TOPICS IN BIODIVERSITY AND CONSERVATION

Volume 6

The titles published in this series are listed at the end of this volume.

Plant Conservation and Biodiversity

Plant Conservation and Biodiversity

Edited by

David L. Hawksworth

and

Alan T. Bull

Reprinted from *Biodiversity and Conservation*, volume 16:6 (2007)

 Springer

A C.I.P. Catalogue record for this book is available from the library of Congress.

Pershore College

ISBN 978-1-4020-6443-2 (HB)
ISBN 978-1-4020-6444-2 (e-book)

Published by Springer,
P.O. Box 17, 3300 AA Dordrecht, The Netherlands.

www.springer.com

Cover Photo: *Gustavia augusta* L. (*Lecythidaceae*) a common riverside tree in Amazonia. Photographed at Lago Janauacá, near Manaus, Amazonas Brazil in August 2006. This genus of the Brazil nut family was named by Linnaeus after King Gustav III of Sweden. The photograph was supplied by Professor Sir Ghillean Prance FRS, VMH.

Printed on acid-free paper

Contents
Plant Conservation and Biodiversity

Introduction

Plant conservation and biodiversity

This book brings together a selection of original studies submitted to *Biodiversity and Conservation* addressing aspects of the conservation and biodiversity of plants. Plants are, along with terrestrial vertebrates, the best known organisms on Earth, and so work on them can be a model for that on less known organism groups. Further, plants are crucial to the maintenance of atmospheric composition, nutrient cycling, and other ecosystem processes. In addition they provide habitats and food for myriads of dependent organisms. At the same time, plants are exploited for food and fuel by humans, and forests continue to be felled for the timber trade or to provide more grazing for cattle. As individual plants are not mobile, and often co-exist as parts of complex plant communities, they are also particularly vulnerable to global climate and other ecological changes.

The contributions are drawn mainly from tropical and subtropical countries, especially Central and South America and Asia, and collectively provide a snapshot of the types of issues and concerns in plant conservation in these regions today. The subjects treated range from effects of climate and habitat changes, including effects of alterations in management and major fires, through the exploitation of forests for medicinal plant and other products as well of trees, to genetic variation within endangered or exploited species, and factors affecting seed production and germination.

This series of themed issues aims to provide an indication of current research activities across this wide range of topics, examples of issues of current concern, that will make the book especially valuable for use in conservation biology courses. They can be viewed as a series of case studies that will expose students to primary research being conducted now. As such they will complement the necessarily less-detailed information available in textbooks and review articles.

DAVID L. HAWKSWORTH
Editor-in-Chief, Biodiversity and Conservation
Universidad Complutense de Madrid;
The Natural History Museum, London
25 April 2007

Springer

Biodiversity and Conservation (2007) 16:1575–1592   © Springer 2006
DOI 10.1007/s10531-006-9002-4

Floristic and structural changes related to opportunistic soil tilling and pasture planting in grassland communities of the Flooding Pampa

C.M. GHERSA*, S.B. PERELMAN, S.E. BURKART and R.J.C. LEÓN

*IFEVA (CONICET) – Facultad de Agronomía, Universidad de Buenos Aires, Av. San Martín 4453, C1417DSE, Buenos Aires, Argentina; *Author for correspondence (e-mail: ghersa@agro.uba.ar; fax: +54-11-4514-8730)*

Received 7 December 2004; accepted in revised form 5 May 2005

Key words: Biodiversity, Ecological impact, Grassland communities, Landscape diversity, Old pastures, Soil tilling

Abstract. The Flooding Pampa natural grassland has an intricate pattern of plant communities, related to small topographic differences that determine important changes in soil characteristics. Despite limitations imposed by soil properties and periodic waterlogging, opportunistic tilling is carried out to plant pastures. There is little information on how pasture planting may affect the structure of the grassland communities. In order to document changes caused by cultural activities on structural and functional characteristics of plant communities in this landscape, we made field surveys in grasslands and very old pastures (grassland communities recovered through secondary succession) using transects located across existing topographic gradients.The patchy structure of this landscape was revealed by the multivariate analysis, by means of which four plant communities could be identified in the natural grassland. Species composition of these communities differed from that of the corresponding old pastures. They lost an important number of exclusive species, but also gained species: some new to the landscape and many already present in other environments.Pasture planting reduced the rate of species replacements along the gradient and produced changes in patchiness, but had no effect on the species–area curve at the landscape scale. Neither did we find differences in total number of species, average number of species/site and proportion of functional types. The new grassland created by opportunistic pasture planting has developed into a structural gradient in which important differences occurred in the lower waterlogged-prone stands, whereas the sites of the other communities experienced less structural changes.

Introduction

The Flooding Pampa grasslands in eastern Argentina cover 90,000 km² of an extremely flat area with poorly drained soils, which have been extensively modified by anthropogenic disturbances (Chaneton et al. 2002). This natural grassland is heterogeneous and characterized by well-defined plant communities that are strongly correlated with the small topographic differences determining dramatic changes in soil characteristics (Soriano 1992; Perelman et al. 2001). Cattle husbandry on unfertilized natural grasslands has been the main activity in this area, which increased species richness in most of the plant communities by enhancing invasion of exotics, but also reduced the compositional and functional heterogeneity of the vegetation at landscape scale (Chaneton et al. 2002).

Technological improvements and need to increase the economic revenue of the land is continually pushing ranchers to replace this traditional practice of extensive low input grazing of natural grasslands, by intensive grazing of planted pastures and, in some cases, even cropping of annual species (Cahuepé et al. 1982; Oesterheld and León 1987; De León and Cauhepé 1988; Soriano 1992; Gerschman et al. 2003). For this reason, despite the limitations imposed by soil properties and periodic water logging, opportunistic tilling is carried out to plant pastures, which may remain grazed and without replanting for very long periods of time. Although there are no patterns, these periods frequently extend to 10 years or more, and fields with pastures older than 50 years are not difficult to find. The secondary succession developing after pasture planting in the regions better drained soils has been described in detail (León and Oesterheld, 1982; Oesterheld and León 1987). These authors studied changes in floristic composition, plant soil cover and specific productivity, finding that in a 15–18-year period the community recovers almost all of its original species and the dominance of *Stipa charruana*.Yet there is still very little information on how pasture planting may affect the structure of the grassland communities as a consequence of alterations in soil–plant relationships and plant-to-plant interactions. Tilling causes dramatic changes in the physical and chemical properties of the topsoil. Layers with different pH, salt and organic matter contents are mixed, and tilling machinery produces soil compaction, that may be aggravated later by cattle trampling. Flooding may revert some of these changes in soil properties, and this reversion is followed by the secondary succession that operates once tilling of the soil disappears (Lavado et al. 1992; Chaneton et al. 2002). Changes in the succesional pathways initiated by opportunistic pasture planting may induce important changes in nutrient and water cycle within the ecosystem, strongly modifying its properties (Wedin 1995; Wedin and Tilman 1996; Quinos et al. 1998).

In this study we made field surveys in natural grasslands and very old pastures, using transects located across existing topographic gradients, in order to document the changes caused by cultural activities on the structural and functional characteristics of the plant communities in a landscape of the Flooding Pampa grassland. We expect that opportunistic pasture planting will expand the narrow ecotones separating patches of the natural grassland, which make apparent the discrete limits between stands of the different communities. Therefore, the natural patchy pattern of the grassland will change into a vegetation gradient following the small topographic differences, especially reducing the compositional and functional heterogeneity of the vegetation at landscape scale.

Characteristics of the landscape

The landscape of the area is made up of the divides, slopes and bottoms of watersheds of the tributaries of the Samborombon River in the northern portion of the Flooding Pampa (León et al. 1979), with slopes of less than 0.1% and lying between 10 and 20 m a.s.l. These flat, extensive divides and slopes present an intricate mosaic of three landscape elements (Figure 1) with height differences no greater than 0.30 m, each with differing soil characteristics and plant communities. These elements are: (1) positive areas (60% of the total surface); (2) flat, depressed areas (30%) and (3) negative, concave, elongated areas (microchannels) (10%). The soils of the positive areas are typic paleudols or typic cromuderts; in the lower negative areas, aquentic cromuderts and typic natracualfs are dominant, while on the flat areas, typic natracuols are predominant (INTA 1989). While the positive areas are rarely flooded, the other two elements of this pattern get water-logged during winter and early spring. The vegetation changes according to these soil and topographic characteristics of the landscape. Four different plant communities are found: *Stipa charruana–Cynara cardunculus–Borreria dascycephala* (a mesophytic prairie), or *Stipa papposa–Stenotaphrum secundatum–Distichlis* spp. (a humid mesophytic prairie) communities on the higher parts of the landscape; *Sporobolus pyramidatus–Nostoc* sp.–*Sporobolus indicus* community (an halophytic steppe) on the flats and *Althernantera philoxeroides–Mentha pulegium* community (a meadow) on the depressions (León et al. 1979; Perelman et al. 2001).

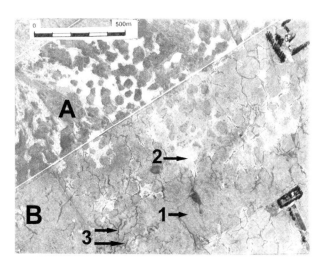

Figure 1. Airphotograph of the studied landscape showing its elements and their arrangement. A: recently tilled field; B: grassland. 1: positive areas, 2: flat, depressed areas, 3: concave micro-channels.

Materials and methods

We selected five pairs of neighboring fields, where each pair consisted of a field of natural grassland and an old pasture. A transect encompassing the entire topographic gradient of each field was located to sample vegetation. Along each transect, samples were taken systematically every 20 m, resulting in 6–9 samples per transect (totalizing 38 samples in natural grasslands and 33 in old pastures). Each vegetation sample consisted of a complete list of the vascular plants present, and an estimate of species cover within an area of 10×2 m located perpendicular to the transect. Sampling was carried out in early summer, when most species are present and easily identifiable through their reproductive organs.

In order to determine and describe the variation in species composition across sites in both, old pastures and grasslands, we classified them in vegetation units by means of a fusion algorithm, which resulted from Sorensen distance measures (Digby and Kempton 1987) used to calculate the farthest neighbor clustering from presence–absence data of all species present in more than 2 samples: 92 species in old pastures and 96 species in grasslands. The frequency (constancy) and the average cover of each species were compared for old pastures and grasslands within vegetation units.

We studied the principal gradients in species composition and compared the amount of habitat heterogeneity present in tilled and non-tilled grasslands with correspondence analysis (Greenacre 1984), applied on the complete matrix of cover data encompassing pastures and grasslands (127 species with presence greater than 2 in the complete inventory = 71 sampling units). Both, for classification and ordination, cultivated pasture species (*Festuca arundinacea, Thinopyrum ponticum, Dactylis glomerata, Phalaris aquatica*) were excluded from the analysis.

We explored the proportional contribution of different plant functional types that may be indicators of changes in ecosystem processes as a result of opportunistic tilling, such as biological invasions or production seasonality: exotics vs. natives, annuals vs. perennials, cool-season vs. warm-season species, testing the hypothesis of homogeneity between grassland and old-pastures within each environment. We applied Fisher exact probability test (Weerahandi 1995) to determine whether each pair of sites differ, significantly, in the proportion of species belonging to each functional group. We estimated average number of species per sample, total number of species and rate of species turnover (beta diversity = total number of species per site/mean number of species per site, Magurran 1988) for the two groups of samples. For each grassland community, we determined the percentage of exclusive species lost in relation to total number of exclusive species for that community. For each old-pasture community, we counted the number of species gained according to their origin, considering 'species new to the landscape' all those not present in the grassland communities and 'from other communities' the rest of species gained. We also calculated beta turnover for each grassland community as (number of species gained + number of species lost)/(total number species in grassland + total number species in old pasture) (Shmida and Wilson 1985). For each transect, we calculated Jaccards percentage similarity index (Digby and Kempton 1987) between sites located at the extreme topographic positions. We compared the average of these values between grassland and old-pasture sites with a *t*-test. Finally, we built a species–area curve for each transect by calculating the cumulative number of species as area increased. Then we averaged the species area curves for old pasture transects and for grassland transects, respectively.

Results

The correspondence analysis revealed the floristic heterogeneity of this grassland, determining four clusters that correspond to four plant communities and confirms the patchy structure of this landscape (Figure 2, Table 1). The two most similar clusters include the stands of the well-drained soil: (G1) associated with the habitat offered by the convex highest topographic positions of the landscape and (G2) constrained to the similar but relatively lower positions. Noteworthy, richness in (G1) is greater than in (G2), which lacks 33 species (Table 2a) including a native shrub (*Baccharis spicata*), some sub-shrubs (*B. pingrae, Margiricarpus pinnatus, Sida rhombifolia, Vernonia rubricaulis*), and some exotic weeds, such as thistles (*Cynara cardunculus, Carduus acanthoides, Cirsium vulgare*). The other two clusters correspond to the stands

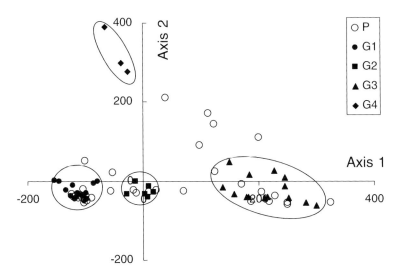

Figure 2. Correspondence analysis: the clusters correspond to four grassland communities. Old pastures sample-points are superimposed and scattered in-between the grassland clusters, the upper extreme along axis 2 missing.

[5]

Table 1. Partial ordered constancy table for grasslands (G) and old pastures (P) in the different vegetation units: MP = mesophitic prairie, HMP = humid mesophitic prairie, HS = halophytic steppe, M = meadow.

ES	Gained NL	GC	Vegetation unit Community Species/No. samples	MP G1 15	P1 10	HMP G2 6	P2 9	HS G3 14	P3 9	M G4 3	P4 5
1*			*Asclepias mellodora*	27							
1*			*Calotheca brizoides*	7							
1*			*Chevreulia acuminata*	7							
1*			*Teucrium cubense*	7							
1*			*Torilis nodosa*	7							
			Carex bonariensis	20				7			
1*			*Panicum sabulorum*	13							
1			*Bromus hordeaceus*	53	30						
1			*Taraxacum officinale*	13	10						
1			*Baccharis coridifolia*	20	10						
1			*Bromus catharticus*	20	10						
1			*Medicago lupulina*	47	30						
	1		*Crepis setosa*		10						
	1		*Digitaria sanguinalis*		10						
	1		*Oxalis conorrhiza*		10						
	1		*Scutellaria racemosa*		10						
	1		*Veronica arvensis*		10						
	1		*Centaurea jacea*		30						
	1		*Cichorium intybus*		20						
	1		*Melilotus albus*		20						
3*		1	*Eryngium elegans*		30			7			
1			*Paspalum notatum*	7	20						
1			*Sonchus asper*	7	20						
1			*Hypochaeris chillensis*	7	30						
1			*Trifolium repens*	13	30						
1			*Trifurcia lahue*	13	40						
1			*Lactuca serriola*	7	10						
1			*Margyricarpus pinnatus*	7	10						
2*			*Briza minor*			17					
2*			*Juncus dicothumus*			17					
2*			*Pfaffia gnaphaloides*			17					
2*			*Trichocline sinuata*			17					
2*			*Richardia stellaris*			33					
			Schizachyrium spicatum	7		33					
2*		1	*Chevreulia sarmentosa*		10	17					
			Leucanthemum vulgare	20	40	17					
			Stenotaphrum secundatum	73	10	33		21			
			Cyperus eragrostis	13		17					
			Cuphea glutinosa	13	10	17					
			Baccharis trimera	33	30	17	11				
			Centaurea calcitrapa	53	30	17	22				
			Piptochaetium stipoides	53	20	33	44				
			Verbena montevidensis	87	80	33	67				

Table 1. (Continued)

ES	Gained NL	GC	Community Species/No. samples	MP G1 15	P1 10	HMP G2 6	P2 9	HS G3 14	P3 9	M G4 3	P4 5
			Piptochaetium bicolor	80	80	67	44				
			Jaegeria hirta	33	40	17	11				
1		2	*Cynara cardunculus*	53	30		11				
1		2	*Melica brasiliana*	73	30		11				
1		2	*Convolvulus hermanniae*	7	60		22				
1		2	*Physalis viscosa*	13	70		33				
1		2	*Anagallis arvensis*	7	30		22				
1		2	*Cirsium vulgare*	47	70		22				
1		2	*Briza subaristata*	13	10		11				
1		2	*Conyza blakei*	73	40		67				
1		2	*Carduus acanthoides*	87	80		67				
1		2	*Vernonia rubricaulis*	33	20		33				
1		2	*Sida rhombifolia*	87	90		11				
1		2	*Ammi majus*	67	90		56				
1		2	*Baccharis pingraea*	13	20		11				
1		2	*Solidago chilensis*	13	20		11				
1		2	*Borreria dasycephala*	80	90		33				
	2		*Hirschfeldia incana*				11				
	2		*Polygala australis*				11				
1*		2	*Baccharis spicata*	7			44				
	2		*Avena* sp.				22				
	2		*Oenothera parodiana*				22				
1*		2	*Glandularia peruviana*	7			33				
1*		2	*Wahlenbergia linarioides*	7			33				
1*		2	*Apodanthera sagittifolia*	7			11				
	1,2		*Dactylis glomerata*		30		44				
	1,2		*Phalaris aquatica*		60		78				
	1,2		*Trifolium pratense*		20		56				
	1,2,3,4		*Festuca arundinacea*		80		22		22		20
	1,2,3,4		*Thinopyrum ponticum*		70		22		78		60
	1,2,3		*Nicotiana longiflora*		10		11		11		
		2	*Stipa charruana*	87	40		11	29	33		
		2	*Cynodon dactylon*	7	50		22	14	11		
2		1,4	*Silene gallica*		10	17	22				20
	3		*Medicago polymorpha*						11		
	3		*Portulaca oleracea*						33		
			Aristida murina	40		100	22	21			
			Noticastrum diffusum	27		33	11	21			
			Carthamus lanatus	53	40	33	22	7	11		
			Gaudinia fragilis	33	40	17	22	29	33		
		3	*Cyclospermum leptophyllum*	73	80	33	56		11		
			Vulpia myuros	60	20	50	11			33	
			Hypochaeris radicata	80	70	17	22			33	20
1		4	*Jaborosa integrifolia*	7	20						20

[7]

Table 1. (Continued)

ES	NL	GC	Community Species/No. samples	MP G1 15	P1 10	HMP G2 6	P2 9	HS G3 14	P3 9	M G4 3	P4 5
1		4	*Oxypetalum solanoides*	40	70						20
		4	*Adesmia bicolor*	47	70	50	67				20
		4	*Conyza primulifolia*	53	20	33	22				20
		4	*Dichondra microcalyx*	87	50	33	56	7			20
			Grindelia discoidea			67	44	64	44		
			Spergula villosa			17	11	14			
			Cypella herberti	33	10	67		7		33	40
			Eryngium echinatum	27	40	50	44	29		67	60
			Juncus imbricatus	67	60	67	11	36		33	40
			Sisyrinchium platense	7		33	33	36		33	20
			Leontodon taraxacoides	7	20	33				67	20
			Mentha pulegium	33	20	17	22			100	20
		2	*Aster squamatus*	13	30		33	14		100	60
			Agalinis communis	13	20	50	11			67	20
			Panicum hians	33	40	50	22	14	22	67	40
			Phyla canescens	100	80	83	56	7	22	67	60
			Lotus glaber	93	100	100	100	36	33	100	80
			Centaureum pulchellum	87	100	100	89	43	11	33	20
			Sporobolus indicus	20	30	100	33	86	44	33	80
			Paspalum distichum			17	11	14	22	33	40
			Cyperus corymbosus	27	10	50	33	14	33	33	80
			Acmella decumbens	13	40	33	78	43	44	67	80
			Paspalum dilatatum	93	100	33	33	7	33	33	40
			Lolium multiflorum	73	80	50	78	21	11	67	60
			Setaria parviflora	60	70	83	89	21	22	33	60
		4	*Conyza bonariensis*	53	60	50	56	14	22		20
		4	*Stipa papposa*	73	60	83	33	86	67		20
		4	*Piptochaetium montevidense*	33	30	67	67				20
		4	*Berroa gnaphalioides*	60	40	83	22	14	22		40
		4	*Eragrostis lugens*	13	30	83	78	14	67		20
		4	*Gamochaeta* sp.	33	60	67	56	14	11		20
		4	*Plantago myosurus*	7	50	50	22	43	22		20
		4	*Stipa neesiana*	53	70	83	78	7	44		20
		4	*Pterocaulon virgatum*	60	50	100	78	71	89		60
		4	*Panicum bergii*	33	40	67	56	57	56		40
		4	*Ambrosia tenuifolia*	80	80	100	100	50	78		40
		3,4	*Bothriochloa laguroides*	100	100	100	22		11		20
		4	*Nostoc* sp.			50		93	100		60
		4	*Chloris berroi*			50	56	79	100		60
		4	*Spergula laevis*			33	33	29	44		40
		1,4	*Eleusine tristachya*		10	33	22	29	78		20
		1,4	*Chaetotropis elongata*		10			21			40
		4	*Distichlis scoparia*			33		57	100		60
3		2,4	*Distichlis spicata*				11	57	11		20

Table 1. (Continued)

ES	NL	GC	Vegetation unit Community Species/No. samples	MP G1 15	P1 10	HMP G2 6	P2 9	HS G3 14	P3 9	M G4 3	P4 5
3		2,4	Sporobolus pyramidatus				11	100	100		40
1		3,4	Juncus capillaceus	13	10				11		20
		2,4	Juncus microcephalus	7	10		11	7			20
3		2,4	Diplachne uninervia				11	43	11		60
		1	Paspalum vaginatum		10			14	22	33	60
3			Puccinelia glaucescens					21	44		
3			Lepidium parodii					21	22		
3			Solanum elaeagnifolium					21	22		
3		4	Lepidium spicatum					36	78		40
			Pappophorum philippinianum			33		64	67		
3		2	Acicarpha procumbens				22	21	22		
3*			Senecio pinnatus					50			
3*			Petunia parviflora					14			
3*			Spergula ramosa					7			
3*		4	Hordeum stenostachys					36			20
4*			Bromidium hygrometricum							67	
4*			Echinochloa sp.							33	
4*			Glyceria multiflora							33	
4*			Juncus sp.							33	
4*			Marsilea ancylopoda							33	
4*			Roripa bonariensis							33	
4*			Rumex sp.							33	
4*			Gratiola peruviana							100	
			Nothoscordum gracile	7						33	
			Stipa philippii	7						33	
4			Alternanthera philoxeroides							100	20
			Eleocharis sp.	7						100	20
4			Leersia hexandra							67	40
		2	Eryngium ebractateum	40			11			100	20
4			Danthonia montevidensis							33	20
4			Pamphalea sp.							33	20
4			Stipa formicarum							33	20
	4		Centella asiatica								20
3*		4	Acicarpha tribuloides					14			20
	2,4		Sida spinosa				11				20

ES: species exclusive to one grassland community,* exclusive species lost.
NL: species gained, new to the landscape, GC: species gained from other communities. Values are constancy (% frequency) in each community. Nomenclature follows Zuloaga et al. (1994), Zuloaga and Morrone (1996, 1999).

occupying the topographic sites with waterlogged-prone soils: (G3) found in the less frequently flooded, saline–alkaline soils habitat, which is characteristic of the mild topographic gradient between uplands and lowlands, and (G4) with the least species richness (Table 2a), distributed on the lowest topographical positions and small channels. In the old pastures, sample-points are

Table 2. Species gained or lost in: (a) the grassland communities; (b) the old pastures.

(a)				
Grassland community	G1	G2	G3	G4
Total number of species	99	66	60	39
Number of exclusive species	40	7	14	13
Number exclusive species lost	10	6	6	8
% exclusive species lost	25	86	43	61
(b)				
Old pasture community	P1	P2	P3	P4
Species gained from other communities	6	28	3	32
Species gained, new to the landscape	14	11	5	2

superimposed and scattered in-between the original grassland clusters, gener-ally displaced downwards and with the upper extreme along axis 2 missing (Figure 2). Both, the grassland communities (G1, G2, G3 and G4) and the corresponding old pastures (P1, P2, P3 and P4) had differences in their species composition (Table 1). G4 sites had the greatest difference with the other grassland communities, with an average similarity of species lists equal to 0.36. The corresponding old pasture (P4) instead, becomes more similar to the old pastures of the other environments with an average similarity of 0.57.

There are 74 species exclusive to anyone of the grassland communities, while only 48 exclusive to anyone of the old pastures. As a result of tillage, the different grassland communities lost an important number of their exclusive species, but also gained species, some new to the landscape while others were already present in the other environments (see Table 2 for total numbers and Table 1 for species identity). The grasslands from the intermediate and lower topographical positions (G2 and G4) lost the most of their exclusive species (86 and 61%, respectively, Tables 1, 2a) and were the ones that gained most species (38 and 34 species, respectively, Table 1). The sites of the richest community (G1) gained the highest number of species new to the landscape, while the one from the poorest (G4) was enriched mainly by species from the landscape species pool (Tables 1, 2b). A total of 20 species were gained in the landscape: 35% of them were established in more than 1 community and 14 species col-onized the sites of the better drained soils (G1) (Table 1).

Other species had different behaviors in response to tilling and pasture planting. Some, generally having greater constancy values than the exclusive species to each of the grassland habitats, were still present in the old pastures, but had constancy values ca. 50% lower than in the untilled fields. *Cynara cardunculus*, *Melica brasiliana* and *Conyza blakei* are within the species that appear to have some sensitivity to pasture establishment, but their constancy reduction was partially compensated for by their presence in the new habitats created by the cultural practices in the formerly G2 stands (P2, Table 1). *Stipa charruanna*, a native, dominant grass was distributed along a wider range of environments than the previous set of species, as it also occupied the habitats of the lower topographic stands with less drained soils (G3, Table 1). Cultural practices for pasture planting reduced by 46% the constancy value for this coarse, perennial grass, only in the convex well-drained soil habitats (G1 vs. P1, Table 1). This reduction, as in the case of other species, was partially compensated for by its occurrence in P2 cultural habitats, located in the intermediate topographic sites, where it was originally absent (G2).

Numerous species disappeared as a consequence of the disturbance. For example, *Sisyrinchium platense*, *Chaetotropis elongatus*, *Stenotaphrum secundatum* and *Eryngium elegans* present in G3, are absent in P3. A number of constant species (*Althernantera philoxeroides*, *Leersia hexandra*, *Eryngium ebracteatum*, *Danthonia montevidensis*, *Pamphalea bupleurifolia* and *Stipa formicarum*) present in G4 decreased their constancy in P4. On the other hand, after pasture planting in G1 stands, five species increase their constancy (*Paspalum notatum*, *Sonchus asper*, *Trifolium repens*, *Hypochaeris chilensis*, *Trifurcia lahue*) and only three decrease (two *Bromus* species and *Medicago lupulina*). Despite the important changes in species composition that resulted from pasture planting, comparisons of plant functional types among the various grassland communities and with the corresponding old pastures yielded no significant differences (Figure 3).

On average, total plant cover of grasslands and old pasture sites was ca. 70% (Figure 4). In both cases, the sites of the higher topographic positions had the highest cover, while the waterlogged and halomorphic sites, the lowest. In sites of intermediate topographic positions, the cover data from the grassland stands had less variability than those from the old pastures. Therefore, despite of the similarity in cover values (G2 = 76%, G4 = 62% and P2 = 72%, P4 = 59%), statistical differences ($p = 0.005$) between higher and lower positions appeared only for grasslands (Figure 4).

At a more detailed scale, pasture planting reduced the rate of species replacements along the studied topographic gradient, since the floristic similarity between contiguous samples was greater in the old-pastures than in the grasslands (Figure 5). However, despite the changes in patchiness, the species–area curve (Figure 6) at landscape scale was not affected. Neither did we find differences between grasslands and old-pastures in total number of species (G = 146; P = 140), average number of species/site (G = 27.9; P = 29.1) and beta diversity (G = 5.2; P = 4.8).

Discussion

The characteristic structure of the grassland communities of this flat intricate landscape is patchy, with well-defined stands corresponding to important edaphic differences that follow the existing subtle relative topographic variation. Soil tilling and pasture planting erased the defined boundaries separating these stands and generated a vegetation gradient. Probably, a great part of this change is caused by tilling operations, which flattens the convex patches as well as the small channels, mixes the organic matter and salts of the surface soil

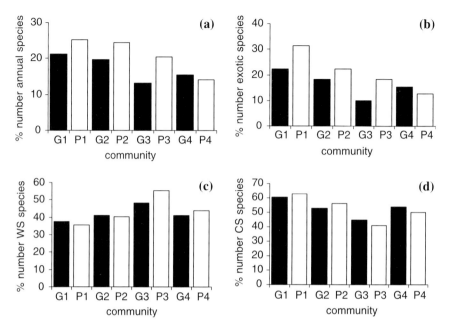

Figure 3. Percentage number of species corresponding to the different functional types yielded no significant differences for grasslands and old pastures: (a) annual species, (b) exotic species, (c) warm-season species, (d) cool-season species.

accumulated in the different layers, distributing them rather homogenously within the plowed profile. Alien species to the landscape appeared in the cultural habitats created by pasture planting, which included not only the sown forage species, but also some weeds and natives to the region (Chaneton et al.

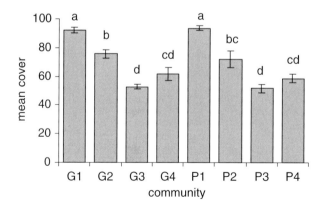

Figure 4. Percentage total plant cover for each grassland and old-pasture community. Vertical bars are standard errors. Different letters represent significant differences between mean values ($p < 0.05$).

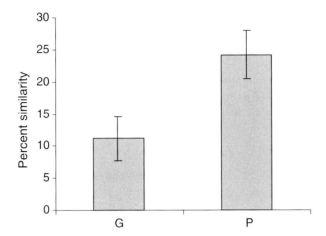

Figure 5. Floristic similarity between contiguous samples (Jaccard Similarity Index) for grasslands and old pastures.

2002). The new grasslands (P) created by the old-pastures had less structural differences along the topographic gradient than the original ones (G). The mesophytic meadows (G1) and the hallophytic steppe (G3) gained mainly species new to the landscape (70 and 62%, respectively) while in the humid mesophytic prairies (G2) and meadows (G4) more than 70% of the incoming species belonged to the regional species pool (70 and 94%, respectively). All communities lost at least 25% of their exclusive species, the humid mesophytic prairies (G2) losing the most (86% of exclusives). The exclusive species that expanded their distribution ranges did so to two or three of the communities of old pastures.

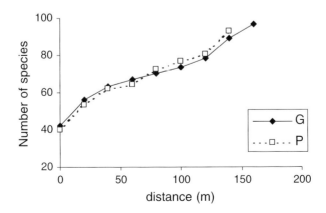

Figure 6. Species–area curves for grasslands and old-pastures at landscape scale were not affected despite changes in patchiness.

Opportunistic pasture planting produced the least impact in the patches of better drained soils covered by mesophytic prairies (G1, beta turnover = 20%), as accounted by loosing 25% of its exclusive species group, and by the addition of 20% of new species, (14% alien to the landscape and 6% redistributed within the landscape), relative to the total of 99 species. It is the richest community and it contributed to 40% of the species redistributions that occurred within the patches of the landscape as a consequence of pasture plantings. As shown for other ecosystems (Levine and D'Antonio 1999; Londsdale 1999; Stohlgren et al. 1999), in these grasslands it is also the richest community the one that gains the highest number of species. Forty three percent of G3 exclusive species were lost in the old pasture stands (P3), and were replaced by the invasion of 8 species, yielding an overall 25% (beta turnover) change, compared to its 60 total species number.

G2 stands that were turned into old pastures increased by 35% their species richness (from 66 to 89 species) and lost 86% of its exclusive species. They underwent a floristic change amounting to 35% (beta turnover value) of its average list (72% of gained species belong to species redistributed within the landscape patches). Seventy seven percent of the species that came to P2 through local species redistribution belonged to G1, indicating that whatever ecological factor/s, biotic or not, curtailing G1 species from expanding their ranges into G2 sites, which occur at subtle topographic differences, disappeared. Contrarily only two species from G2 were able to expand their ranges into P1 environments. Tilling activities increased the floristic similarity of this community with the mesophytic prairie and with the meadows.

The greatest structural impact caused by opportunistic pasture planting occurred in G4 humid community of the grassland. These sites lost 61% of its exclusive species, and had 46% change in its floristic composition (beta turnover value) with 94% redistribution within the landscape. Half of the redistributed species that appeared in the P4 list were also in the G1 and G2 lists, but a significant number (60% of the redistributed species) also were shared with those in G3. Only two species gained by G4 are new to the landscape. This complex of species, fed by all the originally different communities of the landscape, was almost certainly caused by the mixture of soil layers resulting from soil tilling in the less drained lower topographic sites of the landscape, which created a small-grain heterogeneous soil mosaic that allowed the establishment of a diverse group of species. Despite the fact that flooding may revert some of the changes in soil properties caused by tilling, which is followed by a recovery of the original community (Lavado et al. 1992), the patches of the more humid communities surveyed in our study remained opened to the establishment and growth of the other habitat species, notwithstanding the several flooding events that have occurred since the old pastures planting time. It is interesting to point out that not a single species of G4 extended their distribution range to the P1, P2 and P3 cultural habitats. These floristic shifts, caused by immigrations to stands in different topographic positions along the gradient, and local extinctions, could be induced by habitat changes due to pasture planting, but also related to species sampling (Casagrandi and Gatto 1999; Hillerbrand and Blenckner 2002). For example, stands of the better drained soils (G1, G2) covered 60% of the landscapes area and thus, it would be expected that they gain most of the species new to the landscape and have few extinctions, as they did, while the microchannels covering only 10% of the area should have the greatest extinctions and least immigrations.

Chaneton et al. (2002) studied the effects of grazing on heterogeneity and alien plant invasion in the temperate grasslands of the Flooding Pampa. They concluded that grazing promoted exotic plant invasion and generally enhanced community richness, whereas it reduced the compositional and functional heterogeneity of vegetation at landscape scale. That means that grazing effects on floristic heterogeneity were scale dependant. Our data suggests that adding the soil-tilling disturbance to domestic animal grazing did not change the general pattern they described for grazing alone. Noteworthy, apart from the five forage species that were presumably sown into the original habitats and the other 17 that invaded into the landscape old pasture communities, most of the contributions to the patches species richness came from redistribution within the landscape habitats dominated by native species (60 species gained from other communities).

As it was shown by our results, several infrequent species, but with high fidelity to their community and therefore strongly fitted to particular habitats created by subtle topographic variations, appeared to be very sensitive to the ecological changes introduced by opportunistic pasture planting. Within this group, there are perennial grass species that are highly valuable for animal grazing, like *Calotheca brizoides* and *Panicum sabulorum*, and others, like *Asclepias mellodora*, which is one of the few native Asclepiadaceae found in the region that has flowers hosting a lepidopteran *(Danaus emipus)*. Both, the showy flowers and the black and yellow larva of this butterfly, makes this species conspicuous in the stand. Exclusive species loss could alter the grassland functions, as the one just described, as well as others that are unknown because they are less conspicuous. Therefore, these losses should not be underestimated when discussing the overall impact of opportunistic pasture planting on the grassland communities.

Clearly the new grassland created by an opportunistic pasture planting has developed into a structural gradient in which the sites of the well drained soils experienced only small structural change (León and Oesterheld 1982; Oesterheld and León 1987) whereas important differences occurred in the lower waterlogged-prone stands. Some of the steppe species in the saline–alkaline patches of the original grassland were replaced by perennial forage and native species with autumn–winter–spring production peaks, while the community from the more humid acid soil patches was completely changed by the invasion of species from the rest of the habitats.

Plant cover differed only between grassland stands sampled in sites with intermediate positions of the landscapes topographic gradient (Figure 4). The increase in variability associated to pasture planting could be related to differences in management of animal grazing that may have existed between grassland and old pasture fields, which were not controlled in our study. Nevertheless, because grass tussocks contributed the most to plant cover, we believe that other more stable structural changes, related to those factors affecting floristic composition of the old pasture stands were mostly responsible for cover variability (Figure 2, Table 1).

Implications for grassland conservation

The fine grain packaging of species diversity suggests that most of the heterogeneity at the species level could be conserved by establishing protected areas, although the minimum area needed to include a given amount of diversity may not necessarily be enough for its persistence (Perelman et al. 2001). As expected, opportunistic planting of pastures impacted mostly changing the patchy structure of the landscape by reducing its heterogeneity. This may be viewed by ranchers as an improvement for managing grazing of cattle (De León and Cauhepé 1988), yet it may have profound ecological impacts, especially on trophic webs (Bailey et al. 1996; De Vries et al. 1999; Hutchings et al. 2002; Braschler et al. 2003). A large proportion of the Flooding Pampa natural grassland communities are still preserved, but this situation is rapidly changing as the need to improve economic outputs, and availability of new technologies, enhance their replacement to pastures and annual crops. The landscape we studied in this work has an intricate patchiness and a soil type (vertisol, cromudert) that is only found in a relatively small area in the north of this region, which is near to the urban landscapes of the cities of La Plata and Buenos Aires. These characteristics draw attention to its importance for developing conservation activities as soon as possible, not only because of its uniqueness but also, because it is highly jeopardized by the expansion of human activities. Our study provides important information for taking decisions about conservation. On one hand, extensive grazing activity has to be preserved in order to curtail expansion of alternative land uses (Naveh et al. 2001), but most importantly, this activity has to be carried out avoiding soil plowing and pasture planting. Should the landscape's original intricate community pattern be preserved, expansion of soil tilling and pasture planting ought to be stopped. On the other hand, if aims are focused only on preservation of floristic richness, these agronomical practices have little impact.

Acknowledgements

To S. Suárez and R. MacDonough for help during fieldwork. This work was supported by grants from the University of Buenos Aires and the Agencia Nacional para la Promoción Científica y Técnica (Argentina).

References

Bailey D.W., Gross J.E., Laca E.A., Rittenhouse L.R., Coughenour M.B., Swift D.M. and Sims P.L. 1996. Mechanisms that result in large herbivore grazing distribution patterns. J. Range Manage. 49: 386–400.

Braschler B., Lampel G. and Baur B. 2003. Experimental small-scale grassland fragmentation alters aphid population dynamics. Oikos 100: 581–591.

Casagrandi R. and Gatto M. 1999. A mesoscale approach to extinction risk in fragmented habitats. Nature 400: 560–562.

Cauhépé, M.A, León, R.J.C., Sala, O.E. and Soriano, A. (ex aequo). 1982. Pastizales naturales y pasturas cultivadas, dos sistemas complementarios y no opuestos. Revista Facultad de Agronomía (UBA), 3: 1–10.

Chaneton E.J., Perelman S.B., Omacini M. and León R.J.C. 2002. Grazing, environmental heterogeneity, and alien plant invasions in temperate grasslands. Biol. Inv. 4: 7–24.

De León M. and Cauhépé M.A. 1988. Caracterización forrajera y utilización de las comunidades vegetales de la Depresión del Salado. Revista Argentina de Producción Animal 8(Supl. 1): 76–77.

De Vries W.M., Laca E.A. and Demment M.W. 1999. The importance of scale of patchiness for selectivity in grazing herbivores. Oecologia 121: 355–363.

Digby P.G.N. and Kempton R.A. 1987. Multivariate analysis of ecological communities. Chapman & Hall, London.

Greenacre M.J. 1984. Theory and applications of correspondence analysis. Academic Press, London.

Guerschman J.P, Paruelo J.M., Sala O.E. and Burke I.C. 2003. Land use in temperate Argentina: environmental controls and impact on ecosystem functioning. Ecol. Appl. 13: 616–628.

Hutchings M.R., Gordon I.J., Kyriazakis I., Robertson E. and Jackson F. 2002. Grazing in heterogeneous environments: infra- and supra-parasite distributions determine herbivore grazing decisions. Oecologia 132: 453–460.

Hillebrand H. and Blenckner T. 2002. Regional and local impact on species diversity from pattern to processes. Oecologia 132: 479–491.

INTA 1989. Mapa de Suelos de la Provincia de Buenos Aires. SAGyP-INTA, Buenos Aires.

Lavado R., Rubio G. and Alconada M. 1992. Grazing management and soil salinization in two pampean natraqualfs. Turrialba 42: 500–508.

León R.J.C. and Oesterheld M. 1982. Envejecimiento de Pasturas implantadas en el norte de la Depresión del Salado. Un enfoque sucesional. Revista Facultad de Agronomía (UBA) 3: 41–49.

León R.J.C., Burkart S.E. and Movia C.P. 1979. Relevamiento fitosociológico del pastizal del norte de la Depresión del Salado. Serie Fitogeográfica 17. INTA. Buenos Aires.

Levine J.M. and D'Antonio C.M. 1999. Elton revisited: a review of evidence linking diversity and invasibility. Oikos 87: 15–26.

Lonsdale W.M. 1999. Global patterns of plant invasions and the concept of invasibility. Ecology 80: 1522–1536.

Magurran A.E. 1988. Ecological diversity and its measurement. Princeton University Press, Princeton, New Jersey.

Naveh Z., Lieberman A.S., Sarmiento F.O., Ghersa C.M. and León R.J.C. 2001. Ecología de Paisajes. Editorial Facultad de Agronomía U.B.A. Buenos Aires Argentina, pp. 15–18.

Oesterheld M. and León R.J.C. 1987. El envejecimiento de pasturas implantadas: su efecto sobre la productividad primaria. Turrialba 37: 29–35.

Perelman S.B., León R.J.C. and Oesterheld M. 2001. Cross-scale vegetation patterns of the flooding Pampa grassland. J. Ecol. 89: 562–577.

Quinos P.M., Insausti P. and Soriano A. 1998. Facilitative effect of Lotus tenuis on Paspalum dilatatum in a lowland grassland of Argentina. Oecologia 114: 427–431.

Shmida A. and Wilson M.V. 1985. Biological determinants of species diversity. J. Biogeograph. 12: 1–20.

Soriano A. 1992. Río de la Plata Grasslands. In: Coupland R.T. (ed.), Natural Grasslands: Introduction and Western Hemisfere Ecosystems of the World, Vol. 8a. Elsevier, Amsterdam, pp. 367–407.

Stohlgren T.J., Binkley D., Chong G.W., Kalkhan M.A., Schell L.D., Bull K.A., Otsuki Y., Newman G., Bashkin M. and Son Y. 1999. Exotic plant species invade hot spots of native plant diversity. Ecol. Monogr. 69: 25–46.

Wedin D.A. 1995. Species, nitrogen, and grassland dynamics; the constrains of stuff. In: Jones C. and Lawton J.H. (ed.), Linking Species and Ecosystems, Chapman and Hall, London, pp. 253–262.

Wedin D.A. and Tilman D. 1996. Influence of nitrogen loading and species composition on the carbon balance of grasslands. Science 247: 1720–1723.

Weerahandi S. 1995. Exact Statistical Methods for Data Analysis. Springer Verlag, New York.

Zuloaga F., Nicora E., Rúgolo de Agrasar Z., Morrone O., Pensiero J. and Cialdella A. 1994. Catálogo de la Familia de Poáceas en la República Argentina, Monografía no 47. Missouri Botanical Garden Press.

Zuloaga F. and Morrone O. (eds), 1996. Catálogo de la Plantas Vasculares de la República Argentina I, Monografía no 60. Missouri Botanical Garden Press.

Zuloaga F. and Morrone O. (eds), 1999. Catálogo de la Plantas Vasculares de la República Argentina II, Monografía no 74. Missouri Botanical Garden Press.

Biodivers Conserv (2007) 16:1593–1602
DOI 10.1007/s10531-006-9007-z

ORIGINAL PAPER

Decreased frugivory and seed germination rate do not reduce seedling recruitment rates of *Aristotelia chilensis* in a fragmented forest

Carlos E. Valdivia · Javier A. Simonetti

Received: 30 May 2005 / Accepted: 6 February 2006 / Published online: 12 May 2006
© Springer Science+Business Media B.V. 2006

Abstract Habitat fragmentation reduces frugivorous bird abundance. Such a reduction may lead to a reduction in seed dispersal, thereby compromising seedling recruitment rate with far reaching consequences for plant population persistence. We assessed frugivory, seed germination, and seedling recruitment rates in a fragmented forest of central Chile by comparing a continuous forest with four forest fragments surrounded by pine plantations. Frugivory was 2.4 times higher in continuous forest than in forest fragments. Seeds eaten by birds germinated 1.7 and 3.7 times higher than non-eaten seeds from continuous forest and fragments respectively. Non-eaten seeds from continuous forest germinated 2.2 times higher than those from forest fragments, suggesting inbreeding depression. However, seedling recruitment rates at forest fragments were far higher than in continuous forest where no seedling recruited in the five years analysed. Therefore, despite forest fragmentation negatively affected frugivory, it did not translate into a decreased fitness of plants, thus highlighting the importance of considering the overall processes leading the reproductive success of plants following anthropogenic disturbances.

Keywords Forest fragmentation · Avian frugivory · Seed quality · Seedling recruitment rates

Introduction

Habitat fragmentation can have profound effects on frugivory and seed dispersal with negative consequences for plant fitness (Santos and Telleria 1994; Galetti et al. 2003; Şekercioğlu et al. 2004). The reduction in habitat size and increment in isolation can reduce the diversity and abundance of the remaining frugivores. Such reductions may lead

C. E. Valdivia (✉) · J. A. Simonetti
Departamento de Ciencias Ecológicas, Facultad de Ciencias, Universidad de Chile, Casilla 653, Santiago, Chile
e-mail: cvaldiviap@yahoo.com

[19]

 Springer

to a reduction in the intensity of frugivory, and consequently, in seed dispersal and upon seedling recruitment (Galetti et al. 2003; Cordeiro and Howe 2001, 2003; Traveset and Riera 2005).

Small patches of fruiting plants may be less attractive to frugivores as they offer a lower food reward (Saracco et al. 2004). Because sugar concentration of pulp fruits depends on environmental conditions (Ito et al. 1999), forest fragmentation might then affect the attractiveness of plants to frugivores by increasing sugar concentration given the lower environmental humidity of forest fragments. Frugivores prefer concentrated pulps, which may in turn trigger an increased frugivory rate (Stanley et al. 2002). Because habitat fragmentation reduces atmospheric humidity (Camargo and Kapos 1995), the attractiveness of fruits, in terms of sugar concentration, might be higher in fragmented populations. Therefore, habitat fragmentation could lead to contrasting tendencies: a reduction of frugivory, since small and isolated patches are less attractive and harbour a depauperated fruit-feeding animal abundance; or increased frugivory rates, because fruits in fragments might produce more concentrated pulps, leading to some sort of compensation, or even overcompensation, of the lowered population sizes, and hence similar rates of frugivory in continuous forests and forest fragments.

A failure in seed dispersal triggered by a lowered frugivory might lead to a clumped pattern of tree recruitment which may in turn favour a higher mating ratio among close relatives (Bleher and Böhning-Gaese 2001). Therefore, failures in frugivory coupled to a reduction in plant population size and increased isolation triggered by habitat fragmentation may lead to inbreeding depression (Barrett and Khon 1991; Young et al. 1996). Inbreeding depression in turn, may produce a reduction in seed quantity i.e., seed per fruit) and seed quality (i.e., germination rate) (Barrett and Khon 1991; Young et al. 1996; Henríquez 2004). Certainly, the reduction in frugivory coupled to the lowered fitness of plants in terms of the seed quality and quantity may reduce the seedling recruitment rates, which might compromise the long-term population persistence (Barrett and Khon 1991; Galetti 2003).

Although frugivorous vertebrates are rather scarce in the temperate rainforests of southern Chile, the proportion of vertebrate seed dispersal is roughly comparable to tropical forests (Armesto and Rozzi 1989; Willson 1991). In these temperate forests, birds are the most important frugivores being more than three-quarters of the whole vertebrate dispersers (Willson 1991; Aizen et al. 2002). Therefore, changes in frugivorous birds may lead to a changes in frugivory.

Currently, the southern temperate rainforest is severely fragmented which may affect animal-dispersed plants (Bustamante and Castor 1998; Bustamante et al. 2005). In these forests, assessments on the effects of forest fragmentation on birds have only recently been considered, reporting negative effects on richness and abundance of fruit-feeding birds (e.g., Willson et al. 1994; Vergara and Simonetti 2004). Despite this fact, however, there is a great knowledge vacuum concerning the potential effects of such reductions on seed dispersal and seedling recruitment rates. Here, we assess the effect of forest fragmentation on avian frugivory, seed quality, and seedling recruitment rates on the bird-dispersed tree *Aristotelia chilensis* (Elaeocarpaceae), a broadly distributed understory tree in temperate forest of southern South America (Rodríguez et al. 1983). If forest fragmentation negatively affect the frugivorous assemblage, we expect a negative effect on frugivory at the forest fragments with regard to continuous forest. Similarly, because forest fragmentation may also negatively affect seed quality, we expect a lower germination rate of seeds from forest fragments than from continuous forest. Therefore, because we expect a lowered seed dispersal rate coupled to a lowered seed quality, it is reasonable to expect a lowered seedling recruitment rate in fragments than in the continuous forests.

🖄 Springer

Materials and methods

Study site and species

The study was conducted in Maulino forest in the northernmost zone of the Southern temperate rainforest (35°59′S, 72°41′W; San Martín and Donoso 1996). Specifically, we worked in Los Queules National Reserve and four neighbouring forest fragments. Distance between continuous forest and fragments, and between fragments, ranges from 1 to 4 km (see Donoso et al. 2003 for a map). Los Queules is a protected area of 145 ha embedded in a large tract of 600 ha of continuous forest. Forest fragments, ranging from 1 to 6 ha, are patches surrounded by commercial plantations of *Pinus radiata*. Both in the forest fragments and in the continuous forest composition and abundance of adult trees are similar, the most abundant trees being *Aetoxicon punctatum, Cryptocarya alba, Gevuina avellana*, and *Persea lingue* among others (Bustamante et al. 2005).

Aristotelia chilensis (Elaeoearpaceae) is a dioecious tree of up to 4 m tall that inhabits the southern temperate rainforests (Rodríguez et al. 1983). While the adult abundance is roughly the same at both continuous forest and forest fragments, seedling abundance is higher at forest fragments (Bustamante et al. 2005). It bears black-coloured fleshy berries which are eaten and dispersed by fruit-feeding birds. Fruiting occurs from October to December, while fruit ripeness and seed dispersal occur from mid-December to January in the austral spring-summer season (Rodríguez et al. 1983). In the Maulino forest, potential dispersers are *Anairetes parulus, Elania albiceps, Xolmis pyrope* (Tyrannidae), *Aphrastura spinicauda* (Furnariidae), and *Turdus falklandii* (Muscicapidae) (Vergara and Simonetii 2004). In the study site, *A. parulus, E. albiceps* and *X. pyrope* are less abundant in forest fragments than in continuous forest, whereas *A. spinicauda* and *T. falklandii* are equally abundant in both forest fragments and continuous forest (Vergara and Simonetti 2004; González-Gómez 2004).

Fruit characteristics

To determine fruit characteristics, in January 2002, we randomly selected trees and branches and then collected ripe fruits ($n=30$ fruits from six trees in the continuous forest, $n=35$ fruits from six trees in the forest fragment, 1–2 trees per fragment) for assessing fruit size, quantity of seeds, and sugar concentration. Fruit size was estimated as the diameter of each fruit, while sugar concentration (mass percentage) was assessed with a hand-held temperature-compensated refractometer. All measurements were made in fresh fruits immediately after being collected.

Frugivory

The intensity of frugivory was experimentally assessed. We mimicked fruits making red-coloured wire branches bearing nine black-coloured plasticine fruits ($n=96$ branches with 864 fruits placed in the continuous forest, $n=100$ branches with 900 fruits in the forest fragments). The artificial infructescences were attached to individual fruiting trees (2–8 wire branches per tree depending on tree size; $n=16$ trees at continuous forest, $n=16$ trees at forest fragments, four per fragment) leaving them for the action of fruit-feeding birds during 30 days, from December 2001 to January 2002, coinciding with the period of seed dispersal. The only dispersers known for *A. chilensis* are fruit-feeding birds; hence, fruits

🕭 Springer

removed from infructescences were considered as dispersed by frugivorous birds, assumption reinforced by bill impressions on the picked fruits. This experimental array, mimicking natural infructescences of *A. chilensis*, allowed for a control of numerous traits which might otherwise bias the fruit-feeding animal choice, and thus the effect of forest fragmentation on frugivory (see Alves-Costa and Lopes 2001 on methodological details).

Seed germination

To determine the effect of consumption on seed germination, we placed for germination seeds randomly collected in January 2002 from faeces placed on the ground in the continuous forest and forest fragments (n=150 seeds from ca. 100 faeces), together with seeds collected from ripe fruits on trees (n=150 seeds from six trees in the continuous forest, and n=150 seeds from six trees in the forest fragments, 1–2 trees per fragment). This experimental design allowed to assess the effect of frugivores (eaten versus non-eaten seeds) as well as the origin of seeds (continuous forest versus forest fragments) on the seed germination rate.

Germination trials were carried out through a common garden experiment. Because we were incapable of assigning bird-eaten seeds to a specific site, they were only considered as dispersed seeds with mixed origin (i.e., from both continuous forest and forest fragments), thereby rendering conservative the comparisons with non-dispersed seeds from both types of sites. Seed germination was carried out during 250 days by placing 10 seeds from all individual trees into Petri capsules (n=15 capsules per treatment) placed inside germination chambers. Photoperiod of germination chambers was 12-h day/12-h night throughout all time of seed germination, whereas the temperature was modified according to the month when germination naturally occurs, thus mimicking natural conditions for germination (monthly temperature from March to November: 13.3, 10.9, 9.7, 8.5, 8.6, 8.8, 10.6, 12.3, and 13.6°C). Seeds were weekly irrigated with water. We considered a seed as germinated when the emerged root presented 2 mm elongation.

Seed rain and seedling recruitment rates

Forty seed traps were erected within the continuous forest, while other 40 seed traps were placed in forest fragments (10 at each fragment). Each trap consisted of a 0.25 m², open-topped, 1 mm wire-mesh bag held 0.8 m above the ground on a PVC frame. All seeds falling into the traps were counted and identified to the species level each month from January 2002 to December 2002. Because we were incapable of determining if seeds corresponded to bird-dispersed seeds or not, they were only considered as seed rain remaining the dispersal mechanism unknown.

To assess seedling recruitment rates we randomly placed plots of 2 m² (n = 20 in the continuous forest, and 20 in forest fragments, five per fragment). In these plots, we registered all woody plant less than 50 cm height (hereafter seedlings). Each seedling was individually tagged and identified to the species level from 1998 to 2004. Sampling was performed in September of each year, and survivors were recorded in January of the next year for registering all new recruits of the same recruitment season.

At both continuous and forest fragments, we placed two seed traps 2-m away from each plot. Thereafter, each plot and its two adjacent seed trap constituted a census station. Census stations were roughly linearly arranged at each site. The average distance (±1 SE) from each census station to the next census station was 21.13±2.14 m at continuous forest,

Table 1 Fruit characteristics of *Aristotelia chilensis* in the continuous forest and forest fragments (Mean ± 1se)

Fruit characteristics	Continuous forest	Forest fragments
Fruit size (mm)	5.8±0.1[a]	6.2±0.1[a]
Number of seeds per fruit	2.9±0.2[a]	2.9±0.2[a]
Sugar concentration (%)	15.4±1.5[a]	11.0±0.7[b]

Different letters depict significant differences (*P*<0.05)

and 20.83±3.14 m at forest fragments, which did not differ significantly (Mann–Whitney test: U =121, P = 0.305).

At each census station we determined the seedling recruitment rate as $r = (N_t/N_0)^{1/t}$, where N_0 is the number of seedlings at the beginning of the interval, N_t is the number of seedlings at the end of the interval, and t is the number of years (Sheil et al. 1995). The time interval assessed was five years, from 1998 to 2004, thus diminishing a possible overestimation of seedling recruitment rate associated to a narrow time interval (Lewis et al. 2004).

Results

Fruit characteristics

Fruit size was marginally bigger at forest fragments than at continuous forest (Mann–Whitney Test: U = 387.5, P = 0.07; Table 1). The number of seeds per fruit did not differ between both types of sites (U = 520.0, P = 0.948; Table 1). Nevertheless, sugar concentration of fruits was 1.4 times higher in the continuous forest than in the forest fragments (U = 327.5, P = 0.047; Table 1).

Frugivory

Frugivory was 2.4 times higher in the continuous forest than in forest fragments (Mean ± SE: 0.47±0.05 fruits preyed/infructescence in the continuous forest, 0.20±0.04 fruits preyed/infructescence in the forest fragments; U = 3510.0, P= 0.001; Fig. 1b). Such a reduction was 2.2 times greater than the reduction in frugivorous bird abundance previously reported by Vergara and Simonetti (2004) (Fig. 1a).

Seed germination

Overall, both consumption by birds and the origin of seeds (i.e., continuous forest and forest fragments) affected seed germination rate (Table 2). In fact, dispersed seeds germinated 1.7 (Tukey HSD test for balanced data: P<0.01) and 3.7 (P<0.01) times higher than non-dispersed seeds from continuous and forest fragments respectively, while seeds from continuous forest germinated 2.2 times higher than seeds from forest fragments (P=0.02) (Fig. 2).

Seed rain and seedling recruitment rates

Seed rain was 5.4 times higher in the forest fragments than in the continuous forest (U = 54, P<0.001; Fig. 3a). At forest fragments, 21 seedlings were tagged at the onset of the

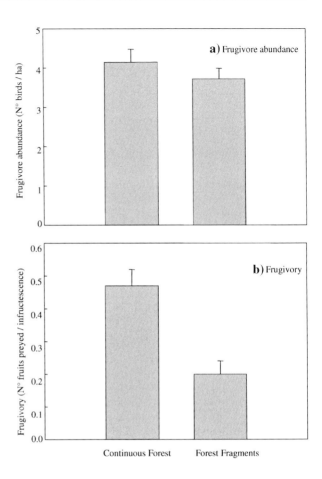

Fig. 1 Frugivore abundance (**a**) (modified from Vergara and Simonetti 2004) and frugivory (**b**) of *Aristotelia chilensis* in a continuous forest and forest fragments (Mean ± 1se)

Table 2 Summary of repeated measures ANOVA testing the effect of seed type (non-eaten seeds from continuous forest and from forest fragments, and eaten seeds from both continuous forest and forest fragments) and time on the seed germination of *Aristotelia chilensis*

Source	df	MS	F	P
(a) Within-subjects:				
Type of seeds	2	169.39	12.39	<0.001
Error	42	13.67		
(b) Between-subjects:				
Time	36	44.73	70.13	<0.001
Time × type of seeds	72	6.69	10.49	<0.001
Error	1512	0.64		

experimental assessment. Of these, only one stayed alive until 2004. On the contrary, at continuous forest no seedling was recorded neither at the onset of the experimental assessment nor in the time interval evaluated (1998–2004). Consequently, the seedling recruitment rate was far higher at the forest fragments with respect to continuous forest (Fig. 3b).

Fig. 2 Effects of bird consumption and forest fragmentation on seed germination rate of *Aristotelia chilensis* (points correspond to Mean ± 1 SE and are presented at fortnight intervals)

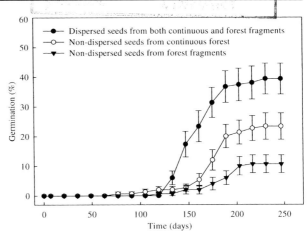

Fig. 3 Seed rain (**a**) and seedling recruitment rates (**b**) of *Aristotelia chilensis* in a continuous forest and forest fragments (Mean ± 1SE)

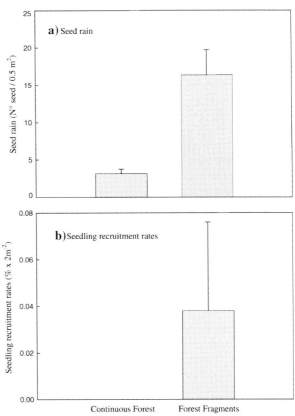

Discussion

Aristotelia chilensis faces a seemingly contradictory scenario in the fragmented Maulino forest. It faces a lowered frugivory and consequently a reduced probability for seed dispersal, yet higher seedling recruitment rates as the final outcome. Forest fragmentation significantly reduced frugivory coupled to the lower abundance of frugivorous birds (Vergara and Simonetti 2004). Contrary to our expectations, forest fragmentation reduced sugar concentration of fruits. However, because we used artificial fruits for the frugivory assessments, thus avoiding the effects of sugar concentration, the lowered frugivory at fragments herein reported may be even more reduced regarding that furgivores prefer more concentrated nectar (Stanley et al. 2002).

A failure in seed dispersal triggered by a lowered frugivory may lead to a clumped pattern of tree recruitment which may in turn favour a higher mating ratio among close relatives, and hence to a likely frugivore-mediated inbreeding depression (Bleher and Böhning-Gaese 2001; Şekercioğlu et al. 2004). Frugivores in the forest fragments, besides providing a lowered seed dispersal service, might carry also seeds with a reduced germination rate, suggesting a likely evidence for inbreeding depression. In fact, perennial and dioecious plants may suffer most strongly the effects of forest fragmentation, in term of inbreeding depression, on account of these reproductive traits (sensu Murcia 1996). Because *A. chilensis* is pollinated by the small-sized bees *Cadeguala albopilosa* and *Ruizantheda mutabilis* (Aizen et al. 2002), which may be strongly restricted for moving across fragments, the possibility of inbreeding depression in *A. chilensis* is indeed to be expected. For instance, in the same forest, the self-compatible vine *Lapageria rosea*, despite exhibiting a suite of traits that render less prone to express inbreeding depression of compared with *A. chilensis*, exhibits a lowered genetic diversity which has triggered inbreeding depression expressed through a lower pollen quality and seed germination capability (Henríquez 2002).

Therefore, the lowered seed dispersal service coupled to a reduced germination rate should lead to *A. chilensis* towards a lowered seedling recruitment rate, and hence towards a likely threat in the long-term population persistence in fragments. Nevertheless, seedling recruitment rates were far greater in fragments than in continuous forest. This fact may be a result of the great number of seeds falling into the fragments, which could exert a ''mass effect'', overcompensating the lowered seed quality and reduced possibilities for seed dispersal. Even though we were incapable of determining the exact origin of the seed rain, the great number of seeds falling into the fragments may be a consequence of the in situ seed production further those seeds produced outside. In fact, the Maulino forest fragments seem to be part of a source-sink system, where there would be an active colonisation process by bird-dispersed trees, since birds would be able of moving across sites with no restriction (sensu Bustamante et al. 2005).

A second aspect to be considered deals with the latter mechanisms leading the seedling recruitment following the seed rain. Population of *A. chilensis* inhabiting the remaining forest fragments do not exhibit the negative density-dependent mechanisms for seedling recruitment exhibited by a great proportion of trees in the continuous forest (Valdivia 2004). For this reason, the high number of seeds falling into the fragments would be not under the mechanisms for population regulation, which increases the probability for seedling recruitment (Valdivia 2004; Bustamante et al. 2005).

Additionally, *A. chilensis* is a short-lived tree with a great phenotypic plasticity being capable to survive in forests with a great variability of environmental light (Lusk and del Pozo 2002). However, *A. chilensis* exhibits a higher rate of growth in light than in darkness of compared with other species of the temperate rainforest of Chile (Lusk and del Pozo 2002). Because light is higher in fragments than in continuous forest (Simonetti, to be published), fragments would be a more suitable habitat for *A. chilensis* regarding the continuous forest, which in turn would increase the likelihood of growth and survival.

As far as seed dispersal by birds is concerned, the scenario faced by *A. chilensis* may be representative of numerous other bird-dispersed trees of the highly fragmented temperate rainforest of South America, as for instance, *Aetoxicon punctatunn, Amomyrtus luma, Drymis winteri*, and *Raphithamnus spinosus*, all of them occurring at the Maulino forest as well (Armesto and Rozzi 1989; Aizen et al. 2002). Concerning the niche regeneration and probabilities for recruitment, however, the same species might respond in many different ways (Lusk and del Pozo 2002; Bustamante et al. 2005). Therefore, several functional responses are to be expected in trees inhabiting the temperate rainforest of southern South America after suffering anthropogenic disturbances such as forest fragmentation. Accordingly, despite some reports have provided evidence that disrupted mutualisms may lead to an almost irrevocable decline in populations (e.g., Cordeiro and Howe 2001, 2003; Traveset and Riera 2005), our report points to consider the importance of assessing the overall mechanisms leading the long-term population persistence of trees, and thus the fate of such forests, before rigorous biological interpretations can be offered.

Acknowledgements   We are indebted to the Chilean Forestry Service (CONAF) and Forestal Millalemu for permission to work in their lands. Fernando Campos kindly helped us with fieldwork. Florencia Prats, Carlos O. Valdivia and Sandra Valdivia also provided invaluable logistic support. This research was funded by Fondecyt 1010852 and 1050745.

References

Aizen MA, Vázquez DP, Smith-Ramírez C (2002) Historia natural y conservación de los mutualismos planta-animal delbosque templado de Sudamérica austral. Rev Chil Hist Nat 75:79–97

Alves-Costa CP, Lopes AV (2001) Using artificial fruits to evaluate fruit selection by birds in the field. Biotropica 33:713–717

Armesto JJ, Rozzi R (1989) Seed dispersal syndromes in the rain forest of Chiloé: evidence for the importance of biotic dispersal in a temperate rain forest. J Biog 16:219–226

Barrett SCH, Kohn JR (1991) Genetic and evolutionary consequences of small population size in plants: implications for conservation. In: Falk DA, Holsinger KE (eds) Genetics and conservation of rare plants: 3–30. Oxford University Press, Oxford, UK, pp 3–30

Bleher B, Böhning-Gaese K (2001) Consequences of frugivore diversity for seed dispersal, seedling establishment and the spatial pattern of seedlings and trees. Oecologia 129:385–394

Bustamante RO, Castor C (1998) The decline of an endangered temperate ecosystem: the ruil (*Nothofagus alessandrii*) forest in central Chile. Biod Cons 7:1607–1626

Bustamante RO, Simonetti JA, Grez AA, San Martín J (2005) Fragmentación y su dinámica de regeneración del bosque Maulino: diagnóstico actual y perspectivas futuras. In: Smith-Ramírez CS, Armesto JJ, Valdovinos C (eds) Historia, biodiversidad y ecología de los bosques costeros de Chile. Editorial Universitaria, Santiago, pp 555–564

Camargo JLC, Kapos V (1995) Complex edge effects on soil moisture and microclimate in central Amazonia forest. J Trop Ecol 11:205–221

Cordeiro NJ, Howe HF (2001) Low recruitment of trees dispersed by animals in African forest fragments. Cons Biol 15:1733–1741

Cordeiro NJ, Howe HF (2003) Forest fragmentation severs mutualism between seed dispersers and an endemic African tree. PNAS 100:14052–14056

Donoso DS, Grez AA, Simonett JA (2003) Effects of forest fragmentation on the granivory of differently sized seeds. Biol Cons 115:63–70

Galetti M, Alves-Costa CP, Cazetta E (2003) Effects of forest fragmentation, anthropogenic edges and fruit colour on the consumption of ornithocoric fruits. Biol Cons 111:269–273

González-Gómez PL, (2004) Intensificación de la insectivoría en un bosque templado fragmentado. M.Sc. Thesis, Facultad de Ciencias, Universidad de Chile, Santiago, Chile

Henríquez CA (2002) El dilema de *Lapageria rosea* en bosques fragmentados: cantidad o calidad de la progenie? Ph.D. Thesis, Facultad de Ciencias, Universidad de Chile, Santiago, Chile

Henríquez CA (2004) Efecto de la fragmentación del hábitat sobre la calidad de las semillas en *Lapageria rosea*. Rev Chil Hist Nat 77:177–184

 Springer

1602 Biodivers Conserv (2007) 16:1593-1602

Ito J, Hasegawa S, Fujita K, Ogasawara S, Fujiwara T (1999) Real time diagnosis of environmental stress by micromorphometric method −1. Effect of air temperature during fruitlet stage of fruit on stem and fruit diameters, and fruit growth in Japanese pear tree (*Pyrus serotina reheder* cv. Kosui). Soil Sci Plant Nut 45:395–402

Lewis SL, Phillips OL, Sheil D, Vinceti B, Baker TR, Brown S, Graham AW, Higuchi N, Hilbert DW, Laurance WF, Lovejoy J, Malhi Y, Monteagudo A, Núñez-Vargas P, Sonké B, Supardi N, Terborgh JW, Vásquez-Martínez R (2004) Tropical forest tree mortality, recruitment and turover rates: calculation, interpretation and comparison when census intervals vary. J Ecol 92:929–944

Lusk CH, del Pozo A (2002) Survival and growth of seedlings of 12 Chilean rainforest trees in two light environments: Gas exchange and biomass distribution correlates. Aust Ecol 27:173–182

Murcia C (1996) Forest fragmentation and the pollination of Neotropical plants. In: Schelhas J, Greenberg R (eds) Forest patches in tropical landscapes. Island Press, Washington, District of Columbia, USA, pp 19–36

Rodríguez R, Matthei O, Quezada M (1983) Flora arbórea de Chile. Editorial de la Universidad de Concepción, Concepción, Chile

San Martín J, Donoso C (1996) Estructura florística e impacto antrópico en el bosque Maulino de Chile. In: Armesto JJ, Villagrán C, Arroyo MK (eds) Ecología de los bosques nativos de Chile. Editorial Universitaria, Santiago, Chile, pp 153–168

Santos T, Telleria JL (1994) Influence of forest fragmentation on seed consumption and dispersal of Spanish juniper *Juniperus thurifera*. Biol Cons 70:129–134

Saracco JF, Collazo JA, Groom MJ (2004) How do frugivores track resources? Insights from spatial analyses of bird foraging in a tropical forest. Oecologia 139:235–245

Şekercioğlu Ç, Gretchen CD, Ehrlich PR (2004) Ecosystem consequences of bird decline. PNAS, USA 101:18042–18047

Sheil D, Burslem D, Alder D (1995) The interpretation and missinterpretation of mortality rates measures. J Ecol 83:331–333

Stanley MC, Smallwood E, Lill A (2002) The response of captive silvereyes (*Zosterops lateralis*) to the colour and size of fruit. Aust J Zool 50:205–213

Traveset A, Riera N (2005) Disruption of a plan-lizard seed dispersal systems and its ecological effects on a threatened endemic plant in the Balearic Islands. Cons Biol 19:421–431

Valdivia CE (2004) Reclutamiento denso-dependiente en un bosque fragmentado. M.Sc. Thesis, Facultad de Ciencias, Universidad de Chile, Santiago, Chile

Vergara PM, Simonetti JA (2004) Avian responses to fragmentation of the Maulino forest in central Chile. Oryx 38:1–6

Willson MF (1991) Dispersal of seeds by frugivorous animals in temperate forests. Rev Chil Hist Nat 64:537–554

Willson MF, de Santo TL, Sabag C, Armesto JJ (1994) Avian communities of fragmented Southern-temperate rainforest in Chile. Cons Biol 8:508–520

Young A, Boyle T, Brown T (1996) The population genetic consequences of habitat fragmentation for plants. TREE 11:414–418

Springer

Biodivers Conserv (2007) 16:1603–1615
DOI 10.1007/s10531-006-9021-1

ORIGINAL PAPER

Vascular plant diversity and climate change in the alpine zone of the Lefka Ori, Crete

**G. Kazakis · D. Ghosn · I. N. Vogiatzakis ·
V. P. Papanastasis**

Received: 24 October 2005 / Accepted: 27 February 2006 / Published online: 16 May 2006
© Springer Science+Business Media, Inc. 2006

Abstract   The aim of this study is to analyse the vascular flora and the local climate along an altitudinal gradient in the Lefka Ori massif Crete and to evaluate the potential effects of climate change on the plant diversity of the sub-alpine and alpine zones. It provides a quantitative/qualitative analysis of vegetation-environment relationships for four summits along an altitude gradient on the Lefka Ori massif Crete (1664–2339 m). The GLORIA multi-summit approach was used to provide vegetation and floristic data together with temperature records for every summit. Species richness and species turnover was calculated together with floristic similarity between the summits. 70 species were recorded, 20 of which were endemic, belonging to 23 different families. Cretan endemics dominate at these high altitudes. Species richness and turnover decreased with altitude. The two highest summits showed greater floristic similarity. Only 20% of the total flora recorded reaches the highest summit while 10% is common among summits. Overall there was a 4.96°C decrease in temperature along the 675 m gradient. Given a scenario of temperature increase the ecotone between the sub-alpine and alpine zone would be likely to have the greatest species turnover. Southern exposures are likely to be invaded first by thermophilous species while northern exposures are likely to be more resistant to changes. Species distribution shifts will also depend on habitat availability. Many, already threatened, local endemic species will be affected first.

G. Kazakis · D. Ghosn
Department of Environmental Management, Mediterranean Agronomic Institute of Chania,
P.O. Box 85, 73100 Chania, Greece

I. N. Vogiatzakis (✉)
Centre for Agri-Environmental Research, School of Agriculture Policy and Development,
University of Reading, Earley Gate, RG6 6AR Reading, UK
e-mail: i.n.vogiatzakis@rdg.ac.uk

V. P. Papanastasis
Laboratory of Rangeland Ecology, Faculty of Forestry and Natural Environment, Aristotle University
of Thessaloniki, GR 54006 Thessaloniki, Greece

 Springer

Keywords   Islands · Global change · Greece · Mountains · Species richness · Species turnover

Abbreviations
GLORIA   Global Observation Research Initiative in Alpine Environments
GCM   General circulation models
LOW   Lowest Summit
CHO   Chorafas
SEK   South-East Kakovoli
STR   Sternes

Introduction

Variations in plant community structure and composition have been attributed to the characteristics of the physical environment, environmental history and past and current human activities (Tivy 1993). However, it is the gradients of these that determine the levels of biodiversity (Gaston and Spicer 1998; Huston 1994). Altitude is an indirect gradient which is correlated with resources and regulators of plant growth (Austin 1980; Austin and Smith 1989). Although it is known that richness of vascular plant species decreases with increases in altitude (Begon et al. 1996; Odland and Birks 1999), the patterns response are variable (Rahbek 1995). Since mountains are characterised by steep climate gradients, they provide ideal settings to study changes in species richness over relatively short distances, as most species have upper altitudinal limits that are set by various climatic parameters and by limitation of resources (Grabherr et al. 1995; Lomolino 2001; Theurillat et al. 2003).

Therefore, understanding changes in plant species richness along altitudinal transects is of great importance in the study of global climate change (Grabherr et al. 1995; Sætersdal et al. 1998). Along these gradients, higher mountain summits contain a strong endemic element since they have been less threatened by human activities and have acted in geological times as natural laboratories (Strid and Papanikolaou 1985; Körner and Spehn 2002). This is exemplified by the Lefka Ori massif, Crete, where the presence of many endemic species supports the argument that the current landscape has been in existence for most of the Pleistocene (Rackham Moody 1996).

The importance of the Lefka Ori massif in the context of national and international biodiversity is well documented (Strid 1993; Phitos et al. 1996; Dafis et al. 1996). The vegetation formations of the Lefka Ori above the tree line are typical of the oro-medi-terranean and alti-mediterranean zone (Quézel 1981a, b) of the Mediterranean mountains. These two zones, which correspond to the sub-alpine and alpine zones of the Central European mountains, comprise either low prickly scrub formations or communities of spiny cushion-shaped dwarf shrubs (Quézel 1981a; Zaffran 1990). Previous research undertaken in Lefka Ori has resulted in several studies of the vegetation formations (Barbero and Quézel 1980; Zaffran 1990; Bergmeier 2002; Vogiatzakis et al. 2003). The majority of these studies agree upon the need for more work to be done in the massif to improve knowledge of species distribution patterns. This will result in improved protection and management and will provide insights on species behaviour in response to future climate change. High mountain zones are considered more vulnerable to climate change

                                        [30]

compared to those of lower altitudes (Körner 1994, 1999) and are also good sites to study the impact of climate change due to low anthropogenic pressure (Beniston 1994).

In recent years, as a result of increasing concern about the effects of climate change, there has been an increase of international initiatives on mountains and climate change (Becker and Bugmann 2001; Grabherr et al. 2001; Huber et al. 2005). In Europe the southern part will be more vulnerable to climate change according to IPCC (2001). Mooney et al. (2001) suggest that in the Mediterranean Basin a warmer climate will push the closed canopy evergreen-oak forest northward while forest communities at the southern margins will be increasingly replaced by low-cover desert shrubland. In mountain regions higher temperatures will lead to an upward shift of biotic zones with possible decrease in the numbers and abundance of endemic species (IPCC 2001; Mooney et al. 2001). Nevertheless, evidence is still scarce on the effects of climate change on the Mediterranean mountain flora with few exceptions (Stanisci et al. 2005; Peñuelas and Boada 2003).

Despite the importance of the Greek flora worldwide (Davis et al. 1994) and the fact that threats to plant diversity are well documented (see Phitos et al. 1996), climate change is not yet high up in the national conservation or political agenda. Recent research suggests that by 2100 temperature in Greece will increase between 3.1 and 5.1°C (Akylas et al. 2005). However, there have been no attempts so far to evaluate the impact of climate change on biodiversity in Greece.

The paper is part of the GLORIA project (http://www.gloria.ac.at/res/gloria_home/) an initiative towards an international research network to assess climate change impacts on mountain environments. This project involved 18 mountain regions from 13 European countries, one of which was Lefka Ori in Crete, Greece. Therefore, the focus herein is on the vegetation and species patterns of the high summits of the Lefka Ori. The aim of this paper is to analyse the vascular flora and the local climate along an altitudinal gradient in the Lefka Ori and evaluate the potential effects of climate change on plant species.

Methods

Study area

The study area is the Lefka Ori or White Mountains of Crete, Greece (Fig.1). Situated on the western part of Crete, Lefka Ori covers 38,500 ha above 1,000 m with 15 peaks above 2,200 m, reaching 2,453 m at Pachnes summit. The massif is the wettest place on the island with 1,900–2,000 mm mean annual precipitation (Rackham and Moody, 1996). It is a rugged marble and dolomite massif rich in rock debris and karstic formations. The soils, typical of heavily eroded hard crystalline limestones and dolomites are calcaric lithosols. They are poor in organic matter, very shallow and stony while calcareous scree formations resulting from limestone weathering are also abundant above 1,900 m. Above the tree line, the vegetation formations comprise either low prickly scrub formations or communities of spiny cushion-shaped dwarf shrubs dominated by *Berberis cretica* L., *Euphorbia acanthothamnos* Heldr. & Sart. ex Boiss., *Juniperus oxycedrus* L. *subsp oxycedrus Acantholimon androcaceum* (Jaub. & Spach) Boiss., and *Astragalus angustifolius* Lam. Grazing, by far the major human impact on Lefka Ori, decreases in intensity along altitude. The four selected summits were Lowest summit (1664 m), Chorafas (1965 m), South-East Kakovoli (2160 m) and Sternes (2339 m), reported hereafter as LOW, CHO, SEK and STR, respectively.

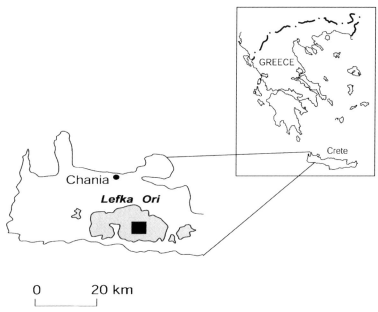

Fig. 1 Greece and the study area: Lefka Ori massif Crete

Field data

Sampling was undertaken from June until August 2001 and was based on the multi-summit approach which is described in detail in the field handbook of the GLORIA project (http://www.gloria.ac.at/res/gloria_home/). Accordingly the summits selected are based on the following criteria: (1) gentle and round morphology with slopes, along the four principal exposures, with an average steepness between 5 and 25°; (2) co-location along the altitude gradient which includes the subalpine, alpine and nival belts; (3) reduced or absent human disturbance.

The survey area in each summit was defined as a polygon with four corners at 10 m lower from the summit top where a complete list of plants was compiled. Within each of these survey areas a smaller one was defined in a similar manner at 5 m lower from the summit top in order to evaluate the quantitative floristic composition for each principal exposure (N–E–S–W). For that purpose, four 9 m² grids (permanent plots) one for each principal exposure were placed in each summit. The list of plants was drawn in each grid and the cover value for every species was determined visually and expressed as a percentage of the total ground cover. Shoot frequency was also counted for every species by measuring as presence any part of the plant within the quadrat. The position of every plot was recorded using a GPS receiver, accompanied by photodocumentation.

The floristic similarity between summits was calculated using the Sørensen index while the chorologic (biogeographic) and life form spectra according to Raunkaier (1937) were also evaluated. For species identification the Flora Europaea (1964–1980) and Mountain Flora of Greece (Strid and Tan 1991) were used. Nomenclature of the plant taxa given in this paper is according to Turland et al. (1993) and Chilton and Turland (1997). For the nomenclature on chorotypes we refer to Jahn and Schönfelder (1995) while bioclimatic zonation is according to Quézel (1981a) and Rivas-Mártinez (1997). Syntaxonomically the area sampled belongs to the *Dapho- Festucetea* class for the lower altitudes and *Thlas-pietea rotundifolia* class for the vegetation on screes (Bergmeier 2002). Voucher speci-mens were collected and kept in the herbarium at the Mediterranean Agronomic Institute at Chania (MAICh).

In addition to the detailed quantitative species description, a data logger was installed at a depth of 10 cm into the soil in the center of each grid in order to record the temperature at 1h intervals. The data loggers installed in 2001 have been removed and substituted with others at the end of 2002 in order to allow readings to be collected for a first analysis. From the temperature data (August 2001–July 2002), the mean annual temperature and the mean daily temperature for each summit was calculated, whereas for each exposure the mean year temperature has been considered. The assumption made herein is that the defrosting period starts when the mean daily temperature are constantly above 0°C while the freezing period starts in the autumn when the mean daily temperatures are constantly below 0°C.

Analysis

Species richness was measured as the total number of species at each altitudinal interval. In addition, the Shannon–Wiener index was used to express **species richness** weighted by species evenness (Krebs 1999). **Species turnover** (beta diversity), was calculated as the gain and loss of species between altitudes according to the formula proposed by Wilson and Shmida (1984):

$$\beta = \frac{g(H) + l(H)}{a(H) + a(H - 1)}$$

where *g(H)* and *l(H)* are the number of species gained and lost, respectively, from altitude $H - 1$ to altitude H and $\alpha(H)$ is the species richness at altitude H. The *Sørensen* similarity coefficient S_s was employed in order to measure floristic similarity between the summits (Kent and Coker 1992):

$$S_s = \frac{2a}{2a + b + c}$$

where a is the number of common species between two summits, b is the number of species unique to the first of the two summits and c is the number of species unique to the second summit.

Results

Temperature data from the data loggers allowed for the analysis of the temperature gra-dient in the soil surface for each summit. According to Fig.2 there is a temperature decrease along the altitude gradient which is followed by a similar decrease in species richness (Table 1). Overall, there is a 4.96°C decrease in temperature from the lowest to the highest summit (a range of 675 m). There is a drop of 2.71°C in 301 m (first lag), a drop of 1.20°C in 195 m (second lag) and 1.05°C in 179 m (third lag).

There is virtually no frost period in the LOW summit since only on the 10th and 11th January 2002 temperatures recorded were below 0 °C. In CHO summit, periods of frost and defrosting alternate (c. week's duration) from 6 December until 27 January, when temperature rises constantly above 0°C. In SEK summit, the defrosting period starts in January since below and above 0°C temperatures alternate. After 28 January, the values were above 0°C with a week in spring i.e. 28 March to 3 April when the temperature were below 0°C. In STR, although it seems that after 28 February frost ends, there is a short period 23 March until 7 April when temperatures fall below and then rise again above 0 °C.

The number of species recorded in the four summits amounts to 70, 20 of which are endemic, belonging to 23 different families. The dominant families are *Cruciferae* (15%), *Graminae* (14%) and *Compositae* (12%), followed by *Caryophyllaceae* (8%) and *Labiatae* (8%). Species number per summit decreased with altitude: 58 species were recorded in the LOW summit, 32 in CHO, 18 in SEK and 14 in STR. The proportion of endemics increases with the elevation gradient starting from 33% for the LOW summit, to 31% for CHO, 33% for SEK and 36% for STR (Table 1).

The diversity index decreased with elevation. The average respective diversity indices for species richness were 1.62 for the LOW summit, 1.21 for CHO, 1.04 for SEK and 0.94 for STR (Table 1). Beta diversity was also found to decrease along the elevation gradient (Table 1). In particular species turnover showed the highest values between 1664 and 1965 m and between 1965 and 2160 m.

The pair STR–SEK showed the highest similarity among the pairs of summits examined (0.69). The LOW summit (tree line) showed the lowest similarity with the other three summits. The highest floristic similarity between STR and SEK is due to common species

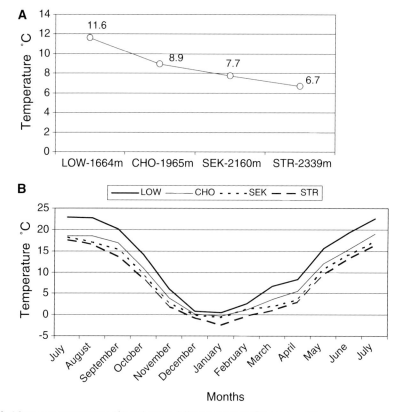

Fig. 2 Mean year temperature (**A**) and mean daily temperature (**B**) per summit

[34]

Table 1 Summary of the summits' properties*

Summit altitude (m)	T (°C)[a]	Species No.[b]	Endemism (%)[c]	α diversity[d]	β diversity[e]
Lowest (1664)	11.6	58	33	1.62	
Chorafas (1965)	8.9	32	31	1.21	0.44
SE Kakovoli (2160)	7.7	18	33	1.04	0.38
Sternes (2339)	6.7	14	36	0.94	0.28

*For details see text: [a]Mean annual temperature, [b]Total species number, [c]Proportion of endemics, [d]Species richness, [e]Species turnover

with a wide altitudinal range such as *Prunus prostrata, Aubrietta deltoidea, Asperula idaea* but also species confined to the highest zones such as *Alyssum sphacioticum* and *Alyssum fragillimum*. More than half of the total species recorded (54%) were found to occur on a single summit. The majority of them were species occurring only on the LOW summit which had the highest species richness. Only 10% of the total recorded flora was shared by the studied summits and comprised the following species: *Prunus prostrata*, a Euro-Mediterranean species, *Aubrieta deltoidea* and *Paracaryum lithospermifolium*, both Balkan species, and the endemics *Acantholimon androsaceum, Draba cretica, Asperula idaea* and *Alyssum fragillimum*.

The chorologic analysis shows that the Cretan endemics had the highest percentage (26%) of the total flora recorded followed by the Euro-Mediterranean element with 18%. These are followed by the East Mediterranean and the Aegean element (both with 12%) and the Balkan element (11%). The distribution of the chorotypes by summit is presented in Fig.3.

The Balkan element shows an increase by altitude from 9% in the LOW summit to 29% in STR, mainly due to the presence of many orophytes, while the Euro-Mediterranean species shows a decrease along the altitudinal gradient from 18% in the LOW summit to 9% in STR. This could possibly be due to the fact that generalist species cannot cope with the adverse environmental conditions of the higher summits. There is also an increase along altitude of the Lefka Ori endemics from 3–9% while Cretan endemics are constantly high in all four summits ranging from 27–30% of the total flora recorded in every summit.

All growth forms were present in all summits i.e. herbaceous perennials, low stature or prostrate woody shrubs, graminoids, annuals and biennials, succulents, geophytes, cushion plants except trees which were present only in the LOW summit. The life form spectrum shows that hemicryptophytes dominate over chamaephytes and therophytes (Fig.3).

The vegetation structure of the LOW summit is typical of low prickly scrub formations dominated by *Juniperus oxycedrus* subsp. *oxycedrus, Acer sempervirens, Thymus capitatus, Prunus prostrata, Euphorbia acanthothamnos*. The CHO summit with *Astragalus angustifolius, Berberis cretica, Prunus prostrata* is typical of cushion heath formations. The two higher summits SEK and STR have typical scree vegetation with *Paracaryum lithospermifolium Peucedanum alpinum, Alyssum fragillimum, Alyssum sphacioticum* and *Euphorbia herniariifolia*.

There seems to be no direct relation between species richness and principal exposure. For example, while the northern exposure is the coldest at the soil level for all summits it was not the poorest one in terms of species (Fig.4). The southern exposure is consistently warm for all summits but it was not in all four summits the richest in species. Vegetation cover was highest in the LOW summit and lowest in the STR summit. However, there is no apparent relation between vegetation cover and aspect (Fig.4).

🌀 Springer

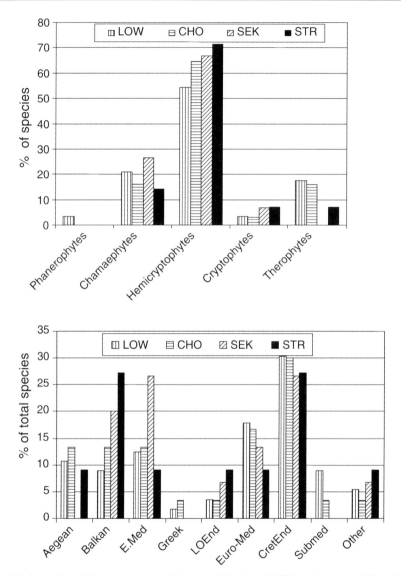

Fig. 3 Life form (**A**) and chorotypes spectra (biogeographic element) (**B**) 1 = Aegean, 2: Balkan, 3: East Mediterranean, 4: Greek, 5: Lefka Ori endemic, 6: Euro-Mediterranean 7: Cretan endemic 8: Submediterranean, 9: Other

Discussion

Species responses to climate change (Harrison et al. 2001; Austin and Rehfisch 2003) will result in direct changes in the composition and structure of plant communities but also indirect threats (Hódar and Zamora 2004). Current simulations for the Mediterranean Basin

▶

Fig. 4 Species richness (**A**), Vegetation cover (**B**) and Mean annual temperature (**C**) for each of the main aspects of each summit

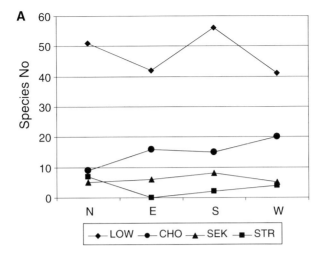

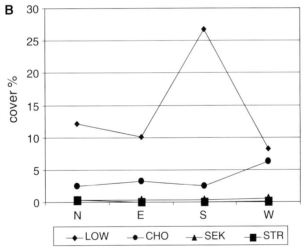

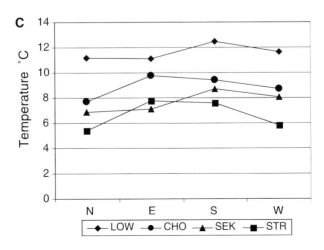

predict a warming of about 2°C in winter and between 2 and 3°C in summer by 2100 (Palutikof and Wigley 1996). Although lower than the rest of the Mediterranean, the increase in mean annual temperature in Crete is estimated to be 1.3–1.4°C with temperatures in the summer increasing between 2.0 and 2.2°C (Akylas et al. 2005). How will this warming scenario affect vegetation structure and composition in the higher zones of Lefka Ori?

The first process to be recorded will concern the shift of more thermophilous species towards higher bioclimatic zones; therefore there will be an increase of species richness in higher altitudes. For many alpine species chilling is required for germination, therefore any changes in frost pattern will in turn affect their reproductive abilities. The results from Lefka Ori indicate that in the highest zones the southern exposures will probably be affected first by the arrival of new species, whereas the northern exposures will be the most resistant to changes. These findings differ somewhat from those reported by Stanisci et al. (2005) in a similar study in Majella massif in Italy, who suggested that the eastward exposures will be the first affected by new arrivals.

The decrease in species number along the elevation gradient studied support the hypothesis that species richness decreases with altitude (e.g. Woodward 1987; Körner 1999). This decrease reflects limited habitat diversity and harsh conditions due to environmental stress. Although the number of endemics follow the same pattern, their percentage contribution to the overall flora shows an increase (Table 1) which is in agreement with other studies for Lefka Ori (Bergmeier 2002; Vogiatzakis et al. 2003). This confirms the hypothesis that such taxa are more susceptible to bioclimatic zone shifting and therefore extinction (Stanisci et al. 2005; Pauli et al. 2001). Future changes will first affect local endemics of the massif's highest alpine zone. These comprise in particular the populations of already endangered species (Phitos et al. 1996) such as *Nepeta sphaciotica*, *Ranunculus radinotrichus* and *Alyssum sphacioticum*. *Nepeta sphaciotica* in particular is confined to the northern slopes of Svourichti mountain in Central Lefka Ori at an altitude of 2200–2300 m (Phitos et al. 1996). Although under a warming scenario it might face extinction since it occurs at a northern aspect, it might resist changes longer than other species on more susceptible exposures as temperature data for the area indicate. Other non-endemic species with a distribution restricted to the uppermost parts of the massif such as *Peucedanum alpinum, Festuca sipylea, Arabis alpina*, will be equally affected.

Species turnover in this study showed the highest values between 1664 and 1965 m a.s.l. This altitude overlaps more or less with the sub-alpine zone, which is the transition between montane-mediterranean and the alpine zone. This high turnover should be attributed to the loss of montane-mediterranean species and the gain of more alpine species, in combination with the relatively low species richness in these altitudes. Another possibility could be the patchy vegetation occurring generally in mountain areas (Sklénar and Ramsay 2001). Therefore, in terms of climate change these middle zones (ecotones) are of great importance for identifying possible future boundary shifts and predicting the fate of species in the higher altitudes.

Although changes in vegetation physiognomy are among the most common responses expected (Harrisson et al. 2001; Mooney et al. 2001) in Lefka Ori the geomorphology of the highest zones, which comprises mainly screes, has also a role to play. Plants occurring on rocks and screes are light-requiring species which would not be able to compete for resources under a dense growth (Ellenberg 1988). Most of these plants have developed various adaptive strategies in this hostile environment (Ellenberg 1988; Vogiatzakis et al. 2003). Therefore, it is more likely that vegetation composition will change with species which are more thermophilous but at the same time able to adapt to life in screes.

🖄 Springer

The current conservation objectives in Lefka Ori under its designation as National Park and Natura 2000 site focus exclusively on in situ protection with priorities based on existing information. The main challenges facing present conservation policy and practice concerns prediction of the patterns and rates of change in species' abundance, distributions and phenology in response to climate change (Hosell et al. 2003). This has placed conservation effectiveness as it stands currently under scrutiny (Araujo et al. 2004). The problem is aggravated in mountainous regions, where interaction of complex gradients has created a mosaic of habitat conditions. Despite criticisms, the development of climate change impact models and scenarios at national or even regional level in the Mediterranean could assist with such an immense task. There are many examples worldwide that can be followed (e.g. Harrison et al. 2001; Bartlein et al. 1997; Scott et al. 2002).

Monitoring studies such as the current one are of utmost importance and provide insights into the responses of mountain ecosystems to climate change. These studies coupled with in situ conservation, bioclimatic modeling and, where necessary, ex situ measures will allow safeguarding of the species and ecosystems subject to threat.

Acknowledgements   The research was funded by the project EU-GLORIA (2001–2003) No. EVK2-CT-2000-00056. We are thankful to Mrs Christina Fournaraki at MAICH for her assistance with plant identification, Dr A.M. Mannion and Dr G.H. Griffiths at the University of Reading for their comments on an earlier draft of this paper.

References

Akyllas E, Lykoudis S, Lalas D (2005) Climate Change in Greece. National Observatory Report, Athens (in Greek)

Araújo MB, Williams PH, Cabeza M, Thuiller W, Hannah L (2004) Would climate change drive species out of reserves? An assessment of existing reserve-selection methods. Glob Change Biol 10:1618–1626

Austin MP (1980) Searching for a model for vegetation analysis. Vegetatio 43:11–21

Austin MP, Smith TM (1989) A new model for the continuum concept. Vegetatio 83:35–47

Austin GE, Rehfisch MM (2003) The likely impact of sea level rise on waders (Charadrii) wintering on estuaries. J Nat Conserv 11:43–58

Barbero M, Quézel P (1980) La végétation forestière de Crète. Ecol Mediterr 5:175–210

Bartlein PJ, Whitlock C, Shafer SL (1997) Future climate change in the Yellowstone National Park region and its potential impact on vegetation. Conserv Biol 11:782–792

Becker A, Bugmann H (eds) (2001) Global change and mountain regions: The Mountain Research Initiative. IGBP Report 49, IGBP Secretariat, Stockholm, Sweden

Begon M, Harper JL, Townsend CR (1996) Ecology: individuals, populations and communities, 2nd edn. Blackwell Scientific, Oxford

Beniston M (1994) Mountain environments in changing climates. Routledge, London

Bergmeier E (2002) The vegetation of the high mountains of Crete: a revision and multivariate analysis. Phytocoenologia 32:205–249

Chilton L, Turland NJ (1997) Flora of Crete: a supplement. Marengo Publications, Retford

Dafis S, Papastergiadou E, Georghiou K, Babalonas D, Georgiadis T, Papageorgiou M, Lazaridou T, Tsiaoussi V (eds) (1996) Directive 92/43/EEC – The Greek "Habitat" project NATURA (2000): an overview. Commission of the European Communities DG XI, The Goulandris Natural History Museum – Greek Biotope/Wetland Center

Davis SD, Heywood VH, Hamilton AC (eds) (1994) Centres of Plant Diversity. WWF/IUCN, Cambridge

Ellenberg H (1988) Vegetation ecology of Central Europe, 4th edn. Cambridge University Press, Cambridge

Gaston KJ, Spicer JI (1998) Biodiversity: an introduction. Blackwell Science, Oxford

Grabherr G, Gottfried M, Gruber A, Pauli H (1995). Patterns and current change in alpine plant diversity. In: Chapin FS, Körner C (eds) Arctic and Alpine biodiversity: pattern causes and ecosystems consequence ecological studies, vol 113. Springer, Heidelberg, Germany, pp 167–181

Grabherr G, Gottfried, M, Gruber A, Pauli, H (2001) Aspects of climate change in the Alps and the high arctic region. Long term monitoring of mountain peaks in the Alps. In: Burga CA, Kratochwil A (eds) Biomonitoring: general and applied aspects on regional and global scales. Kluwer Academic Publishers, Dordrecht, The Netherlands, pp 153–177

 Springer

Harrison PA, Berry PM, Dawson TE (eds) (2001) Climate change and nature conservation in Britain and Ireland: modelling natural resource responses to climate change (the MONARCH project). UKCIP Technical Report, Oxford

Hódar JA, Zamora R (2004) Herbivory and climatic warming: a Mediterranean outbreaking of caterpillar attacks on a relict boreal, pine species. Biodivers Conserv 13:493–500

Hossell JE, Ellis NE, Harley MJ, Hepburn IR (2003) Climate change and nature conservation: implications for policy and practice in Britain and Ireland. J Nat Conserv 11:67–73

Huber UM, Bugmann HKM, Reasoner MA (eds) (2005) Global change and mountain regions: an overview of current knowledge. Advances in global change research, Series vol 23. Springer

Huston MA (1994) Biological diversity: the co-existence of species in changing landscapes. Cambridge University Press, Cambridge

IPCC (International Panel on Climatic Change) (2001) Climate change 2001. Impacts adaptation and vulnerability. CUP

Jahn R, Schönfelder P (1995) Exkursionsflora für Creta. Ulmer, Stuttgart

Kent M, Coker P (1992) Vegetation description and analysis: a practical approach, 2nd edn. Belhaven, London

Körner C (1994). Impact of atmospheric changes on high mountain vegetation. In: Beniston M (eds) Mountain environments in changing climates. Routledge, London, pp 155–166

Körner C (1999) Alpine plant life: functional plant ecology of high mountain ecosystems. Springer, Berlin, 338 pp

Körner C, Spehn EM (2002) Mountain biodiversity: a global assessment. CRC Press, 350pp

Krebs CJ (1999) Ecological methodology, 2nd edn. Addison-Welsey

Lomolino MV (2001) Elevation gradients of species-diversity: historical and prospective views. Glob Ecol Biogeogr 10:3–13

Mooney HA, Kalin Arroyo MT, Bond WJ, Canadell J, Hobbs RJ, Lavorel S and Neilson RP (2001) Mediterranean climate ecosystems. In: Chapin FS, Sala OE, Huber-Saanwald E (eds) Global diversity in a changing environment. Scenarios for the 21st century ecological studies, vol 152. Springer, New York, pp 157–199

Odland A, Birks HJB (1999) The altitudinal gradient of vascular plant richness in Aurland, Western Norway. Ecography 22:548–566

Palutikof JP, Wigley TML (1996). Developing climate change scenarios for the Mediterranean region. In: Jeftic L, Keckes S, Pernetta JC (eds) Climate Change and the Mediterranean, vol 2. Edward Arnold, London, pp 27–56

Pauli H, Gottfried G, Grabherr G (2001) High summits of the Alps in a changing climate. In: Walter A, Burga, A, Edwards PJ (eds) Fingerprints of climate change, adapted behaviour and shifting species range. Kluwer Academic Publishers, Dordrecht, The Netherlands, pp 139–149

Peñuelas J, Boada M (2003) A global change-induced biome shift in the Montseny mountains (NE Spain). Glob Change Biol 9:131–140

Phitos D, Strid A, Snogerup S, Greuter W (1996) The red data book of rare and threatened plants of Greece. WWF, Athens

Quézel P (1981a) Les hautes montagnes du maghreb et du proche-orient:essai de mise en parallèle des charactères phytogéographiques. An Jard Bot Madr 37(2):353–372

Quézel P (1981b) The study of groupings in the countries surrounding the Mediterranean: some methodological aspects. In: Di Castri F, Goodall DW, Sprecht RL (eds) Mediterranean-type shrublands. Elsevier, Amsterdam, pp 87–93

Rackham O, Moody JA (1996) The making of the cretan landscape. Manchester University Press, Manchester

Rahbek C (1995) The elevation gradient of species richness: a uniform pattern? Ecography 18:200–205

Raunkaier C (1937) Plant life forms. Clarendon Press, Oxford

Rivas-Martínez S (1997) Bioclimatic map of Europe. Servicio Cartográfico. Universidad de León, León

Sætersdal M, Birks HJB, Peglar SM (1998) Predicting changes in Fennoscandian vascular plant species richness as a result of future climatic change. J Biogeogr 25:111–122

Scott D, Malcolm JR, Lemieux C (2002) Climate change and modelled biome representation in Canada's national park system: implications for system planning and park mandates. Glob Ecol Biogeogr 11:475–484

Sklenár P, Ramsay PM (2001) Diversity of zonal páramo plant communities in Ecuador. Divers Distrib 7:113–124

Stanisci A, Pelino G, Blasi C (2005) Vascular plant diversity and climate change in the alpine belt of the central Apennines (Italy). Biodivers Conserv 14:1301–1318

Strid A (1993) Phytogeographical aspects of the Greek mountain flora. Frag Floristica Geobot Suppl 2:411–433

Strid A, Papanikolaou K (1985) The Greek mountains. In: Gomez-Campo C (ed) Plant Conservation in the Mediterranean area. Dr. Junk, Dordrecht, pp 89–111

Strid A, Tan K (eds) (1991) Mountain flora of Greece, vol 2. Edinburgh University Press, Edinburgh

Theurillat JP, Schlüssel A, Geissler P, Guisan A, Velluti, C, Wiget L (2003) Vascular plants bryophyte diversity along elevation gradients in the Alps. In: Nagy L, Grabherr G, Körner CH, Thompson DBA (eds). Alpine biodiversity in Europe. Ecological studies, vol 167. Springer, Berlin, pp 185–193

Tivy J (1993) Biogeography: a study of the plants in the ecosphere, 3rd edn. Longman Scientific & Technical, Harlow

Turland NJ, Chilton L, Press JR (1993) Flora of the Cretan area. Annotated checklist and Atlas. HMSO, London

Vogiatzakis IN, Griffiths GH, Mannion AM (2003) Environmental factors and vegetation composition, Lefka Ori massif Crete, S.Aegean. Glob Ecol Biogeogr 12:131–146

Wilson MV, Shmida A (1984) Measuring beta diversity with presence–absence data. J Ecol 72:1055–1064

Woodward FI (1987) Climate and plant distribution. Cambridge studies in ecology. Cambridge University Press, Cambridge

Zaffran, J (1990) Contributions à la Flore et à Vegétation de la Crète. Universitè de Provence, Aix en Provence

🍋 Springer

Biodivers Conserv (2007) 16:1617–1631
DOI 10.1007/s10531-006-9024-y

ORIGINAL PAPER

RAPD variation among North Vietnamese *Flemingia macrophylla* (Willd.) Kuntze ex Merr. accessions

Bettina Heider · Meike S. Andersson · Rainer Schultze-Kraft

Received: 28 February 2005 / Accepted: 20 February 2006 / Published online: 20 May 2006
© Springer Science+Business Media B.V. 2006

Abstract *Flemingia macrophylla* (Willd.) Kuntze ex Merr., a multi-purpose legume with potential as dry-season forage crop, mainly occurs in subhumid to humid environments of tropical and subtropical Asia. Despite increasing interest in conservation of germplasm suitable for low-input production systems information on the genetic diversity of *F. macrophylla* is extremely scarce. The creation of baseline data is supposed to contribute to more efficient conservation management and to identify collecting strategies of novel germplasm. Random amplified polymorphic (RAPD) markers were used to investigate the genetic variation among 37 *F. macrophylla* accessions. Germplasm analysed in this study originated from Bac Kan province, Northeast Vietnam. Eight primers generated a total of 47 amplified RAPD loci of which 38 were polymorphic. Jaccard's similarity coefficients among accessions ranged from 0.069 to 1 with a mean of 0.67. The UPGMA dendrogram revealed three clusters along with three outliers. No correspondence between geographic and genetic distance was found (Mantel test: $R = 0.21$; $P = 0.016$). Analysis of molecular variance (AMOVA) revealed significant ($P < 0.001$) differentiation between accessions collected in lowland and upland regions. Results of UPGMA clustering were confirmed by the pattern of principle coordinates analysis (PCO) plotting. Future collecting strategies should target populations at large distances and along the altitudinal range. Ex situ conservation should encompass those accessions that showed genetic divergence. In situ conservation may consist of establishing a system of interconnected population fragments to guarantee continuing genetic exchange via corridors and of rehabilitating degraded habitats.

B. Heider (✉) · R. Schultze-Kraft
Department of Biodiversity and Land Rehabilitation, Institute of Plant Production and
Agroecology in the Tropics and Subtropics, University of Hohenheim, Garbenstrasse 13,
D-70599 Stuttgart, Germany
e-mail: heider@uni-hohenheim.de

M. S. Andersson
Centro Internacional de Agricultura Tropical (CIAT), Forages, A.A. 6713, Cali, Colombia

🖉 Springer

Keywords   Dry season feed · Ex situ conservation · *Flemingia macrophylla* ·
Genetic diversity · In situ conservation · Multi-purpose legume · Northeast Vietnam ·
RAPD · Soil conservation

Introduction

Flemingia macrophylla (Willd.) Kuntze ex Merr. (Fabaceae) is a woody, perennial
shrub of up to 3 m in height. It naturally occurs in subhumid to humid environments
of tropical and subtropical regions, usually below 2000 m a.s.l. Due to its profound
tap root system it withstands drought periods of 3 to 4 months while retaining green
leaves (Asare 1985; Godefroy 1988). *F. macrophylla* tolerates poor soil drainage or
waterlogging (Budelman and Siregar 1997; Shelton 2001), resists fires, and is well-
adapted to acid, infertile soils with high soluble Al-saturation (Budelman 1989;
Schultze-Kraft 1996).

F. *macrophylla* is a multi-purpose legume with a potential as dry-season forage
crop. Due to its soil conserving and even rehabilitating quality, the species is used as
nematode control, cover crop, mulch, contour hedgerows, erosion control and as
green manure increasing the yield of associated food crops. Moreover, *F. macrophylla* has significance as shade plant in young coffee or cocoa cultivations, medicinal plant, silk dye, host for lac producing insects, and as fuelwood (e.g. Perry and
Metzger 1980; Budelman and Siregar 1997; Andersson et al. 2002; 2006b).

Adaptation to depauperate soils, tolerance to water stress and regrowth after
cutting along with multi-purpose characteristics constitute a promising suitability for
low-input production systems in the subhumid and humid tropics. *F. macrophylla* is a
good example of a formerly neglected species that became interesting as the demand
for soil improving germplasm adapted to degraded sites (Hanson and Maass 1997)
and drought arose (Andersson et al. 2003). Irrespective of this, *F. macrophylla* is a
poorly researched species. The genus is in need for further studies including
taxonomic revision (van der Maesen 2003). Precise taxonomic classification and
documentation of species and species relationships as well as an understanding of the
genetic variation in a species are imperative for collecting, selection, genetic
improvement, sustainable germplasm management and effective conservation. The
existing taxonomic confusion within the genus *Flemingia* and the lack of available
genetic diversity data has most likely impeded the comprehensive inclusion of
utmost genetic diversity in plant improvement and conservation programs
(Valdecasas and Camacho 2003).

Despite the considerable efforts devoted to collect legume forage species in
Thailand, Malaysia, Indonesia, tropical China, and Vietnam, the genetic base of
F. *macrophylla* in genebanks is still narrow (Schultze-Kraft et al. 1987; 1989a, b;
Schultze-Kraft and Pattanavibul 1990; Schultze-Kraft et al. 1993; Hanson and Maass
1997). This is partly due to the fact that during former expeditions the sampling of
F. macrophylla germplasm was not a priority and partly because some areas were not
visited. Vietnam is ranked among the world's ten biologically most diverse countries
(World Bank 2002). Located within Southeast Asia, Vietnam is also considered part
of a legume diversity centre (Williams 1983). At the same time, the fragile mountain
ecosystems in North Vietnam are threatened by genetic erosion as a consequence of
ecosystem transformation and resource overexploitation during the last decades
(Jamieson et al. 1998; Quy 1998). In order to complement previous initiatives the
mountainous North of Vietnam was chosen as collecting area in this study targeting

at herbaceous and shrubby legume species with forage or soil improving potential (Heider et al. 2002).

Regardless of North Vietnam's significance as part of a centre of origin for legume species, there exists no information on genetic diversity among *F. macrophylla* populations from North Vietnam, though the wide geographical and environmental range in which the species occurs, suggests a similar diversity dimension at the genome level. The classical methodology of assessing genetic diversity is to study phenological and morphological characteristics which, however, might be influenced by environmental factors. Molecular markers provide a more reliable and accurate estimation of genetic diversity. Among the different DNA based approaches Random Amplified Polymorphic DNA (RAPD) and Amplified Fragment Length Polymorphism (AFLP) markers are widely used to assess genetic variation in plants. Preliminary trials were conducted with AFLPs due to their good reproducibility, but owing to partial DNA digestion—probably caused by high tannin contents in *F. macrophylla*—the approach did not succeed and finally studies using this methodology could not be completed (Andersson et al. 2003). In contrast, for RAPDs neither previous information on DNA sequence of the species under investigation, nor enzyme digestion is required. RAPD markers have been successfully applied to analyse population structures, conduct taxonomic and phylogenetic studies, and assess intra- and interspecific genetic variation in legume shrubs or trees (e.g. Harris 1995; Liu 1997; Casiva et al. 2002).

The objectives of this study were to determine genetic variation among accessions of *F. macrophylla* originating from Bac Kan province, North Vietnam, using RAPDs to contribute to conservation management of this germplasm for potential future use, and to identify collecting strategies for future germplasm.

Materials and methods

Plant material

Germplasm analysed in this study consisted of 37 *Flemingia macrophylla* accessions which were originally collected in Bac Kan province, Northeast Vietnam (Table 1; Heider et al. 2002). During the collecting process passport data were recorded and the collecting site classified as either lowland or upland. Any collecting site of 300 m a.s.l. or above was regarded as upland and below as lowland. Each accession represents a bulked seed sample deriving from a natural population at a single site. Seedlings were raised under greenhouse conditions before being transplanted into the field at the Quilichao Experimental Station of the Centro Internacional de Agricultura Tropical (CIAT), Cali, Colombia. A randomised complete block design with three replicates was used as experimental set-up (Andersson et al. 2006a).

DNA isolation and amplification

Three individual plants per accession, each providing five young leaves, formed each bulk. Bulk samples were shock-frozen and homogenised in liquid nitrogen. Template DNA extracted from 50 mg of leaf tissue was processed according to a small scale DNA extraction method by Qiagen (DNeasy Plant Mini Kit) with the following modifications: 600 instead of 400µl Buffer AP1, 2 instead of 4µl RNAse, and 150

Table 1 Passport data as recorded at collecting sites of 37 *F. macrophylla* accessions

Accession no.	Latitude	Longitude	Altitude (m) a.s.l.	Site	Soil fertility	Soil characteristics
394	N 22°29′21″	E 106°04′24″	510	Slope	Poor	Compacted
400	N 22°25′26″	E 105°59′05″	660	Slope	Good	Good
403	N 22°24′40″	E 106°00′37″	625	Depression	Poor	Eroded
421	N 22°18′45″	E 105°55′27″	410	Depression	Poor	Compacted
442	N 22°10′24″	E 105°50′04″	240	Hillfoot	Poor	Degraded
464	N 22°07′29″	E 105°45′56″	150	Plain	Moderate	Compacted
473	N 22°10′23″	E 105°39′20″	310	Slope	Poor	Degraded
542	N 22°29′43″	E 106°01′31″	425	Slope	Poor	Degraded
562	N 22°29′52″	E 106°04′22″	650	Slope	Poor	Poor drainage
576	N 22°25′46″	E 105°51′55″	300	Slope	Excellent	Humid
582	N 22°19′39″	E 105°57′51″	521	Slope	Good	Shallow
591	N 22°20′24″	E 106°00′36″	450	Slope	Good	Profound
595	N 22°26′02″	E 106°03′03″	710	Slope	Poor	Very humid
601	N 22°26′17″	E 106°02′37″	675	Slope	Poor	Degraded
607	N 22°23′04″	E 105°55′44″	340	Plain	Poor	Humid
614	N 22°36′52″	E 105°38′22″	350	Slope	Good	Humid
615	N 22°36′52″	E 105°38′22″	350	Plain	Excellent	Humid
629	N 22°06′21″	E 105°40′23″	280	Plain	Moderate	Humid
632	N 22°08′07″	E 105°34′25″	290	Hillfoot	Moderate	
659	N 22°09′43″	E 105°46′19″	240	Plain	Good	Humid
682	N 22°03′13″	E 105°45′06″	190	Plain	Moderate	Humid
712	N 22°16′48″	E 106°12′23″	330	Slope	Poor	Compacted
740	N 22°19′01″	E 106°10′00″	300	Slope	Poor	Degraded
753	N 22°08′03″	E 106°06′07″	300	Slope	Roadside	Shallow
780	N 22°19′15″	E 106°01′12″	400	Plain	Good	Humid
804	N 22°01′29″	E 106°04′32″	300	Slope	Poor	Compacted
816	N 22°02′47″	E 105°50′48″	180	Plain	Good	Humid
821	N 22°09′51″	E 105°55′15″	170	Plain	Good	Profound
843	N 22°15′43″	E 105°57′38″	620	Plain		Humid
857	N 22°01′11″	E 105°52′14″	200	Plain		Humid
870	N 21°58′34″	E 105°49′01″	110	Plain	Poor	Degraded
881	N 22°04′33″	E 106°00′40″	350	Slope	Moderate	Compacted
896	N 21°56′30″	E 105°57′26″	360	Slope	Poor	Degraded
906	N 21°51′11″	E 105°48′46″	100	Terraced slope	Excellent	Humid
914	N 21°55′08″	E 105°51′18″	200	Terraced slope	Poor	Degraded
923	N 22°33′29″	E 105°42′21″	550	Plain	Poor	Humid
924	N 22°19′36″	E 105°54′12″	270	Slope	Poor	Degraded

instead of 130μl Buffer AP2. DNA concentration was measured by a fluorometer (DyNA Quant 200, Hoefer Scientific Instruments) and determined against known standards of λ-DNA concentrations by agarose gel electrophoresis.

An initial screening of 47 decamer primers (Operon Technologies, Alameda, CA, USA) resulted in a selection of eight primers that yielded clear and consistently reproducible banding patterns (Table 2).

PCR reactions were performed in a final volume of 25μl (5μg DNA μl^{-1}). Each reaction mixture consisted of 25 ng template DNA, 0.2μM primer, 800μM dNTPs, 25 mM PCR buffer, 2.5 mM MgCl$_2$, and 1 unit *Taq*-polymerase. Reaction mixtures were placed in a PTC-100 thermal cycler (MJ Research Inc.) and subjected to the following amplification conditions: initial denaturation of 5 min at 94°C; 40 cycles of 30 s at 94°C (denaturation), 30 s at 38°C for annealing and 1 min at 72°C for primer extension concluded by a final extension step of 5 min at 72°C. Amplification

Table 2 Oligonucleotide primers employed in RAPD analysis of *F. macrophylla*: sequence, number of bands obtained and percentage of polymorphic bands per species (% PBS)

Primer code	Sequence (5′ to 3′)	Number of bands	
		Polymorphic	Monomorphic
D 01	ACCGCGAAGG	6	0
D 04	TCTGGTGAGG	2	0
D 15	CATCCGTGCT	10	0
I 07	CAGCGACAAG	2	0
J 04	CCGAACACGG	2	1
J 06	TCGTTCCGCA	8	0
J 07	CCTCTCGACA	6	1
J 12	GTCCCGTGGT	2	0
Total		38	2
PBS (%)		95	

products were electrophorised through 1.4% agarose gel run at 20 V cm^{-1} in 0.5 × TBE buffer. Gels were finally stained with ethidium bromide, exposed to UV-light and photographed using a digital camera (Kodak DC 120).

Data analysis

Presence (1) or absence (0) of reproducible fragments were scored for each analysed accession generating a binary matrix. For each pairwise comparison, the Jaccard's similarity coefficient (JSC) was calculated using SIMQUAL (Similarity for Qualitative Data) routine. Based on JSCs, cluster analysis was performed. A dendrogram was constructed by means of the unweighted pair-group method with arithmetic averages (UPGMA) employing SAHN (Sequential, Agglomerative, Hierarchical, and Nested clustering). These calculations were conducted using NTSYS-pc, version 2.0 (Rohlf 1998). Based on the calculation of pairwise binary genetic distances for dominant data (Excoffier et al. 1992; Huff et al. 1993), the subsequent analyses were performed using GenAlEx V5 (Peakall and Smouse 2001). Analysis of molecular variance (AMOVA) was implemented to estimate variance components and their significance levels for variation between highland and lowland regions of Bac Kan province and among accessions within regions. PCO (Principal Coordinates Analysis) was computed to investigate the overall variation in the present data set and to provide an additional representation of genetic similarity. Finally, Mantel test (999 permutations) was applied to analyse correlations between genetic and geographic distances for each accession (Mantel 1967).

Results

Eight primers selected to detect polymorphism among 37 *F. macrophylla* accessions generated a total of 40 amplified DNA bands of which 38 were polymorphic, i.e. 95% of polymorphism was obtained (Table 2). The number of fragments scored per primer ranged between 2 (D 04, I 07, and J 12) and 10 (D 15) with an average of five while the percentage of polymorphic bands produced per primer ranged between 100% (D 01, D 04, D 15, I 07, J 06 and J 12) and 50% (J 04). Molecular mass of bands varied from 450 to 2500 bp in size. Seven unique markers were detected. In accession 712, primer D 01 yielded three and primer D 15 two unique bands. Primer I 07 and primer J 07 produced one unique band each in accessions 562 and 712, respectively. Examples of RAPD profiles due to primer J 07 are shown in Fig. 1.

JSCs among all analysed *F. macrophylla* accessions ranged from 0.069 to 1 with a mean of 0.67. Average JSC among the 35 *F. macrophylla* accessions, excluding the cluster formed by accession 562 and 712, amounted to 0.73. Perfect similarity (JSC = 1) occurred between accession 464 and 629, as well as between 601, 607, and 780. The lowest JSC was found in comparisons of accession 712 with accessions 629, 914, and 923 (Appendix 1).

The UPGMA dendrogram denoted three principle clusters along with three outliers (Fig. 2). Following an initial division at JSC = 0.13, accessions 562 and 712 represented the most distinct cluster. The two accessions comprising this cluster shared 17% of all alleles with each other and only 13% with the rest of the investigated genotypes indicating a high level of dissimilarity. Subsequent clustering at JSCs of 0.59, 0.66, and 0.70 encompassed three outliers (accessions 421, 615, and 403) which originated from different sites in the northern part of the collecting region sharing a tendency to occur in humid but heavily degraded environments. The two main clusters are composed of 86% of all analysed samples. These clusters corresponded to eco-geographic conditions of collecting sites of the respective accessions. The larger cluster containing 18 accessions coincided with sloping collecting sites at altitudes of ≥300 m a.s.l. (uplands), while the smaller cluster consisted of 14 accessions which were collected below 300 m a.s.l. in mainly plain habitats (lowlands).

AMOVA was performed in order to test the influence of lowland and upland habitats on genetic diversity of *F. macrophylla* accessions (Table 3). For this, a subsample of the 32 accessions belonging to the two main clusters was included. Differentiation between lowland and upland regions was highly significant ($P \leq 0.001$). Of the total genetic diversity covered by the two main clusters 15% was due to variation between regions while 85% of variation was distributed among accessions within regions.

Genetic distances among accessions were not significantly correlated to geographic distances, i.e. geographic distance did not sufficiently explain genetic diversity among accessions. A Mantel test resulted in a coefficient of correlation of $R = 0.21$ ($P = 0.016$).

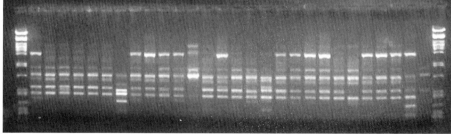

Fig. 1 Example of a RAPD profile among *F. macrophylla* accessions (lanes 1–27 show accession 582–924 in chronological order, lane 28 stands for CIAT acc. no. 18438) detected by primer J 07 (Operon Technologies). Lanes M describe the pattern of a DNA-ladder 10 bp λ-DNA digested with PstI (Gibco-BRL)

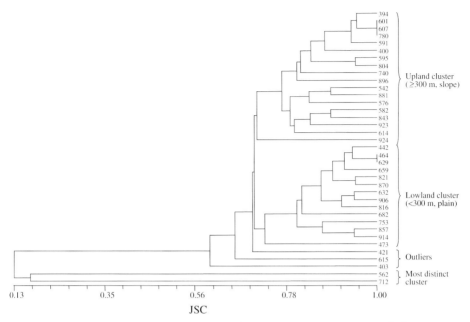

Fig. 2 UPGMA dendrogram visualising the genetic diversity among 37 *F. macrophylla* accessions based on JSCs

Table 3 Analysis of molecular variance (AMOVA) for RAPD variation in a subsample of 32 *F. macrophylla* accessions from the upland and lowland regions of Bac Kan province, North Vietnam, comprising the two main clusters

Source of variation	df	MS	Variance component	% Total	*P*-value
Among regions	1	3.869	0.180	15	<0.001
Accessions within regions	30	0.993	0.993	85	<0.001

Data include degrees of freedom (df), mean squared deviation (MS), variance component estimates, percentage of total variance (% total) contributed by each component, and significance of variance (*P*-value)

Results of UPGMA clustering were confirmed by the pattern of PCO plotting (Fig. 3). A total of 35.2% of the variation among *F. macrophylla* samples was explained by the first three principal coordinates. The first two principal coordinates accounted for 16.7 and 11.8% of total variation, respectively. The first principal coordinate clearly divided accessions analysed into two groups while the second principal coordinate separated outliers and presumed taxonomic misclassifications from the two main clusters, forming an upland and a lowland cluster. This pattern corresponded entirely to clusters of the UPGMA dendrogram.

Discussion

Analysis of variation in RAPD fragments revealed a moderate level of variation among *F. macrophylla* accessions (average JSC = 0.67) and a high level of overall diversity represented by the percentage of polymorphic loci found. The level of

🖄 Springer

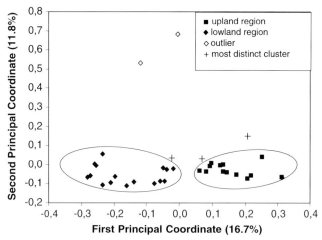

Fig. 3 Principle coordinates analysis of 37 *F. macrophylla* accessions. First and second principal coordinates explain 28.5% of the total variation

polymorphism (95%) across all *F. macrophylla* accessions collected in North Vietnam was considerably higher than the variation observed in previous RAPD studies of tropical legume trees and shrubs. Casiva et al. (2002) reported 19.1% averaged across populations of Argentinian *Acacia* species, Juárez-Muñoz et al. (2002) found an average of 28.3% in *Prosopis* species, Schierenbeck et al. (1997) quantified 43.3% in *Inga thibaudiana*, and Lacerda et al. (2001) detected 70.8% polymorphism in *Plathymenia reticulata*. Andersson et al. (2006c) reported a level of 91.8% of polymorphism for *Cratylia argentea* while RAPD marker analyses conducted by Hawkins and Harris (1998) resulted in moderate levels of variation in *Leucaena leucocephala* (14.3%), *L. esculenta* (17.2%), *Parkinsonia aculeata* (12.1%), and *Cercidium praecox* (23.6%). The high percentage of polymorphism found in our study is consistent with an analysis of genetic diversity in ex situ collections of *F. macrophylla* held by CIAT (94.2%; Andersson et al. 2006d). The CIAT collection comprises accessions from several countries in which the North Vietnamese accessions were recently included. The average number of polymorphic fragments scored per primer was higher in *F. macrophylla* (5) than compared to other legume trees (Schierenbeck et al. 1997; Juárez-Muñoz et al. 2002); equally high in *P. reticulata* (Lacerda et al. 2001) but lower than in *C. argentea* (9.3) (Andersson et al. 2006c).

While the high level of genetic variability represented by a large percentage of polymorphic loci was in agreement with the wide range of JSCs (0.069–1), it contrasted with moderate levels of diversity among accessions (average JSC = 0.67). This phenomenon might be attributable to a higher than expected degree of outcrossing. Inbreeding species tend to be less diverse within but more differentiated among populations while outbreeding species are characterised by high levels of intra-population diversity and low differentiation among populations (Hamrick and Godt 1996). A number of recent studies support the findings of Hamrick and Loveless (1989) who showed that tropical woody plants tend to exhibit high levels of diversity, residing mainly within populations. Schierenbeck et al. (1997) found a JSC (0.68) among *I. thibaudiana* individuals that was comparable to the average similarity values detected among *F. macrophylla* accessions in the present study.

Unfortunately most studies on tropical legume tree and shrub genetic structure used different diversity measures and thus cannot be compared directly with each other. A range of 0.301–0.367 (Shannon index) was considered to be substantial within *P. reticulata* populations while populations were moderately differentiated (Lacerda et al. 2001). Equally, genetic variation was mostly distributed within *Prosopis* populations while 7.15% of the total variation occurred among populations (Juárez-Muñoz et al. 2002). In opposition to these studies, intra-accession variation was considerably low ($H_S = 0.028$; Nei estimates of genetic diversity) in a selection of *F. macrophylla* and *F. stricta* accessions accounting for only 16% of the total genetic diversity ($H_S = 0.179$) (Andersson et al. 2006d). However, as our research interest focussed on the assessment of diversity among accessions, data in hand may not be sufficiently resolved to reflect within accession variation.

Summarising plant isozyme literature, Hamrick et al. (1991) identified a combination of species characteristics including breeding system, seed dispersal, life form and geographic distribution range being decisive factors for the level of genetic variation and its partitioning among and within populations. Of these factors, breeding biology had the most profound effect on the genetic structure of plant populations (Hamrick and Godt 1989). Despite the importance of *F. macrophylla*, there is a total lack of information in literature about its mating system. However, morphological uniformity and lack of segregation observed by one of us (Schultze-Kraft) in field nurseries of several *F. macrophylla* generations suggest a high degree of self-pollination. *Cajanus cajan* (pigeonpea), a distant relative of *F. macrophylla* is generally self-pollinating, but a range of <1–70% outcrossing has been reported (Saxena et al. 1990; 1994; Shiying et al. 2002) while *F. strobilifera* is equipped with large bracts that hide small flowers, thereby suggesting self-pollination (van der Maesen personal communication 2004). Even though *F. macrophylla* flowers are small and therefore likely to be unattractive for insects, field plots were visited by pollinators such as bees, flies, bumblebees and wasps. Natural populations in Vietnam were usually inhabited by aggressive ant species which made seed collecting sometimes difficult. These ant-plant mutualisms are well documented (Beattie 1985). Plants provide food for ants and in turn receive protection from herbivores. However, extrafloral nectaries are not associated with pollination, on the contrary the metapleural secretion of ants may deteriorate pollen (Beattie et al. 1985). So all in all, it appears that *F. macrophylla* is mainly selfing but—as is the case with most tropical legumes considered to be self-pollinated—some degree of outcrossing should be assumed.

A number of pairwise comparisons resulted in perfect similarity thereby identifying three redundant accessions (780 = 601 = 607 and 464 = 629). The remaining 33 accessions were unique genotypes. In order to guarantee efficient germplasm conservation, duplicates should be excluded from ex situ collections provided agronomic or morphological data indicate that low RAPD diversity is correlated with low diversity at quantitative trait loci which is not necessarily the case (Virk et al. 1996; Wouw et al. 1999). Otherwise coding regions for important agronomic features might be lost.

Within the most distinct cluster (Fig. 2), one taxonomical mismatch was detected (562). Based on greenhouse observations, field book notes and passport data it was possible to reconstruct a mislabelling of bags during collecting activities which proved that one *F. strobilifera* accession was accidentally swapped with *F. macrophylla*. Thus, accession 562 was re-classified as *F. strobilifera* (L.) R. Br. It was therefore suspected that accession 712, due to its considerable genetic distinctiveness compared to all other accessions, might as well represent another *F. macrophylla* species or variety but consultation of available documentation did not validate this suspicion. Soil analyses of growing sites of accession 562 and 712 resulted in almost identical values for a group of mineral nutrients, i.e. low MgO content but high contents of CaO, K_2O, and Na_2O. But unfortunately, seedlings of both accessions died prematurely so that exact taxonomic classification could not be scrutinised. However, these findings corroborate that RAPDs are an appropriate methodological approach to identify genotype duplications and taxonomical mismatches in germplasm collections.

Statistical evidence based on a Mantel test showed that groupings of the UPGMA dendrogram did not correspond to geographic distance (Fig. 2). Thus, an isolation-by-distance model does not explain diversity patterns detected in this study. Given the often adjacent location of lowland and upland collecting sites, the pattern of variation found seems to reflect ecological variation: The two main clusters grouped in accordance with a factor combination of altitude and site inclination with site inclination being the most decisive factor, clearly separating a lowland and an upland cluster (Fig. 3). Both clusters, the PCO and the UPGMA dendrogram showed that lowland and upland clusters were genetically distinct (Figs. 2 and 3). Accessions in the upland cluster grew at altitudes ≥300 m a.s.l. at heavily degraded sloping sites whereas accessions of the lowland cluster were collected at altitudes below 300 m a.s.l. on better soils and at plain sites or depressions, less prone to drought. As a result, it appears that microclimatic selection forces based on environmental stress created a moderate but significant differentiation between lowland and upland regions. This differentiation accounted for 15% of the total variation detected between accessions from lowland and upland regions comprising the two main clusters (Table 3). The fact that within a rather small collecting area differentiation among uplands and lowlands was found suggests that gene flow among accessions is limited. This does not surprise as Bac Kan province is characterised by a diverse eco-geographical endowment. Within this environment, *F. macrophylla* populations occurred as isolated patches, being frequently separated by natural barriers such as rivers, forests, and mountain ranges. However, the lack of correlation between genetic diversity and geographic distance found in this study does not generally rule out a correlation if longer distances and a wider geographic range are concerned, but could not be proved by this data set.

It is not clear whether the disjunct occurrence of *F. macrophylla* populations was due to natural distribution patterns or to habitat fragmentation interrupting population continuums. Yet, the high percentage of polymorphic loci along with moderate JSCs suggests that the partitioning of genetic diversity among *F. macrophylla* accessions was not strongly enough differentiated to suspect population genetic bottlenecks (Lynch and Milligan 1994). Current gene flow among *F. macrophylla* populations might be high enough to prevent stronger differentiation though the high level of variability detected might reflect historical population variation indicating that the populations from which *F. macrophylla* accessions were collected have not been isolated long enough yet for genetic drift or selection to affect allelic composition (Ellstrand 1992). As no historic data is available it is not possible to determine if and to which degree genetic erosion due to human interference has already had impact on the genetic diversity of *F. macrophylla* in North Vietnam. However, the fragmentation of previously interlinked habitats results in population isolation and decreasing gene flow (Young et al. 1996). As the mountainous North of Vietnam is affected by overexploitation of natural resources and consequently

[52]

by environmental degradation (Jamieson et al. 1998; Quy 1998), it is very likely that *F. macrophylla* populations underwent fragmentation and reduction of population sizes.

F. macrophylla presently exhibits a relatively high degree of genetic variation at genotype level. Moderate differentiation observed among accessions indicates that the detrimental effects of genetic erosion are not of immediate concern. *F. macrophylla* is currently not threatened by extinction in North Vietnam. However, habitat fragmentation and further reduction of genetic diversity by genetic drift should be avoided to ensure the maintenance of genetic diversity over time.

In order to capture ecological and geographic ranges of *F. macrophylla*, future collecting missions should target populations at large distances and along the altitudinal gradients (Brown and Marshall 1995). If no particular conservation aim is defined, a common principle of it ex situ conservation is to maximise genetic diversity within the stored collection. For the accessions analysed in this study, this encompasses those accessions that showed genetic divergence such as outliers (403, 421, 615), and a representative subsample of different ecotypes from lowland and upland regions.

The ideal in situ protection concept covering the range of genetic variation represented in the target taxon would involve large areas comprising the natural geographic and ecological range of *F. macrophylla* including lowland and upland regions (Margules and Pressey 2000). However, in the case of *F. macrophylla* such an ideal concept would hardly be feasible in an intensively cultivated environment such as Northern Vietnam where land is scarce. In order to restore or maintain gene flow among disrupted populations, a more feasible in situ conservation approach could consist of establishing a system of interconnected population fragments and of rehabilitating degraded habitats as well as of increasing population sizes (Simberloff 1988; Templeton et al. 1990). For tropical tree populations, it was demonstrated that population fragments serve as reservoirs of genetic variation and as sanctuaries of pollen and seed vectors, and therefore deserve special protection (Shafer 1995; Turner and Corlett 1996; Nason et al. 1997).

Conclusions

RAPD analysis of genetic variation among North Vietnamese *F. macrophylla* accessions revealed a high level of genetic variability and moderate levels of differentiation among accessions while detecting a significant differentiation between accessions collected in lowland and upland regions. The study provides genetic information for further and more detailed analysis of genetic diversity of this species. It serves as guidance for the delineation of ex situ and in situ conservation approaches. The generated baseline data are valuable for follow-up research and for future decision making processes associated with management and conservation of genetic resources. RAPDs proved to be highly discriminant but co-dominant molecular markers such as AFLPs should be used for more in-depth investigations of population genetic structure.

Acknowledgements Research was kindly funded by the Vater und Sohn Eiselen-Stiftung, Ulm, Germany. The authors are indebted to the Vietnam Agricultural Science Institute, Hanoi, Vietnam, especially to Prof. Dr. Tran Dinh Long and Mr. Ha Dinh Tuan for their valuable collaboration. Thanks also to Gerardo Ramírez (CIAT) for his statistical advice.

Appendix   Matrix of Jaccard's similarity coefficients of 37 *F. macrophylla* accessions

	394	400	403	421	442	464	473	542	562	576	582	591	595	601	607	614	615	629	632	659	682	712	740	753	780	804	816	821	843	857	870	881	896	906	914	923
400	0.90																																			
403	0.73	0.65																																		
421	0.75	0.75	0.59																																	
442	0.67	0.67	0.46	0.68																																
464	0.71	0.71	0.50	0.74	0.94																															
473	0.55	0.55	0.36	0.55	0.82	0.78																														
542	0.75	0.67	0.68	0.68	0.65	0.55	0.55																													
562	0.13	0.13	0.18	0.15	0.20	0.16	0.10	0.10																												
576	0.84	0.84	0.67	0.78	0.68	0.65	0.55	0.88	0.10																											
582	0.80	0.80	0.64	0.83	0.74	0.79	0.60	0.83	0.20	0.83																										
591	0.90	0.90	0.73	0.75	0.67	0.71	0.55	0.84	0.13	0.84	0.80																									
595	0.85	0.85	0.76	0.68	0.62	0.70	0.57	0.79	0.14	0.70	0.75	0.85																								
601	0.95	0.95	0.86	0.77	0.71	0.64	0.52	0.80	0.13	0.80	0.76	0.95	0.90																							
607	0.95	0.95	0.86	0.77	0.71	0.64	0.52	0.80	0.13	0.80	0.76	0.95	0.90	1.00																						
614	0.84	0.84	0.59	0.78	0.68	0.74	0.55	0.78	0.10	0.78	0.83	0.84	0.79	0.80	0.80																					
615	0.75	0.75	0.52	0.60	0.57	0.72	0.68	0.68	0.15	0.68	0.65	0.75	0.62	0.71	0.71	0.68																				
629	0.71	0.71	0.50	0.74	0.94	1.00	0.78	0.65	0.20	0.65	0.79	0.71	0.75	0.68	0.68	0.74	0.57																			
632	0.71	0.71	0.50	0.65	0.94	0.89	0.78	0.71	0.13	0.71	0.74	0.71	0.75	0.68	0.68	0.74	0.65	0.89																		
659	0.67	0.67	0.52	0.78	0.88	0.94	0.72	0.64	0.21	0.68	0.83	0.67	0.70	0.64	0.64	0.68	0.52	0.94	0.83																	
682	0.59	0.59	0.46	0.68	0.83	0.72	0.68	0.15	0.68	0.74	0.59	0.62	0.57	0.57	0.60	0.52	0.83	0.83	0.88																	
712	0.10	0.14	0.18	0.11	0.07	0.11	0.07	0.11	0.17	0.07	0.11	0.14	0.14	0.13	0.13	0.11	0.07	0.11	0.07	0.11	0.07															
740	0.77	0.86	0.63	0.64	0.68	0.52	0.71	0.13	0.71	0.68	0.86	0.81	0.82	0.82	0.71	0.64	0.68	0.68	0.64	0.57	0.17															
753	0.76	0.76	0.61	0.79	0.84	0.65	0.10	0.19	0.75	0.85	0.80	0.81	0.81	0.70	0.62	0.84	0.75	0.79	0.70	0.14	0.81															
780	0.95	0.86	0.77	0.71	0.64	0.52	0.80	0.13	0.80	0.76	0.95	0.90	1.00	1.00	0.80	0.71	0.68	0.68	0.64	0.57	0.13	0.82	0.81													
804	0.81	0.73	0.65	0.59	0.67	0.71	0.55	0.13	0.67	0.71	0.81	0.95	0.86	0.86	0.75	0.59	0.71	0.71	0.67	0.59	0.18	0.77	0.76	0.86												
816	0.75	0.75	0.52	0.60	0.88	0.83	0.72	0.78	0.10	0.78	0.74	0.75	0.79	0.71	0.71	0.78	0.74	0.83	0.94	0.78	0.07	0.71	0.70	0.73	0.84											
821	0.73	0.73	0.52	0.67	0.84	0.89	0.70	0.67	0.24	0.67	0.80	0.73	0.76	0.70	0.70	0.75	0.59	0.89	0.89	0.84	0.10	0.70	0.68	0.76	0.80	0.74	0.80									
843	0.80	0.71	0.64	0.74	0.74	0.60	0.70	0.20	0.70	0.74	0.79	0.85	0.81	0.81	0.79	0.57	0.84	0.84	0.79	0.70	0.10	0.74	0.74	0.76	0.71	0.85	0.84	0.85								
857	0.85	0.85	0.61	0.79	0.79	0.84	0.65	0.70	0.19	0.70	0.84	0.85	0.80	0.81	0.81	0.84	0.62	0.94	0.94	0.84	0.13	0.79	0.89	0.81	0.76	0.79	0.85	0.84	0.89							
870	0.76	0.76	0.54	0.70	0.89	0.94	0.74	0.70	0.19	0.70	0.76	0.76	0.80	0.73	0.73	0.62	0.94	0.94	0.89	0.10	0.71	0.80	0.85	0.89	0.95	0.84	0.89	0.75	0.71	0.71						
881	0.85	0.76	0.68	0.62	0.70	0.57	0.89	0.09	0.79	0.75	0.85	0.89	0.90	0.90	0.75	0.67	0.75	0.75	0.67	0.62	0.10	0.90	0.71	0.73	0.80	0.68	0.75	0.59	0.58	0.64	0.68	0.61	0.76			
896	0.81	0.81	0.65	0.81	0.84	0.70	0.64	0.08	0.67	0.64	0.81	0.76	0.86	0.86	0.67	0.64	0.57	0.57	0.46	0.14	0.77	0.68	0.80	0.73	0.68	0.89	0.85	0.75	0.89	0.71	0.61					
906	0.76	0.76	0.54	0.70	0.89	0.84	0.70	0.14	0.70	0.73	0.76	0.71	0.76	0.76	0.74	0.74	0.83	0.70	0.68	0.07	0.73	0.80	0.84	0.76	0.71	0.83	0.80	0.79	0.94	0.84	0.75	0.64	0.94			
914	0.80	0.80	0.57	0.74	0.83	0.79	0.68	0.14	0.68	0.79	0.80	0.75	0.76	0.76	0.74	0.74	0.89	0.89	0.74	0.07	0.68	0.74	0.74	0.74	0.74	0.74	0.74	0.83	0.80	0.79	0.94	0.84	0.84	0.89		
923	0.80	0.80	0.57	0.74	0.83	0.79	0.68	0.14	0.68	0.83	0.80	0.75	0.76	0.76	0.83	0.74	0.89	0.89	0.79	0.07	0.68	0.74	0.74	0.74	0.74	0.74	0.74	0.61	0.68	0.57	0.13	0.74	0.73	0.77	0.73	
924	0.70	0.77	0.56	0.64	0.64	0.61	0.52	0.71	0.08	0.71	0.65	0.77	0.65	0.74	0.74	0.64	0.64	0.61	0.68	0.57	0.13	0.74	0.73	0.74	0.63	0.64	0.63	0.61	0.3	0.65	0.73	0.77	0.73	0.65	0.73	0.76

References

Andersson MS, Schultze-Kraft R, Peters M (2002) *Flemingia macrophylla* (Willd.) Merrill. FAO Grassland Index, Rome, Italy. Available online: http://www.fao.org/ag/AGP/AGPC/doc/GBASE/data/pf000154.htm

Andersson MS, Schultze-Kraft R, Peters M, Tohme J, Franco LH, Avila P, Gallego G, Hincapié B, Ramírez G, Lascano CE (2003) Shrub legumes with adaptation to drought. In: CIAT Annual Report 2003. CIAT, Cali, Colombia, pp 99–107

Andersson MS, Schultze-Kraft R, Peters M, Hincapié B, Lascano CE (2006a) Morphological, agronomic and forage quality diversity of the *Flemingia macrophylla* world collection. Field Crops Res 96: 387–406

Andersson MS, Lascano CE, Schultze-Kraft R, Peters M (2006b) Forage quality and tannin concentration and composition of a collection of the tropical shrub legume *Flemingia macrophylla*. J Sci Food Agric DOI: 10.1002/jsfa.2433 (published online 20 March 2006)

Andersson MS, Schultze-Kraft R, Peters M, Duque MC, Gallego G (2006c) Extent and structure of genetic diversity in a collection of the tropical multipurpose shrub legume *Cratylia argentea* (Desv.) O. Kuntze as revealed by RAPD markers. Plant Genetic Resources: Characterization and Utilization (accepted)

Andersson MS, Peters M, Schultze-Kraft R, Gallego G, Duque MC (2006d) Molecular characterization of a collection of the tropical multipurpose shrub legume *Flemingia macrophylla*. Agrofores Syst (accepted)

Asare EO (1985) Effects of frequency and height of defoliation on forage yield and crude protein content of *Flemingia macrophylla*. Proceedings of the XV International Grassland Congress, 24–31 August 1985, Kyoto, Japan. The Science Council of Japan and the Japanese Society of Grassland Science, Tochigi-ken, Japan, pp 164–165

Beattie AJ (1985) The evolutionary ecology of ant-plant mutualisms. Cambridge University Press, Cambridge UK, p 192

Beattie AJ, Turnbull C, Hough T, Jobson S, Knox RB (1985) The vulnerability of pollen and fungal spores to ant secretions: evidence and some evolutionary implications. Am J Bot 72:606–614

Brown AHD, Marshall DR (1995) A basic sampling strategy: theory and practice. In: Guarino L, Ramanatha Rao V, Reid R (eds) Collecting plant genetic diversity—technical guidelines. CAB International, Wallingford, UK, pp 75–91

Budelman A (1989) *Flemingia macrophylla*—a valuable species in soil conservation. NFT Highlights, NFTA Hawaii, p 2

Budelman A, Siregar ME (1997) *Flemingia macrophylla* (Willd.) Merrill. In: Faridah Hanum I, van der Maesen LJG (eds) Auxiliary plants. PROSEA (Plant Resources of South-East Asia) No. 11. Backhuys Publishers, Leiden Netherlands, pp 144–147

Casiva PV, Saidman BO, Vilardi JC, Cialdella AM (2002) First comparative phenetic studies of Argentinean species of *Acacia* (Fabaceae), using morphometric, isozymal, and RAPD approaches. Am J Bot 89:843–853

Ellstrand NC (1992) Gene flow among seed plant populations. New Forest 6:217–228

Excoffier L, Smouse PE, Quattro JM (1992) Analysis of molecular variance inferred from metric distances among DNA haplotypes: application to human mitochondrial DNA restriction sites. Genetics 131:479–491

Godefroy J (1988) Observations de l'enracinement du stylosanthes, de la crotalaire et du flémingia dans un sol volcanique du Cameroun. Fruits 43:79–86

Hamrick JL, Godt MJW (1989) Allozyme diversity in plant species. In: Brown AHD, Clegg MT, Kahler AL, Weir BS (eds) Plant population genetics, breeding, and genetic resources. Sinauer Associates, Sunderland, Massachussets, USA, pp 43–63

Hamrick JL, Godt MJW (1996) Effects of life-history traits on genetic diversity in plant species. Philos Trans Roy Soc London, Ser B 351:1291–1298

Hamrick JL, Loveless MD (1989) The genetic structure of tropical tree populations: associations with reproductive biology. In: Bock JH, Linhart YB (eds) The evolution ecology of plants. Westview Press, Boulder, Colorado, USA, pp 129–146

Hamrick JL, Godt MJW, Muraswski DA, Loveless MD (1991) Correlations between species traits and allozyme diversity: implications for conservation biology. In: Falk D, Holsinger K (eds) Genetics and conservation of rare plants. Oxford Press, London UK, pp 75–86

Hanson J, Maass BL (1997) Conservation of tropical forage genetic resources. Proceedings of the XVIII International Grassland Congress, 8–19 June 1997, Winnipeg, Saskatoon, Canada. The Canadian Forage Council, the Canadian Society of Agronomy and the Canadian Society of Animal Science, Calgary, Alberta, Canada, pp 31–35

Harris SA (1995) Systematics and randomly amplified polymorphic DNA in the genus *Leucaena* (Leguminosae, Mimosoideae). Plant System Evol 197:195–208

Hawkins JA, Harris SA (1998) RAPD characterisation of two neotropical hybrid legumes. Plant System Evol 213:43–55

Heider B, Schmidt A, Schultze-Kraft R, Long TD (2002) Assessment of legume diversity in the highlands of North Vietnam—an on-going research project. Plant Genet Resources Newslett 131:73

Huff DR, Peakall R, Smouse PE (1993) RAPD variation within and among populations of out-crossing buffalograss (*Buchloë dactyloides* (Nutt.) Engelm.). Theor Appl Genet 86:927–934

Jamieson NL, Le Trong C, Rambo AT (1998) The development crisis in Vietnam's mountains. Special Report no. 6. East–West Center, Honolulu, Hawaii, USA

Juárez-Muñoz J, Carrillo-Castañeda G, Arreguin R, Rubluo A (2002). Inter- and intra-genetic variation of four wild populations of *Prosopis* using RAPD-PCR fingerprints. Biodivers Conserv 11:921–930

Lacerda DR, Acedo MDP, Lemos Filho JP, Lovato MB (2001) Genetic diversity and structure of natural populations of *Plathymenia reticulata* (Mimosoideae), a tropical tree from the Brazilian Cerrado. Mol Ecol 10:1143–1152

Liu CJ (1997). Geographical distribution of genetic variation in *Stylosanthes scabra* revealed by RAPD analysis. Euphytica 98:21–27

Lynch M, Milligan BG (1994) Analysis of population genetic structure with RAPD markers. Mol Ecol 3:91–99

van der Maesen LJG (2003). Cajaninae of Australia (Leguminosae: Papilionoideae). Aust Syst Bot 16:219–227

Mantel NA (1967) The detection of disease clustering and a generalized regression approach. Cancer Res 27:209–220

Margules CR, Pressey RL (2000) Systematic conservation planning. Nature 405:243–253

Nason JD, Aldrich PR, Hamrick JL (1997) Dispersal and the dynamics of genetic structure in fragmented tropical tree populations. In: Laurence WF, Bierregaard RO (eds) Tropical forest remnants: ecology, management and conservation of fragmented communities. University Press of Chicago, Chicago Illinois USA, pp 304–320

Peakall R, Smouse PE (2001) GenAlEx V5: Genetic Analysis in Excel. Population genetic software for teaching and research, Australian National University, Canberra, Australia. Available online: http://www.anu.edu.au/BoZo/GenAlEx/

Perry LM, Metzger J (1980) Medicinal plants of East and Southeast Asia: attributed properties. The MIT Press, Cambridge, Massachusetts, USA

Rohlf FJ (1998) NTSYS-pc numerical taxonomy and multivariate analysis system. Version 2.0. Exeter Software, Setauket, New York, USA

Saxena KB, Singh L, Gupta MD (1990) Variation for natural out-crossing in pigeonpea. Euphytica 46:143–148

Saxena KB, Jayasekera SJBA, Ariyaratne HP, Ariyanayagam RP, Fonseka HHD (1994) Frequency of natural out-crossing in partially cleistogamous pigeonpea lines in diverse environments. Crop Sci 34:660–662

Schierenbeck KA, Skupski M, Lieberman D, Lieberman M (1997) Population structure and genetic diversity in four tropical tree species in Costa Rica. Mol Ecol 6:137–144

Schultze-Kraft R (1996) Leguminous forage shrubs for acid soils in the tropics. Wageningen Agricultural University Papers 96–4:67–81

Schultze-Kraft R, Pattanavibul S (1990) Germplasm collection of tropical forage legumes in Thailand. Thai J Agric Sci 23:359–369

Schultze-Kraft R, Gani A, Siregar ME (1987). Collection of native forage legumes in Sumatra, Indonesia. IBPGR Regional Committee Southeast Asia Newslett 11:4–6

Schultze-Kraft R, Lascano C, Benavides G, Gómez JM (1989a) Relative palatability of some little-known tropical forage legumes. Proceedings of the XVI International Grassland Congress, 4–11 October 1989, Nice, France. Association Française pour la Production Fourragère, Nice, France, pp 785–786

Schultze-Kraft R, Pattanavibul S, Gani A, He C, Wong CC (1989b) Collection of native germplasm resources of tropical forage legumes in Southeast Asia. Proceedings of the XVI International Grassland Congress, 4–11 October 1989, Nice, France. Association Française pour la Production Fourragère, Nice, France, pp 271–273

Schultze-Kraft R, Ha Dinh Tuan, Ngyuen Phung Ha, Pattanavibul S, Sornprasitti P (1993) Collecting tropical forage legume germplasm in Vietnam and Thailand. FAO-IBPGR Plant Genet Resources Newslett 91/92:50–52

Shafer CL (1995) Values and shortcomings of small reserves. Bioscience 45:80–88

Shelton HM (2001) Advances in forage legumes: shrub legumes. Proceedings of the XIX International Grassland Congress, 11–21 February 2001, São Pedro, São Paulo, Brazil. The Brazilian Society of Animal Husbandry, Piracicaba, São Paulo, Brazil, pp 549–556

Shiying Y, Saxena KB, Pang Wen, Guangtian W, Ziping H (2002) First report of natural outcrossing in pigeonpea from China. Int Chickpea Pigeonpea Newslett 9:37–38

Simberloff D (1988) The contribution of population and community biology to conservation science. Ann Rev Ecol System 19:473–511

Templeton AR, Shaw K, Routman E, Davis SK (1990) The genetic consequences of habitat fragmentation. Annal Missouri Bot Garden 77:13–27

Turner IM, Corlett RT (1996) The conservation value of small, isolated fragments of lowland tropical rain forest. Trends Ecol Evol 11:330–333

Valdecasas AG, Camacho AI (2003) Conservation to the rescue of taxonomy. Biodivers Conserv 12:1113–1117

Virk PS, Ford-Lloyd BV, Jackson MT, Pooni HS, Clemeno TP, Newbury HJ (1996) Predicting quantitative variation within rice germplasm using molecular markers. Heredity 76:296–304

Quy V (1998) An overview of the environmental problems in Vietnam. Vietnamese Studies 3(129):7–32

Williams RJ (1983) Tropical legumes. In: McIvor JG, Bray RA (eds) Genetic resources of forage plants. CSIRO, Melbourne, Australia, pp 17–31

World Bank (2002) Vietnam environment monitor 2002. World Bank, Hanoi, Vietnam

van de Wouw M, Hanson J, Luethi S (1999) Morphological and agronomic characterization of a collection of Napier grass (*Pennisetum purpureum*) and *P. purpureum* × *P. glaucum*. Trop Grasslands 33:150–158

Young A, Boyle T, Brown T (1996). The population genetic consequences of habitat fragmentation for plants. Tree 11:413–418

🍃 Springer

Biodivers Conserv (2007) 16:1633–1651
DOI 10.1007/s10531-006-9030-0

ORIGINAL PAPER

Non-timber forest product harvesting in alien-dominated forests: effects of frond-harvest and rainfall on the demography of two native Hawaiian ferns

**Tamara Ticktin · Hō'ala Fraiola ·
A. Nāmaka Whitehead**

Received: 12 July 2005 / Accepted: 6 March 2006 / Published online: 12 May 2006
© Springer Science+Business Media B.V. 2006

Abstract Non-timber forest products (NTFP) represent culturally and economically important resources for millions of people worldwide. Although many NTFP are harvested from disturbed habitats and therefore subject to multiple pressures, few quantitative studies have addressed this issue. Similarly few NTFP studies have assessed seasonal variation in demographic rates even though this can confound harvest effects. In Hawai'i, the wild-gathered ferns, *Microlepia strigosa* and *Sphenomeris chinensis*, represent highly important cultural resources but declining populations have led to conservation concerns. Both ferns are harvested from disturbed, alien-dominated forests and contemporary Hawaiian gathering practices often consist of harvest and concurrent weeding of alien invasive species. We assessed the effects of concurrent frond-harvest and alien species weeding on frond structure, density, and rates of production by comparing experimentally harvested vs. control plots, and documented relationships between frond demographic patterns and precipitation. Gathering practices had no impact on frond density of either species or on most other demographic parameters over the short term. Exceptions included a significant decrease in the density of the longest *S. chinensis* fronds and a significant decrease in *M. strigosa* frond production when fronds were gathered without alien weeding. However, seasonal and annual changes in frond density and production occurred across all plots of both species and were significantly correlated with precipitation. The relatively low harvest effects for both species are likely due to several factors including short frond longevity and the strict criteria used by gatherers to select harvestable fronds. The potential for sustainable harvest in the context of alien-dominated forests is discussed.

Keywords Demography · Ferns · Hawai'i · Invasive species · *Microlepia strigosa* · Non-timber forest products · Sustainable harvest · *Sphenomeris chinensis*

T. Ticktin (✉) · H. Fraiola · A. N. Whitehead
Botany Department, University of Hawai'i at Mānoa, 3190 Maile Way, Honolulu, HI 96822, USA
e-mail: Ticktin@hawaii.edu

A. N. Whitehead
Kamehameha Schools, 78-6831 Ali'i Drive, Suite 232, Kailua-Kona, HI 96740, USA

Introduction

Non-timber forest products (NTFP) hold economic and cultural significance for millions of people across the globe. For instance, hundreds of millions of rural peoples currently derive a significant portion of both their subsistence needs and cash income from gathered plant and animal products (Iqbal 1993; Vedeld et al. 2004). Thousands of wild NTFP are also harvested for local use in other cultural and religious activities, such as rituals, ceremonies and dances (e.g. FAO 1991).

A growing body of literature has illustrated that NTFP harvest can have ecological consequences at multiple ecological scales, and that demographic responses to harvest are heavily influenced by harvesting practices (see review by Ticktin 2004). However, although many NTFP are harvested from disturbed habitats and subject to multiple pressures (Cunningham 2001), few quantitative studies have addressed this issue. For example, a small number of studies have assessed the demographic effects of NTFP management practices in the context of browsing or grazing (Endress et al. 2004b; Ghimire et al. 2005), or human-induced fire (Ticktin 2005). Nonetheless, the demographic consequences of NTFP harvest and management in the context of other disturbance factors, such as alien invasive species, have not been addressed to date. This is surprising given that invasive species have invaded many areas where NTFP harvesting represent highly important cultural and economic activities, such as in parts of South Asia, North America and the Pacific Islands; and that invasive species can have major negative impacts on the demographics of native plant populations (e.g. Williamson 1996; Cox 1999).

Seasonal variation in demographic rates of NTFP species is another consideration that has received little attention in the NTFP literature. While many NTFP studies have documented interannual variation in demographic rates (e.g. Nantel et al. 1996; Ticktin et al. 2002; Endress et al. 2004a), few have assessed how demographic parameters may vary seasonally with environmental factors such as precipitation, light or other variables (e.g. Joyal 1996; Ticktin et al. 2003). Seasonal and annual variation in demographic rates can influence and/or confound the impacts of harvest, as well as provide information on optimal timing for harvest. The latter can have important implications for ecological sustainability as well as for local peoples' livelihoods.

The native Hawaiian ferns, *Microlepia strigosa* (Thunb.) C. Presl. and *Sphenomeris chinensis* (L.) Maxon, are two culturally important NTFP that are frequently gathered from disturbed habitats, and over which conservation concerns have emerged. Their fronds are collected to make *lei* (garlands), especially those used in the art of *hula*, as they represent *kinolau* or physical manifestations of Hawaiian gods. Hawaiian cultural practitioners gather these species from local forests, many of which are now dominated by alien invasive species. Many cultural practitioners have adapted their practices to include the weeding of invasive species with which the fern species grow (Ticktin et al. in review).

M. strigosa and *S. chinensis* are also among multiple fern species worldwide that are harvested for their fronds (Chin 1997). Despite the significant economic importance of various ferns and the conservation concerns generated as a result of their harvest (Chamberlain et al. 1998; Milton and Moll 1988), there is still surprisingly little information on the impacts of frond harvest on fern populations (see Milton 1987; Geldenhuys and van der Merwe 1988). In fact, demographic patterns in tropical herbaceous ferns remain poorly studied and have only been documented for a few species (Windisch and Pereira-Noronha 1983; Sharpe and Jernstadt 1990; Sharpe 1997; Mehltreter and Palacios-Rios 2003). Notably, for most of these few species, rates of frond production and growth have been found to vary significantly with rainfall (Sharpe 1997; Mehltreter and Palacios-Rios 2003).

In this paper, we present a quantitative analysis of the impacts of NTFP gathering practices in a disturbed habitat, through a case-study of the effects of frond harvest on fern populations in Hawai'i's alien-dominated forests. We employ a short-term experimental approach that, given the lack of previous research on NTFP harvest in alien-dominated habitats, provides guidance for longer-term studies both in Hawai'i and elsewhere. Specifically, the objectives were to

(1) Experimentally assess the impacts of a contemporary Hawaiian gathering practice—concurrent harvest and alien-species removal—on demographic parameters (frond structure, frond density, rates of frond production, mortality) of *S. chinensis* and *M. strigosa*. To do so, we also quantify rates and patterns of frond-harvest.
(2) Assess the relationships, if any, between demographic parameters in both species and rainfall patterns.
(3) Discuss the implications of the above for the sustainable use and conservation of *S. chinensis* and *M. strigosa* in the context of Hawai'i's disturbed, alien-dominated forests; and for NTFP research in alien-invaded forests elsewhere.

Methods

Study species and sites

Microlepia strigosa (Dennstaedtiaceae), known locally and in Hawaiian as *palapalai*, is a terrestrial fern that is indigenous to all the main Hawaiian Islands, and distributed across parts of South Asia, Southeast Asia and Polynesia (Palmer 2003). It's fronds are less than one meter long and it has long-creeping rhizomes. In Hawai'i, it is found in dry to moderately wet habitat, from near sea level to 1,770 m (Palmer 2003). *M. strigosa* is sacred to Laka, the principle goddess of *hula*, and is therefore used to decorate the hula altar and to make *lei*, especially for adorning *hula* dancers. It is reported to be the most commonly collected *lei* plant (Hawaii Volcanoes National Park Database).

Sphenomeris chinensis (Lindsaeaceae) or *pala'ā*, is also a terrestrial fern, indigenous to all the main Hawaiian Islands, and widely distributed across Asia, Polynesia and Madagascar (Palmer 2003). *S. chinensis* grows in mesic to wet forests and is commonly found in forest openings and other disturbed areas where there is adequate moisture, such as trails, roads, grasslands and streamsides from 40 to 1,310 m (Palmer 2003). It has fronds that range from 15 to 80 cm and short-creeping rhizomes. *S. chinensis*'s lacy, delicate fronds make it one of the most popular *lei* plants. In addition, the dead fronds and rhizomes were and still are used to make a browny-reddish dye for bark cloth.

Lei-giving and *hula* are Native Hawaiian traditions that continue to persist and flourish in the culturally-diverse population of the Hawaiian Islands today. Neck *lei* are presented as gifts and are commonly used by all sectors of society for a variety of occasions such as birthdays, graduations, political events, etc. *Hula*, a sacred and ceremonial art composed of chants and dance, plays an important role in the lives of tens of thousands of people of all ages living in Hawai'i today. It represents one of the important symbols of the Hawaiian cultural renaissance and today Hawai'i boasts over 100 *hālau hula* (hula schools), many with several hundred dancers each (Josephson 1998).

Like most of Hawai'i's native plants, *M. strigosa* and *S. chinensis* populations are reported to be declining or disappeared from most of the accessible forest areas on all the main Hawaiian islands, and gatherers have expressed concern that they are increasingly difficult to find (Timmons 1996). Both species are almost exclusively gathered from the wild. Although conservation organizations have promoted culti-vation in home-gardens (Timmons 1996), gatherers prefer to harvest from forest populations (Vieth et al. 1999).

We carried out experimental frond harvests at two sites on the island of O'ahu (Table 1). Note that we were unable to replicate our experiments for either species at any other sites due to the lack of availability of other accessible populations that were large enough for experimental manipulation and free from uncontrolled gathering pressure.

Sphenomeris chinensis experiments

To assess the effects of Hawaiian gathering practices on *S. chinensis* frond demog-raphy, we randomly established five pairs of 0.5 m × 0.5 m plots (total of 10 plots) in

Table 1 Description of study sites for *Sphenomeris chinensis* and *Microlepia strigosa* experiments

Species	*Sphenomeris chinensis*	*Microlepia strigosa*
Study site	Mānana-Waimano tract of the Ewa Forest Reserve	Kahanahāiki tract of the Mākua Military Reserve
Location	21.43′ N, 157. 94′ W	21.54′ N, 158.20′ W
Elevation	520 m	550 m
Public access	Yes, but populations not easily accessible and showed no signs of harvest by others before or during experiment	No
Dominant overstory species	*Psidium cattleianum*,[1] *Eucalyptus* spp.,[1] *Metrosideros polymorpha, Acacia koa*	*Psidium cattleianum*,[1] *Schinus terebinthifolius*,[1] *Aleurites moluccana*[1]
Dominant understory species	*P. cattleianum*,[1] *Clidemia hirta*,[1] alien grasses	Mix of alien grasses and native & alien ferns
Understory alien cover[2]	15%	10%
Treatments	Control vs. Harvest & Weeding	Control vs. Harvest vs. Harvest & Weeding
Replicate plots per treatment	5	4
Plot size[3]	0.5 m × 0.5 m	Variable: 1–2.25 m² (see text)

[1] Alien invasive species

[2] Estimated visually and for *M. strigosa* using the pole intercept method as well

[3] There were no significant differences in frond density across plots for either species

an extended *S. chinensis* population in October 2001 (Table 1). All fronds in each plot (total of 1,266 fronds) were labeled, numbered, and measured for frond length (excluding stipe). No significant differences were found in frond density among plots, and one half of the plots were then randomly assigned for harvest. We invited a cultural practitioner (*hula* dancer) to harvest the fronds in December 2001, since her gathering techniques were characteristic of traditional Hawaiian gathering practices (Ticktin et al. in review). These consisted of harvesting the longest fronds at about 10 cm above the base, removing many of the dead stipes, and weeding the alien invasive species in the plots. The number and length of each frond harvested (the length that was removed) was recorded. Plots were re-censused every 6 months over a period of 2 years. In each census all fronds were counted, measured and new fronds marked. Harvest was carried out once per year, so that the second harvest was in January 2003. Note that we were unable to include a 'harvest only' treatment (with no weeding) in this experiment due to insufficient numbers of *S. chinensis* fronds.

Microlepia strigosa experiments

In August 2002, we initiated a second experiment to test the effects of Hawaiian gathering practices on the demography of *M. strigosa* (Table 1). In this experiment we added a third treatment, so that we could compare the effects of (1) frond-harvest alone vs. (2) frond-harvest and alien weeding combined vs. (3) control (no harvest or weeding). We randomly established four large square plots that ranged in size from 4 to 9 m^2 so that each plot had a similar number of fronds. (Note that for the *S. chinensis* experiments, we were able to use equal-sized plots since they contained similar frond densities). We then divided each plot into four equal-sized subplots. Those subplots that did not differ significantly in frond density or alien cover were then selected for the experiment; this ranged from 3 to 4 subplots per plot. One subplot from each plot was randomly assigned to each of the three treatments mentioned above, so that for each of the three treatments we had four replicate subplots (Table 1).

All plants (of all species) in each plot were labeled, measured and their spatial coordinates within the plot were recorded. For *M. strigosa*, the number of fronds and length of longest frond (excluding stipe) for each plant was recorded (for a total of 665 fronds). Unlike *S. chinensis*, whose fronds grow very densely so that it is very difficult to distinguish individuals, *M. strigosa* individual plants are usually distinguished by clusters of fronds. When there was doubt as to whether rhizomes of different clusters were connected, we attempted to assess connectivity by gently prodding the soil with our fingers to trace the rhizomes.

Harvest and weeding treatments were carried out by a *hula* dancer who was also trained in gathering protocols in her *hālau hula* in August 2002, right after the first census. Plots were re-censused every 3 months for 1 year, and the length and number of all *M. strigosa* fronds, plant mortality and number of new plants were recorded. In addition, for a subset of four to six plants in each subplot, the newest frond was measured and marked with non-toxic acrylic paint and all of the surviving, previously-marked fronds were measured. For the harvest, and harvest and weeding treatments, these plants consisted solely of those from which fronds had been harvested. For the control plots, plants with fronds of similar lengths to those that were

🖄 Springer

harvested were randomly selected. The number of selected plants per plot was consistent across all treatments (21 plants/treatment).

Rainfall data for both sites was obtained from NOAA (National Oceanic and Atmospheric Administration) rain gages located nearest to the study sites, which were HI-11 for *S. chinensis* and HI-02 for *M. strigosa*.

Data analysis

One-way repeated measures ANOVAs, where time was the repeated factor, were used to assess changes in *S. chinensis* and *M. strigosa* frond density, density of specific size-classes, *M. strigosa* plant density, mean frond size and number of *S. chinensis* fronds harvested. Changes in the proportion of new *S. chinensis* fronds produced over time were analyzed with repeated measures ANOVAs with both year and time as repeated factors. All the data fit repeated-measures ANOVA assumptions, including normality, homogeneity of variances and when circularity was not met, the Greenhouse-Geisser corrected probability was used.

Changes in rates of *M. strigosa* frond production among plants subject to different treatments were assessed using one-way ANOVAs and SNK tests ($N = 21$ plants/treatment). The data was square-root transformed. Differences in final lengths of *M. strigosa* fronds that were buds (still furled) at the time of harvest were assessed using a two-sample Wilcoxon test, since the data were not normally distributed. Differences in *M. strigosa* plant mortality among treatments were analysed using loglinear analyses (Sokal and Rolf 1995), where we tested whether plant fate (alive, dead) was dependent on treatment (harvest; harvest and weeding; control). We carried out one analysis in which, for the harvest treatments, only those plants actually harvested were considered; and a second analysis in which all plants (harvested and non-harvested) in all plots were considered. Loglinear analyses were also used to test whether size-class structure of fronds was independent of harvest treatment for both species. For *S. chinensis*, size-class distribution was based on the sizes of all fronds, where as for *M. strigosa*, it was based on the length of the longest frond per plant. Mean frond longevity was calculated by dividing the mean number of living fronds by the mean rate of leaf production over the study period for each species, as in Tanner (1983). Correlations between demographic variables and pooled precipitation values were assessed using the Pearson correlation coefficient.

Results

Effects of gathering on *Sphenomeris chinensis*

Contemporary Hawaiian harvest practices (combined frond-harvest and alien weeding) had no significant effects on the density of *S. chinensis* fronds (Fig. 1), or on the proportion of new *S. chinensis* fronds produced (Fig. 2) over the 2-year study period. However, frond size-class distribution was dependent on the treatment ($N = 3,121$, d.f. $= 9$, $F = 49.47$, $P = 0.0001$), and by the end of the second year, the number of fronds in the largest size-category (>25 cm) was significantly lower in harvested plots than in control plots (Fig. 3; d.f. $= 1,8$, $F = 5.93$, $P = 0.04$). Mean frond length in harvested plots was lower than that of control plots one year after

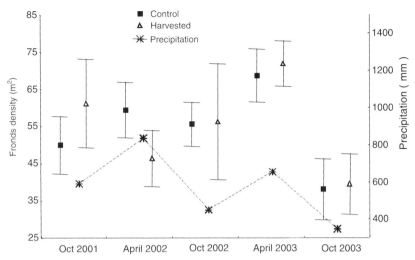

Fig. 1 Changes in density of mature (unfurled) *S. chinensis* fronds over time in harvested vs. control plots and in relation to precipitation (precipitation measurements represent the amount of rain that fell between census periods. For instance the rainfall value for April 2002 is the cumulative amount of precipitation that fell between census dates in October 2001 and April 2002. The value for October 2001 represents the cumulative total between April 2001 and October 2001). Means ±1 SE. Frond-harvest and alien weeding took place in December 2001 and January 2003. There were no significant differences in frond density between control and harvested plots, but frond density varied significantly over time ($P > 0.001$) and was correlated with annual precipitation ($r = 0.99$, $P = 0.067$)

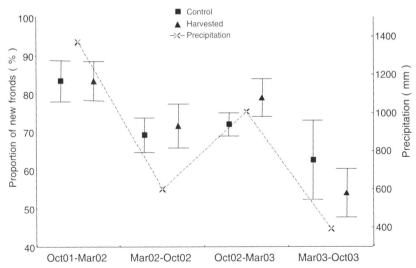

Fig. 2 Changes in the proportion of new *S. chinensis* fronds produced between October 2001 and October 2003, in harvested vs. control plots. Means ±1 SE. The proportion of new fronds is calculated as the number of new fronds divided by the number of mature fronds in each plot at the time of the last survey. Frond harvest and alien weeding took place in December 2001 and January 2003. There were no significant differences in frond production between control and harvested plots. There were significant effects of season ($P = 0.02$) and year ($P = 0.01$) on frond production, and seasonal rates of precipitation were significantly correlated with rates of frond production ($r = 0.99$, $P = 0.03$)

(a) 2001 (before harvest and alien-weeding)

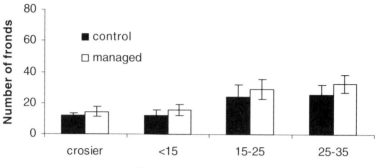

(b) 2002

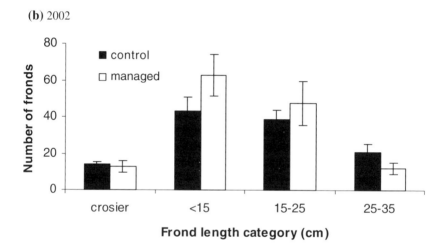

(c) 2003

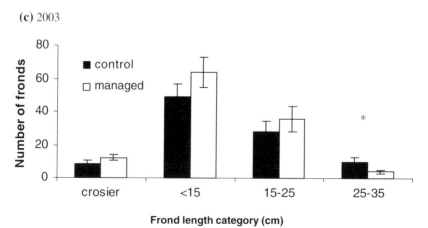

◄

Fig. 3 Change in size-class distribution of *S. chinensis* fronds in harvested vs. control plots, from 2001 to 2003. Means ±1 SE. Size-categories represent frond length, excluding stipe. Asterisks indicate significant differences ($P < 0.05$) between control and harvested plots. Density of the longest fronds (25–35 cm) decreased over time ($P < 0.001$) while that of the shortest fronds (<15 cm) increased over time ($P < 0.001$)

both annual harvests, although this differences was only marginally significant (ANOVA, d.f. = 1,8, $F = 4.72$, $P = 0.06$; Fig. 4).

All of the above variables changed significantly over time in both harvest and control plots. Repeated measures ANOVA of frond density showed significant time effects (d.f. = 3,24, $F = 11.02$, $P < 0.0001$; Fig. 1), and there was a tendency towards decreased frond density between the first and last censuses (d.f. = 1,8, $F = 4.33$, $P = 0.0711$). The proportion of new fronds produced was also decreased significantly from the first year to the second year of monitoring (d.f. = 1,8, $F = 11.95$, $P < 0.01$; Fig. 2). In addition, there were significant seasonal effects, as the proportion of new fronds produced was significantly greater from the end of October through the beginning of March, than from the end of March through the beginning of October (d.f. = 1,8, $F = 7.72$, $P = 0.02$; Fig. 2).

Frond size-class distribution was dependent on the census year for both control ($N = 1,424$, d.f. = 6, $F = 150.54$, $P = 0.0001$) and harvested plots ($N = 1,717$, d.f. = 6, $F = 307.72$, $P = 0.0001$). Mean frond length decreased significantly over time (d.f. = 2,16, $F = 27.60$, $P < 0.0001$; Fig. 4). The density of fronds in the largest size-category (>25 cm) decreased significantly over time (d.f. = 2,16, $F = 27.21$, $P < 0.001$), while the density of shortest fronds (<15 cm) increased significantly (d.f. = 2,16, $F = 15.69$, $P < 0.001$; Fig. 3).

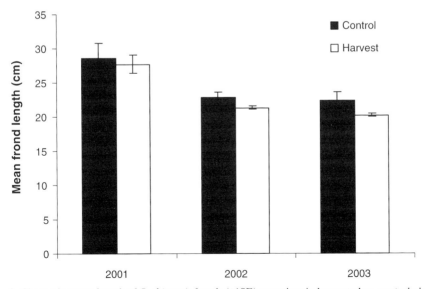

Fig. 4 Change in mean length of *S. chinensis* fronds (±1SE) over time in harvested vs. control plots. The difference in mean frond length between harvested and control plots was marginally significant in 2002 ($P = 0.06$). Mean frond length decreased significantly over time ($P < 0.0001$)

 Springer

The changes in frond density and rates of frond production over time coincide closely with changes in precipitation levels over time (Figs. 1, 2). Seasonal rates of precipitation were significantly correlated with rates of frond production ($r = 0.99$, $P = 0.03$) and annual rates of precipitation were strongly correlated with frond density ($r = 0.99$, $P = 0.067$). In addition, mean frond length and density of the longest fronds (>25 cm) recorded during the wet season were also strongly correlated with the precipitation levels during the proceeding 6-month census period ($r = 0.77$, $P = 0.07$; $r = 0.92$, $P = 0.01$ respectively).

Mean frond longevity of *S. chinensis* was estimated to be 5.1 months in control plots and 5.4 months in harvested plots, and harvested plots had significantly more fronds that survived after 6 months than did non-harvested plots (d.f. = 1,4, $F = 7.26$, $P = 0.039$).

The cultural practitioner harvested fronds selectively with respect to size and frond appearance, preferring longer fronds and avoiding those that were browning or wilting. The number of *S. chinensis* fronds gathered decreased from 289 in December 2001 (43.8% of all fronds) to 42 one year later (6.5% of all fronds), as few fronds met the selection criteria of the cultural practitioner in the second year. The size-distribution of harvested fronds also differed significantly between years ($N = 334$, d.f. = 3, $F = 12.9$, $\chi2 = 0.0016$), as the first harvest consisted of a greater percentage of longer fronds than did the second harvest (Fig. 5). A higher proportion of fronds (number of fronds harvested/total number of fronds) was also harvested in the first harvest than in the second harvest, this difference was significant for fronds in the smallest size-class, and marginally significant ($P = 0.06$) for those in the largest size-class (Fig. 5).

Effects of gathering on *Microlepia strigosa*

Harvest of *M. strigosa* fronds had no significant effects on frond density, plant density, or size-class distribution over the 1-year study period (Figs. 6–8). Similarly, mean annual mortality rate of *M. strigosa* control plants (18.2 ± 10%) did not differ significantly from that of harvested (20 ± 7%) or harvested and weeded plants (25.2 ± 16%). The latter analysis includes all plants within all plots, whether they were harvested or not. In addition, comparison of only the harvested plants also showed that harvest and harvest and weeding had no significant effects on plant mortality.

M. strigosa plants in control plots produced an average of 5.6 fronds per years, and produced significantly more new fronds per year than frond-harvested plants (mean = 3.8 fronds). However, there was no difference in rates of frond production between control plants and those plants subject to both frond-harvest and weeding (Fig 9). In addition, in the census period right after harvest (August–November 2002), the buds of harvested plants showed a tendency to produce shorter fronds (25.9 ± 4.2 cm) than those of non-harvested plants (37.4 ± 8.4 cm) (Wilcoxon two sample test $z = 1.60$, $P = 0.055$).

Frond density in both harvest and control plots changed significantly over time, increasing from February to May (d.f. = 1,9, $F = 7.46$, $P = 0.02$) and then decreasing again in August 2003 (d.f. = 1,9, $F = 21.8$, $P < 0.001$; Fig. 6). These seasonal changes in frond density were significantly correlated with patterns of rainfall (Fig. 6; $r = 0.87$, $P = 0.05$). In addition, total frond density decreased significantly from August 2002 to August 2003 (d.f. = 1,9, $F = 37.82$, $P = 0.0002$), coinciding with a 50% drop in annual precipitation between the 2 years (from 1109 mm between August 2001 and August

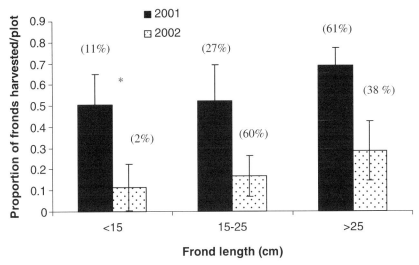

Fig. 5 Mean proportion of *S. chinensis* fronds harvested per plot (±1SE) in each size-class in 2001 vs. 2002. Numbers in parentheses indicate the percentage of harvested fronds in each size-class (number of harvested fronds in each size-class/total number of harvested fronds). Asterisks indicate significant differences (*P* < 0.05) between 2001 and 2002

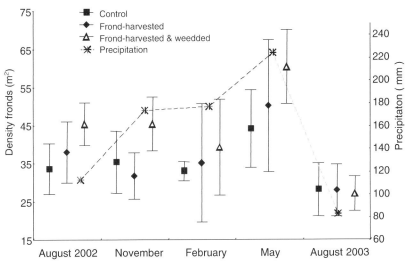

Fig. 6 Changes in density of *M. strigosa* fronds over time and in relation to precipitation (rainfall measurements represent the amount of rain that fell between census periods. For instance the rainfall value for November 2002 is the amount of precipitation between August 2002 and November 2002. For May–August 2003, rainfall values from the nearest rain gage were not available and the next-nearest rain gage was used—see text), in plots subject to three different management treatments. Means ±1 SE. Harvesting and alien-weeding treatments took place after the first census, in August 2002. There were no significant differences in frond density among treatments. Frond density increased significantly from February to May (*P* = 0.02) and decreased significantly from May to August (*P* < 0.001). Seasonal changes in frond density were significantly correlated with precipitation (*r* = 0.87, *P* = 0.05)

🖄 Springer

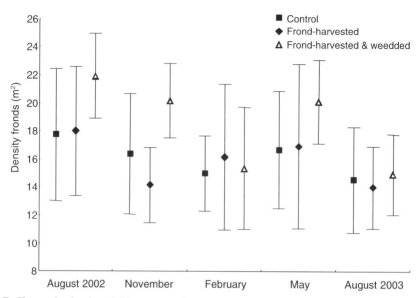

Fig. 7 Change in density of *M. strigosa* plants in plots subject to three different management treatments. Means ±1 SE. Harvesting and alien-weeding treatments took place after the first census, in August 2002. There were no significant differences among treatments or over time

2002, to 598 between August 2002 and August 2003). Similarly, the density of plants with fronds in the largest size-classes decreased significantly over this time period (d.f. $= 1,9, F = 11.05, P = 0.009$; Fig. 8), as did the mean length of the longest fronds of those plants in the largest two size-categories. The latter decreased from 35.9±0.8 cm to 30.7 ± 0.9 cm ($N = 57$, d.f. $= 1,54, F = 41.96, P<0.0001$).

Note that while data for the rain gage nearest our study site was not available for May-August 2003, there is very strong support for a great decrease in rainfall at our site during these months for the following reasons: (1) data from the next nearest gage (Poamoho), which is located near our study area but at a lower elevation, showed a great drop in precipitation from May to August and we found consistency in rainfall patterns between these two sites in the long-term rainfall record; (2) the historical record at our site illustrates that these three months are consistently much drier than the previous three; for instance over the past 3 years, rainfall during May–August was only 28–60% of the levels of March–May; and (3) rainfall across the island dropped during this period, and data from the past 30 years show that June–August tend on average to receive only 36.7% as much rain as during March–May.

The number of fronds per *M. strigosa* plant ranged from one to six, with most plants having one to three fronds. Frond longevity was calculated to be 4.1 months for the control plants, 4.3 months for plants in harvested and weeded plots, and 5.9 months for those in frond-harvested plots.

The cultural practitioner used the same selection criteria discussed for *S. chinensis* to gather *M. strigosa* fronds. In addition, she tended to harvest only one frond per plant: of the 53 fronds that were gathered from the plots, there was only one plant from which more than one frond (two fronds) was gathered. In addition, in almost all cases, the longest frond was gathered. Of those fronds harvested, 70% were greater than 35 cm long, 26% of harvested fronds were between 25–35 cm long and the rest

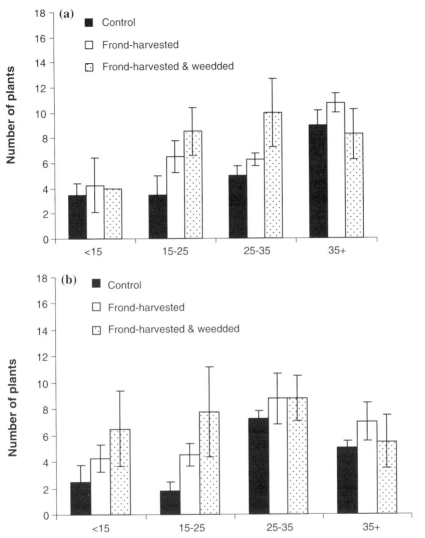

Fig. 8 Size-class structure of *M. strigosa* plants in plots subject to different types of management (a) before harvest (August 2002); and (b) 1 year after harvest (August 2003). Means ±1 SE. Size class structure is based on the length (cm) of the longest frond (excluding stipe) of each plant. There were no significant differences among treatments. The density of fronds in the largest size class (>35 cm) decreased significantly over time ($P = 0.009$)

(4%) fell between 15 and 25 cm. Overall, an average of 11.8 ± 3% of all fronds were gathered.

Discussion

The Hawaiian ferns, *Microlepia strigosa* and *Sphenomeris chinensis*, represent highly important cultural resources over which conservation concerns have emerged. Like

⊘ Springer

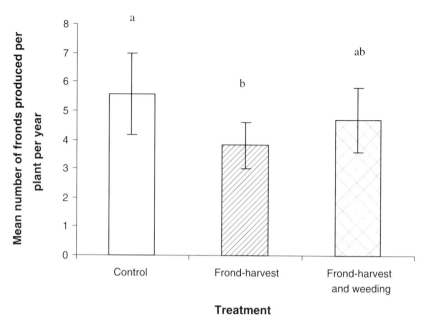

Fig. 9 Number of *M. strigosa* fronds produced per year (August 2002–August 2003), for *M. strigosa* plants subject to three different treatments (control vs. frond-harvest vs. frond-harvest and alien-weeding). Means ±1 SE. Different letters indicate significant differences among treatments (*P* < 0.05)

many non-timber forest products across the globe, they are harvested from disturbed forests and are therefore subject to multiple pressures. The results of our study suggest that the demographics of fern populations subject to the adaptive, contemporary Hawaiian gathering practice of concurrent harvesting and alien weeding differ little from control populations over the short-term, and that their demographic trends appear to be more strongly influenced by temporal variation in precipitation.

Impacts of gathering practices on frond density, production and size-structure

Frond-harvest can have significant impacts on the productivity and size-structure of fern populations. Frond-harvest of the South African ferns, *Blechnum giganteum*, *B. punculatum*, *Polystichum lucidum*, and *Rumohra adiantifolium* growing in the indigenous forests of the Southern Cape, resulted in highly significant decreases in frond density, frond production and fertility; as well as in frond size for the latter two (Milton 1987; Milton and Moll 1988; Geldenhuys and van der Merwe 1988). Similarly, for some species of palms, partial defoliation of fronds has also been illustrated to decrease the size of new fronds (Endress et al. 2004a; Ratsirarson et al. 1996; Joyal 1996).

Our results illustrated that concurrent frond-harvest and alien weeding had no significant impacts on *S. chinensis* frond density or annual rates of frond production, nor on *M. strigosa* frond density, plant density, size-class structure, mortality rates or frond production. While our study design did not allow us to determine the precise time it took populations to recuperate to pre-harvest density levels, our results indicate that *S. chinensis* recuperation took less than six months and *M. strigosa* recovery took less than three months (these were the census periods for each species). We did not quantitatively assess differences in fertility, but all mature ferns in our plots appeared to be fertile, suggesting that harvest had no apparent effect on fertility.

However, frond harvest of *S. chinensis* did lead to a significant decrease in density of the longest fronds by the end of the study, and showed a tendency to decrease mean frond length. *M. strigosa* fronds produced in the first three months after harvest were also significantly shorter than those produced in control plots, though these differences disappeared after the first three months. In addition, *M. strigosa* frond-harvest with *no* weeding significantly decreased rates of frond production. The fact that *S. chinensis* fronds that were crosiers in harvested plots at the time of harvest survived longer than those in control plots, suggests that growth of crosiers may have been initially delayed. This is also likely why the frond longevity calculations for both species of ferns in harvested plots were longer than those in control plots.

The above impacts could be due to a combination of factors, including depletion of nutrient reserves (Geldenhuys and van der Merwe 1988), reduced nutrient uptake due to reduced root growth and depletion of photosynthates (Crawley 1983), or decreases in photosynthetic material. Studies of palm frond harvest have documented decreases in photosynthetic capacity after harvesting through changes in nutrient ratios (O'Hara 1999).

The relatively low impacts of *S. chinensis* and *M. strigosa* frond harvest on demographic parameters, especially as compared to the South African fern species (Milton 1987; Milton and Moll 1988; Geldenhuys and van Mervwe 1988—the only other frond-harvested ferns for which we could find demographic data), is likely due to at least three factors. First, and perhaps most importantly, frond longevity of *M. strigosa* and *S. chinensis* is much shorter (~4–5 months) than that of the South African ferns (>2.8 years, Milton 1987; Milton and Moll 1988). Removal of longer-lived fronds clearly represents a much greater cost to the plant than does the removal of short-lived fronds. Second, in the case of *M. strigosa*, harvest intensity was lower than that of the South African species. Moreover, for both species harvest levels were limited by culturally appropriate selection criteria. Third, our experiment was carried out a relatively short time period. It is possible that consecutive annual harvests of the Hawaiian ferns may have impacts that only emerge over the longer term, especially if harvest significantly reduces nutrient reserves or if initial low responses to harvest are due re-allocation of stored photosynthates (Whitham et al. 1991) that can be depleted over the longer term. Longer-term studies, as well as ecophysiological studies, will be needed to evaluate this.

The fact that these species, like the harvested South African ferns, are rhizomatous, may also allow them to better withstand leaf loss than non-rhizomatous species, since they may benefit from translocation of nutrients from non-harvested plants or fronds (e.g. Schmid et al. 1988). However, harvest of a greater proportion of fronds or plants would then be expected to increase negative impacts.

Relationships between demographic patterns and precipitation

The strong correlations we found between precipitation and *S. chinensis* and *M. strigosa* frond density, production and size structure suggest that these variables are more heavily influenced by rainfall patterns, and/or potentially other related

Springer

environmental factors, than by gathering practices. These results are consistent with other studies of ferns, which have found positive, significant relationships between precipitation levels and frond size, frond production rates and/or frond development rates (Geldenhuys and van der Merwe 1988; Milton 1988; Sharpe 1997; Mehltreter and Palacios-Rios 2003).

These results clearly illustrate how the effects of harvest on demographic patterns can easily be compounded by and/or confounded with, the effects of temporal variation in environmental factors. They point therefore, to the importance of assessing responses to this kind of variation when conducting studies of NTFP harvest impact.

It is important to point out that our results are based on experimental harvests at single sites. We were unable to replicate our experiments at other sites due to the lack of availability of other accessible populations that were large enough for experimental manipulation and free from uncontrolled gathering pressure. Therefore we are unable to assess the impacts or extent of spatial variation in demographic parameters and their relationships to precipitation, nor how this could interact with harvest effects; these factors warrant further investigation.

Implications for sustainable harvest and conservation

In the context of disturbed habitats, NTFP harvest is often thought to exacerbate the problem of diminishing native plant populations. In Hawai'i, the wild-gathering of *S. chinensis* and *M. strigosa* is perceived as an additional threat to populations already declining due to habitat destruction and degradation by invasive species. Consequently, there have been attempts to curtail native plant use by cultural practitioners, ranging from a proposed Hawai'i state senate bill (that was rejected after protests by cultural practitioners), to workshops with *hula* schools on *lei* plant cultivation, to *lei* celebrations that prohibit the use of native species.

The gathering of *S. chinensis* and *M. strigosa* populations invaded by alien species in the forest sites we assessed appear to have fairly small impacts on the demographics of both species, at least over the short-term. However, higher rates or more frequent gathering will likely increase the magnitude of the negative impacts that we did document. This presents a conservation concern for two reasons. One is that some gatherers not trained in Hawaiian gathering protocols do currently harvest in greater quantities than we assessed here. For example, experienced gatherers harvest only those fronds that are fully mature and not wilted or browned, which greatly limits rates of harvest. In our experiment, a significant portion of fronds in the longest size-class did not meet these criteria and were therefore left unharvested (22% and 51% of the longest fronds were left unharvested during the first harvest in *S. chinensis* and *M. strigosa* respectively, and a full 73% for *S. chinensis* in the second year). A second reason is that the growing difficulty of finding accessible populations due to habitat destruction and disturbance by invasive species, has meant that multiple gatherers are increasingly forced to gather from the same few, remaining populations, and therefore that the frequency of harvest is increasing.

Cultivation and enrichment plantings are two options that have been implemented elsewhere to increase the abundance of highly-valued NTFP species (e.g. Sugandhi and Sugandhi 1995; Carpentier et al. 2000; Ticktin 2005). Cultivation of *S. chinensis* and *M. strigosa* has been encouraged and initiated, but most cultural practitioners prefer to gather from the wild for spiritual reasons (Vieth et al. 1999).

Another strategy may therefore be to focus on increasing the number of accessible populations though outplanting and transplanting into accessible, forest areas. In this way, *hula* schools, *lei* makers and other cultural practitioners could gather from, and help maintain, their own populations. Indeed, many *hula* teachers have expressed interest in this. Initial outplanting experiments with *M. strigosa* illustrate that transplanting to forest patches can be successful (H. Fraiola, unpublished data). However, this would require care that the harvest of propagules for outplanting does not negatively impact the remaining wild populations, as has happened with other NTFP (e.g. Ticktin 2005).

Our results point to some simple practices that could help ensure harvest sustainability in both wild and outplanted populations. Since both species of ferns appear to respond to stress by reducing the size of new fronds, frond-size could be used as proxy to assess the vigor of a population and its potential to regenerate after harvest. Gatherers could avoid harvesting from populations with small fronds or with fronds that appear smaller than they used to. Similarly, given that fern productivity appears to be dependent on rainfall, populations are likely able to recuperate most rapidly from frond-harvest during or after periods of high rainfall. Gathering could be focused during these periods and avoided during dry seasons or periods. This has been recognized by a few traditional gatherers, who maintain that one should only gather a few weeks after rains (Ticktin et al. in review).

Future research on NTFP harvest in disturbed habitats

The results of this short term, experimental study of *M. strigosa* and *S. chinensis* point to several lines of future research for NTFP in disturbed habitats, and alien-dominated forests in particular. First, our study suggests that there could be a relationship between frond-harvest, fern demographics and invasive species that warrants further investigation. In this case, *M. strigosa* plants subject to frond-harvest and concurrent weeding of alien species had rates of frond production equal to control plots, while frond-harvested only plants showed significantly decreased frond production. It is possible that weeding may have compensated for the negative effects of harvest by providing more light and/or liberating more nutrient resources for new fronds. While we were unable to test the effects of harvest and weeding separately for *S. chinensis* due to the lack of available comparable plots, other research has illustrated that alien cover increases immediately after *S. chinensis* harvest without weeding (Ticktin et al. in review). Thus, for NTFP growing in alien-invaded habitats, longer-term studies that assess the effects of harvest on the dynamics of both the harvested and the invasive species populations, and at differing levels of alien cover, will be needed to better elucidate their relationships.

Second, since demographic patterns of some species (such as the ferns in this study) vary according to seasonal variation in environmental factors, and since germination patterns of invasive species often vary seasonally as well, it will be very important to assess how the interactions between harvested populations and invasive species populations (or other disturbances) differ when harvested at differing times of the year.

Third, as our study illustrates, local gathering practices can adapt to deal with disturbance such as alien species, and there is therefore a need to understand the various ways in which local gathering practices can adapt to disturbance, and to quantitatively assess their impacts. As NTFP harvest in Hawai'i and elsewhere continues in forest habitats increasingly threatened by invasive species or other disturbances, a better understanding of these kinds of relationships will become increasingly important.

Acknowledgements   We would like to thank the people who provided enormous help to us with the fieldwork for this research: Laurette Boulet, Arlene Sison, Kaleo Wong, Patricia Tannahill, Kylene Bargamento, Anya Schiller, Elena Morgan, Jen Rodwell, Rory O'Connor, and Gustavo de la Peña Valencia. Mahalo nui loa to kumu hula Vicky Holt Takamine for encouraging and supporting this work, and without her this would not have been possible. Thanks also to Kapua Kawelo and Joby Rohrer from the Army Environmental Corps for granting us permission to use Mākua Military Reserve for this research; and to Mehanaokalā Hind for gathering for the *S. chinensis* experiments. Thanks to Sarah Dalle, Isabelle Schmidt and two anonymous reviewers for their help in greatly improving this paper. Funding and support for this research was generously provided by 'Ilio'ula-okalani Coalition Inc., Environmental Defense Fund, and the Hawai'i Conservation Alliance.

References

Carpentier CL, Vosti SA, Witcover J (2000) Intensified production systems on western Brazilian Amazon settlement farms: could they save the forest? Agriculture Ecosystems and Environment 82:73–88

Chamberlain J, Bush R, Hammett AL (1998) Non-timber forest products: the other forest products. Forest Products Journal 48:10–19

Chin WY (1997) Ferns of the tropics. Timber Press, Portland, OR

Cox G (1999) Alien species in North America and Hawai'i: impacts on natural ecosystems. Island Press, Washington, DC

Crawley MJ (1983) Herbivory: the dynamics of plant–animal interactions. Studies in ecology 10. Blackwell Scientific Publications, Oxford, pp 21–110

Cunningham AB (2001) Applied ethnobotany: People, wild plant use and conservation. Earthscan Publications Ltd, London, UK

Endress B, Gorchov D, Peterson M, Padrón Serrano E (2004a) Harvest of the palm *Chamaedorea radicalis*, its effects on leaf production, and implications for sustainable management. Conservation Biology 18:822–830

Endress B, Gorchov D, Noble RB (2004b) Non-timber forest production extraction: effects of harvest and browsing in an understory palm. Ecological Applications 14:1139–1153

FAO (1991) Non-wood forest products: the way ahead. Food and Agriculture Organization of the United Nations, Italy

Geldenhuys CJ, van der Merwe CJ (1988) Population structure and growth of the fern *Rumohra adiantiformis* in relation to frond harvesting in the southern Cape forests. South African Journal of Botany 54:351–362

Ghimire SK, McKey D, Aumeeruddy-Thomas Y (2005) Conservation of Himalayan medicinal plants: harvesting patterns and ecology of two threatened species, *Nardostachys grandiflora* DC. and *Neopicrorhiza scrophulariiflora* (Pennell) Hong. Biological Conservation 124:463–475

Iqbal M (1993) International trade in non-wood forest products. An overview. FAO, Rome

Josephson M (1998) Evaluating two options for increasing the availability of lei plant materials for use by hālau hula : an application of conjoint analysis. M.S. Thesis, University of Hawai'i, Honolulu

Joyal E (1996) The palm has its time: an ethnoecology of *Sabal uresana* in Sonora, Mexico. Economic Botany 50:446–462

Mehltreter K, Palacios-Rios M (2003) Phenological studies of *Acrostichum danaeifolium* (Pteridaceae, Pteridophyta) at a mangrove site on the Gulf of Mexico. Journal of Tropical Ecology 19:155–162

Milton S (1987) Effects of harvesting on four species of forest ferns in South Africa. Biological Conservation 41:133–146

Milton SJ, Moll EJ (1988) Effects of harvesting on frond production of *Rumohra adiantiformis* (Pteridophyta: Aspidiaceae) in South Africa. Journal of Applied Ecology 25:735–743

Nantel P, Gagnon D, Nault A (1996) Population viability analysis of American ginseng and wild leek harvested in stochastic environments. Conservation Biology 10:608–621

O'Hara JL (1999) An ecosystem approach to monitoring non-timber forest products harvest: the case study of Bayleaf palm (*Sabal mauritiiformis*) in the Rio Bravo Conservation and management Area, Belize. Ph.D. Thesis, Yale University, New Haven, Connecticut

Palmer D (2003) Hawai'i's ferns and fern allies. University of Hawai'i Press, Honolulu

Ratsirarson J, Silander JAJ, Richard AF (1996) Conservation and managements of threatened Madagascar palm species, *Neodypsis decary*, Jumelle. Conservation Biology 10:40–52

Schmid B, Puttick GM, Burgess KH, Bazzaz FA (1988) Clonal integration and effects of simulated herbivory in old-field perennials. Oecologia 75:456–471

Sharpe JM (1997) Leaf growth and demography of the rheophytic fern *Thelypteris angustifolia* (Willdenow) Proctor in a Puerto Rican rainforest. Plant Ecology 130:203–212

Sharpe JM, Jernstadt JA (1990) Leaf growth and phenology of the dimorphic herbaceous layer fern *Danaea wendlandii* (Maratiaceae) in a Costa Rican rainforest. American Journal of Botany 77:1040–1049

Sokal RR, Rolf FJ (1995) Biometry, 3rd edn. W.H. Freeman and Company, New York, NY

Sugandhi R, Sugandhi M (1995) Conservation and cultivation of MFP and their potential for rural development in India. Journal of Non-Timber Forest Products 2:83–85

Tanner EVJ (1983) Leaf demography and growth of the tree-fern *Cyathea pubescens* Mett. Ex Kuhn in Jamaica. Botanica Journal of the Linnean Society 87:213–227

Ticktin T (2005) Applying a metapopulation framework to management and conservation of non-timber forest species. Forest Ecology and Management 206:249–261

Ticktin T (2004) The ecological implications of harvesting non-timber forest products. Journal of Applied Ecology 41:11–21

Ticktin T, Johns T, Chapol Xoca V (2003) Patterns of growth in *Aechmea magdalenae* and its potential as a forest crop and conservation strategy. Agriculture, Ecosystems and Environment 94:123–139

Ticktin T, Nantel P, Ramírez F, Johns T (2002) Effects of variation on harvest limits for nontimber forest species in Mexico. Conservation Biology 16:691–705

Ticktin T, Whitehead AN, Fraiola H (in review) Traditional gathering of native hula plants in alien-invaded Hawaiian forests: adaptive practices, impacts on alien invasive species and implications for conservation. Environmental Conservation

Timmons G (1996) Hālau and the forest. The Nature Conservancy of Hawai'i Membership Newsletter, Spring 1996

Whitham TG, Maschinski J, Larson KC, Page KN (1991) Plant responses to herbivory: the continuum from negative to positive and underlying physiological mechanisms. In: Price DW, Lewinsohn TM, Fernandes GW, Benson WW (eds) Plant–animal interactions. John Wiley & Sons, New York, pp 227–256

Williamson M (1996) Biological invasisions. Chapman & Hall, London, UK

Windisch PG, Pereira-Noronha M (1983) Notes on the ecology and development of *Plagiogyria fialhoi*. American Fern Journal 73:79–84

Vedeld P, Angelsen A, Sjaastad E, Kobugabe, Berg G (2004) Counting on the environment, forest incomes and the rural poor, environment economics series paper 98. World Bank, Washington, DC

Vieth GR, Cox LJ, Josephson M, Hollyer J (1999) Alternatives to forest gathering of plant materials for hula lei adornment. College of Tropical Agriculture and Human Resources, Resource Management Fact Sheet 2, Honolulu, Hawai'i

Biodivers Conserv (2007) 16:1653–1667
DOI 10.1007/s10531-006-9031-z

ORIGINAL PAPER

Tequila and other *Agave* spirits from west-central Mexico: current germplasm diversity, conservation and origin

Patricia Colunga-GarcíaMarín ·
Daniel Zizumbo-Villarreal

Received: 4 August 2005 / Accepted: 6 March 2006 / Published online: 9 July 2006
© Springer Science+Business Media B.V. 2006

Abstract   Current germplasm diversity used in the production of *Agave* spirits in west-central México is in danger of erosion due to an expansion in the cultivation of the clone *A. tequilana* Weber var. azul, used for the elaboration of the famous drink "Tequila". In order to define critical areas of in situ conservation and to determine the role of local native and mestizo cultures in the generation and maintenance of diversity, an ethnobotanical exploration was conducted in the center and south of the state of Jalisco. Results situate the nucleus of greatest diversity at present in the south of Jalisco and indicate that this is a result of a continuous process of selection initiated by the indigenous population for the production of food and fermented drinks, which continued into the final years of the 16th century but with a new objective: distillation using the Filipino technology introduced to west-central Mexico through Colima. More than 20 variants were found to be cultivated by the traditional farmers, the majority relating to the *A. angustifolia* Haw. complex. We discuss the possibilities of in situ germplasm conservation and its legal protection.

Keywords   *Agave tequilana* · *Agave angustifolia* · Mezcal · Tequila ·
Intraspecific diversity · In situ conservation · Genetic resources · Domestication

Introduction

Mesoamerica is one of the three most important areas of origin and diversity of cultivated plants in the world. In the case of the agaves, it was here that "The great genetic diversity in a genus rich in use potential came into the hands of several peoples who developed the main agricultural center of the Americas" (Gentry 1982, p. 3), allowing this area to become the center of an explosive diversification through human selection. A group of species of this genus, collectively known in many

P. Colunga-GarcíaMarín (✉) · D. Zizumbo-Villarreal
Unidad de Recursos Naturales, Centro de Investigación Científica de Yucatán, Calle 43 No 130.
Col. Chuburná de Hidalgo, Mérida, Yucatán CP 97070, México
e-mail: pcolunga@cicy.mx

🍎 Springer

regions of Mexico as "mezcals" (from the Nahuatl "metl" = *Agave* and "ixcal-li" = cooked or baked), is among the most important and widely used prehispanic food plants in the seasonally dry regions of Mexico and Central America, where it is naturally distributed (Bruman 1940). Its history of diversification under cultivation and human selection can be divided into three important periods: its use as food, dating back at least 11,000 years, its use in the elaboration of fermented drinks, and its use in the elaboration of spirits. This last stage, which began towards the end of the 16th century, has registered a considerable commercial growth over the last 50 years, with characteristics that are endangering the current diversity of this group of species.

In order to maintain the current diversity of cultivars and their genetic and productive improvement, along with legal protection of the germplasm and derived products, special attention must be given to the role that local and indigenous cultures have played in these processes, since it is within these cultures that the most drastic diversification and genetic improvement of cultivars has been occurring, i.e. the passage from wild populations to domesticated ones. In the case of Mexico, it is within the indigenous and mestizo cultures that this diversity continues to show a dynamic development, maintenance and improvement (Hernández-Xolocotzi 1978), thus research in these areas is of vital importance. The rapid socio-economic and cultural changes involved in the globalization of markets represent a threat to the cultural diversity which has maintained the agricultural biodiversity. However, globalization of markets offers, at the same time, new opportunities to maintain and spread the diversity of its cultivars.

Before the development of agriculture, the group of *Agave* species known as "mezcales" represented a basic food source for the gatherer populations living in the arid and semi-arid areas to the north of the Isthmus of Tehuantepec and as far as Rio Gila in Arizona. The "quiotes" (floral peduncles) and in particular the "heads" (stalks and leaf bases that remain attached to the stalk after the leaves are removed), both cooked in pits, have been used in the same way since 9000 BC approximately (Fig. 1) (Bruman 1940; Callen 1965; Smith 1986). This group of species was one of the basic food sources to which other plants, such as maize, beans and squash were eventually added. In time these crops gradually displaced the agaves, their importance declined and they became a food source for times of scarcity.

The use of cooked "heads" in the elaboration of fermented drinks did not extend as far north as their use as food, but it was quite extended at the time of the Spanish Conquest, forming what Bruman (1940, pp. 17–18) called the great cultural region of "Mezcal wine" (Fig. 1). Within this great region, this author defined five areas according to the alcoholic drinks predominating in them. The area in which drinks based on agaves and *Spondia purpurea* L. predominated, he called "Mezcal-Jocote" (Fig. 1), delimiting it as the area occupied by ethnic groups inhabiting part of the modern states of Sinaloa, Nayarit, Jalisco, Colima, Michoacan and Guerrero in the west of Mexico (Bruman 1940, p. 142–147).

The crucial event for the origin of *Agave* spirits was the introduction of the still in Colonial times (Bruman 1940). Within the "Mezcal-Jocote" cultural area, this introduction derived originally from the Filipinos brought by the Spaniards to the coasts of Colima and Jalisco in the Manila Galleons in the late 16th century (Bruman 1940, 1944, 2000). The Filipinos introduced the elaboration of fermented and distilled drinks from coconuts (*Cocos nucifera* L.) with their own technology,

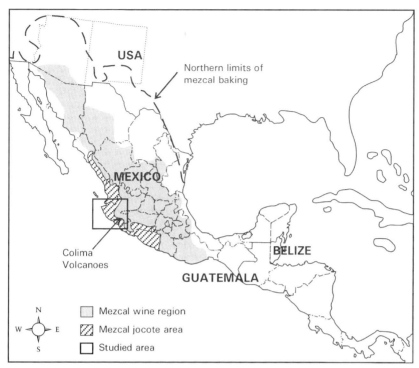

Fig. 1 The great cultural region where agave is used as food, the "Mezcal wine" region where fermented agave beverages were very important in prehispanic times, and the "Mezcal-Jocote" cultural area, in which fermented drinks were derived mainly from agaves and from *Spondias purpurea* L. Modified from Bruman (1940, 2000)

using a still unlike any in Europe and which could be built with local materials, very different from the Arabian still made of copper, used at that time by the Spaniards.

> The introduction of the process of distillation was so early, the product of the stills so attractive to the natives, and stills, once seen, so easily fashioned from native materials, that the mezcal wine soon come to be looked upon almost everywhere, an intermediate stage (Bruman 1940, pp. 18–19).

Wherever an *Agave* distilling industry arose in the colonial period, as a result of the introduction of the Asian or European stills, it is possible to confirm, according to Bruman (1940, p. 18; 2000, p. 20), that an undistilled *Agave* fermented beverage was already known in pre-Columbian times, which in turn was preceded by the knowledge of a plentiful source of wild *Agave* plants for food.

Walton (1977), synthesizing the findings of Bruman (1940, 1944), while adding new data, proposes that the 400-year-old history of Tequila and the other *Agave* spirits has been brought forward from its beginnings in the 16th century when the indigenous population living in the foothills of the volcanoes of Colima first submitted the fermented *Agave* beverage to the Filipino still. Its subsequent diffusion in the 17th and 18th centuries, according to this author, was directed north, towards the district of Tequila and the region of "Los Altos" in the present-day state of Jalisco;

to the south in the modern states of Michoacan, Guerrero and Oaxaca, after which it then spread to the mining centers of the north: Sonora, San Luis Potosi and Tamaulipas.

During the 19th century, one of these agave spirits drinks, the "mezcal" produced in the city of Tequila, in the state of Jalisco, became known by the name of the city in which it was elaborated, giving rise to the famous "Tequila" (Walton 1977). Legally recognized as a Denomination of Origin in 1974 (Diario Oficial de la Federación 1974, 1997), this drink has become enormously popular at national and international levels, which has led to a dramatic increase in its areas of cultivation. Given the fact that, since 1949, the Official Mexican Standard for Tequila stipulates that only the distilled drinks produced with the clone known as *A. tequilana* Weber var. azul or "blue agave" can use this name, its explosive increase has had a negative effect on the cultivation of other species and varieties of agaves used traditionally in the region of Tequila for the same purpose (Valenzuela-Zapata 1994, 1997). Its asexual propagation by shoots and, more recently, its cloning propagation in vitro, has also led to a drastic reduction in its genetic diversity (Gil-Vega et al. 2001). *Agave tequilana* Weber was recognized by Gentry (1982) as a clone derived from the complex *A. angustifolia* Haw.

This work presents the results of an ethnobotanical exploration of the current diversity of germplasm in relation to the production of Tequila and other *Agave* spirits in west central Mexico. The objectives were (1) Define critical areas for in situ conservation, and (2) Determine the role played by local native and mestizo cultures in the generation and maintenance of this diversity. Based on the hypothesis that if the foothills of the volcanoes of Colima was the area in which the Filipino still was adapted to the distillation of traditional fermented drinks from agaves in west central Mexico, then it is precisely in this area that the primary nucleus can be found from which the traditional farmers initiated the selection of germplasm for this purpose. Therefore, it is in this area that the ancestral populations of the native cultivars and their greater diversity can be found. This is the hypothesis which will direct a series of studies commencing with the present work.

Methodology

The ethnobotanical exploration was carried out in September 2003 and April 2004, following the principles of Hernández-Xolocotzi (1971). It was conducted in two regions: (1) In the south of Jalisco, including the area surrounding the River Armería-Ayuquila, which flows into the valley of Colima, and the area around the River Cohuayana-Naranjo-Tuxpan which flows through the valley of Alima (Fig. 2). Both river basins are situated in the foothills of the volcanoes of Colima. Since the production of spirits requires an abundant supply of cold water, the use of the still could have been extended along the banks of these rivers, passing through these valleys to the foothills of the volcanoes. (2) In the center of the state of Jalisco, the exploration was conducted in the region of Tequila-Amatitán (Fig. 2), the area in which the Tequila mezcal became famous, and which is now the most important area of production.

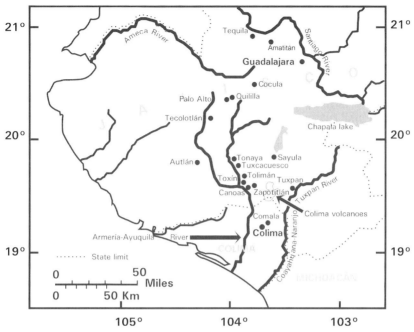

Fig. 2 Studied area. Localities explored in the Center and South of Jalisco, México

In the south of Jalisco, we explored from Cocula to Tuxpan, including Palo Alto, Tecolotlán, Quililla, Sayula, Tonaya, San Pedro Toxín, Tolimán, Tuxcacuesco, Zapotitlán and Canoas (Fig. 2). We interviewed one or two farmers from each town in their fields, asking them about the diversity of the agaves used, their biological and productive differences, the selection criteria and the elaboration processes for the spirits. We collected live plants and herbarium specimens in their areas of cultivation and in the natural habitats. We also included the wild populations of *A. angustifolia* Haw. growing on the semi-arid slopes between Cocula and Tecolotlán, and which Gentry (1982, pp. 583–584) mentioned as being very similar to those of *A. tequilana*, and from which this author believes the farmers of Tequila may have made their original selections. In the center of Jalisco, the exploration was carried out in the region of Tequila-Amatitán (Fig. 2), where we interviewed eight farmers in their fields, collecting samples from the cultivated varieties. One of the farmers accompanied us on an inspection of the natural areas of potential distribution of the wild populations.

The live samples were deposited in the Collection of *Agave* Germplasm in the Centro de Investigación Científica de Yucatán (CICY) (Colunga-GarcíaMarín 2004). The samples of herbarium were also deposited in the CICY.

Specific historical information of the area under study, used to discuss the ethnobotanical information, was obtained from sources that describe the living and production conditions of the ethnic groups inhabiting the Bishoprics of Michoacan and Nueva Galicia during the 16th and 17th centuries. These bishoprics were the political-religious entities to which the area under study belonged at that time. The works consulted were Acuña (1987, 1988), Sevilla del Río (1977) and Tello (1637–1653).

🎄 Springer

Results

South of Jalisco. Valley of the river Armería-Ayuquila

We found natural populations of *A. angustifolia*, from Cocula to Tuxpan (Fig. 2), including those highlighted by Gentry (1982, pp. 583–584) for their morphological similarity with *A. tequilana*. In Palo Alto, near the locality mentioned by Gentry (1982), we found peasant farmers maintaining encouraged populations of this species for spirit elaboration, beside their traditional crops. In Quililla, also near this locality, we found people who collect the floral peduncles to be cooked and sold.

In the towns of Zapotitlán and Canoas, the most isolated of the populations visited, we found 15 cultivated variants, most of them belonging to the *A. angustifolia* complex sensu Gentry (1982): "azul", "cimarrón cenizo", "cimarrón verde", "cimarrón negro", "chancuella", "cuaquesoca", "cuchara", "de brocha", "ixtero amarillo", "ixtero verde", "lineño", "perempis o siriaco", "prieto", "telcruz" and "pencudo". One producer, alone, Macario Partida, cultivates 11 of these in association with cattle and the traditional food crops: maize, beans and squash, a veritable in situ germplasm bank (Fig. 3). Macario, in addition to the production of food for his family, is also engaged in the elaboration of *Agave* spirits. He frequently roams the foothills of the Colima volcanoes and the "Sierra de Manantlán" searching for wild *Agave* populations and, in his own words, "of any plant that he finds interesting, he brings home two to four samples". The ones that give good spirit, he propagates and improves through the selection of the best individuals. His criteria for selection are: sweeter juice, a larger fleshy stem which is softer and therefore easier to grind, a more precocious plant with a greater production of shoots and resistance to plagues, diseases and grazing. He also maintains older variants from his town and some that have been selected in others.

The pre-hispanic tradition of elaborating fermented drinks from the *Agave* in this region was recorded by the Spanish chroniclers of the 16th century, who, in the "Relación de Zapotitlán" say:

Fig. 3 Macario Partida's plot where he grows 11 agave cultivars for elaboration of spirits (mezcals). Cultivars are in association with staple food crops and cattle breeding. Zapotitlán, Jalisco. (**A**) General view. At the background, the Biosphere Reserve "Sierra de Manantlán". (**B**) Detail showing the cultivation of *Zea mays* L. and *Cucurbita* sp

There is in this province a tree named "MEXCATL", that Spaniards named "maguey", they produced with it, wine, vinegar, syrup, rope, fabric, timber, needles, nails, and a very proven balsam for injuries. (Acuña 1988, p. 68).

In this region, most of the *Agave* spirits are still elaborated with the same procedures described in the 16th and 17th centuries. In places called "taverns", close to rivers and streams, the agave "heads" are baked in ground pits and pounded on rock outcrops with wooden mallets (Fig. 4). The crushed material is fermented in pits hand-hewn in the rock below ground level (Fig. 5). The fermented beverage is, even to this day, called "tuba", a Filipino term referring to traditional fermented and distilled coconut drinks (Bruman 1944). Although the cooking of the *Agave* "heads" is often carried out in groups among several farmers, the raw material is labeled so the fermentation process and subsequent distillation is done individually, allowing each farmer to evaluate his own genotypes.

The similarity of the stills used today (Fig. 6), with the stills used in the 16th and 17th century by the Filipinos in the elaboration of "coconut wine", is perfectly clear in the description provided by Fray Antonio de Tello for Colima in 1623:

The stills are hollow trunks, the thickness of a man, covered by a copper encasing full of water, which is changed as it is heated, and in the middle of the hollow part there is a round fitted board, with a pipe protruding from one side, through which the distillation occurs (Tello 1637–1653, p 650 front).

Fig. 4 (**A**) *Agave* "heads" (leaf bases and stems) being carried to the ground pit. (**B**) Heads being baked. (**C**) Heads pounded on rock outcrops with wooden mallets. Canoas, Jalisco

Fig. 5 Fermentation pits hand-hewn in the rock below ground level. Río Armeria. Canoas, Jalisco

This similarity was noted by Bruman (1944) through the comparison between the stills he found in use during 1938 in Bolaños, Tuxcacuesco, Tolimán and Tuxpan, Jalisco, and the Filipino stills described by Feliciano (1926).

In Zapotitlán and Tolimán (Fig. 2), we found coconut plantations apparently introduced since the 16th century. In Zapotitlán, we found the two oldest "taverns" of the locality, now abandoned: "Los Chinos tavern" and "Del Campanario tavern". These "taverns", according to the oldest *Agave* spirit makers of the community, were built long ago by "outsiders" (probably Filipinos or "Chinese Indians", as they were named by the Spaniards). They report that at the turn of the 20th century the "taverns" were still used. The fermentation pits are very small, both in diameter and in depth. When they were not being used, these pits were covered with earth in order not to be detected. These facts suggest that the "taverns" were originally used by the Filipinos who evaded the colonial laws, running away upriver, to continue with the production of coconut spirit (Gálvez 1785; Sevilla del Río 1977). The enormous influence of the "Chinese Indians" in this region has been demonstrated in several studies (Zizumbo-Villarreal 1996; Fuchigami 1990; Gómez-Amador 2000).

In Tonaya (Fig. 2), a less isolated town where commercial production of "Mezcal Tonaya" has developed, we found six cultivated variants. One of these, the "green mezcal" or "lineño", is the preferred for the elaboration of this drink.

Fig. 6 A traditional "tavern". Zapotitlán, Jalisco. (**A**) Filipino type still made out of a trunk of *Enterolobium cyclocarpum* (Jacq.) Griseb. The external receiver is an agave leaf. (**B**) Upper portion showing the condenser made with a copper pot and running water from the stream. (**C**) Inner receiver ("spoon") made of wood

South of Jalisco. Valley of the Río Coahuayana-Naranjo-Tuxpan

In the locality of Tuxpan we found two *Agave* cultivated variants to produce spirits: "garabato" and "peruano". *Agave* spirits from "garabato" are very distinctive of this town.

The farmers from the south of Jalisco are greatly concerned by the expansion of the "blue agave" (*A. tequilana* var. azul) used for the production of Tequila. They think that this expansion will result in: (1) the substitution of traditional varieties, (2) a reduction of the area destined to the traditional food crops, such as maize, beans and squash, all of which can be cultivated in association with the traditional varieties of *Agave*, but not with the "blue agave", and (3) soil erosion propitiated by monoculture, since this species is planted in the direction of the slope, it is not associated with other crops that protect the soil, such as squash; furthermore, herbicides are applied which eliminate all the plant cover causing exposure of the soil for longer periods (Fig. 7). The main mechanisms of expansion of the "blue agave" are land leasing by the Tequila companies and the pressures they exert on the local farmers to stop cultivating their traditional cultivars so as to prevent them from

Fig. 7 Tequila (*Agave tequilana* Weber var. azul) plantations in the "Barranca de Amatitán", Jalisco. Soil erosion is propitiated by this monoculture, since it is planted in the direction of the slope and without association with crops that could protect the soil, such as squash. Herbicides are applied to eliminate all the plant cover causing soil exposure for long periods

mixing with the "blue agave" sold to the companies. The farmers are also concerned about the elimination of wild populations as a result of purchases made by Tequila companies when "blue agave" is scarce.

The interest shown by the farmers in conserving their own varieties resides in their better adaptation to the productive conditions of the area, and to the local market flavor preferences. Since they are more resistant to plagues and diseases, they can be cultivated in association with traditional food crops or with grasses for cattle breeding and without using herbicides and pesticides. The local population prefers the flavor of these varieties over that of the "blue agave".

Center of Jalisco. Tequila-Amatitán region

We did not find wild populations of "blue agave" nor of *A. angustifolia*, and there was no evidence of any botanical collection having been carried out in this region. This fact indicates that this region is probably outside the natural distribution of the species or that it has disappeared from the area. As for the cultivated variants, already reported by Valenzuela-Zapata (1994, 1997), we found, predominantly, the "blue agave", and a few examples of "sigüin", "chato or saguayo", "bermejo", "pata de mula" and "listado". The farmers referred to the "zopilote" and the "moraleño" but no examples were found. The erosion of the germplasm of agaves used in the production of spirits in this region has already been documented by Valenzuela-Zapata (1994, 1997) as a consequence of the marked tendency of 19th century producers to select the "blue agave" because of its short maturing cycle, its higher industrial characteristics and its greater production of shoots, on the one hand, and, the ordinance by the Official Mexican Standard of Tequila since 1949 requiring the exclusive use of this variety, on the other.

In spite of current pressures on the farmers of this area to stop cultivating the other Tequila varieties so as to comply with the current Official Standard it is still possible to find producers who maintain them in special plots. Their main motivation being their appreciation of the flavor of the spirits derived from them.

Discussion and conclusions

The ethnobotanical exploration confirmed the hypothesis to the effect that the traditional farmers inhabiting the foothills of the volcanoes of Colima, in the south of the present day state of Jalisco, maintain a wealth of local varieties of *Agave* for the traditional production of spirits (more than 20), more in fact than the nine reported for the region of Tequila-Amatitán at the end of the 19th century by Pérez (1887, pp. 132–136). The exploration also showed that the still used at present in this process is the same as that described in the begining of the 17th century for the production of coconut spirits by the Filipino population (Tello 1637–1653), and similar to the elaboration process of *Agave* spirits described by Lumholtz (1902) in the 19th century among the indigenous population of the north and south of Jalisco, and in the first half of the 20th century by Bruman (1940, 1944) in the same area. This evidence suggests that present day varieties involved in the elaboration of Tequila and the other *Agave* spirits in west-central Mexico are the result of a continuous selection process initiated by the native population for the early use of agaves for food and fermented beverages that continued at the end of the 16th century with a new objective: the elaboration of spirits using the Filipino technology introduced through Colima. The selection of varieties for distillation continues today among the descendents of those human populations, now mestizos, who follow this local tradition.

The main characteristics of the three important stages in the history of diversification and evolution under cultivation and human selection of these varieties could have been the following:

Diversification and evolution under human selection

Agaves as food

During this stage, early human selection pressures seem to have focused on the taste of the fresh floral peduncle, determined by their quantity of sugars, water, and saponines (pungency). The localization and harvesting of the peduncles must have favored individuals with taller and thicker floral peduncles. The selection would have initiated for a great number of species in a wide geographical area, from Arizona to the Peninsula of Yucatan (Colunga-GarcíaMarín and Zizumbo-Villarreal 1986; Colunga-GarcíaMarín et al. 1993). The cutting of the peduncle could have indirectly favored the species and individuals with greater capacity for young shoot production, both in the sexual reproductive stage and as juvenile. The selection of taller peduncles could also have favored plants with greater size.

The later baking and consumption of the stalks could have reinforced the selection of plants with a greater capacity to store sugars, less fiber and pungency, and greater size. This type of use, which involves cutting the plant at ground level, at the sexual reproduction stage, favored the species and individuals with a production of young shoots in the juvenile stage. This form of consumption, which increased human interest in the agaves, led to a marked nutritional dependency on them in the semi-arid areas. This, combined with the natural capacity for vegetative propagation of many *Agave* species, favored the recurrent selection of clones with anthropocentric characteristics. Their cultivation under favorable environmental conditions, such as those prevailing around human settlements, allowed selection of individuals

🖉 Springer

with less resistance to plagues, diseases and grazing, as well as the survival of polyploids and other mutants which presented a reduced capacity for sexual reproduction, generating a reciprocal human-plant dependency.

Agaves for fermented beverages

This use probably involved new selection pressures: the success of fermentation and the flavor derived, through the indirect selection of specific associations with bacteria and yeasts. The reduction of the area in which strong selection pressures on the agaves for food and fermented beverages had been occurring, as a consequence of agriculture development, probably also involved a reduction of the species under human selection for these purposes. The populations of *A. angustifolia* of west-central Mexico maintained a relevant role for these objectives.

Agaves for spirits

The Filipino still, adapted in west-central Mexico to the elaboration of *Agave* spirits, became a key selective instrument to exert new selection pressures on the germplasm already selected during previous stages, which included cultivars adapted to the agro-ecosystems of the region, with high production of shoots and sugars stored in the stem, with specific successful associations with bacteria and yeasts for fermentation and which, therefore, produced fermented beverages with different flavors. The versatile design and relatively small size and weight of the Filipino still, which allows distillations on a small scale, made it possible to continue the individual selection of the genotypes most suitable for the new use. We found in the south of Jalisco a great diversity of genotypes selected with this technique. The introduction of cattle and new agricultural systems, both extensive and intensive, made it necessary to also select variants suitable for these agro-ecosystems.

The diffusion of the idea of producing *Agave* spirits in other areas, some of them with different agro-climatic characteristics, once again increased the diversity of species subjected to human selection and management which decreased when their nutritional importance declined. Such is the case of *A. salmiana* Otto ex Salm in San Luis Potosi (Aguirre et al. 2001); *A. cupreata* Trel & Berger; *A. potatorum* Zucc. and *A. hookeri* Jacobi in Guerrero and Michoacán and *A. marmorata* Roezl and *A. karwinskii* Zucc. in Oaxaca. There are at least 11 species used in Mexico to produce *Agave* spirits nowadays (García-Mendoza 2003).

A very rich and diverse Mexican culture of agave spirits (mezcals) emerged with the different adaptations of the Filipino still, the later adoption of the Arab still, introduced by Spanish colonists for sugar cane distillation, to produce *Agave* spirits, and the blending of both the Asian and the Arab techniques. All these technical alternatives were adapted to the native raw materials and the indigenous and mestizo cultural practices. Another adaptation of the Filipino still may be observed in the description of Bourke (1893) in Michoacan. This adaptation is in use at present. An example of the blending of both techniques may be found in the description of Bahre and Bradbury (1980) in Sonora.

Evidence regarding the early introduction of the Filipino still and its role in the origin of *Agave* spirits presented by Bruman (1940, 1944) and Walton (1977), and confirmed in this work, is in marked contrast with recent postulations regarding the origin of Tequila as an idea of Spaniards, originated in Tequila, Jalisco, based on the technological model for the fabrication of rum using the arab still, and as a process out of the hands of the indigenous community (Valenzuela-Zapata 1997, p. 36–37; CRT, 2002; Luna-Zamora 2002, p. 33; Muriá 2003, p. 4). Valenzuela-Zapata and Nabhan (2003, p. 9), based on a wrong assignation of affirmations in the works of Bruman and Walton, state that

> The Filipinos brought sugar cane to the coasts of Jalisco and Oaxaca, along with the distillation of rum or aguardiente (Bruman 1935[1]; Walton 1977). The backyard technology for making cane sugar into firewater became the model for transforming pit-roasted and fermented mezcal into the first "vinos de mescal" as they were originally called.

The hypothesis for Prehispanic distillation of *Agave*, first proposed by Lumholtz (1902) upon his finding of Huichol and Cora stills, and proposed again by Hernández (2002) due to similarities between the archeological ovens found in Tlaxcala and those used at present in Oaxaca has yet to be proven (Barrios-Ruiz 2004).

Agaves for Tequila

Finally, to paraphrase Gentry (1982), just as this genus rich in use potential came into the hands of those human groups who developed the main agricultural center of the Americas, propitiating its explosive diversification; towards the end of the 17th century it fell into the hands of a commercial plantation system which has been drastically reducing the diversity generated over the previous 10 centuries.

Germplasm conservation and legal protection

The isolated conditions in which traditional agriculture has remained in the south of Jalisco and its socio-economic productive conditions have allowed it to continue being a dynamic scenario for the selection, diversification and conservation of germplasm. This situation, however, is being increasingly endangered by the expansion of the cultivation of the "blue agave" clone, that has been reducing the wild populations and the areas cultivated with other traditional varieties.

It is necessary to implement an in situ conservation program in the surrounding areas of the Armería-Ayuquila and Coahuayana-Naranjo-Tuxpan rivers and their tributaries for both the genetic resources of these cultivates and the cultural processes that originated them. This program must support and encourage farmers in their desire to conserve their local varieties and to continue generating new germplasm, through the legal protection of their germplasm and the products derived from it, in order to benefit them directly. These actions may be carried out in accordance to the "International Treaty on Plant Genetic Resources for Food and Agriculture" (FAO 2004) and the "Payment Schemes for Environmental Services" that the Mexican Government has been trying to stimulate. As this area is located within the Biosphere Reserve "Sierra de Manantlán", these actions may be facilitated.

---

[1] The correct year of publication for the paper cited, is 1944. In this work, and in that of Walton (1977) no mention is made to the effect that the Filipinos introduced sugar cane on the coasts of Jalisco and Oaxaca along with the distillation of rum. Quite the opposite in fact, it says they introduced the coconut and its distillation on the coasts of Colima and Jalisco and from there it passed to the indigenous population for the distillation of Agave fermented beverages.

Germplasm conservation and the generation of new germplasm will be of great importance for future genetic and productive improvements in the *Agave* spirits industry, especially if there is a change in the productive and legal focus that aims at diversification rather than homogenization. The globalization of markets offers opportunities for diversification, due to a worldwide increase in the interest for local agricultural products, the protection of their areas of origin and the diffusion of their cultural dimensions.

Acknowledgements We thank the Consejo Regulador del Tequila for their facilities to carry out the ethnobotanical exploration within the Tequila-Amatitán region, especially to Ismael Vicente, Jesús Macías and Fabián Rodríguez for their kindness. To the traditional mezcal producers from the south of Jalisco for their willingness to share their knowledge and their courage to preserve their genetic resources, especially to Macario and Apolinar Partida. To Luis Eguiarte, Jorge Larson, Catarina Illsley, Janet Long, Gerardo Gutierrez-Mendoza and Teresa Rojas for their comments on a previous version of this manuscript.

References

Acuña R (ed) (1987) Relaciones geográficas del siglo XVI: Michoacán. Instituto de Investigaciones Antropológicas, Universidad Nacional Autónoma de México, México D.F

Acuña R (eds) (1988) Relaciones geográficas del siglo XVI: Nueva Galicia. Instituto de Investigaciones Antropológicas, Universidad Nacional Autónoma de México, México D.F

Aguirre-Rivera J, Charcas-Salazar H, Flores-Flores J (2001) El maguey mezcalero Potosino. Universidad Autónoma de San Luis Potosí, San Luis Potosí

Barrios-Ruíz AA (2004) Estudio comparativo de muestras de hornos arqueológicos del sitio Nativitas en Tlaxcala, y hornos actuales del estado de Oaxaca destinados a la cocción de agave, por medio de microscopía, espectroscopia y cromatografía. Bachelor Thesis. Universidad Nacional Autónoma de México. México D.F

Bahre CJ, Bradbury D (1980) Manufacture of Mescal in Sonora, Mexico. Econ Bot 34:391–400

Bourke JG (1893) Primitive distillation among the Tarascoes. Am Anthropol 6:65–70

Bruman HJ (1940) Aboriginal drink areas of New Spain. Ph.D. Dissertation. University of California, Berkeley

Bruman HJ (1944) The Asiatic origin of the Huichol Still. Geogr Rev 34:418–427

Bruman HJ (2000) Alcohol in Ancient Mexico. The University of Utah Press, Utah

Callen EO (1965) Food habits of some Pre-Columbian Mexican Indians. Econ Bot 19:335–343

Colunga-GarcíaMarín P (2004) Colección mexicana de germoplasma de Agave spp. In: Carnevali G, Sosa V, León de la Luz JL, León Cortés J (eds) Colecciones Biológicas. Centros de Investigación CONACyT, Consejo Nacional de Ciencia y Tecnología, México D.F., pp 18–19

Colunga-GarcíaMarín P, Coello-Coello J, Espejo-Peniche L, Fuente-Moreno L (1993) Agave studies in Yucatan, Mexico II. Nutritional value of the inflorescence peduncle and incipient domestication. Econ Bot 47:328–334

Colunga-GarcíaMarín P, Zizumbo-Villarreal D (1986) Diversidad y uso alimenticio del henequén: implicaciones para su proceso evolutivo y perspectivas de aprovechamiento. Boletín de la Escuela de Ciencias Antropológicas de la Universidad Autónoma de Yucatán 13:30–41

Consejo Regulador del Tequila (2002) The History of Tequila. http://www.crt.org.mx

de Gálvez Viceroy M (1785) Bebidas prohibidas. In: Orozco y Berra M (ed) (1855) Apéndice al Diccionario Universal de Historia y de Geografía. Colección de artículos relativos a la República Mexicana. Tomo I, VIII. Imprenta de JM Andrade y F Escalante. México, pp 354–362

Diario Oficial de la Federación (1974) Declaración General de Protección a la Denominación de Origen Tequila. December 9. Modified on October 13, 1977, and October 26, 1999. México, D.F.

Diario Oficial de la Federación (1997) Norma Oficial Mexicana Bebidas Alcohólicas Tequila-Especificaciones. NOM-006-SCFI-1994. September 3, Modified on December 24, February 1 and March 1, 2000. México, D.F.

FAO (2004) International Treaty on Plant Genetic Resources for Food and Agriculture.http://www.fao.org/ag/cgrfa/itpgr.htm

Fuchigami E (1990) Indios Chinos en Colima, siglos XVI y XVII. Archivo Municipal de Colima. Documento inédito AH-178. Colima

García-Mendoza A (2003) Sistemática y distribución actual de los Agave spp. mezcaleros. Final Technical Report. Project V029. Comisión Nacional para el conocimiento y uso de la Biodiversidad. CONABIO. México D.F

Gentry SH (1982) Agaves of Continental North America. University of Arizona Press, Tucson

Gil-Vega K, González ChM, Martínez VO, Simpson J, Vandemark G (2001) Analysis of genetic diversity in Agave tequilana var. azul using RAPD markers. Euphytica 119:335–341

Gómez-Amador A (2000) La palma de cocos en la arquitectura de la Mar del Sur. Ph. D. thesis. Facultad de Arquitectura. Universidad Nacional Autónoma de México. México D.F

Hernández S (2002) El uso de los hornos asociados a las unidades habitacionales del Formativo terminal (300 a.C. – 100 d.C.) del sitio Nativitas, Tlaxcala. Un estudio etnoarqueológico. Bachelor Thesis. Escuela Nacional de Antropología e Historia. México, D.F

Hernández-Xolocotzi E (1971) Apuntes sobre la exploración etnobotánica y su metodología. Colegio de Postgraduados, Escuela Nacional de Agricultura, Chapingo México

Hernández-Xolocotzi E (1978) Exploración etnobotánica para la obtención de plasma germinal para México. In: Cervantes T (ed) Sociedad Mexicana de Fitogenética. Chapingo, México, pp 3–12

Lumholtz C (1902) Unknown Mexico: a record of five years' exploration. 2 vols. New York

Luna-Zamora R (2002) La historia del tequila, de sus regiones y sus hombres. CONACULTA, México D.F

Muriá JM (2003) Una bebida llamada tequila. El Colegio de Jalisco, Zapopan Jalisco

Pérez L (1887) Estudio sobre el maguey llamado mezcal en el estado de Jalisco. Imprenta Ancira y hermano, Guadalajara México

Sevilla del Río F (1977) La Provanca de la Villa de Colima: En su defensa ante un mandamiento de la Real Audiencia de México, que ordenaba la tala total de los palmares colimenses, año 1612. Editorial Jus, México D.F

Smith Jr CE (1986) Preceramic Plant Remains from Guilá Naquitz. In: Flannery KV (ed) Guilá Naquitz. Archaic Foraging and Early Agriculture in Oaxaca, México. Academic Press. New York, pp 265–301

Tello AF (1637–1635) Crónica Miscelánea de la Sancta Provincia de Xalisco: Libro Segundo. Instituto Jaliscience de Antroplogía e Historia. Serie de Historia 9 Vol. III. (1984) Gobierno del Estado de Jalisco. Guadalajara, México.

Valenzuela-Zapata AG (1994) El agave tequilero: Su cultivo e industrialización. Monsanto, Guadalajara México

Valenzuela-Zapata AG (1997) El agave tequilero, su cultivo e industria. 2nd. ed. Monsato, Guadalajara México

Valenzuela-Zapata AG, Nabhan GP (2003) Tequila! A natural and cultural history. The University of Arizona Press, Tucson

Walton MK (1977) The evolution and localization of mezcal and tequila in Mexico. Geografica 85:113–132

Zizumbo-Villarreal D (1996) History of coconut (Cocos nucifera L.) in México: 1539-1810. Gene Resour Crop Evol 43:505–515

Springer

Biodivers Conserv (2007) 16:1669–1677
DOI 10.1007/s10531-006-9037-6

ORIGINAL PAPER

Spatial patterns of dicot diversity in Argentina

**Alejandra Juárez · Pablo Ortega-Baes · Silvia Sühring ·
Walter Martin · Guadalupe Galíndez**

Received: 2 December 2005 / Accepted: 7 April 2006 / Published online: 3 May 2006
© Springer Science+Business Media B.V. 2006

Abstract In this paper, we analyzed the taxonomic diversity of the Argentine dicots
to evaluate their relationships with area, latitude, and longitude. We also evaluated
species diversity and higher taxa diversity relationships. The families, genera and
species diversity in Argentine dicots was not explained by the area of each province
but it varied through latitudinal and longitudinal gradients. The taxonomic diversity
of these plants increased from high to low latitudes and west–east longitudes. These
patterns would explain why the main diversity centers are located in the North
region of this country. As we expected the species diversity and higher taxa diversity
showed a positive relationship. At this scale, higher taxa diversity could be use as
surrogate for species diversity.

Keywords Argentina · Dicots · Plant diversity · Spatial gradients ·
Species richness

Introduction

Several studies have reported that species richness can vary with spatial scale and
through latitudinal, longitudinal and/or altitudinal gradients. It has been also reported that the number of higher taxonomic units (i.e. orders, families, and genera)
vary spatially. Positive relationships between the number of higher taxonomic units
and the species richness have been registered (Gaston and Williams 1993; Williams
and Gaston 1994; Gaston 1996a, 1996b, 2000).

A. Juárez · P. Ortega-Baes · S. Sühring · W. Martin · G. Galíndez
Laboratorio de Investigaciones Botánicas (LABIBO), Facultad de Ciencias Naturales,
Universidad Nacional de Salta, Buenos Aires 177, 4400 Salta, Argentina

P. Ortega-Baes (✉)
Departamento de Ecología de la Biodiversidad, Instituto de Ecología, Universidad Nacional
Autónoma de México, Apdo. Postal 70-275, 04510 México, D.F, México
e-mail: portega@miranda.ecologia.unam.mx

 Springer

Species richness tends to increase as the area becomes larger (Gaston 1996a; Brown and Lomolino 1998; Connor and McCoy 2001). This pattern was documented at different scales, with larger areas supporting more species than smaller ones (Ceballos and Brown 1995; Gaston 1996b; Brown and Lomolino 1998; Brown 2003). On the other hand, the variation in species richness with latitudinal gradients is one of the most widely cited patterns, where the number of species increases according to latitude decrease. This pattern has been described for mammals, birds, amphibians, fishes, insects and plants in several regions in the world (Gaston and Williams 1996; Gaston 2000; Brown 2003; Willig et al. 2003).

Argentina is one of the 25 most diverse countries in the world due to their species richness and endemism (Caldecott et al. 1996). For vascular plants there have been cited 248 families, 1927 genera and 9690 species from which 71.84 % are dicots. In this country, the Northeast and Northwest regions have the highest number of families, genera and species; Misiones and Salta being the most important provinces (Zuloaga et al. 1999). The fact that the Northern provinces show higher vascular plant diversity in Argentina could be related with their area and/or to latitudinal and longitudinal variations. However, at the present, the factors that explain these patterns are unknown.

In this paper, we analyzed the species diversity of the Argentine dicots to evaluate their relationships with spatial variables and higher taxa diversity. Taking into account the patterns described by Zuloaga et al. (1999), we expected that the species diversity: (1) would be explained by the area of each province (Ceballos and Brown 1995; Ortega-Baes and Godínez-Álvarez 2006), (2) increase as latitude decrease (Willig et al. 2003), (3) increase west-eastwards due to the effects of air masses flow (Gaston and Williams 1996) and (4) correlate positively with higher taxon diversity (Gaston 1996b, 2000).

Methods

The checklist of Argentine dicots was elaborated according to *Catálogo de las Plantas Vasculares de la República Argentina II* (Zuloaga and Morrone 1999a, 1999b). This is the only taxonomic diversity and geographical distribution database available for these plants. Based on these data, we constructed presence–absence matrixes for families, genera and species for each province. We included all the provinces of Argentina, except Ciudad Autónoma de Buenos Aires (CABA; Fig. 1). For each province we determined the number of families, genera and species. Total area, latitude and longitude of each province were obtained from INDEC (2005).

We analyzed the relationship among area and the number of families, genera and species through a lineal regression analysis with area as independent variable and taxa richness as dependent variable. Both variables were transformed (Zar 1984).

To analyze taxonomic diversity versus latitude, taxonomic diversity versus longitude and species diversity versus higher taxa diversity relationships, we used the Pearson correlation analysis (Zar 1984). Species richness (ln transformed) and Taxonomic Biodiversity Index (TBI) were used as diversity measures. The TBI was estimated as: $TBI = n/\ln (A)$, where n is the taxa richness and $\ln A$ is the natural logarithm of each province area (Squeo et al. 1998). For each province, the latitude was determined as the average of north and south latitude limits, whereas the longitude as the average of east and west limits. With these variables, we evaluated

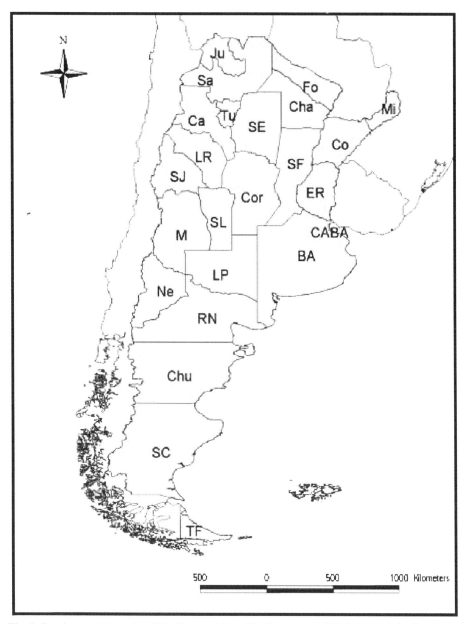

Fig. 1 Provinces of Argentina. BA: Buenos Aires, Ca: Catamarca, CABA: Ciudad Autónoma de Buenos Aires, Cha: Chaco, Chu: Chubut, Co: Corrientes, Cor: Córdoba, ER: Entre Ríos, Fo: Formosa, Ju: Jujuy, LP: La Pampa, LR: La Rioja, M: Mendoza, Mi: Misiones, Ne: Neuquén, RN: Río Negro, Sa: Salta, SC: Santa Cruz, SE: Santiago del Estero, SF: Santa Fe, SJ: San Juan, SL: San Luis, TF: Tierra del Fuego and Tu: Tucumán

the following relationships: (1) family richness-latitude; (2) genus richness-latitude; (3) species richness-latitude; (4) family richness-longitude; (5) genus richness-longitude; (6) species richness-longitude; (7) families TBI-latitude; (8) genera

🖉 Springer

TBI-latitude; (9) species TBI-latitude; (10) families TBI-longitude; (11) genera TBI-longitude; (12) species TBI-longitude; (13) species richness-family richness; (14) species richness-genus richness; (15) species TBI-families TBI; and (16) species TBI-genera TBI.

Results

In Argentina, the dicot plants were represented by 167 families, 1205 genera and 6395 species.

The richness of the different taxa considered in this study could not be explained by the area of each province. No significant relations were found between the family ($F = 0.73$, $P = 0.40$; Fig. 2a), genus ($F = 0.99$, $P = 0.33$; Fig. 2b) and species ($F = 0.85$, $P = 0.37$; Fig. 2c) richness and the area of the provinces.

We found a significant negative correlation between family ($r = -0.75, P < 0.001$), genus ($r = -0.84, P < 0.001$) and species ($r = -0.72, P < 0.001$) richness and latitude of the provinces (Fig. 3a). The same pattern was found between families TBI ($r = -0.73$, $P < 0.001$), genera TBI ($r = -0.78$, $P < 0.001$) and species TBI ($r = -0.67, P < 0.001$; Fig. 3b).

Similarly, a significant negative correlation between family ($r = -0.63, P < 0.001$), genus ($r = -0.53, P = 0.01$) and species ($r = -0.42, P = 0.04$) richness and longitude of the provinces were registered (Fig. 4a). Furthermore, we registered a significant correlation to families TBI ($r = -0.62, P < 0.001$), genera TBI ($r = -0.49, P = 0.02$) and species TBI ($r = -0.42, P = 0.05$; Fig. 4b).

We found significant positive correlations between species richness and family ($r = 0.93, P < 0.001$; Fig. 5a) and genus ($r = 0.97, P < 0.001$; Fig. 5c) richness. In addition, we met a significant positive correlation between species TBI and families TBI ($r = 0.93, P<0.001$; Fig. 5b) and genera TBI ($r = 0.98, P < 0.001$; Fig. 5d).

Discussion

Several studies have reported that the number of species increases with the increment of area; which means that there is a positive relation between species richness and the area size (Gaston 1996b; Brown and Lomolino 1998; Connor and McCoy 2001). At global scale, some authors found a positive relationship between country areas and species richness; thus the species richness would be explained, at least in part, by the area size of each geopolitical unit (Ceballos and Brown 1995; Ortega-Baes and Godinez-Alvarez 2006). In the present study, no relation was found between taxa richness (families, genera and species) and the area of each province. Though there is a wide variations in the size of Argentine provinces (maximum: 307,571; minimum: 22,524; INDEC 2005), we can find both high and low diversity in any of them, independently of its size (Fig. 2). According to these results, it seems that the location is more important than the province size to explain the taxonomic diversity. Gaston (1996b), suggested that if the area range considered varies broadly through a latitudinal gradient, as it occurs for Argentine provinces, the interaction between the area and the species richness can be weak or non significant.

The most widely documented pattern about species richness is their variation through latitudinal gradient, thus the species richness increase with the latitude

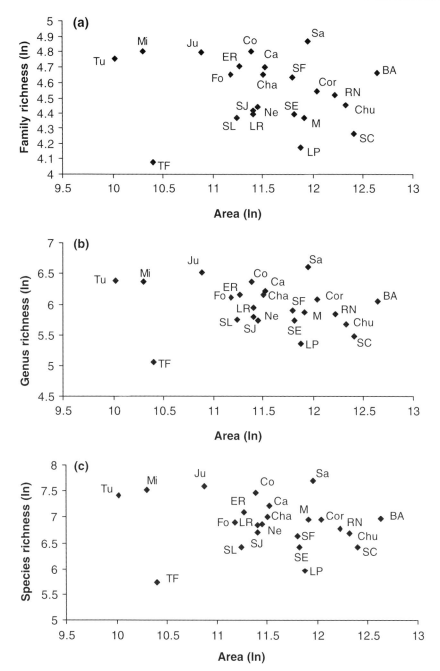

Fig. 2 Relationship between Argentine dicot diversity and province area. (a) Family richness, (b) Genus richness, and (c) Species richness. References are the same as in Fig. 1

diminution (Gaston and Williams 1996; Willig et al. 2003). This pattern has been registered for a wide range of organisms in several regions in the world (Willig et al. 2003). In all cases, the latitude has been considered a surrogate of environmental

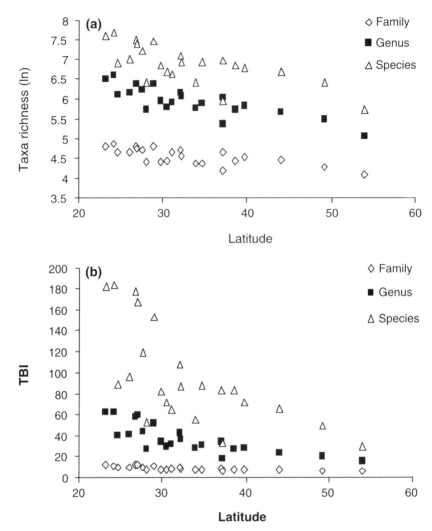

Fig. 3 Relationship between Argentine dicot diversity and latitude. (a) Family, genus and species richness. (b) Families, genera and species TBI. The TBI was estimated as: TBI = n/ln (A), where n is the taxa richness and ln A is the natural logarithm of each province area

gradients such as temperature, solar energy and seasonality which can interact and sometimes they can be correlated between them (Gaston 1996a; Willig et al. 2003). In this work, both the taxa richness and the Taxonomic Biodiversity Index (TBI) for each one of the taxa considered (families, genera and species) showed a negative correlation with latitude, where provinces of low latitude showed higher value of taxonomic diversity. The same pattern has been registered for the taxonomy diversity of plant at global and regional scale (Williams et al. 1994; Willig et al. 2003; Kier et al. 2005). However, there is little evidence of latitudinal variations in plant diversity in tropical regions of South America (Gentry 1988; Prance 1994), as we registered for a country almost completely temperate as Argentina, probably due to the little climatic variation observed. The latitudinal pattern registered for Argentine

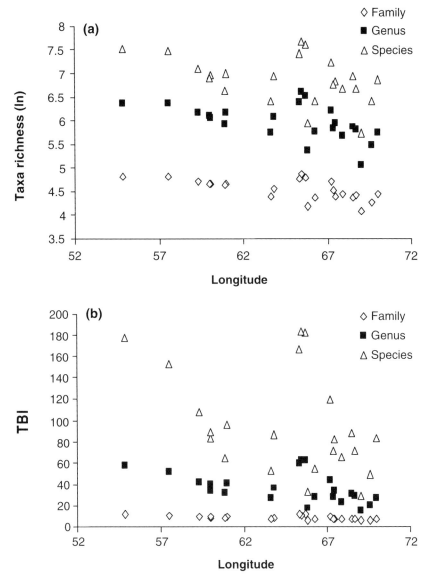

Fig. 4 Relationship between Argentine dicot diversity and longitude. (a) Family, genus and species richness. (b) Families, genera and species TBI. The TBI was estimated as in Fig. 3

dicot diversity is similar to other temperate regions of the world where the climate is considered as the main explicative factor (Silvertown 1985). Real et al. (2003) found the same pattern for land mammals of Argentina, where spatial (e.g. latitude) and environmental (e.g. climate) variables explain a high proportion of mammals species diversity variation.

Gaston and Williams (1996) mentioned for some taxa of Northern Hemisphere that longitudinal variations in species richness can be the results of air flow effects which generate less extreme seasonal climatic variations. According to our results,

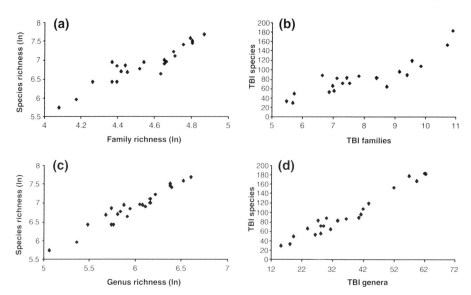

Fig. 5 Relationship between species diversity and higher taxa diversity for Argentine dicots. (a) Species richness–Family richness, (b) Species TBI–Families TBI, (c) Species richness–Genus richness, (d) Species TBI–Genera TBI. The TBI was estimated as in Fig. 3

the family, genus and species richness showed a negative correlation with the longitude, it means that the taxa diversity increase in west-east direction. In Argentina, air flow from the Pacific Ocean cannot go into the continent due to the presence of the Andean Cordillera which is a natural physical barrier, mainly in the center and north regions in this country. Polar, Continental Tropical, Maritime Tropical, and Continental Equatorial air masses have important effects on the climate of this country (Bianchi and Yañez 1992).

A positive relationship among species richness and higher taxa diversity (i.e. families, tribes, genera) has been documented for a wide range of organisms in different regions in the world (Gaston and Williams 1993; Williams and Gaston 1994; Williams et al. 1994; Gaston 1996b). For Argentine dicots we found a significant correlation between species richness and family and genera richness. This means that higher taxa could be used as a surrogate of specific richness (Gaston 1996b). However, if we consider the scale used in this paper (provinces), the utilization of these taxa as surrogate at local scale should be evaluated in more details. Contrary as we registered for Argentina, for the South America tropic it has been suggested an absence of relationship between the higher taxonomic units and the species richness (Prance 1994).

To sum up, the taxonomic diversity in Argentine dicots was not explained by the area of provinces, but it varied through latitudinal and longitudinal gradients. The taxonomic diversity of these plants increases from high to low latitudes and west–east longitudes. As we expected the species diversity and the higher taxa diversity showed a positive relationship. The spatial patterns of dicot diversity in Argentina would explain why the main diversity centers are located in the Northeast region of this country. Finally, the high diversity of the Northwest provinces (e.i. Salta, Jujuy, Tucumán) could be explained by the latitudinal gradient. This could be due to that tropical eco-regions have their limits of distribution in this region of Argentina, which added to the others eco-regions generate a high environmental heterogeneity.

Acknowledgements We are grateful to M. Alonso-Pedano by collaborate with database and H. Zarza by elaborate figure one. We are also grateful to C. Moraga and N. Boetsch for their assistant with English version. POB and SS thanks to CIUNSalta for partial supporting.

References

Bianchi AR, Yañez CE (1992) Las precipitaciones en el noroeste Argentino. INTA

Brown JH, Lomolino MV (1998) Biogeography. Sinauer Associates, INC. Publishers

Brown JH (2003) Macroecología. Fondo de Cultura Económica, México

Caldecott JO, Jenkins MD, Johnson TH, Groombridge B (1996) Priorities for conserving global species richness and endemism. Biodivers Conserv 5:699–727

Ceballos G, Brown JH (1995) Global patterns of mammalian diversity, endemism, and endangerment. Conserv Biol 9:559–568

Connor EF, McCoy ED (2001) Species-area relationships. In: Levin SA (ed) Encyclopedia of biodiversity, San Diego Academics Press, pp 397–411

Gaston KJ (1996a) What is biodiversity? In: Gaston KJ (ed) Biodiversity: a biology of numbers and difference, Blackwell, pp 1–9

Gaston KJ (1996b) Species richness: measure and measurement. In: Gaston KJ (ed) Biodiversity: a biology of numbers and difference, Blackwell, pp 77–113

Gaston KJ (2000) Global patterns in biodiversity. Nature 405:220–227

Gaston KJ, Williams PH (1993) Mapping the world's species the higher taxon approach. Biodivers Lett 1:2–8

Gaston KJ, Williams PH (1996) Spatial patterns in taxonomic diversity. In: Gaston KJ (ed) Biodiversity: a biology of numbers and difference, Blackwell, pp 202–229

Gentry A (1988) Changes in plant community diversity and floristic composition on environmental and geographical gradients. Ann Missouri Bot Gard 75:1–34

INDEC (2005) Geografía. http://www.indec.mecon.ar/

Kier G, Mutke J, Dinerstein E, Ricketts TH, Küper W, Kreft H, Barthlott W (2005) Global patterns of plants diversity and floristic knowledge. J Biogeogr 32:1107–1116

Ortega-Baes P, Godínez-Alvarez H (2006) Global diversity and conservation priorities in the Cactaceae. Biodivers Conserv 15:817–827

Prance GT (1994) A comparison of the efficacy of higher taxa and species numbers in the assessment of biodiversity in the neotropics. Phil Trans R Soc Lond B 345:89–99

Real R, Barbosa AM, Porras D, Kin MS, Marquez AL, Guerrero JC, Palomo LJ, Justo ER, Vargas JM (2003) Relative importance of environment, human activity and spatial situation in determining the distribution of terrestrial mammal diversity in Argentina. J Biogeogr 30:939–947

Silvertown J (1985) History of a latitudinal diversity gradient: woody plants in Europe 13,000–1000 years b.p. J Biogeogr 12:519–525

Squeo FA, Cavieres LA, Arancio G, Novoa JE, Matthei O, Marticorena C, Rodríquez R, Arroyo MTK, Muñoz M (1998) Biodiversidad de la flora vascular en la región de Antofagasta, Chile. Revista Chilena Hist Nat 71:571–591

Williams PH, Gaston KJ (1994) Measuring more of biodiversity can higher-taxon richness predict wholesale species richness? Biol Conserv 67:2111–2117

Williams PH, Humphries CJ, Gaston KJ (1994) Centres of seed-plant diversity: the family way. Proc R Soc Lond B 256:67–70

Willig MR, Kaufman DM, Stevens RD (2003) Latitudinal gradients of biodiversity: pattern, process, scale, and synthesis. Annu Rev Ecol Evol Syst 34:273–309

Zar JH (1984) Biostatistical Analysis. Prentice-Hall, INC

Zuloaga FO, Morrone O (eds) (1999a) Catálogo de las Plantas Vasculares de la República Argentina II. A-E. Missouri Botanical Garden, St. Louis, Missouri

Zuloaga FO, Morrone O (eds) (1999b) Catálogo de las Plantas Vasculares de la República Argentina II. F-Z. Missouri Botanical Garden, St. Louis, Missouri

Zuloaga FO, Morrone O, Rodríguez D (1999). Análisis de la biodiversidad en plantas vasculares de la Argentina. Kurtziana 27:17–167

ⓓ Springer

Biodivers Conserv (2007) 16:1679–1697
DOI 10.1007/s10531-006-9039-4

ORIGINAL PAPER

Unsustainable collection and unfair trade? Uncovering and assessing assumptions regarding Central Himalayan medicinal plant conservation

Helle Overgaard Larsen · Carsten Smith Olsen

Received: 27 June 2005 / Accepted: 24 March 2006 / Published online: 11 November 2006
© Springer Science+Business Media B.V. 2006

Abstract The trade in medicinal plants for herbal remedies is large and probably increasing. The trade has attracted the attention of scientists and development planners interested in the impact on plant populations and the potential to improve rural livelihoods through community based management and conservation. This has resulted in a large number of publications and development activities, ranging from small NGO projects to new government policies. Through a review of 119 references from Nepal, 4 common assumptions regarding the medicinal plant collection and trade have been identified: I. The commercial medicinal plant resource base is becoming ever more degraded as a consequence of collection; II. The medicinal plants are an open-access resource; III. Cultivation can contribute to conservation of commercially collected medicinal plant species; and IV. Medicinal plant harvesters are cheated by middlemen. The frequency of the assumptions is documented, their empirical support is evaluated, and the consequences of their presence for conservation and rural livelihoods are discussed. It is concluded that the empirical backing for the assumptions is weak, and that some reviewed references use logically flawed argumentation. It is argued that the assumptions are leading to misguided conservation efforts, and an inclusive approach to conservation of commercial central Himalayan medicinal plant species is briefly outlined.

Keywords Commercialization · Community-based conservation · Endangered species · Medicinal plants · Nepal · Non-timber forest products

Electronic Supplementary Material Supplementary material is available in the online version of this article at http://dx.doi.org/10.1007/s10531-006-9039-4 and is accessible for authorized users.

H. O. Larsen · C. S. Olsen (✉)
Danish Centre for Forest, Landscape and Planning, The Royal Veterinary and Agricultural University, Rolighedsvej 23, Frederiksberg C, DK-1958, Denmark
e-mail: cso@kvl.dk

H. O. Larsen
e-mail: hol@kvl.dk

 Springer

Introduction

Millions of people rely on, or choose to use, medicinal plants to cover all or part of their health care needs (IUCN 1993; WHO 2002). Tens of thousands of species are used, and medicinal usage may constitute the most common human use of biodiversity (Hamilton 2004). The annual world market for botanical medicines is estimated at USD 20–40 billion, with an annual growth rate of 10–20% (Kate and Laird 1999). Most medicinal plant species are harvested in the wild (Schippmann et al. 2002) and the extent of use, including the exchange of plant material across national borders and continents, has led to conservation concerns spurring recommendation of extensive cultivation programs (Akerele et al. 1991; IDRC 2002).

Medicinal plants are important to urban and rural populations throughout the Indian sub-continent (CSIR 1986). The traditional medical systems practiced in the region are dependent on input from a wide range of habitats and there is large-scale trade in medicinal plants. From Nepal alone, thousands of tons are annually harvested and exported (Olsen 2005, 2006), and the trade is known to have taken place for centuries (Regmi 1972; Olsen and Helles 1997b). Today, products from hundreds of species are gathered by rural harvesters in forests and other vegetation types, and pass through well-established market channels from remote Nepalese villages to cities in India (Olsen 1998). Quantitative end-use consumption estimates are not available for any products, but regional and international consumer demand for traditional plant-based medicines is argued to be increasing rapidly (Lambert et al. 1997; Hamilton 2004). In the past decades, trade has attracted the attention of scientists and development planners due to concerns over the impact of trade on plant populations and the potential for using trade to improve rural livelihoods through community based management and conservation (HMG 1988a). This has resulted in a large number of publications and development activities, ranging from small NGO projects to new government policies.

In our research and work with commercial medicinal plant conservation and trade in Nepal since 1992, we have had the opportunity to study the literature and interact with many policy makers, traders, and development agents. We have become increasingly aware that many of the actors in this sector, including scientists, government officials, and development workers, seem to implicitly operate with a number of major assumptions that influence their way of thinking and mode of action. Uncovering and discussing these assumptions is not just an academic exercise: the assumptions shape the public and academic discussions, and influence the use of scarce research funds and the formation of national and local decisions. Without an explicit discussion and assessment of the assumptions, the risk is that both conservation and development efforts in relation to commercial medicinal plants will be misguided. Through formulating and discussing what we perceive to be four unfounded assumptions, the present paper seeks to initiate a critical discussion of sustainable ways forward for commercial medicinal plant conservation.

Methodology

This literature review was conducted with the criteria for assessment of reviews of medical science in mind (Gates 2002). Though not directly comparable, issues such as selection and publication biases, as well as the formulation of a review protocol are valuable. In contrast to reviews of medical science this review did not assess the quality of the included references; the only parameter noted was whether primary research had been carried out or not.

Literature was compiled from libraries, project offices, and government agencies. Criteria for inclusion were that references should include aspects of commercial medicinal plant conservation and/or trade in Nepal. Accordingly, e.g., purely ethnobotanical or pharmacological studies were not included. A total of 119 references dating from 1976 to 2004 were included in the review. There were contributions from 123 authors; the average number of references per author was 1.6 with a maximum of 7. The review included published but not peer-reviewed material (conference proceedings), unpublished material – e.g. consultancy reports, peer-reviewed material, policy documents, and journalistic material. An overview of reference types, main subjects addressed, and methods applied in the reviewed references is presented in Table 1.

This literature review does not claim to be exhaustive. Peer-reviewed material is relatively easy to find and access, while books containing non-peer-reviewed material and proceedings are often published regionally in small impressions. Consultancy reports usually have to be collected from the organisation commissioning it, and policy documents from the relevant ministry. The large number of non-peer-reviewed references included in the review were collected during periods in which the authors stayed in Nepal. If project offices were visited again more references could be added to the list, but the analysed literature probably includes all major opinions and views on medicinal plant conservation and trade in Nepal. Journalistic references were only included if they were frequently referred to in other publications. Only references in English were included; references in Nepali were only included if translated into English.

Careful reading of the literature resulted in the identification and formulation of four assumptions related to commercial medicinal plant conservation and trade. Identification of the assumptions was assisted by information gathered from informal and formal discussions and interviews held with researchers, NGO and donor staff, government officers, harvesters, and traders over the past ten years (Olsen 1995, 1998, 2006; Larsen 2002; Olsen and Larsen 2003). The identification and formulation

Table 1 An overview of types, main subjects addressed, and methods applied in reviewed references

Type	No.	Subject	No.	Method	No.[a]
Non-peer reviewed	60	General status	39	Interview	59
Unpublished	30	Sustainable management	12	Literature exclusively	33
Peer-reviewed	21	Trade analysis	12	Expert consultation	10
Policy document	6	Economic potential	12	Inventory	7
Journalistic	2	Resource assessment	9	Questionnaire	6
		Other	9	Not specified	6
		Conservation	7	Plant enumeration	5
		Government policy	6	Workshop	5
		Policy analysis	5	Cultivation trial	2
		Ethnobotany	3	Herbarium	1
		Biological study	3	Population biology	1
		Economic importance	2		
Total	119	Total	119	Total	135

[a] Sums to 135 as some studies apply more than one method

 Springer

of the four assumptions in this paper of course reflect the positions of the authors, and more assumptions could probably be formulated if other aspects than the state, management, and trade of medicinal plants were focused on. While the formulation of assumptions is necessarily a qualitative endeavour, their presence or absence from a given reference can be objectively verified, ensuring scientific rigour.

All references were categorized in terms of whether they featured or disagreed with one or more of the four assumptions. A database was constructed where the paragraphs featuring an assumption and the page number were entered. This was followed by a three-step analysis: (i) the prevalence of the identified assumptions was estimated by counting the frequency of occurrence among references, (ii) the degree to which each assumption is supported by empirical evidence was assessed, and (iii) when assumptions contain elements that cannot be assessed objectively, e.g. whether cultivation is likely to lead to conservation of wild populations, the provided arguments were analysed.

Results

Four assumptions were identified in the reviewed literature. Three assumptions are related to the state and management of the commercial medicinal plant resource base, while the last is related to trade. Information on the frequency of assumptions and the argumentation provided by the authors in the reviewed references are found in Tables 2 and 3.

Assumption I: degradation

The commercial medicinal plant resource base is becoming ever more degraded as a consequence of collection

In most of the studies reviewed, it is stated that the commercial medicinal plant resource base is being overexploited and degraded: "The main problems with the trade are that the resource is being depleted" (DeCoursey 1993, p. 51). A total of 25 different arguments and indicators supporting the assumption of overexploitation were found. The frequency of the three most common arguments (one reference may contain more than one argument) exceeds 50%. Fourteen references conclude that the present level of information renders impossible conclusions on whether the resource is degraded or not.

Assumption II: open-access

The medicinal plants are an open-access resource

Twenty-six studies state that the medicinal plant resource base is subject to open-access conditions: "Non-timber forest products, especially medicinal and aromatic

Table 2 Number of reviewed references featuring or disagreeing with the assumptions

Assumption	I. Degradation	II. Open-access	III. Cultivation	IV. Middleman
Featuring the assumption	77	26	48	37
Disagreeing with the assumption	3	10	3	5

Table 3 Most frequent arguments provided to support the four assumptions on medicinal plant collection and trade

Assumptions and arguments	No. of references
Assumption I: The commercial medicinal plant resource base is becoming ever more degraded as a consequence of collection	77
Collection is indiscriminate	51
Increasing demand for raw materials in India spurs unsustainable harvest	26
Poverty compels harvesters to destruction	15
Assumption II: The medicinal plants are an open-access resource	26
Previous systems disappeared due to nationalisation or increasing market demand	4
Monitoring is difficult	2
Assumption III: Cultivation can contribute to conservation of commercially collected medicinal plant species	48
Reduction of pressure on wild populations	17
Assurance of adequate supplies of raw materials for the industry	16
Provision of local benefits, e.g. income	10
Assumption IV: Medicinal plant harvesters are cheated by middlemen	37
Cash advances mean harvesters have to accept low prices	7
There are few traders	7
Lack of price information	6

plants, are regarded as a free commodity to be collected from nature" (Bhattarai 1997a, p. 82). Any systems in place to regulate collection are said to have disappeared after the Pasture Nationalisation Act in 1974, where traditional local rights to alpine pastures were transferred to the government (DeCoursey 1993). Ten references document some form of community management and/or argue that open-access cannot automatically be assumed. Hertog (1995, p. 24) observed both community management and open-access situations during a fieldtrip to Dolpa District: "Most of the valuable NTFPs are collected on public, government land, access to which is open to a group of villagers, Forest Users Group or all".

Assumption III: cultivation

Cultivation can contribute to conservation of commercially collected medicinal plant species

Linked to the assumption of resource degradation is the assumption that cultivation can play a significant role in commercial medicinal plant conservation. This is expressed in 48 references across all reference types and major methods. Of these, 38 also feature the assumption of resource degradation. Assumptions I and II are related to what could theoretically be observed today (the state of the resource and absence or presence of local management systems), while Assumption III is related to a hypothesised future effect of cultivation.

Two possible drawbacks of cultivation are mentioned by three references: (i) promotion of cultivation on private land will not benefit farmers who cannot afford the necessary land, labour, and capital, and (ii) cultivation may not reduce collection from wild sources which may be carried out, in parallel, by a different sector of the community. In addition, concern over the content of chemical substances of cultivated plants is mentioned (Hertog 1995).

🍎 Springer

Assumption IV: middleman

Medicinal plant harvesters are cheated by middlemen

The fourth assumption is that the middlemen directing the dried medicinal plant products to India pay unreasonably low prices to harvesters, thereby earning more than what is considered fair: "...the mountain peasants who are the primary collectors are getting just the crumbs that fall from a sumptuous table" (Aryal 1993, p. 9). The assumption is expressed by 37 references. Five references argue that the middlemen are not exploiting the harvesters, but rather provide services such as transport and cash advances that would otherwise not be available. Four references do not state whether traders are exploitative, but note the services that traders provide.

There is general agreement that Terai based traders, the exporting central wholesalers, control market information and are able to capture very large profit margins. Some authors also state that village-based traders capture disproportionate margins.

Discussion

The four assumptions presented above can all be concluded relatively common on the basis of the number of references expressing them. In the following, they will be evaluated in terms of their empirical support, the line of argumentation used by their proponents, and the conservation consequences of their prevalence in scientific literature and policy documents.

Is the commercial medicinal plant resource base under threat from collection?

No explicit definition of degradation is provided in any of the reviewed sources, except Bhattarai et al. (2002). However, most references expressing the assumption of degradation refer to overexploitation (e.g. HMG 1988b) and species extinction (e.g. Chaudhary 1998). A narrow definition of degradation is used in this discussion: a significant and measurable decrease in the supply of medicinal plants.

The empirical basis for concluding that the commercial medicinal plant resource base is being degraded is very weak. Of the 77 references expressing the degradation assumption, 20 did so without any empirical support data, and another 29 studies were based on interviews only. Nor can the literature focusing on resource assessment be used to draw general conclusions. Of the 9 references related to resource assessment, three are literature studies discussing methods, two are reports of the National Forest Survey including presence–absence of a small number of medicinal plants in forest inventory plots, two are small-scale inventories based on purposeful sampling recording frequency and biomass, and two are small to medium-scale inventories based on random sampling recording frequency and biomass. Only two studies specifically attempted to evaluate the effect of commercial harvest on plant populations (Ghimire et al. 2002; Larsen 2002), but without data on general collection intensities these studies can offer little in terms of a general resource assessment.

A Conservation Assessment and Management Plan (CAMP) workshop conducted in 2001 assigned threat categories to 51 medicinal plant species using the

IUCN Red List criteria (Bhattarai et al. 2002). This exercise was essentially based on participants' perceptions of changes in various species parameters, such as population size. While useful, this should not be confused with, or equalled to, empirical evidence for the status of any medicinal plant species.

The assumption of degradation seems to a large extent based on beliefs that commercialization automatically translates into pressures on the resource (e.g. Shrestha and Joshi 1996). Though not explicitly stated, it seems that degradation is assumed to stem from increasing market demand leading to a break-down of local management systems because rising prices lead to indiscriminate harvest (levels and techniques) by poor, ignorant harvesters. The general belief that commercialization and increasing demand leads to overexploitation of natural resources is widely supported (Redford and Stearman 1993; Freese 1998), but so is the counterargument that the more valuable the resource, the stronger incentives people will have to manage it sustainably (Plotkin and Famolare 1992; Godoy and Bawa 1993).

Does commercialization lead to degradation of the medicinal plant resource base?

It is uncertain whether demand is increasing. However, there is supporting circumstantial evidence: Lambert et al. (1997) and Holley and Cherla (1998) claim increasing demand for natural medicines in western countries, while Olsen (1998) speculates that Indian demand is increasing. As the principal end-uses are not known for any species, predicting demand is impossible. While official trade data is very inaccurate (Malla et al. 1995), a 43% increase in district-wise annual average medicinal plant export from Nepal from the fiscal years 1989/90 to 2003/04 may indicate increasing demand for some species (Malla et al. 1995; HMG 2004). Longitudinal price information is almost completely absent. Olsen and Bhattarai (2000) found significant price increases for *Nardostachys grandiflora* from 1994/95 to 1997/98, but also that harvester prices were constant. This indicates that increasing demand and wholesaler prices do not necessarily directly affect the harvesters' incentive to collect. Furthermore, both common pool management and open-access regimes are found in Nepal, and evidence that the former are not robust and breaking down in the face of increasing market pressure is missing, as discussed below.

How common is open-access?

This review reveals confusion over the difference between open-access and common pool resources, e.g. 'Community pasturelands are considered as common property. As a result, there is no identifiable entity to accept management responsibility and most of the community pasturelands are overgrazed and deteriorating' (HMG 2002, p. 57). Equating open-access and community management means that exclusion of the harvesters from formal management decisions is a decision foregone. The confusion indicates an implicit adherence among the references to the Tragedy of the Commons line of thinking, where sustainable use of a common pool resource requires either management by the state or privatization (Hardin 1968). It also indicates that more than 15 years of research on collective action building on communication, trust, and reciprocity is disregarded. In this paper, we talk about open-access when no property rights define who can use a common pool resource and how uses are regulated (Dietz et al. 2002). Local institutions are documented to hold the potential, under certain conditions, to keep common-pool resource extraction within sustainable levels (Ostrom 1990). These are, among other things, that the resource is considered valuable, that limits on access exist and can be enforced, and that co-operation is considered rational (Ostrom 1998). Furthermore, the presence of local institutions has been shown to have a positive effect on regulating harvesting practices, an effect argued to be more important than demographic changes (Agrawal and Yadama 1997).

Little emphasis has been placed specifically on studying local institutions in relation to commercial medicinal plants. Only 7 of the 26 references expressing the assumption on open-access were based on local level studies; none of these focused explicitly on local institutions. Eight references documented the presence of local institutions regulating access to, and withdrawal of, medicinal plants. Thus, local institutions for regulating commercial medicinal plant harvest are present in some, but not all locations, and open-access regimes cannot generally be assumed. This is not to say that local institutions ensure sustainable harvest. The local management systems documented in the reviewed references are both old (Larsen 2002), and newly arisen in the face of resource pressure (Edwards 1996b). The specific attributes of a given local system, including an assessment of the resource availability, need to be studied for an evaluation of the regulation potential.

Is cultivation a feasible conservation approach?

The need for cultivation has been voiced for decades (Dobremez 1976). As mentioned above, this is argued to lead to conservation of the wild plant populations, assured supply of raw materials to the industry, and local benefits. However, cultivation is rare. The Government of Nepal has established structures for growing and processing medicinal plants (Rajbhandary and Bajracharya 1994), but their focus is on exotic traditional species such as lemongrass rather than wild harvested and traded medicinal plant species. None of the traded species are commercially cultivated, though cultivation techniques are being developed for a few species (Maharjan 1994; Dhakal 2000). Likewise, medicinal plant cultivation in community forestry is rare as is cultivation on private land (Edwards 1996a). There are not yet any official reports of medicinal plants sold from community forests (HMG 2004).

Given the perceived benefits of, and official emphasis on cultivation, why is it so rare? Since no extensive official cultivation program has been initiated, it must be the assumption that market forces will induce individual farmers to cultivate medicinal plants. But seen from the farmers' perspective, cultivation apparently is not attractive. Several reasons can account for this: First, if there are relatively abundant resources, the unit cost of extraction will be low compared to cultivation. Second, the required investment may be too high in terms of land, labour, and/or knowledge acquisition. Third, other crops may be more profitable.

In the short term, collection of medicinal plants from the wild will continue whether the resource is declining or not: demand is likely to persist and other income-generating activities are unlikely to become available to any significant degree in the rural areas. Additionally, medicinal plant harvesters typically have less land, and will not engage in cultivation. Depending on the resource availability in the wild, cultivation may be seen at small scales. In the medium term, if demand persists, the availability of wild resources will depend on the ability of local systems to balance harvest and growth, at least for products of which harvest is necessarily destructive. This may be tempered by new income-generating activities and a substitution away from extraction (Neumann and Hirsch 2000). If no substitutes are found, and the supply curve is assumed inelastic (Homma 1996), cultivation may eventually take place to a larger degree. Extraction and cultivation can probably take place simultaneously for some time, although the steady supply by cultivation would make prices fall (Homma 1996). Local benefit distribution effects would then depend on whether communal cultivation is initiated.

Thus, it appears that whether cultivation is initiated or not depends on the investment required by the individual farmer. This could to some degree be lowered through facilitating knowledge dissemination through NGOs, as is currently attempted (e.g. CECI 1998). However, cultivation will only substitute wild harvesting in the longer term when financially competitive, and may not serve to protect currently exploited plant populations. A more feasible approach to achieving conservation in the short term appears to be in situ management.

Is trade unfair to harvesters?

The large majority of references expressing the assumption on middlemen's exploitative behaviour provide no data on marketing margins. None of the references reviewed provide suggestions as to what a fair margin is. In the seven studies containing empirical estimates of marketing margins, 23% of Indian retail price (Malla et al. 1999), 9–37% of Terai seller's price (Hertog 1995; Sharma 1995), 7–27% of Indian wholesaler price (CECI 1999), and 12–54% of Indian wholesaler price (Rawal and Poudyal 1999) are perceived to be unfair. On the other hand, 50% of Indian retail price is consider fair (Edwards 1996b), as is 47% of Indian wholesaler price (Olsen 1998). The studies are not directly comparable due to differences in reporting net/gross benefits. The review revealed a need for more stringent price data collection and reporting: prices are published and quoted without specification of time (month and year) and location (physically as well as in the marketing chain). It has been documented that harvesters' margins for a species can change by 100% within the same season (Olsen and Helles 1997a), and that local traders can have negative margins in one year and positive in another (Olsen and Bhattarai 2000). This indicates that observations and recommendations based on data from just one year should be treated with much caution.

To assess whether harvesters of medicinal plants receive too low prices and returns to labour, three issues need consideration: (i) what should harvester prices be compared to, (ii) what are the returns to labour, and (iii) how does price share and return to labour compare with other products such as agricultural produce. Regarding the first, Olsen (2005) showed that almost 93% of the volume of commercial medicinal plants traded in 1997/98 went to India. As end uses and prices beyond the regional wholesaler level for all Nepalese medicinal plant species remain unknown, it seems reasonable to compare harvester prices with regional wholesaler prices in India. Regarding the second, there have been very few studies of return to labour used in commercial medicinal plant harvesting. In a study from western Nepal in 1994–95, Olsen (1998) found returns to labour input in dedicated medicinal plant harvesting comparable to other available income options. Regarding the third, Crawford (1997) calculated the gross percentage of the final retail price accruing to Nepalese farmers for rice and wheat at 56–69%. This is higher than the share of regional wholesaler price realized for medicinal plant harvesters. However, it is difficult to compare the two figures, especially as farmers' costs are not known for the agricultural produce. In collection of medicinal plants in the wild, harvesters' gross and net margins will be almost similar as products can be collected free of charge and the activity does not require capital investments.

Thus available evidence does not indicate very low prices or returns to labour. Empirical data does not paint a clear picture of unfair harvester exploitation. This does not, of course, mean that income and/or return to labour cannot be increased.

Consequences of the assumptions

The four assumptions are common in the scientific, development, and policy communities working with commercial medicinal plants in Nepal. Empirical evidence on the state of the medicinal plant resource is not available, and yet perceptions of overexploitation and rural irresponsibility dominate the political agenda. The commonness of not-too-critical studies based entirely on literature means that assumptions get repeated and perpetuated. The perception of an impending ecological crisis in relation to the commercially harvested medicinal plants has lead to strict rules on collection and export (HMG 1995), and harvesters are frequently regarded as ignorant or ruthless individuals incapable of organising for the management of a valuable common good. At the same time, the present authoritarian approach to commercial medicinal plant conservation and trade has lead to widespread illegal rent-seeking (CECI 1999), while implementation of rules on resource monitoring are virtually non-existent (Larsen et al. 2005). This dire situation is further exacerbated by the attention given to cultivation. Many plans for domestication and cultivation of commercial medicinal plants have been made (HMG 1988a), but few have been implemented (HMG 2002). The cultivation focus is contributing to directing attention towards unknown future benefits, and away from the issue of sustainable collection from the wild today. It may thus be speculated that the eagerness to acknowledge degradation of the commercial medicinal plant resource base, on weak scientific grounds, has contributed to create a situation where real and possibly pressing conservation issues are not being identified or addressed. If the national objective of sustainable management of the commercial medicinal plant resource base (HMG 1988a) is to be achieved, a different approach to conservation is required.

An inclusive approach to conservation of commercial medicinal plants

The current top-down approach to commercial medicinal plant conservation emphasises control of harvesters and establishment of cultivation to relieve pressures on natural plant populations. As noted above, this approach has failed. We suggest an inclusive approach building on three key components: (i) scientific input, (ii) local communities, and (iii) regional co-operation. Consider these in turn. First, scientific input is required to establish a valid and reliable overview of conservation issues. Initial work should focus on conducting a national-level inventory for key commercial species as a basis for a species level empirical threat assessment. Much research and work has already been done that would provide input to and lower the cost of such an inventory, including identification of key commercial species (Olsen 2006), a GIS database with potential vegetation coverage (Lillesø et al. 2005) and reviews of NTFP inventory techniques (Wong et al. 2001). Subsequent research could focus on increasing our meagre understanding of medicinal plant–people dynamics, e.g. how local communities respond to supply and demand changes. Second, conservation should build on involving local communities in resource management. Enforcement of official regulations is currently not possible, nor is any resource monitoring taking place (Larsen et al. 2005). Through involving local communities, several generations' knowledge on medicinal plant ecology could potentially be used for location-specific management decisions. The need for the participation of local communities in in-situ conservation is increasingly recognised. If granted a share in management power and responsibility, local institutions could be part of adaptive co-management contributing traditional ecological knowledge and local institutions to an on-going process of trial-and-error (Berkes 2004). Third, Nepal and other countries in the region should strive to move from the current national conservation approach to a regional approach. Given the structure of the medicinal plant market, and the fact that almost all traded species are found across several countries, there is a risk that current national level regulations may simply, if implemented and enforced, shift demand and thus plant population pressure between countries.

Are assumptions found elsewhere?

The situation in Nepal resembles previous assumptions such as the 'Himalayan environmental degradation theory' (Forsyth 1998) and 'west African deforestation' (Fairhead and Leach 1995). These myths served to justify authoritarian policy interventions. In Nepal, there is a risk that harvesters of commercial medicinal plants will be announced culprits, responsible for the degradation of the medicinal plant resource base, just like the upland farmers of Nepal who were accused of causing erosion because they were said to cultivate ever steeper slopes without concern for lowland livelihoods (Forsyth 1998). A preliminary survey of 13 peer-reviewed studies related to commercial medicinal plant conservation and trade elsewhere indicates that some of the assumptions are also found in India and Pakistan (e.g. Sheikh et al. 2002; Uniyal et al. 2002). Assumptions of degradation and the potential of cultivation for conservation were common in the studies, while assumptions on open-access and middleman exploitation were found in Nepalese references mainly. Conclusions would have to include a thorough review of all types of references, as well as in-depth knowledge on local contexts, but the examined sources suggest that further investigations might be in place.

Conclusion

Four common assumptions on commercial medicinal plant conservation and trade in the central Himalaya were identified and formulated. The empirical support for the assumptions is weak and many arguments used to support the assumptions, such as the nature of the linkage between commercialisation and degradation of the medicinal plant resource base, are found to be accepted and used too uncritically. The need for additional empirical data to support conservation and development initiatives is evident. The present dominance of unfounded assumptions may be a barrier to identification of conservation priorities and may be diverting attention from development of in-situ adaptive co-management systems to ex-situ cultivation

Springer

issues. If sustainable management of commercial central Himalayan medicinal plants is to be achieved, future conservation efforts need to be based on scientifically valid data, build on community participation, and take place at the regional, rather than national, level.

Acknowledgements We wish to thank N.K. Bhattarai and many other colleagues who have participated in discussions on commercial medicinal plants in Nepal. Thorsten Treue and Niels Strange are thanked for their comments on the draft manuscript. The Council for Development Research of the Danish Ministry of Foreign Affairs and the Royal Veterinary and Agricultural University, Copenhagen, provided funding.

References

All 119 reviewed references from Nepal and 13 references from other countries in the region are included in the list although they are not cited in the text. Reviewed references are marked by an asterix.

*Acharya M (2003) Non-timber forest products in Sindhuli District: a review. In: Anonymous (ed) Proceedings of seminar on non-timber forest products, Central Development Region, Kathmandu, Nepal, August 29, 2002. NTFP Database Project, Institute of Forestry, Pokhara, pp 86–107

*Adhikary PM (1993) Medicinal and aromatic plants in Nepal. In: Chomchalow N, Henle HV (eds) Medicinal and aromatic plants in Asia: breeding and improvement. Oxford and FAO, New Delhi, pp 138–144

Agrawal A, Yadama GN (1997) How do local institutions mediate market and population pressures on resources? Forest Panchayats in Kumaon, India. Dev Change 28:435–465

*Airi S, Rawal RS, Dhar U, Purohit AN (2002) Assessment of availability and habitat: preference of Jatamansi – a critically endangered medicinal plant of west Himalaya. Curr Sci 79:1467–1471

Akerele O, Heywood V, Synge HE (1991) The conservation of medicinal plants. Proceedings of an International Consultation, 21–27 March 1988, held at Chiang Mai, Thailand. Cambridge University Press, Cambridge

*Amatya G (1995) A preliminary study of medicinal plants in the Bhotkhola and Tamku regions of the Sankhuwasabha District for commercial scale cultivation. Department of National Parks & Wildlife Conservation and The Mountain Institute, Kathmandu

*Amatya G, Sthapit VM (1991) Rural extension service for raw material procurement in Nepal. Entwick Ländlich Raum 4:12–13

*Amatya G, Sthapit VM (1994) A note on *Nardostachys jatamansi*. J Herbs, Spices, Med Plants 2:39–47

*Anonymous (1994) A feasibility study on establishing a processing plant for medicinal herbs at Chaudabisa, Jumla. Karnali Institute for Local Development and Nature Conservation, Kathmandu

*ANSAB (1997a) Environment and forest enterprise activity. Forest products market/enterprise options study. Final report. Asia Network for Small scale Agricultural Bioresources, Kathmandu

*ANSAB (1997b) Non-timber forest products, final technical report. Asia Network for Small scale Agricultural Bioresources, Kathmandu

*Aryal M (1993) Diverted wealth: the trade in Himalayan herbs. Himal Jan/Feb 1993:9–18

*Baral SR (1999) Use of non-timber forest products in Nepal's mid-western Himalayan region adjoining Tibet (China). In: Mathema P, Dutta IC, Balla MK, Adhikary SN (eds) Sustainable forest management. Proceedings of an international seminar, 31 August–2 September 1998, Pokhara, Nepal. Institute of Forestry, Pokhara, pp 167–171

*Basukala KR (2003) Status of non-timber forest products in Kathmandu District. In: Anonymous (ed) Proceedings of seminar on non-timber forest products, Central Development Region, Kathmandu, Nepal, August 29, 2002. NTFP Database Project, Institute of Forestry, Pokhara, pp 58–63

Berkes F (2004) Rethinking community-based conservation. Conserv Biol 18:621–630

*Bhandari B (1997) Prospects for tourism in Chhekampar. IUCN Nepal, Kathmandu
*Bhandari BR (2003) Status of non-timber forest products in Makwanpur District. In: Anonymous
    (ed) Proceedings of seminar on non-timber forest products, Central Development Region,
    Kathmandu, Nepal, August 29, 2002. NTFP Database Project, Institute of Forestry, Pokhara,
    pp 141–146
*Bhattarai DR (2003) An overview of non-timber forest products of Nepal. In: Anonymous (ed)
    Proceedings of seminar on non-timber forest products, Central Development Region,
    Kathmandu, Nepal, August 29, 2002. NTFP Database Project, Institute of Forestry, Pokhara,
    pp 147–152
*Bhattarai NK (1995) Prospects of community- based NTFP enterprise development for promoting
    biodiversity conservation and economic development in Gorkha District, Nepal. Proceedings of
    the seminar on medicinal and aromatic plants in Gorkha District: how to promote their util-
    isation and marketing. German Development Service, District Forest Office, Gorkha Bazaar,
    pp 14–18
*Bhattarai NK (1997a) Biodiversity – people interface in Nepal. Non-Wood Forest Products 11:78–
    86
*Bhattarai NK (1997b) Medicinal and aromatic plants of Nepal. INBAR Technical Report No. 15.
    International Network for Bamboo and Rattan, Beijing, pp 162–173
*Bhattarai NK, Karki M, Tandon V (2002) Report of the CAMP workshop in Nepal. Med Plant
    Conserv 8:28–30
*Bhojvaid PP, Sharma AK, Khali RP, Gargya GR (2000) Ecological aspects of conservation and
    cultivation of *Taxus baccata* L. and *Nardostachys jatamansi* DC in Garhwal Himalaya.
    In: Amatya SM (ed) Proceedings of the third regional workshop on community based NTFP
    management, South East Asian countries NTFP Network (SEANN) 8–9 April, 2000. Institute of
    Forestry, Pokhara, pp 94–119
*Bist HR (2003) Non-timber forest products in Sarlahi District. In: Anonymous (ed) Proceedings of
    seminar on non-timber forest products, Central Development Region, Kathmandu, Nepal,
    August 29, 2002. NTFP Database Project, Institute of Forestry, Pokhara, pp 80–85
*Burbage MB (1981) Report on a visit to Nepal: the medicinal plant trade in the KHARDEP area –
    a study on the development potential. Natural Resources Institute, Overseas Development
    Administration, London
*CECI (1997a) Inventory of four high value non-timber forest products in Jumla (*Nardostachys
    grandiflora, Picrorhiza scrophulariiflora, Rheum australe* and *Valeriana jatamansii*). Canadian
    Centre for International Studies and Cooperation, Kathmandu
*CECI (1997b) Non-timber forest products in Nepal: opportunities for sustainable harvesting and
    income generation in Jumla. Canadian Centre for International Studies and Cooperation,
    Kathmandu
*CECI (1998) Training on management, marketing and cultivation of non-timber forest products in
    Jumla. Training manual (revised). Community Based Economic Development Project, Canadian
    Centre for International Studies and Cooperation, Kathmandu
CECI (1999) Subsector analysis of high-altitude NTFPs in the Karnali zone. Draft version. Canadian
    Centre for International Studies and Cooperation, Kathmandu
*Chandrasekharan D (1998) NTFPs, institutions and income generation in Nepal: lessons for
    community forestry. Discussion Paper Series MNR 98/1. International Centre for Integrated
    Mountain Development, Kathmandu
*Chaudhary RP (1998) Biodiversity in Nepal: status and conservation. S. Devi, Saharanpur
Crawford IM (1997) Agricultural and food marketing management. Marketing and Agribusiness
    Texts 2. FAO, Rome
*CSIR (1986) The useful plants of India. Council of Scientific and Industrial Research, Publications
    and Information Directorate, New Delhi
*DeCoursey M (1993) Report on the status of selected non-timber forest products (NTFPs) in
    Sindhu Palchok and Kabhre Palanchok Districts. Nepal Australia Community Forestry Project,
    Kathmandu
*Dhakal R (2000) Preliminary study on germination behaviour of some NTFPs for propagation in
    marginal and khoria land of Makalu Barun Conservation Area (Buffer Zone). In: Amatya SM
    (ed) Proceedings of the third regional workshop on community based NTFP management, South
    East Asian countries NTFP Network (SEANN) 8–9 April, 2000. Institute of Forestry, Pokhara,
    pp 315–328
*Dhar U, Manjkhola S, Joshi M, Bhatt A, Bisht AK, Joshi M (2002) Current status and future
    strategy for development of medicinal plants sector in Uttaranchal, India. Curr Sci 83:956–964

🖉 Springer

*Dhar U, Rawal RS, Upreti J (2000) Setting priorities for conservation of medicinal plants – a case study in the Indian Himalaya. Biol Conserv 95:57–65

*Dhital Y. 2003. Status of non-timber forest products in Ramechhap District. In: Anonymous (ed) Proceedings of seminar on non-timber forest products, Central Development Region, Kathmandu, Nepal, August 29, 2002. NTFP Database Project, Institute of Forestry, Pokhara, pp 127–132

Dietz T, Dolsak N, Ostrom E, Stern PC (2002) The drama of the commons. In: Ostrom E, Dietz T, Dolsak N, Stern PC, Stonich S, Weber EU (eds) The drama of the commons. National Research Council, National Academy Press, Washington DC, pp 3–35

*Dobremez JF (1976) Exploitation and prospects of medicinal plants in Eastern Nepal. In: Anonymous (ed) Mountain environment and development. Sahayogi Press, Kathmandu, pp 97–107

*Duwadi VR, Parajuli DP, Lamichhane HP (2000) SAFE-concern's approach to and experience of domestication, cultivation and harvesting of chiraito (*Swertia chirata*) in the mid-hills of Nepal. In: Amatya SM (ed) Proceedings of the third regional workshop on community based NTFP management, South East Asian countries NTFP Network (SEANN) 8–9 April, 2000. Institute of Forestry, Pokhara, pp 128–131

*Edwards DM (1993) The marketing of non-timber forest products from the Himalayas: the trade between east Nepal and India. Rural Development Forestry Network Paper 15b. Overseas Development Institute, London

*Edwards DM (1996a) Non-timber forest products from Nepal: aspects of the trade in medicinal and aromatic plants. Forest Research and Survey Centre, Ministry of Forests and Soil Conservation, Kathmandu

*Edwards DM (1996b) The trade in non-timber forest products from Nepal. Mt Res Dev 16:383–394

Fairhead J, Leach M (1995) False forest history, complicit social analysis – rethinking some west-African environmental narratives. World Dev 23:1023–1035

*Farooquee NA, Saxena KG (1996) Conservation and utilization of medicinal plants in high hills of the central Himalayas. Environ Conserv 23:75–80

Forsyth T (1998) Mountain myths revisited: integrating natural and social environmental science. Mt Res Dev 18:107–116

Freese CH (1998) Wild species as commodities: managing markets and ecosystems for sustainability. Island Press, Washington DC

*FRSC (1997) Forest resources of the Eastern Development Region 1996, Publication 70. Forest Research and Survey Centre, Ministry of Forests and Soil Conservation, Kathmandu

Gates S (2002) Review of methodology of quantitative reviews using meta-analysis in ecology. J Anim Ecol 71:547–557

*Gautam KC (2003) Status of NTFPs in Sindhupalchok District. In Anonymous (ed) Proceedings of seminar on non-timber forest products, Central Development Region, Kathmandu, Nepal, August 29, 2002. NTFP Database Project, Institute of Forestry, Pokhara, pp 9–24

*Gautam M (1995) Non-timber forest product development: new directions for Nepal's community forestry program. In: Fox J, Donovan D, DeCoursey M (eds) Voices from the field: sixth workshop on community management of forest lands. Program on Environment, East-West Center, Honolulu, pp 53–63

*Ghimire SK, Shrestha KK, Gurung TN, Lama YC, Aumeeruddy-Thomas Y (2002) Community-based conservation of medicinal plants and development in the Shey Phoksundo National Park, Nepal. In: Anonymous (ed) Medicinal plants: a global heritage. International Development Research Centre, Delhi, pp 239–246

Godoy R, Bawa KS (1993) The economic value and sustainable harvest of plants and animals from the tropical forest: assumptions, hypotheses, and methods. Econ Bot 47:215–219

Hamilton AC (2004) Medicinal plants, conservation and livelihoods. Biodivers Conserv 13:1477–1517

Hardin G (1968) The tragedy of the commons. Science 162:1243–1248

*Hertog W (1995) Hidden values: non-timber forest products in Dolpa District. Karnali Local Development Programme, Kathmandu

*HMG (1988a) Forest-based industries development plan, part II: medicinal and aromatic plants and other minor forest products. Ministry of Forests and Soil Conservation, Kathmandu

*HMG (1988b) Master plan for the forestry sector Nepal. Main report. His Majesty's Government, the Agricultural Development Bank, and FINNIDA, Kathmandu

*HMG (1993) Nepal environmental policy and action plan: integrating environment and development. Environment Protection Council, His Majesty's Government, Kathmandu

HMG (1995) Forest Regulation, 2051 (official English translation). Ministry of Forests and Soil Conservation, Kathmandu
*HMG (1998) Environmental strategies and policies for industry, forestry and water resource sectors. Ministry of Population and Environment, Kathmandu
*HMG (2002) Nepal biodiversity strategy. Ministry of Forests and Soil Conservation, Kathmandu
HMG (2004) Our forest. Annual report 2003/04. [Hamro ban 2059/60] (In Nepalese). Ministry of Forests and Soil Conservation, Kathmandu
*HMG and IUCN (1988) The national conservation strategy for Nepal. His Majesty's Government and the International Union for Conservation of Nature and Natural Resources, Kathmandu
Holley J, Cherla K (1998) The medicinal plants sector in India: a review. Medicinal and Aromatic Plants Program in Asia, International Development Research Centre, New Delhi
Homma AKO (1996) Modernisation and technological dualism in the extractive economy of Amazonia. In: Pérez MR, Arnold JEM (eds) Current issues in non-timber forest products research. Center for International Forestry Research, Bogor, pp 59–81
IDRC (2002) Medicinal plants: a global heritage. International Development Research Centre, Delhi
*IUCN (1993) Guidelines on the conservation of medicinal plants. IUCN, Gland
*Jayaswal ML (2001) Experiences of the non-timber forest products (NTFP) based enterprise development in Nepal. Harvesting of non-wood forest products. Seminar proceedings, Menemen-Izmir, Turkey, 2–8 October 2000. Ministry of Forestry, Ankara, pp 91–96
*Joshi SP (2003) Status of non-timber forest products in Bara District. In: Anonymous (ed) Proceedings of seminar on non-timber forest products, Central Development Region, Kathmandu, Nepal, August 29, 2002. NTFP Database Project, Institute of Forestry, Pokhara, pp 137–140
*Kadel PN (1998) An interaction between forest and people in Jumla, Nepal. District Forest Office, Jumla
*Kafley GP (2003) Status of non-timber forest products in Lalitpur District. In: Anonymous (ed) Proceedings of seminar on non-timber forest products, Central Development Region, Kathmandu, Nepal, August 29, 2002. NTFP Database Project, Institute of Forestry, Pokhara, pp 36–43
*Kala CP (2000) Status and conservation of rare and endangered medicinal plants in the Indian trans-Himalaya. Biol Conserv 93:371–379
*Kala CP, Farooquee NA, Dhar U (2004) Prioritization of medicinal plants on the basis of available knowledge, existing practices and use value status in Uttaranchal, India. Biodivers Conserv 13:453–469
*Kampen-van Dijken L, Thapa N (1998) Jumla District: description and baseline data 1998. Karnali Local Development Programme, SNV/Netherlands Development Organisation, Kathmandu
*Kanel KR (1999) Policy-related issues in non-timber forest product business. In: Rawal RB, Bhatta R, Paudyal A (eds) Non-timber forest products: production, collection and trade in Mid-Western Development Region of Nepal. New Era, Kathmandu, pp 38–42
*Kanel KR (2000) Non-timber forest policy issues in Nepal. In: Amatya SM (ed) Proceedings of the third regional workshop on community based NTFP management, South East Asian countries NTFP Network (SEANN) 8–9 April, 2000. Institute of Forestry, Pokhara, pp 23–33
*Kanel KR (2002) Policy and institutional bottlenecks: possibilities for NTFP development in Nepal. In: Bhattarai NK, Karki M (eds) Sharing local and national experience in conservation of medicinal and aromatic plants in South Asia. International Development Research Centre, Delhi, pp 54–61
*Kaphle N (2003) Status of non-timber forest products in Rasuwa District. In: Anonymous (ed) Proceedings of seminar on non-timber forest products, Central Development Region, Kathmandu, Nepal, August 29, 2002. NTFP Database Project, Institute of Forestry, Pokhara, pp 50–53
*Karki JBS (1995) Use, availability, and marketing of non-timber forest products in Eastern Nepal. In: Fox J, Donovan D, DeCoursey M (eds) Voices from the field: sixth workshop on community management of forest lands. Program on Environment, East-West Center, Honolulu, pp 64–82
*Karki M (2003) Present status and distribution of NTFPs in Bhaktapur District. In: Anonymous (ed) Proceedings of seminar on non-timber forest products, Central Development Region, Kathmandu, Nepal, August 29, 2002. NTFP Database Project, Institute of Forestry, Pokhara, pp 32–43
Kate Kt, Laird SA (1999) The commercial use of biodiversity. Earthscan Publications Ltd., London

*Kleinn C (1994) Forest resources inventories in Nepal: status quo, needs, recommendations. Forest Resource and Information System, Ministry of Forests and Soil Conservation, His Majesty's Government and FINNIDA, Kathmandu

*Kleinn C, Laamanen R, Malla SB (1996) Integrating the assessment of non-wood forest products into the forest inventory of a large area: experiences from Nepal. Non-Wood For Prod 3:23–30

*KMTNC (1998) Project proposal for Manaslu conservation area. King Mahendra Trust for Nature Conservation, Lalitpur

*Koirala J (2003) Status of non-timber forest products in Rautahat District. In: Anonymous (ed) Proceedings of seminar on non-timber forest products, Central Development Region, Kathmandu, Nepal, August 29, 2002. NTFP Database Project, Institute of Forestry, Pokhara, pp 117–126

*Lama YC, Ghimire SK, Aumeeruddy-Thomas Y (2001) Medicinal plants of Dolpo: amchis' knowledge and conservation. People and Plants Initiative, WWF Nepal, Kathmandu

*Lambert J, Srivastava J, Vietmeyer N (1997) Medicinal plants: rescuing a global heritage. World Bank Technical Paper 355. The World Bank, Washington DC

*Larsen HO (2002) Commercial medicinal plant extraction in the hills of Nepal: local management systems and sustainability. Environ Manage 29:88–101

*Larsen HO, Olsen CS, Boon TE (2000) The non-timber forest policy process in Nepal: actors, objectives and power. For Policy and Econ 1:267–281

*Larsen HO, Smith PD (2004) Stakeholder perspectives on commercial medicinal plant collection in Nepal: poverty and resource degradation. Mt Res Dev 24:241–248

Larsen HO, Smith PD, Olsen CS (2005) Nepal's conservation policy options for commercial medicinal plant harvesting: stakeholder views. Oryx 39:435–441

Lillesø J-PB, Shrestha TB, Dhakal LP, Nayaju RP, Shrestha R (2005) The map of potential vegetation of Nepal – a forestry/agro-ecological/biodiversity classification system. Forest & Landscape Development and Environment Series 2-2005, Centre for Forest, Landscape and Planning, The Royal Veterinary and Agricultural University, Copenhagen

*Maharjan MR (1994) Chiraito cultivation in community forestry. B/NUKCFP/13. Nepal-UK Community Forestry Project, Kathmandu

*Malla SB (1994) Medicinal plants in the Bagmati Zone. ADPI Series 8. International Centre for Integrated Mountain Development, Kathmandu

*Malla SB, Shakya PR, Karki BR, Mortensen TF, Subedi NR (1999) A study on non-timber forest products in Bajura District. CARE Nepal, Kathmandu

*Malla SB, Shakya PR, Rajbhandari KR, Bhattarai NK, Subedi MN (1995) Minor forest products of Nepal: general status and trade. FRIS Project Paper 4. Forest Research and Information System Project, Ministry of Forests and Soil Conservation, His Majesty's Government and FINNIDA, Kathmandu

*Manjkhola S, Dhar U (2002) Conservation and utilization of *Arnebia benthamii* (Wall. ex G. Don) Johnston – a high value Himalayan medicinal plant. Curr Sci 83:484–488

Neumann RP, Hirsch E (2000) Commercialisation of non-timber forest products: review and analysis of research. Center for International Forestry Research, Bogor

*Neupane RK, Ghimere AJ (2000) Income generation through 'community based non-timber forest product (NTFP) management': reflection of mid- and far-west (Dailekh and Accham) development regions of Nepal. In: Amatya SM (ed) Proceedings of the third regional workshop on community based NTFP management, South East Asian countries NTFP Network (SEANN) 8–9 April, 2000. Institute of Forestry, Pokhara, pp 59–64

*New Era (1992) Non-timber forest products commercialization feasibility study. Country report: Nepal. Appropriate Technology International, Kathmandu

*Nicholson K (1997) Problem analysis and strategy development for non timber forest products in SNV project areas in Nepal. SNV/Netherlands Development Organisation, Kathmandu

Olsen CS (1995) Medicinal plants in Gorkha District – a brief introduction to resources, constraints and possibilities. In: Yadav SK, Stoian D (eds) Medicinal and aromatic plants in Gorkha District: how to promote their utilisation and marketing. District Forest Office, Gorkha, pp 11–16

*Olsen CS (1998) The trade in medicinal and aromatic plants from central Nepal to northern India. Econ Bot 52:279–292

*Olsen CS (2001) Trade in the Himalayan medicinal plant product Kutki – new data. Med Plant Conserv 7:11–13

Olsen CS (2005) Quantification of the trade in medicinal and aromatic plants in and from Nepal. Acta Horticult 678:29–35

Olsen CS (2006) Valuation of commercial Central Himalayan medicinal plants. Ambio 34:607–610

[120]

Olsen CS, Bhattarai NK (2000) Forest resources and human welfare in Himalaya: the contribution of commercial medicinal plants. Paper presented at the XXI IUFRO World Congress 2000, 7–12 August, Kuala Lumpur

*Olsen CS, Helles F (1997a) Medicinal plants, markets and margins in the Nepal Himalaya: trouble in paradise. Mt Res Dev 17:363–374

*Olsen CS, Helles F (1997b) Making the poorest poorer: policies, laws and trade in medicinal plants in Nepal. J World For Res Manage 8:137–158

*Olsen CS, Larsen HO (2003) Alpine medicinal plant trade and Himalayan mountain livelihood strategies. Geogr J 169:243–254

*Olsen CS, Treue T (2003) Analysis of trade in non-timber forest products. In: Helles F, Strange N, Wichmann L (eds) Recent accomplishments in applied forest economics research. Kluwer Academic Publishers, Dordrecht, pp 227–239

*Osti KP (2003) Status of non-timber forest products in Nuwakot District. In: Anonymous (ed) Proceedings of seminar on non-timber forest products, Central Development Region, Kathmandu, Nepal, August 29, 2002. NTFP Database Project, Institute of Forestry, Pokhara, pp 133–136

Ostrom E (1990) Governing the commons: the evolution of institutions for collective action. Cambridge University Press, Cambridge

Ostrom E (1998) A behavioral approach to the rational choice theory of collective action. Am Polit Sci Rev 92:1–22

*Pandey NK, Tewari KC, Tewari RN, Joshi GC, Pande VN, Pandey G (1993) Medicinal plants of Kumaon Himalaya and strategies for conservation. In: Dhar U (eds) Himalayan biodiversity: conservation strategies. G.B. Pant Institute of Himalayan Environment & Development, Almora, pp 293–302

*Pandit BH, Thapa GB (2003) A tragedy of non-timber forest resources in the mountain commons of Nepal. Environ Conserv 30:283–292

*Pandit BH, Thapa GB (2004) Poverty and resource degradation under different common forest resource management systems in the mountains of Nepal. Soc Nat Res 17:1–16

*Pandit MK, Babu CR (1998) Biology and conservation of *Coptis teeta* Wall. – an endemic and endangered medicinal herb of eastern Himalaya. Environ Conserv 25:262–272

*Parajuli DP, Gyawali AR, Chaudhary SD, Shrestha BM, Raisaly NK, Tribedi RC, Pandey RC (1999) *Dactylorhiza hatagirea*: resource assessment and issues for sustainable management in Annapurna Conservation Area of Lamjung District, Nepal. In: Mathema P, Dutta IC, Balla MK, Adhikary SN (eds) Sustainable forest management. Proceedings of an international seminar, 31 August–2 September 1998, Pokhara, Nepal. Institute of Forestry, Pokhara, pp 217–224

*Paudel D, Rosset C (1998) What Hanuman brought was not only 'jadibuti'... Swiss Agency for Development and Cooperation, Kathmandu

*Pikkarainen T, Paudyal SS (1996) Forest resources of the central Development Region 1994/1995. Publication 66, Forest Research and Survey Centre, Ministry of Forests and Soil Conservation, Kathmandu

Plotkin M, Famolare L (1992) Sustainable harvests and marketing of rain forest products. Island Press, Washington DC

*Poudyal AS (2003) Non-timber forest products: pros and cons in context with Dolakha District. In: Anonymous (ed) Proceedings of seminar on non-timber forest products, Central Development Region, Kathmandu, Nepal, August 29, 2002. NTFP Database Project, Institute of Forestry, Pokhara, pp 1–8

*Pyakuryal PN (2003) Status of non-timber forest products in Kavrepalanchok District. In: Anonymous (ed) Proceedings of seminar on non-timber forest products, Central Development Region, Kathmandu, Nepal, August 29, 2002. NTFP Database Project, Institute of Forestry, Pokhara, pp 111–116

*Rajbhandary KR (1987) Preliminary field report of the plant collection in east Nepal in 1985. Newslett Himal Bot 2:8–16

*Rajbhandary TK, Bajracharya JM (1994) National status paper on NTFPs: medicinal and aromatic plants. In: Pradhan J, Maharjan P (eds) Proceedings of the national seminar on non-timber forest products: medicinal and aromatic plants, Kathmandu, September 11–12, 1994. Ministry of Forest and Soil Conservation and Herbs Production and Processing Co. Ltd, Kathmandu, pp 8–15

*Rawal RB (1995) Commercialization of aromatic and medicinal plants in Nepal. In: Durst PB, Bishop A (eds) Beyond timber: social, economic and cultural dimensions of non-wood forest products in Asia and the Pacific. FAO, Bangkok, pp 149–155

⊘ Springer

*Rawal RB (1997) Status of commercialisation of medicinal and aromatic plants in Nepal. INBAR Technical Report No. 15. International Network for Bamboo and Rattan, Beijing, pp 174–188

*Rawal RB (2002) Challenges and opportunities in developing community-based forest enterprises in the higher mountains of Nepal. In: Anonymous (ed) Medicinal plants: a global heritage. International Development Research Centre, Delhi, pp 210–217

*Rawal RB, Poudyal A (1999) Marketing of non-timber products in Nepal. In: Rawal RB, Bhatta R, Paudyal A (eds) Non-timber forest products: production, collection and trade in Mid-Western Development Region of Nepal. New Era, Kathmandu, pp 26–37

*Rawal RB, Pradhan J, Bajracharya JM (1996) Commercial utilization of medicinal and aromatic plants in Nepal. In: Jha PK, Ghimire GPS, Karmacharya SB, Baral SR, Lacoul P (eds) Environment and biodiversity: in the context of South Asia. Ecological Society, Kathmandu, pp 256–259

Redford KH, Stearman AM (1993) Forest-dwelling native Amazonians and the conservation of biodiversity: interests in common or in collision? Conserv Biol 7:248–255

Regmi MC (1972) A study in Nepali economic history (1768–1846). Adroit Publishers, Delhi

*Regmi S, Bista S (2002) Developing methodologies for sustainable management of high value medicinal and aromatic plants in Jumla District, Nepal. In: Bhattarai NK, Karki M (eds) Sharing local and national experience in onservation of medicinal and aromatic plants in South Asia. International Development Research Centre, Delhi, pp 96–100

*Sah SP, Dutta IC (1996) Inventory and future management strategies of multipurpose tree and herb species for non-timber forest products in Nepal. Non-Wood For Prod 3:123–146

Schippmann U, Leaman DJ, Cunningham AB (2002) Impact of cultivation and gathering of medicinal plants on biodiversity: global trends and issues. Biodiversity and the Ecosystem Approach in Agriculture, Forestry and Fisheries. Satellite event on the occasion of the Ninth Regular Session of the Commission on Genetic Resources for Food and Agriculture. Inter-Departmental Working Group on Biological Diversity for Food and Agriculture, Rome

*Sharma P (1995) Non-wood forest products and integrated mountain development: observations from Nepal. Non-Wood For Prod 3:157–166

*Sharma P (1996) Opportunities in and constraints to the sustainable use of non-timber forest resources in the Himalayas. ICIMOD Newslett 25:9–11

*Sheikh K, Ahmad T, Khan MA (2002) Use, exploitation and prospects for conservation: people and plant biodiversity of Naltar Valley, northwestern Karakorums, Pakistan. Biodivers Conserv 11:715–742

*Shiva MP (1989) Suggestions for further studies on minor forest products within the Master Plan framework for the forestry sector, Nepal. FAO, Kathmandu

*Shrestha GL, Joshi RB, Amatya G, Sthapit VM (1994) Domestication and commercial cultivation of medicinal and aromatic plants (MAPs) in Nepal. In: Pradhan J, Maharjan P (eds) National seminar on non-timber forest products: medicinal and aromatic plants, Kathmandu, September 11–12, 1994. Proceedings. Ministry of Forests and Soil Conservation and Herbs Production and Processing Co. Ltd, Kathmandu, pp 35–41

*Shrestha KK (2003) Status of non-timber forest products in Dhanusha District. In: Anonymous (ed) Proceedings of seminar on non-timber forest products, Central Development Region, Kathmandu, Nepal, August 29, 2002. NTFP Database Project, Institute of Forestry, Pokhara, pp 108–110

*Shresta KK, Ghimire SK (1996) Diversity, ethnobotany and conservation strategy of some potential medicinal and aromatic plants of Taplejung (Tamur Valley). Asia Network for Small-scale Bioresources, Kathmandu

*Shresta KN (2003) Status of non-timber forest products in Mahottari District. In: Anonymous (ed) Proceedings of seminar on non-timber forest products, Central Development Region, Kathmandu, Nepal, August 29, 2002. NTFP Database Project, Institute of Forestry, Pokhara, pp 64–79

*Shrestha PM, Dhillion SS (2003) Medicinal plant diversity and use in the highlands of Dolakha district, Nepal. J Ethnopharmacol 86:81–96

*Shrestha RK, Malla SB, Regmi S (1995) Marketing of high-value agricultural products from Nepal to Germany (and the EC). Agricultural Projects Services Centre, Kathmandu

*Shrestha TB (1990) The Makalu-Barun national park and conservation area: management plan. Department of National Parks & Wildlife Conservation, His Majesty's Government and Woodlands Mountain Institute, Kathmandu

*Shrestha TB (1994) Utilisation of non-timber forest products: medicinal and aromatic plants. In: Pradhan J, Maharjan P (eds) National seminar on non-timber forest products: medicinal and

aromatic plants, Kathmandu, September 11–12, 1994. Proceedings. Ministry of Forests and Soil Conservation and Herbs Production and Processing Co. Ltd., Kathmandu, pp 16–22

*Shrestha TB, Joshi RM (1996) Rare, endemic, and endangered plants of Nepal. WWF, Kathmandu

*Shrestha TB, Pokharel S (2000) The potential of medicinal and aromatic plants for sustainable mountain development in Nepal. In: Price M, Butt N (eds) Forests in sustainable mountain development, IUFRO Research Series No. 5, pp 312–318

*Singh AP (2003) Status of non-timber forest products of Makwanpur. In: Anonymous (ed) Proceedings of seminar on non-timber forest products, Central Development Region, Kathmandu, Nepal, August 29, 2002. NTFP Database Project, Institute of Forestry, Pokhara, pp 44–49

*Singh LM (1996) Project report: medicinal plants and traditional medical practice in Gorkha District. Himalayan Ayurveda Research Institute, Kathmandu

*Singh MP, Malla SB, Rajbhandary SB, Manandhar A (1979) Medicinal plants of Nepal – retrospects and prospects. Econ Bot 33:185–198

*Siwakot M, Tiwari S (2002) Conservation and sustainable use of medicinal and aromatic plants: IUCN's efforts in Nepal. In: Bhattarai NK, Karki M (eds) Sharing local and national experience in conservation of medicinal and aromatic plants in South Asia. International Development Research Centre, Delhi, pp 179–184

*Subedi BP (1999) Non-timber forest products sub-sector in Nepal: opportunities and challenges for linking the business with biodiversity conservation. Paper presented at the Workshop on natural resources management for enterprise development in Himalayas, August 19–21, 1999, Nainital

*Subedi BP (2002) Participatory utilisation and conservation of medicinal and aromatic plants: a case from western Nepal Himalaya. In: Anonymous (ed) Medicinal plants: a global heritage. International Development Research Centre, Delhi, pp 186–209

*Subedi BP, Bhattarai NK (2002) Community managed enterprise: participation of rural people in medicinal and aromatic plants conservation and use. In: Anonymous (ed) Medicinal plants: a global heritage. International Development Research Centre, Delhi, pp 251–257

*Subedi BP, Binayee SB (2000) Linking conservation to business and local communities: an approach to sustainable management of in situ biodiversity in Nepal. In: Amatya SM (ed) Proceedings of the third regional workshop on community based NTFP management, South East Asian countries NTFP Network (SEANN) 8–9 April, 2000. Institute of Forestry, Pokhara, pp 34–43

*Subedi RH, Singh LM (1994) Ayurvedic importance of medicinal and aromatic plants: present status, constraints and suggested actions for sustainability. In: Pradhan J, Maharjan P (eds) National seminar on non-timber forest products: medicinal and aromatic plants, Kathmandu, September 11–12, 1994. Proceedings. Ministry of Forests and Soil Conservation and Herbs Production and Processing Co. Ltd., Kathmandu, pp 23–26

*Uniyal SK, Awasthi A, Rawat GS (2002) Current status and distribution of commercially exploited medicinal and aromatic plants in upper Gori valley, Kumaon Himalaya, Uttaranchal. Curr Sci 82:1246–1252

*Vantomme P, Markkula A, Leslie RN (2002) Non-wood forest products in 15 countries of tropical Asia: an overview. FAO, Bangkok

*Ved DK, Mudappa A, Shankar D (1998) Regulating export of endangered medicinal plant species – need for scientific rigour. Curr Sci 75:341–344

Wong JLG, Thornber K, Baker N (2001) Resource assessment of non-wood forest products: experience and biometric principles. Non-Wood Forest Products 13. FAO, Rome

WHO (2002) WHO traditional medicine strategy. World Health Organisation, Geneva

*Yadav BK (2003) Status of non timber forest products in Chitwan District. In: Anonymous (ed) Proceedings of seminar on non-timber forest products, Central Development Region, Kathmandu, Nepal, August 29, 2002. NTFP Database Project, Institute of Forestry, Pokhara, pp 25–31

*Yonzon P (1993) Raiders of the park Himal Jan/Feb: 22–23

Biodivers Conserv (2007) 16:1699–1714
DOI 10.1007/s10531-006-9044-7

ORIGINAL PAPER

Woody plant species richness in the Turvo State park, a large remnant of deciduous Atlantic forest, Brazil

Ademir R. Ruschel · Rubens O. Nodari ·
Bruno M. Moerschbacher

Received: 13 September 2005 / Accepted: 22 March 2006 / Published online: 22 May 2006
© Springer Science+Business Media B.V. 2006

Abstract This paper presents a quantitative inventory of woody plants with DBH ≥ 5 cm in Turvo State park, a large remnant (17,500 ha) of seasonal deciduous Atlantic forest in Southern Brazil. The forest inventory was based on 141 sampling points (point centered quarter method). Seventy-eight species from 37 families were recorded, the density was 879 plants ha^{-1} and the basal area 25.12 m^2 ha^{-1}. Timber species of commercial value made up 35% of the species and 72% of the basal area. The ten most abundant species made up 52% of the species, and the ten species with highest basal area accounted for 48% of the total basal area and 12% of the species. Zoochory was the dominant mode of seed dispersal, representing 63% of species and 67% of the plant density. Nine percent of the species were pioneers, 56% were secondary species accounting for 62% of the total basal area, and 35% were opportunistic-climax species accounting for 54% of the plant density. Turvo park has exceptionally high biodiversity and is the last intact large remnant of the Alto-Uruguai river ecosystem. In addition to species richness and an abundance of high value timber species, this forest is of value for the rich genetic resources, and for medicinal and ornamental plants.

Keywords Ecological groups · Endemic species · Floristic composition ·
Seed dispersion · Timber species

A. R. Ruschel (✉) · R. O. Nodari
Departmento de Fitotecnia, Universidade Federal de Santa Catarina, Caixa Postal 476,
Florianópolis CEP 88040-900 SC, Brazil
e-mail: arruschel@yahoo.com.br

B. M. Moerschbacher
Department of Plant Biochemistry and Biotechnology, Westphalian Wilhelm's University
of Münster, Münster, Germany

Introduction

Subtropical Atlantic Forest, also named Seasonal Deciduous Forest, represents a small ecosystem in the basin of the Uruguai river in Southern Brazil (IBGB 1990). The primary forest of that ecosystem is characterized by a closed canopy dominated by Lauraceae and emergents of Fabales species. These emerging trees are responsible for the typical appearance of the forest, more spectacular in summer due to the abundant blossoming of the dense treetops, and less obvious during winter when they shed their leaves. The subcanopy is dominated by *Sorocea bonplandii*, *Gynnanthes concolor*, and *Trichilia* species, and epiphytes are very poor when compared with the tropical Atlantic forest (Rambo 1956). Besides harboring a rich flora, the region is also shelter for endangered animal species (Wallauer and Albuquerque 1986).

The economic development of the Alto-Uruguai river region was initially based on the exploitation of the noble timber species by the lumber companies or by the colonists themselves (Ruschel et al. 2003). As a consequence, in a short period of time, the exuberant forest originally encompassing an area of 47,000 km^2 was reduced to less than 2,000 km^2. Most of this remaining forest occurs near the Uruguai river in the border region between Brazil and Argentina (Google Earth 2005). Turvo park is the largest undisturbed remnant of Subtropical Atlantic Forest ecosystem preserved in Brazil. Some small fragments of this forest have survived in small areas with prominent relief or high slopes in the west of Santa Catarina State, but not even a single public domain area is under legal protection in this state. Thus, the Subtropical Atlantic Forest is the most threatened forest ecosystem in South Brazil today.

Previous studies of the native flora in the Alto-Uruguai river basin reported high species richness of timber value, and new species to this ecosystem have been identified in each new study (Rambo 1956; Klein 1972; Brack et al. 1985; Dias et al. 1992; Cristóbal and Vera 1999; Daviña et al. 1999; Ruschel et al. 2003). Rambo (1956) reported that epiphytes are almost completely lacking in Subtropical Atlantic Forest, but in a small remnant of this ecosystem, Rogalski and Zanin (2003) found 70 epiphytes species belonging to 30 genera and eight families. Little quantitative data on the abundance and dominance of the species of these forests is available, and almost none for primary forest. Knowledge on the species in primary forest are important as a reference for studies of forest fragmentation effects, and for defining guide lines for sustainable management and protection of this ecosystem. The aim of the current study, thus, was to further analyse the floristic, phytosociological, and ecological structure of the shrub and arboreal species in Turvo park.

Methods

Study site

Turvo park, encompassing an area of 17,500 ha of Subtropical Atlantic Forest, is located in the extreme northwestern section of the southern Brazilian state Rio Grande do Sul (Fig. 1). Its western border is marked by the Uruguai river, and it is complemented by the Moconá park west of the river, in the Argentinian province of Misiones. Turvo park, created in 1947, represents one of few remnants of seasonal deciduous forest of the Uruguai river ecosystem, and it is in many ways similar to the

Fig. 1 Map of the study area, Turvo park, in Southern Brazil. On the other side of Uruguai river, Turvo park is complemented by Moconá Park in the Argentinian state Misiones (adapted from Google Earth 2005)

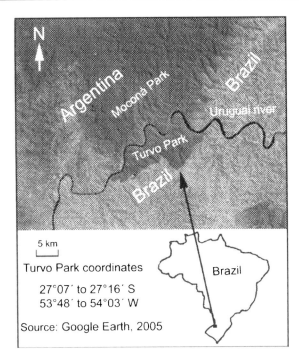

Iguaçú national park, in Paraná-Brazil. According to Rambo (1956) and Klein (1972), the subtropical forest of Alto-Uruguai is an extension of the forest of the Paraná river, through the province of Misiones.

The area of Turvo park reaches altitudes between 200 and 400 m above the sea level. The soil consists of extrusive rocks of basaltic origin, forming an eutrophic red soil, in a hilly relief. The annual average temperature is 18–20°C, with the lowest temperatures in July and the highest in January. Frosts regularly occur between June and August. The annual precipitation (2200–2400 mm) is well distributed, occurring on 80–100 days per year, with a precipitation deficit towards the end of spring and the beginning of summer (IBGE 1990).

Sampling method

The present study was carried out in areas of primary forest during the month of February, 2002. Forest inventories were compiled using the point-centered quarter method (Cottam and Curtis 1956). Three areas were chosen, between 27°11′24″ to 27°12′34″ S and 53°51′03″ to 53°51′33″ W, at altitude of 290–350 m above the sea level. The sample areas were situated 6.5, 9.0 and 11.5 km from the north-eastern border of the park (Fig. 1).

The point-centered quarter method was considered efficient and suitable for floristic inventories in temperate and tropical rain forests, being advantageous in terms of time and costs when compared with the quadrate plots method (Cottam and Curtis 1956; Gibbs et al. 1980; Cavassan et al. 1984; Krebs 1989; Dias et al. 1992; Sparks et al. 2002). According to these studies, sample size is more important for precision in forest inventories than the sampling method used.

[127]   ✌ Springer

In each area, a straight line with a length of 750 m pointing south was marked with sampling points 15 m apart. In each of the quarters defined by a sampling point, the closest living plant (with the exception of lianas) with a DBH ≥ 5 cm (diameter at breast height = 1.3 m) was identified taxonomically, and DBH and total height were taken. Thus, a total of 141 sampling points yielded 564 plants. The classification of the species was performed with assistance of bibliography (Reitz and Klein 1964; Reitz et al. 1978; Lorenzi 1998) and with the help of taxonomists Prof. Marcos Sobral (Federal university of Rio Grande do Sul) and Prof. Ademir Reis (Federal university of Santa Catarina). In the case of plants with more than one living trunk, all trunks with DBH ≥ 5 cm were evaluated. Whenever possible, total height was measured using a forest measure ruler or a clinometer. According to Krebs (1989) the quantitative sampling error was estimated based on the distances of the classified trees from the sampling points, and the qualitative sampling sufficiency was esti- mated by the species area curves (Fig. 2).

The phytosociological parameters basal area, density, and dominance, the index of diversity H' of Shannon and Wiener and the evenness index were determined according to Krebs (1989).

Ecological structure

Species were classified in three groups: (i) Pioneers, the first forest species appearing in the forest succession; (ii) Secondary species, succeeding the previous group and needing greater soil fertility, shade levels, and humidity; and (iii) Climax/Opportu- nists, only occurring in advanced stages of succession when the edaphic conditions (shade, humidity, and fertility) are suitable and agents for their pollination and seed dispersal are present.

Species were also classified according to their method of seed dispersal in three categories: anemochoric; autochoric; and zoochoric (Janson 1983). In the absence of detailed studies for a given species, the classification was based on published information (Tabarelli 1992; Dalling et al. 1998; Nascimento et al. 2000; Traveset et al. 2001).

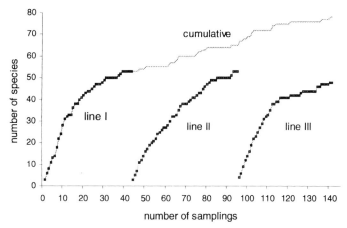

Fig. 2 Species-area curves for the three sampling lines in Turvo park as well as the cumulative species-area curve for all three sampling lines. Woody plants with a DBH ≥ 5 cm were identified using the point-centered quarter method

Results

Species diversity, density and dominance

Using the point-centered quarter method, 564 plants were identified along three lines each of about 750 m total length, in the center of Turvo park. These woody plants with a DBH $\geq$ 5 cm were classified into 78 species from 37 families, 38 of them being timber species of commercial value (Table 1). The diversity index of Shannon (H′) for this set of plants was 3.73 and the index of evenness was 0.86 (Table 2).

In addition to the sampled species shown in Table 1, plants of eight species (*Aralia warmingianus* (March.) Harms, *Carica quercfolia* (St. Hil.) Solms. Lamb., *Maclura tinctoria* (L.) D. Don ex Steud, *Peltophorum dubium* (Spreng.) Taub., *Plinia trunciflora* (Berg.) Kauzal, *Solanum pseudoquina* (A. St.-Hil), *Randia armata* (sw.) DC. and *Xylosma pseudosalzmannii* Sleum. were observed in the area, but not captured in the sampling strategy used. While the number of species sampled was similar (53, 53, and 48), the species compositions of the three sampling lines differed markedly: only about one third of the species (35%) were sampled in all three lines, while more than one third of the species (38%) were sampled in a single line only (Table 1). On average, a new species was identified for every 7.2 sampled plants. The individual species-area curves for the three sampling lines as well as the cumulative species-area curve (Fig. 2) show that an even larger sample size would be required to fully account for the diversity of the area.

In this ecosystem, the density of plants with DBH $\geq$ 5 cm was 879 plants ha^{-1}, adding up to a basal area of 25.12 m^2 ha^{-1} (Table 2). Timber species with commercial value represented approximately one third of the plants (35%) and more than two thirds (72%) of the basal area (Table 1).

Among the families found in Turvo park, some contributed conspicuously more to the number of plants and basal area. Thus, a group of 16 families (42%) comprise 69% of the species, 92% of the plants, and 93% of the basal area. On the other hand, the remaining 21 families were represented each by a single species, with the exception of Myrsinaceae and Annonaceae with two species each (Fig. 3). The Fabales, including the families Caesalpiniaceae, Mimosaceae, and Fabaceae all belonging to the group of 16 dominant families, are represented by 14 species (3, 6, 5, respectively) equaling 18% of all species (Fig. 3). In terms of density, the Fabales account for 13% (2%, 6%, 5%, resp.) of the total number of individual plants; and in terms of dominance, they account for 23% (8%, 8%, 7%, resp.) of the total basal area. Fabales, Euphorbiaceae and Meliaceae together account for 38% of the individual plants, and for 35% of the species in this forest.

The Moraceae and Arecaceae were both represented by a single species only, namely *Sorocea bonplandii* and *Syagrus romanzoffiana*, resp., but the density of both reached 8% (Table 1). While these were the species with the highest density, their dominance in terms of basal area (1% and 5%, resp.) was less pronounced. Similarly, the ten most frequent species, comprising 52% of all plants, accounted for only 26% of the total basal area. On the other hand, the ten most dominant species, comprising 48% of the total basal area, accounted for only 12% of all plants.

Of all sampled plants, 2% showed ramification below DBH, and these belonged to the species *Trichilia claussenii*, *Sorocea bonplandii*, *Pataganula americana*, *Guarea macrophylla*, *Calyptranthes tricona*, *Solanum mauritianum*, *Urera baccifera*, and

Springer

Table 1 List of families and species in sample sites (LI, L2 and L3), absolute and relative density of plants ha^{-1}, relative basal area, value of use (NT, non-timber; T, timber with commercial value), modes of seed dispersion (zoo, zoochoric; aut, autochoric; ane, anemochoric), and succession groups (p, pioneer; s, secondary; c, opportunistic/climax) of the plant species sampled in Turvo park

Family and species	Sampled			Plants ha^{-1} (%)	Basal area (%)	Use value	Seed dispersal	Succession group
	L1	L2	L3					
Achatocarpaceae								
Achatocarpus praecox Griseb.	1	1	–	3 (0.4)	0.3	NT	zoo	c
Agavaceae								
Cordyline dracaenoides Kunth	–	2	–	3 (0.4)	0.1	NT	zoo	p
Annonaceae								
Annona sp.	3	–	1	6 (0.7)	0.1	NT	zoo	s
Rollinia silvatica (A. St.-Hil.) Mart.	1	–	–	2 (0.2)	0	NT	zoo	s
Apocynaceae								
Aspidosperma parvifolium (Müll. Arg.) A. DC.	–	1	–	2(0.2)	0.7	T	ane	c
Tabernaemontana catharinensis A. DC.	–	1	–	2 (0.2)	0.2	NT	zoo	s
Araliaceae								
Schefflera morototoni (Aubl.) Maguire, Steyemark & Frodin	–	–	1	2 (0.2)	0	T	zoo	s
Arecaceae								
Syagrus romanzoffiana (Cham.) Glassm.	9	28	7	69 (7.8)	5.1	NT	zoo	s
Bignoniaceae								
Jacaranda puberula Cham.	–	1	–	2 (0.2)	0	T	ane	s
Bombacaceae								
Chorisia speciosa St.-Hil.	1	3	–	6 (0.7)	0.7	T	ane	s
Boraginaceae								
Cordia ecalyculata Vell.	–	–	2	3 (0.4)	0	T	zoo	s
Cordia trichotoma (Vell.) Arrab. ex. Steud.	1	–	1	3 (0.4)	0.1	T	ane	s
Pataganula americana L.	–	3	–	5 (0.5)	2.4	T	ane	s
Caesalpiniaceae								
Apuleia leiocarpa (Vog.) Macbride	3	2	2	11 (1.2)	4.7	T	ane	s
Bauhinia forficata Link.	2	–	–	3 (0.4)	0.1	NT	aut	p
Holocalyx balansae Mich.	4	–	–	6 (0.7)	3.3	T	zoo	c
Caricaceae								
Jacaratia spinosa (Aubl.) A. DC	–	–	1	2 (0.2)	0.7	NT	zoo	s
Erythroxylaceae								
Erythroxylum deciduum A. St.-Hil.	–	–	1	2 (0.2)	0.4	T	zoo	s
Euphorbiaceae								
Alchornea triplinervia (Spreng.) Müll. Arg.	–	–	2	3 (0.4)	1.8	T	zoo	s
Gynnanthes concolor Spreng.	12	5	19	56 (6.4)	0.7	NT	aut	c
Manihot flabellifolia Pohl	–	1	1	3 (0.4)	0	NT	aut	p
Sapium glandulatum (Vell.) Pax.	–	–	1	2 (0.2)	0.2	NT	zoo	s
Sebastiana brasiliensis Spreng.	4	11	7	34 (3.9)	1.3	NT	aut	c
Sebastiania commersoniana L.B. Smith & R.J. Downs	5	3	–	12 (1.4)	0.3	NT	aut	s

 Springer

Table 1 continued

Family and species	Sampled			Plants ha^{-1} (%)	Basal area (%)	Use value	Seed dispersal	Succession group
	Ll	L2	L3					
Tetrorchidium rubrivenium Poepp. & Endl.	–	–	2	3 (0.4)	0.3	T	aut	c
Flacourtiaceae								
Banara tomentosa Clos	2	2	–	6 (0.7)	0.4	NT	zoo	c
Casearia silvestris SW.	1	2	–	5 (0.5)	0.5	NT	zoo	p
Icacinaceae								
Citronella paniculata (Mart.) Howard	–	–	1	2 (0.2)	0	NT	zoo	c
Lauraceae								
Nectandra megapotamica (Spreng.) Mez	10	10	6	41 (4.6)	4.8	T	zoo	c
Ocotea diospyrifolia (Meisn.) Mez	3	2	1	11(1.2)	5.3	T	zoo	c
Loganiaceae								
Strychnos brasiliensis (Spreng.) Marl	1	–	–	2 (0.2)	0	NT	zoo	c
Meliaceae								
Cabralea canjerana (Vell.) Mart.	2	–	3	8 (0.9)	2.0	T	zoo	s
Cedrela fissilis Vell.	3	3	3	14 (1.6)	7.7	T	ane	s
Guarea macrophylla Vahl	–	2	–	3 (0.4)	0.1	NT	zoo	c
Trichilia catigua A. Juss.	2	7	7	25 (2.8)	0.6	NT	zoo	c
Trichilia claussenii C. DC.	11	8	11	45 (5.1)	3.7	NT	zoo	c
Trichilia elegans A. Juss.	5	2	1	12 (1.4)	0.2	NT	zoo	c
Mimosaceae								
Albizia cf. edwallii (Hoehne) Barneby & J. Grimes	1	–	2	5 (0.5)	1.6	T	ane	s
Calliandra foliolosa Benth.	2	5	2	14 (1.6)	0.2	NT	aut	s
Enterolobium contortisiliquum (Vell.) Morong & Britton	–	1	–	2 (0.2)	1.6	T	aut	s
Inga marginata Willd.	1	5	8	22 (2.5)	1.1	NT	zoo	s
Inga sessilis (Vell.) Mart.	–	–	1	2 (0.2)	0	NT	zoo	s
Parapiptadenia rigida (Benth.) Brenan	1	2	2	8 (0.9)	3.3	T	ane	s
Moraceae								
Sorocea bonplandii (Baill.) Burger. Lanj. & Boer	15	11	20	72 (8.2)	1.0	NT	zoo	c
Myrsinaceae								
Myrsine guianensis (Aubl.) Kuntze	–	1	–	2 (0.2)	0.1	NT	zoo	s
Myrsine loefgrenii (Mez.) Otegui	–	–	1	2 (0.2)	0	NT	zoo	s
Myrtaceae								
Calypthrantes tricona D. Legr.	2	6	14	34 (3.9)	2.5	NT	zoo	c
Campomanesia guazumifolia (Camb.) O. Berg.	–	2	–	3 (0.4)	0.2	NT	zoo	c
Campomanesia xanthocarpa O. Berg	1	5	3	14 (1.6)	0.8	NT	zoo	c
Eugenia uniflora L.	1	1	–	3 (0.4)	0.7	NT	zoo	s
Myrcianthes pungens (O. Berg) D. Legr.	–	–	1	2 (0.2)	0.5	T	zoo	c
Plinia rivularis (Cambess.) Rotman	–	2	3	8 (0.9)	0.7	NT	zoo	c

🖄 Springer

Table 1 continued

Family and species	Sampled			Plants ha^{-1} (%)	Basal area (%)	Use value	Seed dispersal	Succession group
	Ll	L2	L3					
Nyctaginaceae								
Pisonia ambigua Heimerl	2	1	1	6 (0.7)	0.6	NT	zoo	s
Fabaceae								
Dalbergia frutescens Britton	1	1	–	3 (0.4)	0.5	T	ane	s
Lonchocarpus campestris Mart. ex Benth.	2	4	1	11 (1.2)	3.6	T	ane	s
Machaerium paraguariense Hassler	1	2	–	5 (0.5)	0.1	T	ane	s
Machaerium stipitatum Vog.	6	4	2	19 (2.1)	1.9	T	ane	s
Myrocarpus frondosus Fr. Allem.	2	2	–	6 (0.7)	0.9	T	ane	s
Piperaceae								
Piper gaudichaudianum Kunth	–	–	1	2 (0.2)	0	NT	zoo	s
Polygonaceae								
Ruprechtia laxiflora Meisn.	1	–	–	2 (0.2)	0.8	T	ane	s
Rosaceae								
Prunus sellowii Koehn.	–	1	–	2 (0.2)	0.2	T	zoo	s
Rutaceae								
Balfourodendron riedelianum Engl.	2	4	2	12 (1.4)	1.7	T	ane	s
Helietta apiculata Benth.	1	4	–	8 (0.9)	3.4	T	ane	s
Pilocarpus pennatifolius Lem.	4	5	2	17 (2.0)	0.3	NT	ane	c
Zanthoxylum petiolare A. St.-Hil. & Tul.	–	–	1	2 (0.2)	0.1	T	zoo	s
Sapindaceae								
Allophylus puberulus (Cambess.) Radlk.	7	3	5	23 (2.7)	2.6	NT	zoo	s
Cupania vernalis Cambess.	1	–	–	2 (0.2)	0	T	zoo	s
Diatenopteryx sorbifolia Radlk.	4	3	4	17 (2.0)	4.9	T	ane	s
Matayba elaeagnoides Radlk.	1	1	2	6 (0.7)	2.5	T	zoo	s
Sapotaceae								
Chrysophyllum gonocarpum (Mart. & Eichler) Engl.	8	5	11	37 (4.3)	4.3	T	zoo	c
Chrysophyllum marginatum (Hook. & Arn.) Radlk.	4	6	3	20 (2.3)	2.2	T	zoo	c
Simaroubaceae								
Picrasma crenata Engl.	3	1	–	6 (0.7)	0.2	NT	zoo	c
Solanaceae								
Solanum mauritianum Scop.	1	3	–	6 (0.7)	0.8	NT	zoo	p
Styracaceae								
Styrax leprosus Hook. & Arn	3	1	–	6 (0.7)	0.9	T	zoo	s
Tiliaceae								
Luehea divaricata Mart.	1	2	3	9 (1.1)	2.2	T	ane	s
Ulmaceae								
Trema micrantha Blume	1	–	1	3 (0.4)	0.1	NT	zoo	p
Urticaceae								
Urera baccifera (L.) Gaudich.	5	14	7	41 (4.6)	1.6	NT	zoo	p
Verbenaceae								
Vitex megapotamica (Spreng) Mold.	1	–	–	2 (0.2)	0.4	T	zoo	c
Families 37	172	208	184	879	100			
Species 78	53sp	53sp	48sp	(100)				

Achathocarpus praecox. Among them, only *Pataganula americana* belongs to the canopy of the forest. Interestingly, all individuals of *Achathocarpus praecox* showed ramifications below 1.3 m of height (DBH).

Table 2 Important phytosociological studies for ecosystems (FED, Seasonal Deciduous Forest; FMS, Mesophilic Semideciduous Forest; FOD, Ombrophilic Dense Forest) similar or next to Turvo park (RS, Rio Grande do Sul; SC, Santa Catarina; SP, São Paulo), method used (MP, quadrate plots method with area in ha; MQ, point-centered quarter method with number of sampling points), minimum DBH, number of species, density of plants per hectare, basal area (BA) per hectare, index of Shannon and Wiener (H′), and index of evenness of Pielou (J′)

Authors	Ecosystem, State (method used)	DBH (cm)	Number of species	plant ha^{-1}	BA m^2 ha^{-1}	H′	J′
Dias et al. (1992)	FED–RS (MP-1 ha)	≥10	58	399	–	3.47	0.85
	FED–RS (MQ-107 pt)	≥10	60	–	–	3.52	0.86
Present study	FED–RS (MQ-141 pt)	≥5	78	879	25.12	3.73	0.86
Ruschel et al. (2000)[a]	FED–SC (MQ-468 pt)	≥5	51	994	36.5	3.33	0.86
Silva et al. (1982)	FED–SP (MQ-160 pt)	≥10	123	816	–	4.07	0.85
Ivanauskas et al. (1999)	FED–SP (MP-0.42 ha)	≥4.8	97	2,271	31.93	3.77	0.82
Vilela et al. (2000)	FES–MG (MP-1.6 ha)	≥5	116	1,587	53.13	3.79	0.80
Dias et al. (1998)	FOM–PR (MP-1 ha)	≥5	127	1,594	–	3.67	0.76
Longhi (1997)[b]	FOM–RS (MP-1.69 ha)	≥9.5	63	386	22.55	3.26	0.79
Cavassan et al. (1984)	FMS–SP (MQ-129 pt)	≤10	60	643	32.38	3.50	0.85
Jarenkow and Waechter (2001)	FOD–RS (MP-1 ha)	≥5	55	1,855	41.66	2.24	0.56

[a]Total of samplings made in twelve forest fragments

[b]*Nectandra megapotamica association* group only

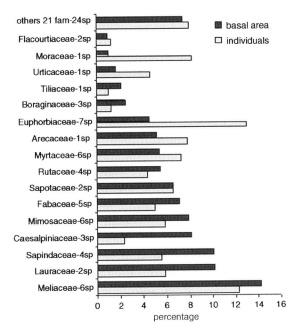

Fig. 3 Relative density (grey bars) and relative dominance (black bars) of the plants belonging to different families, and number of species within each family, as sampled in Turvo park

Dispersion of seeds and succession groups

The majority of species (63%) were zoochoric (Fig. 4A). This type of dispersal was also predominant in terms of number of plants (67%) and basal area (54%). About one quarter of the species (25.6%) were classified as anemochoric, and only few species (11%), including all of the Euphorbiaceae, were classified as autochoric.

🖄 Springer

Interestingly, anemochoric plants exhibited a relative high dominance (42% of total basal area) in spite of their low density (19% of all plants).

The majority of species (56%) were classified as secondary (Fig. 4B), while few species were classified as pioneers (9%). Similarly, species classified as secondary were the most dominant (62% basal area), but species classified as opportunistic/climax reached the highest density (54% of all plants).

Zoochoric species were most prevalent in all succession groups, most notably (74%) in the group of species classified as opportunistic/climax (Fig. 4C). The relative amount of species classified as anemochoric was largest in the group of

Fig. 4 Relative distribution of the species, density of plants (DR), and basal area (BA), subdivided according to the mode of seed dispersal (**A**): autochoric, aut (white bars), anemochoric, ane (black bars) and zoochoric, zoo (gray bars), or according to the succession group (**B**): opportunistic/climax, (white bars), secondary, (black bars) and pioneer (gray bars); and relative distribution of the succession groups, subdivided according to the mode of seed dispersion (**C**): autochoric, aut (white bars), anemochoric, ane (black bars) and zoochoric, zoo (gray bars), as sampled in Turvo park

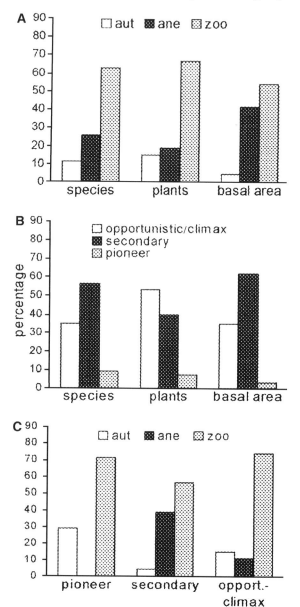

secondary species (39%). In the group of pioneering species, anemochoric species were absent. Autochoric species were relatively rare in all succession groups, and almost absent (5%) in the group of secondary species.

Diameters and total heights

The shrub-arboreal forest plants (DBH ≥ 5 cm) had an average DBH of 14.9 cm (Fig. 5A). In fact, 80% of the plants had a DBH below 20 cm. Larger individuals became increasingly rare, so that only 5% of the plants (44 trees ha^{-1}) reached a DBH ≥ 40 cm, and only 1% (9 trees ha^{-1}) had a DBH above 60 cm. However, trees of commercial interest, i.e. having DBH ≥ 40 cm, represented 43% of the total basal area.

About half of the shrub-arboreal plants (DBH ≥ 5 cm) had a height of below 8 m (Fig. 5B). Taller individuals became increasingly rare, so that only 5% of the plants (39 trees ha^{-1}) reached a height of at least 20 m, and only 0.5% (4.7 tree ha^{-1}) had a height above 26 m. The emerging trees which are characteristic for the ecosystem, i.e. having heights above 20 m, represented 33% of the total basal area.

Discussion

Species richness

The results of the present study indicated a high species diversity in different areas of Turvo park. Similarly, the species identified in previous studies carried out in the

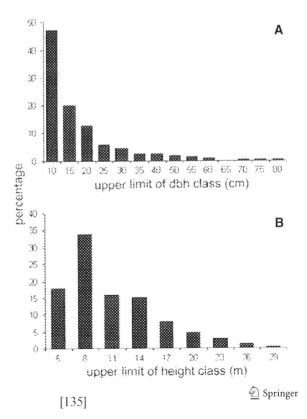

Fig. 5 Relative distribution of the woody plants with DBH ≥ 5 cm classified according to diameter (**A**) or according to height (**B**), as sampled in Turvo park

Alto-Uruguai river ecosystem differed greatly from each other and from the present work (Brack et al. 1985; Dias et al. 1992; Keel et al. 1993; Daviña et al. 1999). The most comprehensive study so far listed a similar number of species as the current one, but more than half of the species listed in the present work were not reported previously by Dias et al. (1992). Thus, the inventory of species diversity in Turvo park still has to be considered incomplete.

While the species identified in different areas of Turvo park differed, the species diversity, as indicated by the relevant indices, was similarly high in all areas studied (Table 1). Brack et al. (1985) observed 727 plant species from 121 families in diverse edaphic environments in Turvo park. Similarly, 411 species from 284 genera of 100 families were reported during successive collections in the neighboring Moconá park (Daviña et al. 1999) which belongs to the same ecosystem. In their detailed studies on shrub-arboreal plants in a single hectare plot in Turvo park, Dias et al. (1992) identified 88 species belonging to 69 genera of 37 families. Similarly, we identified 78 species from 67 genera of 37 families. Three species new to this ecosystem were found in this study, namely *Allophylus puberulus* (Cambess.) Radlk., *Zantoxylum petiolare* A. St.-Hil et Tul., Another new species, *Solanum pseudoquina*, does not show up in the sample data but was collected along the sample lines.

The present work did not investigate possible relations between species richness and edaphic sites, but differences among investigated areas were observed. Generally, one can presume that such differences are most likely due to different edaphic environments, such as litolic grounds and rocky outcrops, canopy gaps, or river springs. As an example, we observed that the occurrence of some species, such as *Parapiptadenia rigida, Helieta apiculata, Pataganula americana, Urera baccifera, Sebastiana brasiliensis, Cordyline dracaenoides, Manihot flabellifolia*, and *Allophylus puberulus*, clearly depended on specific edaphic conditions, being restricted to and dominant at places of litolic ground or rocky outcrops. According to Rambo (1956), 5% of the 400 species described for the ecosystem occurred at river borders, 31% were typical for the edge of the forest, and 64% exclusively occurred inside the forest. It has been emphasized previously that different edaphic environments and other factors are important for the maintenance of species diversity in tropical forests (Budowski 1970; Denslow 1980; Whitmore 1984) and eco-units often constitute niches for endemic species (Budowski 1965).

Phytosociological composition

The woody flora of Turvo park is characterized by three major aspects: (i) Fabales were most prevalent among the trees emerging from the forest canopy, followed by the Euphorbiaceae and Meliaceae. Trees from these few families represented one third of the total plants and more than two thirds of the total basal area; (ii) the Moraceae comprised the highest number of plants, contributing mostly to the subcanopy, followed by Lauraceae, Fabales, and Meliaceae; and (iii) more than one third of all woody plants belonged to species with commercial timber value. Among them, *Nectandra megapotamica* exhibited the highest density, followed by other important timber species with high commercial value such as *Diatenopteryx sorbifolia, Cedrella fissilis, Balfourodendron riedelianum, Apuleia leiocarpa* and many others. Importantly, the seasonal deciduous forest is not only rich in species of a high timber value, but the number of economically valuable individuals with a DBH $\geq$ 40 cm is also high (44 trees ha^{-1} and 72% of the total basal area). In a study

[136]

on the historical use of forest trees in this same ecosystem, though in different forest remnants, prior to the year 1975, Ruschel et al. (2003, 2005) reported an average of 43 trees ha^{-1} with DBH ≥ 40 cm.

Typical understory species were *Sorocea bonplandii*, *Gymmanthes concolor*, and species of the genus *Trichilia*. Other understory species such as *Calyptranthes tricona* and *Plinia rivularis* were found only in areas where the forest canopy was fully developed (personal observations). These rather few species were typically found in large numbers of individuals, often in local groups of a given species, contributing significantly to the high density of the climax forest stage. Consequently, 80% of all plants had a DBH below 20 cm. Similarly, Klein (1972) reported that 50% of the 178 tree species (shrub to arboreal) of the Subtropical Atlantic Forest ecosystem exhibited total heights below 15 m.

Ecological structure

In contrast to the Tropical Atlantic Forest, the Subtropical Atlantic Forest of Turvo park is dominated by secondary species such as *Parapiptadenia rigida*, *Peltophorum dubium*, *Enterolobium contortisiliquum*, *Diatenopteryx sorbifolia*, *Chorisia speciosa*, *Apuleia leiocarpa* and others, representing the emergent trees in the forest canopy. While the secondary species are typically anemochoric (Budowski 1970), zoochoric species were most abundant in all succession groups. This behavior is typical for tropical forests where usually >75% of the species are zoochoric (Foster et al. 1986; Nascimento et al. 2000; Rondon Neto et al. 2001). For this reason, protection of the fauna is extremely important in tropical forests.

Interestingly, most anemochoric species belonged to the secondary succession group, and most belonged to the emergent trees in the forest, reaching heights of 20–40 m. The absence of anemochoric species within the group of pioneers may be explained by the fact that only individuals with a DBH ≥ 5 cm were included in this study. However, in a similar study of small remnants of the same ecosystem, anemochoric pioneers such as *Ateleia glazioviana* Baill and *Aloysia virgata* Juss were identified (Ruschel 2000), probably due to prevailing border effects in these small areas. Clearly, such small remnants do not represent the typical expression of the fully developed ecosystem.

Final consideration

The Atlantic forest ecosystem to which Turvo park belongs has been identified as one of the 25 global hotspots of biodiversity (Myers et al. 2000). With 8,000 of the 20,000 known plant species considered endemic, the Atlantic Forest ranks fourth on the list of biodiversity hotspots, thus being one of the highest priority areas for conservation world wide. Some of the endemic species of the seasonal deciduous forest part of the Atlantic Forest ecosystem described previously (Rambo 1956; Brack et al. 1985; Keel et al. 1993) and in this study are on the list of endangered species of Rio Grande do Sul (CONSEMA 2002), such as *Allophylus puberulus*, *Annona cacans*, *Apuleia leiocarpa*, *Aralia warmingiana*, *Jacaranda puberula*, *Jacaratia spinosa*, *Machaerium nyctitans*, *Myrocarpus frondosus*, and *Picrasma crenata*. Due to our sparse knowledge of the ecosystem and the occurrence of rare

🖄 Springer

species, or of species occurring at low density only, it is quite likely that many more endangered species are present in this ecosystem.

The conservation of the Turvo park provides several benefits. Besides conservation of the landscape physiognomy, the park represents a constant source of diaspore for the colonization and recovery of degraded areas. The Park also provides a valuable resource for research and programs for the improvement of essential forest products, such as timber, medicinal, or ornamental plants. However, protection of the Alto-Uruguai river ecosystem is critical not only because of its uniquely rich biodiversity and, therefore, extremely high economic value, but it also is the last large remnant of this ecosystem in Southern Brazil. The present study, by advancing our knowledge on the biodiversity of Turvo park, which is still not complete, will represent a reference for future studies of species richness and biodiversity in smaller remnants of this ecosystem.

New insights were achieved about the occurrence and prevalence of timber species and genetic diversity of *Sorocea bonplandii*, one of the key species of this ecosystem. Ruschel et al. (2005) found highly similar values of timber volume and species richness between the Turvo park (old grow forest) and, the smaller remnants in its vicinity (secondary forest). In addition, no genetic erosion associated with forest fragmentation was found in *Sorocea bonplandii* (unpublished).

Acknowledgements   We gratefully acknowledge the expert help of Marcos Sobral and Prof. Dr. Ademir Reis in the taxonomic identification of forest species. We thank the Secretaria Estadual do Meio Ambiente (SEMA, RS, Brazil) and the Fundação Estadual de Proteção Ambiental (FEPAM, RS, Brazil) for the research concession, lodging, and support. Financial support came from the Conselho Nacional de Desenvolvimento Científico e Tecnológico (CNPq, Brazil), from the International Programme of the University of Münster, and from the Federal University of Santa Catarina.

References

Brack P, Bueno RM, Falkenberg DB, Paiva MRC, Sobral M, Stehmann JR (1985) Levantamento Florístico do Parque Estadual do Turvo, Tenente Portela, Rio Grande do Sul, Brasil. Roessléria 7:69–94

Budowski G (1965) The distinction between old secondary and climax species in tropical central American lowland forest. Turrialba 15:40–42

Budowski G (1970) Distribution of tropical American rain forest species in the light old successional processes. Trop Ecol 11:44–48

Cavassan O, Cesar O, Martins FR (1984) Fitossociologia da vegetação arbórea da Reserva Estadual de Bauru, Estado de São Paulo. Revista Brasileira de Botânica 7:91–106

Cottam G, Curtis JT (1956) The use of distance measures in phytosociological sampling. Ecology 37:451–460

CONSEMA—Conselho Estadual do Meio Ambiente (2002) Espécies da Flora Ameaçadas de Extinção do Rio Grande do Sul, 2002. http://www.sema.rs.gov.br/sema/html/espec.htm

Cristóbal LL, Vera N (1999) The forest diversity of a native secondary and primary forest of the Guarani reservation area, Misiones, Argentine. Yvyraretá 9:92–99

Dalling JW, Stephen PH, Silveira K (1998) Seed dispersal, seedling establishment and gap partitioning among tropical pioneer trees. J Ecol 86:674–689

Daviña JR, Rodrígues ME, Honfi A, Sijo GJ, Insaurralde I, Guillen R (1999) Floristic studies of the Maconá Park, Misiones, Argentina. Candollea 54:231–249

Denslow JS (1980) Gap partitioning among tropical rainforest trees. Biotropica 12(suppl):45–55

Dias LL, Vasconcellos JMO, Silva CP, Sobral M, Benedeti MHB (1992) Levantamento florístico de uma área de mata subtropical no Parque Estadual do Turvo, Tenente Portela, RS. In: Congresso Nacional Sobre Essências Nativas, 2, 1992, São Paulo. Anais, Instituto Florestal: São Paulo, 1992, pp 339–346

Dias MC, Vieira AOS, Nakajima JN, Pimenta JA, Carneiro Lobo P (1998) Composição floreística e fitossociológica do componente arbóreo das florestas ciliares do rio Iapó, na bacia do rio Tibagi, Tibagi, PR. Revista Brasileira de Botânica 21:183–195

Foster RB, Arce J, Wachter TS (1986) Dispersal and sequencial plant community in Amazonia Peru foodplain. In: Estrada A, Fleming TH (eds) Frugívores and seed dispersal, W. Junk, Dordrecht, pp 357–370

Gibbs PE, Leitão Filho HF, Jabbott RJ (1980) Application of the point-centred quarter method in a florístic survey of an area of gallery forest at Mogi-Guaçu, SP, Brazil. Revista Brasileira de Botânica 3:17–22

Google Earth (2005) Google Earth. http://www.earth.google.com/ Accessed in February 2006.

IBGE, Instituto Brasileiro de Geografia e Estatística (1990) Geografia do Brasil. Região Sul, Rio de Janeiro

Ivanauskas NM, Rodrigues RR, Nave AG (1999) Phytosociology of the semi-deciduous seasonal forest fragment in Itatinga, SP. Scientia Forestalis 56:83–99

Jarenkow JA, Waechter JL (2001) Composição, estrutura e relações florísticas do componente arbóreo de uma floresta estacional no Rio Grande do Sul, Brasil. Revista Brasileira de Botânica 24:263–272

Janson HC (1983) Adaptation of fruit morphology to dispersal agents in a neotropical forest. Science 1:187–188

Keel S, Gentry AH, Spinzi L (1993) Using vegetation analysis to facilitate the selection of conservation sites in eastern Paraguay. Conserv Biol 7:66–75

Klein RM (1972) Árvores nativas da floresta subtropical do Alto-Uruguai. Sellowia 24:9–62

Krebs CJ (1989) Ecological methodology. Harper Collins, New York NY

Longhi SJ (1997) Agrupamentos e Análise Fitossociológica de Comunidades Florestais na Sub-Bacia hidrogárafica do Rio Passo Fundo-RS. PhD thesis, Universidade Federal do Paraná, Curitiba, Brasil

Lorenzi H (1998) Árvores brasileiras: manual de cultivo de plantas arbóreas nativas do Brasil, vol I and II. Plantarum, São Paulo

Myers N, Mittermeier RA, Mittermeier CG, Fonseca GAB, Kent J (2000) Biodiversity hotsposts for conservation priorities. Nature 403:853–858

Nascimento ART, Longhi SJ, Alvarez Filho A, Schmitz Gomes GS (2000) Análise da diversidade florística e dos sistemas de dispersão de semente em um fragmento florestal na região central do Rio Grande do Sul. Napaea 12:49–67

Rambo B (1956) Der Regenwald am oberen Uruguay. Sellowia 7:183–223

Rondon Neto MR, Watzlawick LF, Caldeira MVW (2001). Diversidade florística e síndromes de dispersão de diásporos das espécies arbóreas de um fragmento de floresta ombrófila mista. Revista Ciências Exatas e Naturais 2:209–216

Reitz R, Klein RM (1964) Os nomes populares das plantas de Santa Catarina. Sellowia 16:9–118

Reitz R, Klein RM, Reis A (1978) Projeto madeira de Santa Catarina. Sellowia 28/30:1–320

Rogalski JM, Zanin EM (2003) Floristic composition of the vascular epiphytes of "estreito de Augusto César", Brazilian Semi-Evergreen Forest of Uruguai River, RS, Brazil. Revista Brasileira de Botânica 26(4):551–556

Ruschel AR (2000) Avaliação e Valoração das Espécies Madeiráveis da Floresta Estacional Decidual do Alto-Uruguai. Master Dissertation, Universidade Federal de Santa Catarina, Florianópolis

Ruschel AR, Guerra MP, Moerschbacher BM, Nodari RO (2005) Valuation and characterization of timber species in remnants of the Alto Uruguay River ecosystem, southern Brazil. For Ecol Manage 217:103–116

Ruschel AR, Nodari ES, Guerra MP, Nodari RO (2003) Evolução do uso e valorização das espécies madeiráveis da floresta estacional decidual do Alto-Uruguai-SC. Ciência Florestal 13:153–166

Silva AF, Leitão Filho HF (1982) Composição florística e estrutura de um trecho da mata atlântica de encosta no município de Ubatuba (São Paulo, Brasil). Revista Brasileira de Botânica 5:43–52

Sparks JC, Masters RE, Payton ME (2002) comparative evaluation of accuracy and efficiency of six forest sampling methods. Proc Oklahoma Acad Sci 82:49–56

Tabarelli M (1992) Flora arbórea da floresta estacional baixo montana no Município de Santa Maria, RS. In: Congresso Nacional Sobre Essências Nativas, 2, São Paulo. Anais, Instituto Florestal: São Paulo, pp 260–268

Traveset A, Riera N, Mas RE (2001) Ecology of fruit-colour polymorphism in *Myrtus communis* and differential effects of birds and mammals on seed germination and seedling growth. J Ecol 89:749–760

🖄 Springer

Vilela EA, Oliveira Filho AT, Carvalho DA, Guilhermes FAG, Appolinário V (2000) Caracter-
    ização estrutural de floresta ripária do Alto Rio Grande, em Madre de Deus de Minas, MG.
    Cerne 6:41–54
Wallauer JP, Albuquerque EP (1986) Lista preliminar dos mamíferos observados no Parque Flor-
    estal do Turvo, Tenente Portela, Rio Grande do Sul, Brasil. Roessléria 8:179–185
Whitmore TC (1984) Gap size and species in tropical rain forests. Biotropica 16:767–779

Biodivers Conserv (2007) 16:1715–1730
DOI 10.1007/s10531-006-9048-3

ORIGINAL PAPER

Genetic diversity assessment in Somali sorghum (*Sorghum bicolor* (L.) Moench) accessions using microsatellite markers

Marco Manzelli · Luca Pileri · Nadia Lacerenza ·
Stefano Benedettelli · Vincenzo Vecchio

Received: 14 September 2005 / Accepted: 10 April 2006 / Published online: 20 May 2006
© Springer Science+Business Media B.V. 2006

Abstract In the north-western region of Somalia, bordering Ethiopia, sorghum represents an important resources for human and animal nutrition. The critical situation of Somalia is threatening the preservation of this valuable resource and it becomes urgent to develop a strategy of correct evaluation of the sorghum germplasm in order to promote conservation and preservation programs. Microsatellites, also known as Simple Sequence Repeats (SSRs), are reproducible molecular markers useful in assessing the level of genetic diversity of plants. A total of 5 sorghum SSR-specific primer pairs were used to assess the genetic diversity of Somali sorghum landraces. Extensive variation was found at the microsatellite loci analysed, except for a locus that resulted in a monomorphic for some accessions. Considerable differences were found between total and effective number of alleles indicating non uniform allele frequency. Moreover allele frequency at a single locus significantly changed among accessions. Total gene diversity calculated for each locus ranged from 0.44 to 0.79. Most of the genetic diversity occurred within accessions demonstrating that accessions are not under selection processes and/or there is a continuous exchange of genes between sorghum populations. In any case, the patterns of clustering were significantly affected by the presence/absence of some alleles with high discriminant weight. Accessions Carabi, Abaadiro, Masego Cas and Masego Cad represent distinct genotypes confirming finding observed in previous phenotypic studies. The results highlight the central role of local farmers in maintaining and shaping local germplasm.

Keywords Genetic diversity · Germplasm evaluation · Sorghum ·
Molecular markers

M. Manzelli (✉) · L. Pileri · N. Lacerenza · S. Benedettelli · V. Vecchio
Department of Agronomy and Land Management, Faculty of Agriculture, University of
Florence, Piazzale delle Cascine 18, 50144 Florence, Italy
e-mail: marco.manzelli@unifi.it

[141]

Introduction

Agricultural biodiversity represents the main outcome of thousands of years of farmers activities in selection, breeding and farming and its wideness is highly correlated with duration of domestication and type of farming systems adopted. Those actions have determined a progressive accumulation of genetic variability, much of which is still being maintained and managed by local farmers and probably represents one of the greatest asset ever generated through human activity (de Boef et al. 1996). Local landraces are maintained in a dynamic process and usually tend to be well adapted to the specific environment in which they evolved (de Boef et al. 1996). They specifically exhibit their high potential in local marginal areas, where they appear equal or sometimes superior to modern varieties in terms of yield (Weltzien and Fishbeck, 1990).

Sorghum [*Sorghum bicolor* (L.) Moench] is an important staple food crop serving over 400 million people in the semi-arid regions of the world (Andrew and Bramel-Cox 1993; FAOSTAT 2004). Resistance to harsh environments such droughts and heat and biotic stresses makes this species an important food resources in dryland ecosystems.

According to the literature, the region of the Horn of Africa including Ethiopia, Sudan and Chad is considered the primary centre of origin and domestication of the species *Sorghum bicolor* (Doggett 1988; de Wet et al. 1976). India and China are considered secondary centres of diffusion (Doggett 1988). Previous studies have detected a high level of phenotypic diversity in sorghum landraces coming from many developing countries, highlighting their central role in local subsistence agriculture (Damania and Rao 1980; Prasada Rao et al. 1989; Zongo et al. 1993; Appa Rao et al. 1996, 1998; Teshome et al. 1997; Li and Li 1997; Ayana and Bekele 1998, 1999, 2000; Chivasa et al. 2000; Friis Hansen 2000; Grenier et al. 2004; Geleta and Labuschagne 2005; Manzelli et al. 2005).

In the north-western region of Somalia which borders to Ethiopia, sorghum represents an important resources for human and animal nutrition. Recent events in Somalia (wars, droughts, etc.) are threatening the maintenance of this important genetic resource. Luckily, local farmers have continued to select and preserve sorghum landraces that show a high degree of variability in terms of morphological and agronomic characteristics, as confirmed in our recent studies (Manzelli et al. 2005). The degradation of habitats and the loss of related biodiversity are already leading to irreversible situations responsible for desertification, displacement of local communities and increasing mass poverty. Consequently, it becomes of utmost importance in this area to develop a strategy of characterisation and in situ preservation of this species. The correct evaluation of the germplasm in an area where cultivation is traditionally conducted, without large energy inputs or excessive specialisation, must be a priority in order to safeguard a tremendously valuable resource for this country and others with similar food and agricultural characteristics.

Our previous studies concentrated on the collection and in situ evaluation of sorghum landraces. The main outcome of these efforts was the detection of highly variable germplasm for morphological and agronomic features (Manzelli et al 2005). Purely phenotypic variability is not sufficient to deeply characterise genetic variation because of genotype–environment interaction and multifactorial inheritance of phenotypic traits. In fact, to estimate how much of phenotypic diversity is genetically determined, it is necessary to control the crosses and subdivide the population in families (halsib, fullsib, S1, etc.). A quicker method of evaluating the width of gene pool minimizing environmental influences is represented by the use of molecular markers. DNA molecular markers have been widely utilised to orient plant genetic resource conservation management, providing a significant tool to evaluate the genetic diversity and its distribution within and among populations of several crop species and predict the evolutionary potential of those species (Hamrick and Godt 1997). In the context of ex situ conservation programs, the information provided by molecular markers is used to assess the level of genetic diversity of domesticated plant germplasm collections, define the wideness of core collections and design optimal procedures in field sampling missions for breeding and conservation purposes (e.g. Pecetti et al. 1992; Schoen and Brown 1995; Menkir et al. 1997; Dean et al. 1999; Djè et al. 1999, 2000; Ayana et al. 2000; Dahlberg et al 2002).

Different types of molecular markers have been used to estimate genetic variation in sorghum. Those markers give broad and different ranges of information and substantially differ in terms of practicability and reproducibility (Hamrick and Godt 1997; Uptmoor et al. 2003). We have used the SSR (Simple Sequence Repeat) analysis in order to complete our previous studies on Somali sorghum germplasm. Despite their high initial cost, SSRs offer several advantages compared to other molecular techniques, such as RFLP (Restriction Fragment Length Polymorphisms), AFLPs (Amplified Fragment Length Polymorphisms) and RAPDs (Random Amplified Polymorphic DNAs). In fact, they are highly polymorphic even among closely related individuals (Mazur and Tingey 1995; Tautz and Renz 1984; Tautz and Schlötterer 1994; Morgante and Olivieri 1993; Lagercrantz et al. 1993; Akkaya et al. 1992; Saghai-Maroof et al. 1994), easily reproducible (Jones et al. 1997) and inherited in a co-dominant fashion.

The main objective of this study was to evaluate the extent of genetic variability within and among Somali sorghum accessions in order to promote a program aimed at the preservation and enhancement of this important food and forage resource.

Materials and methods

Plant material

Nine accessions of sorghum [*Sorghum bicolor* (L.) Moench] were used for this study (Table 1). Eight accessions were chosen as representative of local germplasm from the univariate and multivariate analysis obtained by evaluating 16 accessions, collected in Northwest Somalia, for 9 quantitative characters (Manzelli et al. 2005). The ninth accession, even if not assessed in the agronomic trial for lack of seed, has been inserted in this study because of its specific morphological features.

DNA extraction

Twenty seeds per accession have been used for DNA extraction. The seeds were sterilised with 2% Na-hypochlorite for 10 min, rinsed with distilled water and treated with a commercial fungicide (Vitavax) containing carboxin. Seeds were germinated in a climatic chamber at 22°C under dark conditions in order to reduce chlorophyll synthesis. Etiolated coleoptiles, approximately 4 cm in length, were

[143]   ✌ Springer

Table 1 Evaluated sorghum accessions

Accession[1]	Local name	Translated name	Collecting area
1	Elmi Jama Cad	Red Sorghum	Kalabayd
2	Elmi Jama Cas	White Sorghum	Kalabayd
5	Masego Cas	Red Masego	Boodlay
10	Abaadiro	–	Togwajaale
13	Carabi	–	Boodlay
14	Masego Cad	White Masego	Togwajaale
15	Elmi Jama Kuuso Feruur Geely Aalle	Compact camel mouth sweet Sorghum	Togwajaale
16	Elmi Jama Kuuso Feruur Geely	Compact camel mouth Sorghum	Togwajaale
18	Adan Gaab	Small white Sorghum	Boodlay

[1] The original accession numbering is reported in Manzelli et al. 2005

individually sampled, placed in an Eppendorf tube, frozen in liquid nitrogen and stored at –80°C until DNA extraction. Frozen tissue samples were milled by means of a commercial homogenizer (Qiagen Mixer Mill). Plant DNA was extracted according to Macherey–Nagel NucleoSpin® Plant protocol. The DNA quality was assessed by electrophoresis in 0.8% agarose gel stained with ethidium bromide and examinated by an UV transilluminator.

Microsatellites analysis

Five primer pairs of microsatellite loci (Table 2) were chosen from the oligonucleotide sequences indicated by Brown et al. (1996) and Taramino et al. (1997) by a preliminary screening with the DHAMAN software and subsequent electrophoresis in acrylamide gel in order to verify their complementarity and reproducibility.

The PCR reaction was performed in a final volume of 30 μl, consisting of 15 μl of HotStarTaq Master Mix from Qiagen (200 μM dNTPs, 1.5 mM MgCl₂ and 1.5 U of Taq), 0.2 μM of each primer and 25 ng of genomic DNA. PCR conditions have been optimised for each primer pair by adapting the annealing temperature (T_m). PCR cycling condition were: a 15-min initial denaturation at 95°C, 35 cycles of amplification (1 min at 94°C, 1 min at T_m, 1.5-min elongation at 72°C) followed by a final elongation of 8 min at 72°C. All PCR reactions were performed on Hybaid

Table 2 Sequences of microsatellite primers used for estimation of genetic diversity

Name	Author	Flanking sequence		Repeat motif
SbAGE01	Taramino et al. (1997)	forward	GACCGATCTAATGATGCAG	$(AG)_{30}$
		reverse	ACGGTAGAGAAGACCCATC	
SbAGB03	Taramino et al. (1997)	forward	GTGTGTGTAGCTTCTTGGG	$(AG)_{41}$
		reverse	ACGTAGGAGTAGTTTCTAGGATT	
SvPEPCAA	Taramino et al. (1997)	forward	GCAGCTCAGGGACAAATAC	$(AT)_{10}$
		reverse	CTGCTTCAGGTAAGGATCG	
SvHPRGPG	Taramino et al. (1997)	forward	ACTCCGACGCACCCTAAG	$(CGC)_8$
		reverse	CTCCATTCTTGTAGCACGTA	
Sb4-22	Brown et al. (1996)	forward	TGAGCCGAAAACCGTGAG	$(ACGAC)_4/$
		reverse	CCCAAAACCAAGAGGGAAGG	$(AG)_6$

Thermocycler. Detection of PCR products was carried out on 11% non-denaturing polyacrylamide gels (16-cm length) run vertically. Amplified fragments were visualised by an UV transilluminator after ethidium bromide staining.

Data analysis

Each SSR was treated as an independent locus and its alleles were scored as present (1) and absent (0). Three genetic parameters were calculated to estimate the polymorphism degree of each locus and accession: total number of alleles (n_a), effective number of alleles (n_e) and total gene diversity (H_t) computed according to Nei (1972, 1973). The proportion of genetic diversity within and between accessions was estimated for each polymorphic locus by genetic diversity statistics H_s and G_{st} (Nei 1973).

Pairwise genetic similarity matrix was generated among the 9 accessions using Jaccard's similarity coefficient (Jaccard 1908), given as:

$$S_{ij} = \frac{a}{a + b + c}$$

Where a is the number of bands shared between accessions i and j, b is the number of bands present only in accession i, c is the total number of bands present only in accession j. The matrix of similarity was utilised for the cluster analysis by the Unweighted Pair-Group Method Using Arithmetic Average (UPGMA) algorithm (Sheath and Sokal 1973). The relationship among accessions was visualised by a dendrogram. A correspondence analysis (CA) was performed based on allele presence/absence matrix and the relationship between allele and accession distributions was visualised by a scatterplot against the two first principal coordinates. The analysis was done using NTSYS-PC (1992).

Results

Characterisation of microsatellite loci

Table 3 reports the total number of alleles at each locus and the size ranges observed in this study and in the original publications (Brown et al. 1996; Taramino et al. 1997) that first characterised the corresponding microsatellite loci. We scores a total of 30 alleles. All five loci are highly polymorphic except for the locus SvHPRGPG. In particular that locus showed only two allelic forms, one of which resulted as a null

Table 3 Characteristics of the microsatellite loci

| Locus | Original publication | | This study | |
	Allele number	Size range (bp)	Allele number	Size range (bp)
SbAGE01	9	208–240	9 (1–9)	209–235
SbAGB03	5	94–158	8 (10–17)	93–208
SvPEPCAA	7	210–250	7 (18–24)	168–319
SvHPRGPG	3	246–255	2 (25–26)	229
Sb4-22	5	270–300	4 (27–30)	296–306

The numbers in parenthesis indicate the progressive numbering of the alleles allele. A similar result was reported in sorghum by Dean et al. (1999). In all cases the size range of PCR products is different than that reported in the original publications (Brown et al. 1996; Taramino et al. 1997), probably due to the different geographic origin of the accessions used in this study. Similar observations was reported in previous studies (Donini et al. 1998; Djè et al. 2000).

The 9 allelic forms detected in locus SbAGE01 can be subdivided in two main groups: 6 alleles with a unique band size between 209 and 235 bp, 3 alleles with two bands of similar size range (Fig. 1).

Each of the 8 allelic forms of locus SbAGB03 (Fig. 2) was represented by 3 distinct amplified fragments at a distance of 43 bp for allele 1 and 57 bp for the other alleles. The amplification size was wider than reported in the original publications (Taramino et al. 1997) with values ranging from 93 to 208 bp.

The 7 allelic forms of locus SvPEPCAA (Fig. 3) were constituted by two main PCR products, each of which subdivided in one, two or three sub-products of amplification. The size range was wider than reported in the original publications (Taramino et al. 1997).

The locus SvHPRGPG showed only two allelic forms, one of which resulted as a null allele (Fig. 4). The existence of the null allele was confirmed by repeated amplifications.

The locus Sb4-22 showed a lower size range compared to what reported in the literature and 4 allelic forms, each one represented by a single band, respectively of 304, 300, 306 and 296 bp (Fig. 5).

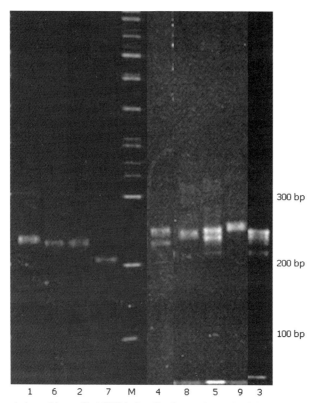

Fig. 1 Allelic variation of locus SbAGE01 visualised on polyacrylamide gel

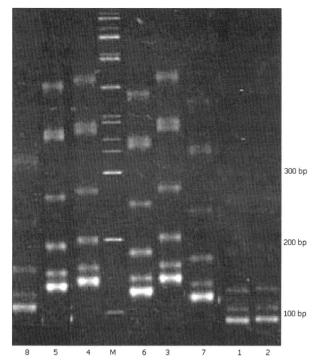

Fig. 2 Allelic variation of locus SbAGB03 visualised on polyacrylamide gel

Genetic diversity analysis

Statistics of genetic diversity over and within accessions are given in Tables 4 and 5. The difference detected between total and effective number of alleles denoted a high differentiation in allele frequencies (Table 4). Referring to locus analysis within accession (Table 5), only the locus Sb4-22 resulted in monomorphic for accessions Masego Cas (2) and Adan Gaab (18), while a high polymorphism degree for the other loci in all the accessions was observed. Moreover allele frequency distribution at single locus markedly changes among accessions.

The highest overall total genetic diversity occurred in loci SbAGE01, SbAGB03 and SvPEPCAA (Table 4). The genetic diversity mostly resides within accessions. A high Gst value was only observed for locus Sb4-22 (42 % of total variance). It indicates the importance of this locus in discriminating between accessions.

Accessions Abaadiro (10), Elmi Jama Kuuso Feruur Geely Aalle (15), Elmi Jama Kuuso Feruur Geely (16) reached higher values of Ht, while accessions Masego Cas (5) and Masego Cad (14) showed lower values (Table 5).

Cluster and correspondence analysis

Cluster analysis based on Jaccard's similarity matrix showed that accessions Masego Cas (5), Masego Cad (14) and Elmi Jama Cas (2) are closely related, whilst accessions Abaadiro (10) and Carabi (13) are isolated (Fig. 6). The other accessions are comprised between those main clusters.

Springer

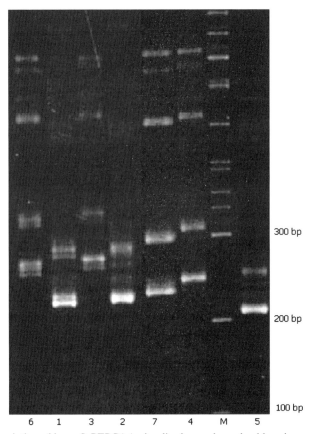

Fig. 3 Allelic variation of locus SvPEPCAA visualised on polyacrylamide gel

Based on the correspondence analysis, allele and accession distributions are given in Figs. 7 and 8. Comparing the two distributions, it is possible to note how some alleles are highly discriminant between accessions. It is the case of alleles 30 (locus Sb4-22), 22 (locus SvPEPCAA) and 17 (locus SbAGB03) for accession Carabi (13), allele 24 (locus SvPEPCAA) present only in some genotypes of accession Adan Gaab (18), alleles 28 (locus Sb4-22) and 3 (locus SbAGE01) for accession Abaadiro (10), alleles 19 (locus SvPEPCAA), 14 (locus SbAGB03) and 4 (locus SbAGE01) observed in the cluster comprising accessions Elmi Jama Cas (2), Masego Cas (5) and Masego Cad (14).

Discussion

According to other studies, this research confirms the usefulness of microsatellite markers in detecting genetic variation in crop species. The level of genetic diversity detected by these molecular markers confers to them a high degree of precision and reliability in the framework of genetic studies, with a particular emphasis on the management of germplasm collection and in the optimisation and orientation of

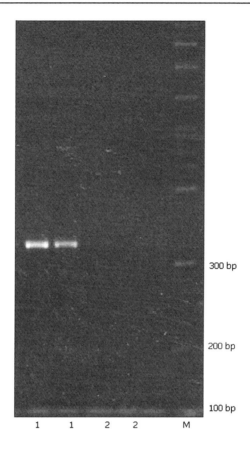

Fig. 4 Allelic variation of locus SvHPRGPG visualised on polyacrylamide gel

300 bp

200 bp

100 bp

1   1   2   2   M

sampling strategies (Schoen and Brown 1995; Djè et al. 2000). In fact SSR analysis allows: the formation of a germplasm collection representative of the region of origin, thus minimizing the number of accessions that need to be kept; the identification of accessions whose genetic distance emphasises characteristics essential to future selection programs; the identification of accessions which, though homogeneous over certain productive or qualitative traits, evidence a wide genetic pool. This kind of evaluations can help in stabilizing output yields in a difficult environment, such as north-western Somalia. Populations that contain a mix of genotypes have the highest capacity for adaptation to marginal areas where climate is not always favourable to optimal crop growth and development. In fact, high input varieties do not always have enough plasticity to adapt to diverse environmental events that characterise tropical and subtropical areas. Thus genetic variability present in landraces serves to avoid crop failure by reducing vulnerability to diseases, pests and environmental stresses (Ceccarelli et al. 1992; de Boef et al. 1996).

To our knowledge this research is the first attempt in estimating genetic diversity of Somali sorghum germplasm based on microsatellite markers. This preliminary investigation showed that Somali sorghum germplasm is highly variable, especially for some loci.

The variability observed at locus SvPEPCAA, characterised by a microsatellite region inserted in the gene encoding for the enzyme phosphoenolpyruvate

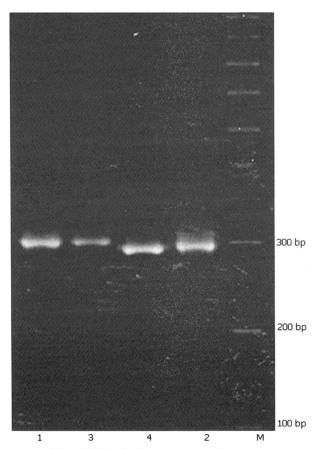

Fig. 5 Allelic variation of locus Sb4-22 visualised on polyacrylamide gel

Table 4 Genetic diversity over accessions

Locus	n_a^a	ne^b	H_t^c	H_s^d	G_{st}^e
SbAGE01	9	3.9	0.74	0.58	0.21
SbAGB03	8	4.9	0.79	0.60	0.23
SvPEPCAA	7	3.6	0.72	0.47	0.34
SvHPRGPG	2	1.8	0.44	0.37	0.14
Sb4-22	4	2.2	0.55	0.31	0.42

[a] Total number of alleles per locus

[b] Effective number of alleles per locus

[c] Total gene diversity

[d] Average gene diversity within accession

[e] Proportion of differentiation among accessions

carboxylase, was high since that region is located in the ninth intron of that gene. This microsatellite can be an useful marker in detecting differences in the expression of PEP carboxylase, an enzyme involved in carbohydrate synthesis in C_4 plant species. Although that microsatellite region is based on the dinucleotide repeat $(AT)_{10}$, its

Table 5 Genetic diversity within accessions

Accession		Locus					Average
		SbAGE01	SbAGB03	SvPEPCAA	SvHPRGPG	Sb4-22	
1 Elmi Jama Cad	Ht	0.50	0.65	0.34	0.50	0.57	0.51
	Na	2	5	3	2	3	3.0
	Ne	2.0	2.8	1.5	2.0	2.4	2.1
2 Elmi Jama Cas	Ht	0.79	0.46	0.34	0.10	0.46	0.43
	Na	6	2	3	2	3	3.2
	Ne	4.7	1.8	1.5	1.1	1.9	2.2
5 Masego Cas	Ht	0.27	0.41	0.38	0.32	0	0.27
	Na	3	4	2	2	1	2.4
	Ne	1.4	1.7	1.6	1.5	1	1.4
10 Abaadiro	Ht	0.79	0.72	0.71	0.48	0.10	0.56
	Na	7	4	5	2	2	4.0
	Ne	4.7	3.5	3.5	1.9	1.1	2.9
13 Carabi	Ht	0.52	0.57	0.27	0.38	0.55	0.45
	Na	3	4	3	2	4	3.2
	Ne	2.1	2.3	1.4	1.6	2.2	1.9
14 Masego Cad	Ht	0.27	0.47	0.44	0.32	0.42	0.38
	Na	4	4	3	2	2	3.0
	Ne	1.4	1.9	1.8	1.5	1.7	1.6
15 Elmi Jama	Ht	0.69	0.60	0.75	0.48	0.60	0.62
Kuuso Feruur	Na	5	6	5	2	3	4.2
Geely Aalle	Ne	3.2	2.5	4	1.9	2.5	2.8
16 Elmi Jama	Ht	0.81	0.78	0.68	0.42	0.18	0.57
Kuuso Feruur	Na	6	7	5	2	2	4.4
Geely	Ne	5.1	4.5	3.1	1.7	1.2	3.1
18 Adan Gaab	Ht	0.68	0.81	0.42	0.42	0	0.46
	Na	5	7	4	2	1	3.8
	Ne	3.1	5.3	1.7	1.7	1	2.6

collocation in the gene intronic region do not determine frameshift mutation as deletion or insertion mutations are eliminated during mRNA maturation phase.

Locus SvHPRGPG is inserted within a gene encoding for glycoproteins rich in hydroxyproline. That primer showed the lowest allele number and value of H_t (0.44) with a percentage of among accession variance of 14%. The presence of a null allele was also observed. The characteristics of microsatellite repeat do not determine frameshift mutations and eventual variations could be due to nucleotide substitutions, induced by a high UV radiation typical of the collecting area. The low variability observed for this locus could be related to a high degree of gene preservation, that's why it would not be possible to accept the presence of null alleles or nonsynonymous mutations. The presence of a null allele could be explained by forming the hypothesis that one of the two sites complementary to the primers is inserted in an intronic sequence of the gene. This position would permit the gene expression but not the amplification of microsatellite sequences because of mutations only at an intronic level. The gene sequence reported on the EMBL database shows that the microsatellite region is located in the terminal part of the gene, adjacent to the sequence of the transcription end. This clearly indicates that, even in the case of mutations in sequences close to the gene end, the gene is expressed while the microsatellite sequence is not amplified without determining a reduction in genotype fitness.

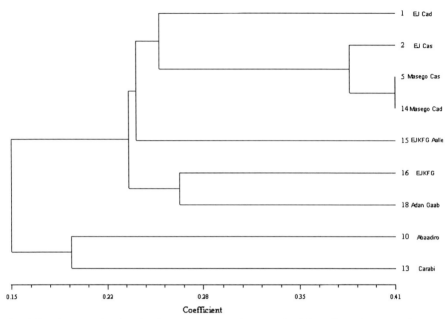

Fig. 6 Dendrogram showing the clustering patterns between the 9 Somali sorghum accessions

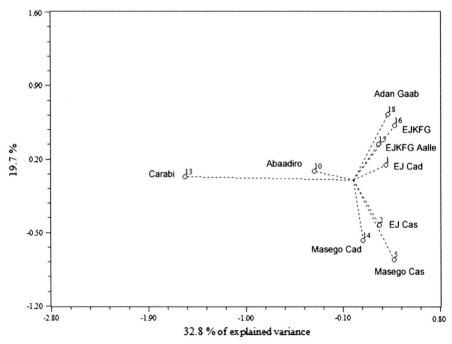

Fig. 7 Scatterplot of sorghum accessions against the two first principal coordinates computed from allele presence/absence matrix

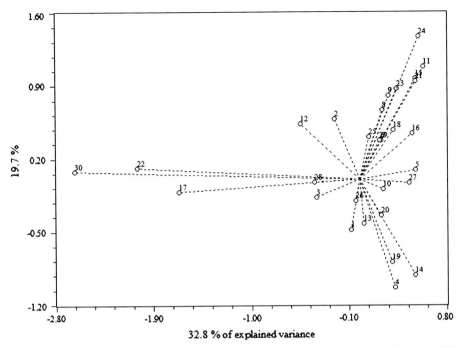

Fig. 8 Scatterplot of alleles against the two first principal coordinates computed from allele presence/absence matrix

Loci SbAGE01 and SbAGB03 showed high value in genetic diversity since they are not related to any gene encoding activity.

Although not inserted in an encoding region, Locus Sb4-22, characterised by a combination of two microsatellite sequences, showed low variability, probably due to its nature of imperfect microsatellites. By analysing the size of observed alleles it is possible to assert that the mutation mechanism *Slipped Strand Mispairing* (Levinson and Gutman 1987), responsible in determining microsatellite variability, is more active in sequences of simple repeats than of more complex repeats.

The high degree of diversification among and within accessions confirmed what was observed at the phenotypic level. Moreover the molecular analysis provided an unbiased estimation of the genetic diversity in absence of environmental interaction. Accessions Carabi (13), Abaadiro (10) and the two Masego (5 and 14) can well represent distinct genotypes. In particular accessions Masego Cas (5) and Masego Cad (14), though collected in different areas, showed a similar gene pool, confirming that names given by farmers to the accessions are consistent. In fact, as reported by Somali farmers, the main difference between the two mentioned accessions resides in the kernel colour.

As previously mentioned, the allele distribution at single locus significantly changes among accessions. Then the patterns of clustering were significantly affected by the presence/absence of some alleles with high discriminant weight (Figs. 7 and 8).

The observed variability is higher within accessions (II_s) than among accessions (G_{st}), demonstrating that accessions are not under selection processes and/or there is a continuous exchange of genes among accessions. In fact in situ conservation allows landrace adaptation and evolution to continue. Undoubtedly, local farmers play a central role in this evolutionary process, shaping and maintaining genetic diversity to meet their needs. Even if in the past farmers' breeding activities have been considered as "primitive" by professional breeders, several researchers have recently changed their points of view and have begun concentrating their attention on those traditional activities that constitute one of the major forces in local farming system development (de Boef et al. 1996). It is now well recognised that landraces and farmers are interdependent and both are in need of each other for their own survival (Appa Rao et al. 1998).

Somalia has been without a government or any form of stable administrative and political structure over the past decade. This has led to severe consequences both on socio-political sphere and even more on the livelihood and life quality of Somali people. The impact on agriculture and rural communities is complex as many causes have affected this traditional agricultural production system. Risk of genetic erosion and the weakening of local seed networks are particularly consistent with unpredictable consequences on rural livelihood.

Nevertheless, the conservation and improvement of sorghum accessions are of practical value in Somalia. The definition of farmer-oriented conservation and improvement programs will lead to a greater production stability of this traditional food crop. The possibility to preserve in situ local sorghum germplasm can be particularly useful in order to define those parameters necessary to the development of cultivars able to grow in marginal and heterogeneous environmental conditions.

This preliminary analysis on the Somali sorghum gene pool attempts to confirm and promote molecular markers as basic tools in the definition of genetic resource conservation and exploitation strategies, and, in general, to recognise the importance of agricultural biodiversity as a strategic resource in subsistence agriculture.

References

Akkaya MS, Bhagwat AA, Cregan PB (1992) Length polymorphisms of simple sequence repeat DNA in soybean. Genetics 132:131–139

Andrews DJ, Bramel-Cox (1993) Breeding cultivars for sustainable crop production in low input dry land agriculture in the tropics. In: Buxton DR, Shibles R, Forsberg RA, Blad BL, Asay KH, Paulsen GM, Wilson RF (eds) International crop science. Crop Sci Soc Amer Inc, Madison, Wisconsin, USA, pp 211–223

Appa Rao S, Mengesha MH, Gopal Reddy V, Prasada Rao KE (1998) Collecting and evaluation of sorghum germplasm from Rwanda. Plant Genet Resour Newslett 114:26–28

Appa Rao S, Prasada Rao KE, Mengesha MH, Gopal Reddy V, (1996) Morphological diversity in sorghum germplasm form India. Genetic Resour Crop Evol 43:559–567

Ayana A, Bekele E (1998) Geographical patterns of morphological variation in sorghum (*Sorghum bicolor* (L.) Moench) germplasm from Ethiopia and Eritrea: qualitative characters. Hereditas 129:195–205

Ayana A, Bekele E (1999) Multivariate analysis of morphological variation in sorghum (*Sorghum bicolor* (L.) Moench) germplasm from Ethiopia and Eritrea. Genet Resour Crop Evol 46:273–284

Ayana A, Bekele E (2000) Geographical patterns of morphological variation in sorghum (*Sorghum bicolor* (L.) Moench) germplasm from Ethiopia and Eritrea: quantitative characters. Euphytica 115:91–104

Ayana A, Bryngelsson T, Bekele E (2000). Genetic variation of Ethiopian and Eritrean Sorghum (*Sorghum bicolor* (L.) Moench) germplasm assessed by random amplified polymorphic DNA (RAPD). Genet Resour Crop Evol 47:471–482

Brown SM, Hopkins MS, Mitchell SE, Senior ML, Wang TY, Duncan RR, Gonzales-Candelas F, Kresovich S (1996) Multiple methods for identification of polymorphic simple sequence repeats (SSRs) in sorghum [*Sorghum bicolor* (L.) Moench]. Theoret Appl Genet 93:190–198

Ceccarelli S, Valkoun J, Erskine W, Weigand S, Miller R, van Leur JAG (1992) Plant genetic resources and plant improvement as tools to develop sustainable agriculture. Exp Agric 28:89–98

Chivasa W, Harris D, Chiduza C, Nyamudeza P, Mashingaidze AB (2000) Biodiversity on farm in semiarid agriculture: a case study from a smallholder farming system in Zimbabwe. Zimbabwe Sci News 34(1):13–18

Dahlberg JA, Zhang X, Hart GE, Mullet JE (2002) Comparative assessment of variation among sorghum germplasm accessions using seed morphology and RAPD measurements. Crop Sci 42:291–296

Damania AB, Rao VA (1980) Collecting sorghum in Somalia. FAO/IBPGR Plant Genet Resour Newslett 40:14–16

de Boef WS, Berg T, Haverkort B (1996) Crop genetic resources. In: Bunders J, Haverkort B, Hiemstra W (eds) Biotechnology: building on farmers' knowledge. Macmillan Education LTD, London and Basingstoke, pp. 103–128

de Wet JMJ, Harlan JR, Price EG (1976) Variability in Sorghum bicolor. In: Harlan JR, de Wet JMJ, Stemler ABL (eds) Origins of African plant domestication. Mouton, The Hague, Paris, pp 453–463

Dean RE, Dahlberg JA, Hopkins MS, Mitchell SE, Kresovich S (1999). Genetic redundancy and diversity among 'orange' accessions in the U.S. national sorghum collection as assessed with simple sequence repeat (SSR) markers. Crop Sci 39:1215–1221

Djè Y, Forcioli S, Ater M, Lefèbvre C, Vekemans X (1999) Assessing population genetic structure of sorghum landraces from North-western Morocco using allozyme and microsatellite markers. Theoret Appl Genet 99:157–163

Djè Y, Heuertz M, Lefèbvre C, Vekemans X (2000) Assessment of genetic diversity within and among germplasm accessions in cultivated sorghum using microsatellite markers. Theoret Appl Genet 100:918–925

Doggett H (1988) Sorghum. 2nd edn. Longman, UK

Donini P, Stephenson P, Bryan GJ, Koebner RMD (1998) The potential of microsatellites for high throughput genetic diversity assessment in wheat and barley. Genet Resour Crop Evol 45:415–421

FAOSTAT data, (2004)

Friis Hansen E (2000) Farmers' management and use of crop genetic diversity in Tanzania. In: Almekinders C (ed) Encouraging diversity: the conservation and development of plant genetic resources, London, UK, pp 66–71

Geleta N, Labuschagne (2005). Qualitative traits variation in sorghum (*Sorghum bicolor* (L.) Moench) germplasm from eastern highlands of Ethiopia. Biodiversity and conservation 14:3055–3064

Grenier C, Bramel PJ, Dahlberf JA, El-Ahmadi A, Mahmoud M, Peterson GC, Rosenow DT, Ejeta G (2004). Sorghums of the Sudan: analysis of regional diversity and distribution. Genet Resour Crop Evol 51:489–500

Hamrick JL, Godt MJW (1997) Allozyme diversity in cultivated crops. Crop Sci 37:26–30

Jaccard P (1908) Nouvelles recherches sur la distribution florale. Bulletin Société Vaudense Sciences Naturelles 44:223–270

Jones CJ, Edwards KJ, Castiglione S, Winfield MO, Sala F, Van de Weil AC, Bredemeijer G, Vosman B, Matthes M, Maly A, Brettschneider R, Bettini P, Buiatti M, Maestri E, Malcevschi A, Marmiroli N, Aert R, Volckaert G, Rueda J, Linaacero R, Vazque A, Karp A (1997) Reproducibility testing of RAPD, AFLP and SSR markers in plants by a network of European laboratories. Mol Breed 3:381–390

Lagercrantz U, Ellegren H, Andersson L (1993) The abundance of various polymorphic microsatellite motifs differs between plants and vertebrates. Nucleic Acids Res 21:1111–1115

Levinson G, Gutman GA (1987) Slipped-strand mispairing: a major mechanism for DNA sequence evolution. Mol Biol Evol 4:203–221

Li Y, and Li C (1997). Phenotypic diversity of sorghum landraces. In: China International Conference on Genetic Improvement of Sorghum and Pearl Millet pp 659–668

Manzelli M, Benedettelli S, Vecchio V (2005). Agricultural biodiversity in Northwest Somalia – an assessment among selected Somali sorghum (*Sorghum bicolor* (L.) Moench) germplasm. Biodivers Conserv 14:3381–3392

Mazur BJ, Tingey SV (1995) Genetic mapping and introgression of genes of agronomic importance. Curr Opin Biotechnol 175:182

Menkir A, Goldsbrough P, Ejeta G (1997) RAPD based assessment of genetic diversity in cultivated races of sorghum. Crop Sci 37:564–569

Morgante M, Olivieri AM (1993) PCR-amplified microsatellites as markers in plant genetics. Plant J 3:175–182

Nei M (1972). Genetic distance between populations. Am Nat 106:283–292

Nei M (1973) Analysis of gene diversity in subdivided populations. Proc Nat Acad Sci USA 70:3321–3323

Pecetti L, Annichiarico P, Damania AB (1992). Biodiversity in a germplasm collection of durum wheat. Euphytica 60:229–238

Prasada Rao KE, Hussein Mao H, Mengesha MH (1989). Collecting sorghum germplasm in Somalia. FAO/IBPGR Plant Genet Resour Newslett 78–79:41

Saghai-Maroof MA, Biyashev RM, Yang GP, Zhang Q, Allard RW (1994). Extraordinarily polymorphic microsatellite DNA in barley: species diversity, chromosomal locations and populations dynamics. Proc Natl Acad Sci USA 91:5466–5470

Schoen DB, Brown AHD (1995). Maximising genetic diversity in core collections of wild relatives of crop species. In: Hodgkin T, Brown AHD, Van Hintum ThJL, Morales EAV (eds) Core collections of plant genetic resources. John Wiley and Sons, Chichester, pp 55–76

Sneath PHA, Sokal RR (1973). Numerical taxonomy. WH. Freeman, San Francisco, CA

Taramino G, Tarchini R, Ferrario S, Lee M, Pé ME (1997). Characterisation and mapping of simple sequence repeats (SSR) in Sorghum bicolor. Theoret Appl Genet 95:66–72

Tautz D, Renz M (1984). Simple sequences are ubiquitous repetitive components of eukaryotic genomes. Nucleic Acids Res 12:4127–4138

Tautz D, Schlötterer C (1994). Simple sequences. Curr Opin Genet Dev 4:832–837

Teshome A, Baum BR, Fahring L, Torrance JK, Arnason TJ, Lambert JD (1997) Sorghum (Sorghum bicolor (L.) Moench) landrace variation and classification in North Shewa and South Welo, Ethiopia. Euphytica 97:255–263

Uptmoor R, Wenzel W, Friedt W, Donaldson G, Ayisi K, Ordon F (2003). Comparative analysis on the genetic relatedness of Sorghum bicolor accessions from Southern Africa by RAPD, AFLPs and SSRs. Theoret Appl Genet 106:1316–1325

Weltzien E, Fishbeck G (1990). Performance and variability of local barley landraces in Near-Eastern environments. Plant Breed 104:58–67

Zongo JD, Gouyon PH, Sandmeier M (1993). Genetic variability among sorghum accessions from the Sahelian agroecological region of Burkina Faso. Biodivers Conserv 2(6):627–636

Biodivers Conserv (2007) 16:1731–1745
DOI 10.1007/s10531-006-9052-7

ORIGINAL PAPER

Demand for rubber is causing the loss of high diversity rain forest in SW China

Hongmei Li · T. Mitchell Aide · Youxin Ma · Wenjun Liu · Min Cao

Received: 21 December 2005 / Accepted: 26 April 2006 / Published online: 20 May 2006
© Springer Science+Business Media B.V. 2006

Abstract As the economies of developing countries grow, and the purchasing power of their inhabitants increases, the pressure on the environment and natural resources will continue to increase. In the specific case of China, impressive economic growth during the last decades exemplifies this process. Specifically, we focus on how changing economic dynamics are influencing land-use and land-cover change in Xishuangbanna, China. Xishuangbanna has the richest flora and fauna of China, but increasing demand for natural rubber and the expansion of rubber plantations is threatening this high-diversity region. We quantified land-use/land-cover change across Xishuangbanna using Landsat images from 1976, 1988, and 2003. The most obvious change was the decrease in forest cover and an increase in rubber plantations. In 1976, forests covered approximately 70% of Xishuangbanna, but by 2003 they covered less than 50%. Tropical seasonal rain forest was the forest type most affect by the expansion of rubber plantations, and a total of 139,576 ha was lost. The increase of rubber plantations below 800 m, shifted agricultural activities to higher elevations, which resulted in deforestation of mountain rain forest and subtropical evergreen broadleaf forest. Although these changes have affected the biodiversity and ecosystem services, we believe that long-term planning and monitoring can achieve a balance between economic and social needs of a growing population and

H. Li · Y. Ma (✉) · W. Liu · M. Cao
Xishuangbanna Tropical Botanical Garden, Chinese Academy of Sciences, 88 Xuefu Road,
Kunming 650223, PR China
e-mail: may@xtbg.ac.cn

T. M. Aide
Department of Biology, University of Puerto Rico, 23360, San Juan, PR 00931-3360, USA

Y. Ma
Institute of Applied Ecology, Chinese Academy of Sciences, 72 Wenhua Road, Shenyang
110016, PR China

Y. Ma
Graduate School of the Chinese Academy of Sciences, 19B Yuquan Road, Beijing 100039,
PR China

the conservation of a highly diverse flora and fauna. Below 800 m , we recommend that no more rubber plantations be established, existing forest fragments should be protected, and riparian forests should be restored to connect fragments. Future rubber plantations should be established in the abandoned arable or shrublands at higher elevations, and tea or other crops should be planted in the understory to improve economic returns and reduce erosion.

Keywords Biodiversity conservation · Economic development · Land-use/land-cover change · Rubber plantations · Xishuangbanna

Introduction

Globally the rate of deforestation has decreased during the last decade, and in more than 50 countries reforestation has surpassed deforestation (FAO 2000). Although these are positive signs, most of the countries that are experiencing forest transition are in the northern hemisphere and are developed countries. In many tropical and developing countries, deforestation continues to exceed reforestation. Logging, mining, slash and burn agriculture, and fire wood collection contribute to deforestation and forest degradation, but by far the most important factor is the conversion of forest to agriculture and pastures. Although technological advances have helped to increase productivity and limit agricultural expansion, the increasing human population and the increase in per capita caloric intake continue to drive agricultural expansion. For example, the demand for soy products has lead to the deforestation of millions of hectares of dry forest (i.e. cerrado, chaco) in South America (Grau et al. 2005a, b), and between 1990 and 2002, an additional 10.7 million ha of tropical forest were converted to oil palm plantations, mainly in SE Asia (Casson 2003). As the economies of developing countries grow, and the purchasing power of their inhabitants increases, the pressure on the environment and natural resources will continue to increase.

The impressive economic growth of China during the last decades exemplifies this process. During the last 25 years, on average, the Gross Domestic Product (GDP) has increased at a rate of more than 8% per year (Wong and Chan 2003). China has shifted from an exporter of soybean to the largest importer (Tuan et al. 2004). It presently uses 40% of the global supply of cement and 20% of the global supply of steel (Chandler and Gwin 2004). Furthermore, hundreds of power plants and dams are in construction around the country (Sinton et al. 2005). What will be the biodiversity consequences of this economic growth?

Within China, the Xishuangbanna region is the most diverse region of China, and it is included in the Indo-Burma biodiversity hotspot (Myers et al. 2000). It represents only 0.2% of the area of China, but it contains approximately 5000 species of higher plants (16% of the nation's total), 102 species of mammals (21.7%), 427 species of birds (36.2%), 98 species of amphibians and reptiles (14.6%), and 100 species of freshwater fish (2.6%) (Zhang and Cao 1995). The ecological and socio-economic context of Xishuangbanna is representative of other tropical regions of Southeast Asia that contain high levels of biodiversity and are threatened with deforestation and environmental degradation.

An important driver of land-use/land-cover change in the Xishuangbanna region has been the transformation of tropical forest to rubber plantations. The first rubber plantations were established in 1956 to meet the needs of national defense and economic development. In 1978, the expansion of rubber plantations was accelerated by the China's Reform and Innovation policy (Edit committee of Xishuangbanna Dai Autonomous Prefecture Difangzhi 2002). Today, the expansion of rubber plantations in Xishuangbanna is continuing, driven by the increase in national consumption (Fig. 1). The increasing demand for natural rubber during the last 10 years, in large part, has been driven by the need for automobile tires due to the dramatic increase in automobile production. Although there has been an expansion of plantations in China and an increase in production (Fig. 1), China produces less than half of what it consumes, and this is promoting the conversion of forest to rubber plantations in other tropical regions.

The Chinese government has enacted different policies in response to the impacts that economic development is having on the environment. At the national scale, new forestry laws and polices were enacted. This included increasing the number and extent of conservation areas, and the implementation of the Natural Forest Protection Plan to control deforestation and increase forest cover (Yin 2001). Within Xishuangbanna, these policies had a mixed impact. On the positive side, a series of nature reserves were established in 1980 that cover 240,000 ha, 12% of Xishuangbanna (Guo et al. 2002). But, the Natural Forest Protection Plan actually lead to a reduction in natural forest cover, because in the plan the definition of forest includes rubber plantations. Furthermore, given that rubber plantations are a major component of the local economy, there continues to be local incentives for establishing new plantations. This policy assumes that rubber plantations would have positive environmental effects by increasing forest cover. However, to assess the net effect, three factors need to be considered: (1) the environmental effects of the plantations in relation to other land cover types, (2) the effects of new plantations on the dynamics of other land uses, in particular if new plantations are replacing agriculture or mature forests, and (3) the spatial distribution of the resulting land cover configuration.

A major limitation for enacting effective development and conservation activities is the lack of spatially explicit information on land-use dynamics. In the past, land-use decisions in Xishuangbanna were often based on government reports that did not even include land-use categories appropriate for the region (Guo et al. 2002). The lack of recent spatially explicit information has led officials to believe that the expansion of rubber plantations was occurring in areas of shifting cultivation, when they were actually replacing the most species diverse forest in China. If local officials

Fig. 1 Production and consumption of natural rubber in China. Data source from Li and Zhang (2004)

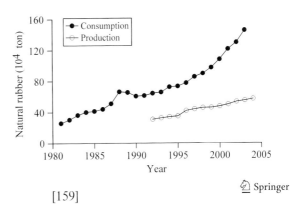

in Xishuangbanna or in other region experiencing rapid transformation are to balance development and conservation needs, they must have access to the best information.

The major objective of this study is to quantify land-use/land-cover changes over the past 27 years (1976–1988 and 1988–2003) in Xishuangbanna, China. Based on our results, we provide suggestions on how to improve land-use policy in a way that balances the need for economic development and biodiversity conservation.

Methods

Study area

Xishuangbanna (21°08′–22°36′ N, 99°56′–101°50′ E), in Yunnan Province, southwest China, covers 19,150 km², includes three counties (Jinghong, Menghai and Mengla), and borders Laos to the south and Myanmar to the southwest (Fig. 2). The region has mountain-valley topography with the Hengduan Mountains running north-south, and about 95% of the region is covered by mountains and hill. The Mekong River flows through the center of Xishuangbanna, and the region contributes more than 20 important tributaries, resulting in many river valleys and small basins (Cao and Zhang 1997). The altitude varies from 2430 to 475 m above sea level. The climate of this region is influenced by warm-wet air masses from the Indian Ocean in summer, including monsoons, and continental air masses of subtropical origin in winter, resulting in a rainy season from May to October, and a dry season from November to April (Zhang 1988). Within Xishuangbanna the annual rainfall ranges from 1100 to 2400 mm, and is lower in center region and higher in the western and eastern regions (Edit committee of Xishuangbanna Dai Autonomous Prefecture Difangzhi 2002). The combination of geography and climate in Xishuangbanna has created a transition zone between the flora and fauna of tropical South East Asia and subtropical and temperate China (Cao et al. 1996), resulting in the region with the highest biodiversity in China (Zhang and Cao 1995; Cao and Zhang 1997). The five primary forest types in Xishuangbanna are: tropical seasonal rain forest, tropical mountain rain forest, evergreen broad-leaved forest, monsoon forest over limestone, and monsoon forest on river banks (Wu et al. 1987).

Data sources

Land-use/land-cover change was determined using two Landsat Multi Spectral Scanner (MSS) images (24 February 1976–#139/45, and 25 April 1975–#140/45), a Landsat Thematic Mapper (TM) image (2 February 1988–#130/45) and a Landsat Enhanced Thematic Mapper (ETM) image (7 March 2003–#130/45). Two images were used to create the 1976 cover, with information from 1975 used to fill in areas with cloud cover in the 1976 image. All images were acquired during the dry season between February and April. Two land-use maps developed by the Xishuangbanna Department of Land and Resource (Xishuangbanna Land-use Status Map 1982, 1991) and a vegetation map developed by the Xishuangbanna Forestry Bureau (Xishuangbanna Vegetation Distribution Map 1993) were used as references for the classification of the MSS and TM images, respectively. Topographic maps (scale = 1:50,000) and digital topographic data with a contour interval of 100 m

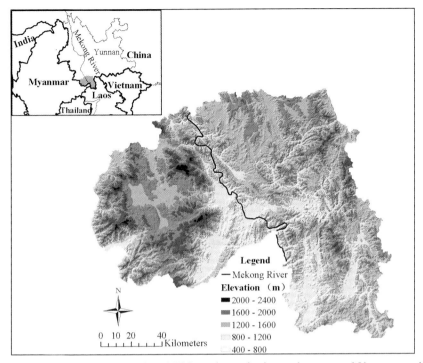

Fig. 2 The location and topography of Xishuangbanna in the southern part of Yunnan province of China

published by the State Bureau of Surveying and Mapping of China were used to build a digital elevation model (DEM).

Geometric correction and classification of satellite images

The TM satellite images were rectified to Albers Conical Equal Area projection system with a 35-m pixel size. The ETM and MSS images were registered to the TM images using an image-to-image registration technique: rectification RMS errors were <0.5 pixels and <1 pixels, respectively.

All non-thermal channels of the TM and ETM images and all channels of the MSS images were used to create class spectral signatures for classification. The images were classified using the supervised maximum likelihood classification method. Training areas for each land-cover class were identified for each image. For the ETM image, training areas were identified in the field during February–March 2003. For the TM and MSS images, training areas were generated from the Department of Land and Resource maps of 1982 and 1991, and the Forestry Bureau's vegetation map of 1993, respectively. We selected large homogeneous areas for the training areas. For each land-use type, we included at least 10 training areas to reflect the variation within a land use due to topography and slope effects.

Initially we used the same 15 land-use classes developed by the National Agricultural Zoning Committee (1984); however, to assure the greatest accuracy we combined these into eight classes (Table 1). Forests were classified into four classes.

🖄 Springer

It was difficult to distinguish the different forest types from the images so tropical seasonal rain forest, mountain rain forest and subtropical evergreen broadleaf forest were separated based on elevation (Guo et al. 1987) (Table 1). Conifers and bamboos dominated the "other forest" class, and they could be distinguished based on differences in texture and spectral characteristics. Rubber plantations were easy to classify because the trees are deciduous during the dry season, and most native forest species are evergreen. Shrubland is a common land-use class, but it is often a transition between abandoned agricultural land and forest or plantations. Arable lands included areas of active agriculture, recent fallow lands, grassland, tea gardens, and paddy rice. All other land-use types (e.g. urban, water) were combined into the "Other land-use" category if the individual land-use cover was <1% of the study area.

Post classification

The classified images were transformed using the clump, elimination, and filter options in Erdas Imagine (Version 8.6, Leica Geosystems). The "clump" option helps maintain spatial coherence by identifying clumps, which are contiguous groups of

Table 1 Land-use classes used in image classification and change detection

Land-use class	General description
Tropical seasonal rain forest (TSRF)	Forested areas with greater than 30% closed canopy dominated by broadleaf trees, and at an altitude less than 800 m
Mountain rain forest (MRF)	Forested areas with greater than 30% closed canopy dominated by broadleaf trees, and at an altitude between 800 and 1000 m
Subtropical evergreen broadleaf forest (SEBF)	Forested areas with greater than 30% closed canopy dominated by broadleaf trees, and at an altitude greater than 1000 m
Other forest (OF)	Forest with greater than 30% closed canopy dominated by conifer trees or bamboo
Rubber plantations (RP)	Forested areas with trees clearly planted in rows and a homogeneous canopy. Deciduous during the dry season
Arable lands (AL)	Shifting cultivation or permanent agriculture (e.g. slash and burn, tea plantation, fallow lands, grassland and paddy rice)
Shrublands (SL)	Land covered by secondary growth or highly degraded forest areas with <30% tree cover
Other land-use (OL)	Other land uses include: urban areas, water and sand. Together these areas accounted for less than 1% of the area

pixels in one thematic class. The "elimination" option removes small clumps (<8 pixels) by replacing them with the value of nearby larger clumps. The "3 × 3 filter" option was used to smooth the classified images.

The transformed images were then exported to ArcView GIS (Version 3.3, Environmental Systems Research Institute, Redlands, USA). In ArcView, the images were converted to a grid format, and then to a shape format. The new polygon themes were exported to ArcGIS (Version 8.3, Environmental Systems Research Institute, Redlands, USA) and polygons <1 ha were eliminated in ArcGIS (e.g. merging the selected polygons with neighboring polygons with the largest shared border). This final transformation was necessary to minimize any classification errors due to differences in the resolution of the three satellite images. These polygon themes were used for the analyses to derive land-use/land-cover changes and spatial distribution of each land-use/land-cover type in various elevations.

Accuracy assessment

The accuracy of our classification was verified by ground-truthing. Specifically, we compared our classification of the 2003 ETM image with field observations in December 2004. A total of 286 points were verified. In each point, we determined the current land-use cover, determined the location using a global positioning system (GPS), and took a photograph of the site. The field observations were then referenced to the classification to assess the overall accuracy and the accuracy of the different land-use categories.

The accuracy of our classification was greater than 90% for all land-use classes, with the exception of shrublands (Table 2). We correctly classified shrublands 66.7% of the time. The errors included classifying shrublands as forest, other land-use, and arable lands. The major explanation for this problem is that in Xishuangbanna shrublands are a transition land-use class between arable land and forests. In the period, between the time the image was taken and field observations a shrubland could be cleared and converted to arable land or the vegetation could have grown sufficiently to be considered forest. Rubber plantations were occasionally classified as grassland (arable lands) because they have similar spectral properties during the dry season when the rubber plantations are deciduous.

Table 2 Accuracy assessment of land-cover classification of the 2003 ETM image. A total of 286 points were verified with field observations

Field observations	Image classification				
	Forest	Rubber plantations	Arable lands	Shrublands	Other land-use
Forest	67			5	
Rubber plantations		64	1		
Arable lands	1	6	77	7	
Shrublands	1	1		24	
Other land-use			1		31
Total	69	71	79	36	31
Accuracy (%)	97.1	90.1	97.5	66.7	100.0

Results

Land-cover area and change

Land cover of Xishuangbanna changed dramatically between 1976, 1988, and 2003 (Fig. 3). The most obvious change was the decrease in forest cover and an increase in rubber plantations. In 1976, forests covered approximately 70% of Xishuangbanna, but by 2003 they covered less than 50% (Figs. 3, 4). During this period, the area of tropical seasonal rain forest was reduced by 67% (139,576 ha). This forest type occurs below 800 m, and covered 10.9% of the area in 1976, but only 3.6% by 2003. A large proportion of this change was due to the conversion of this forest type to rubber plantations. The establishment of rubber plantations began in the 1950s, but by 1976 they only occupied 1.1% of the area; by 2003 they increased to 11.3% (Figs. 3, 4).

The area of mountain rain forest, subtropical evergreen broadleaf forest and other forests also decreased between 1976 and 2003 (Figs. 3, 4). The area covered by mountain rain forest and subtropical evergreen broadleaf forest decreased by approximately 20% (205,847 ha) between 1976 and 2003. The expansion of rubber plantations was not a major cause of the loss of these forests because they occur above the elevation range most appropriate for rubber plantations (i.e. 500–800 m). The major cause of forest loss was the conversion to arable lands and shrublands.

In this region, the land-cover categories of arable lands and shrublands are very dynamic. Between 1976 and 1988, arable lands increased from 17.7% to 22.8%, but between 1988 and 2003, they decreased to 19.8%. More importantly, 60% of the area classified as arable lands in 2003 originated from another land-cover category in 1976. The increase in arable lands during the first period was mainly due to forest conversion to slash and burn agriculture, while the subsequent decrease was primarily due to the Natural Forest Protection Plan (NFPP, 1998). This plan encouraged an increase in forest cover, particularly in areas of steep slopes. The NFPP contributed to an increase in rubber plantations at higher elevation, and an increase in shrublands following the abandonment of slash-and-burn agriculture. Similar to the pattern observed in arable lands, 80% of shrublands in 2003 originated from another land-cover category in 1976. Along with a high turnover rate, shrublands increased from 11.6% to 18.4% between 1976 and 2003, and much of this increase was at the expense of subtropical evergreen broadleaf forest.

The area of "other land-use" also increased from 0.4% to 0.7% between 1976 and 2003, mainly because of increasing urban cover.

Land-cover change at different elevations

Elevation is an important factor influencing land-use change in Xishuangbanna (Fig. 5). Forest, shrublands, and arable lands occurred across a large range of elevation, but rubber plantations were primarily limited to the lower elevations. In 1976, forest was the dominant land cover category at all elevations, but by 2003 rubber plantations dominated areas below 800 m. Between 1976 and 2003, forest area decreased at all elevations, with the largest decreased below 1300 m. For the period 1976 to 1988, the expansion of rubber plantations mainly occurred between 600 and 800 m. The area of rubber plantations continued to expand between 1988 and 2003, and there was a marked increase in areas at higher elevations. For

[164]

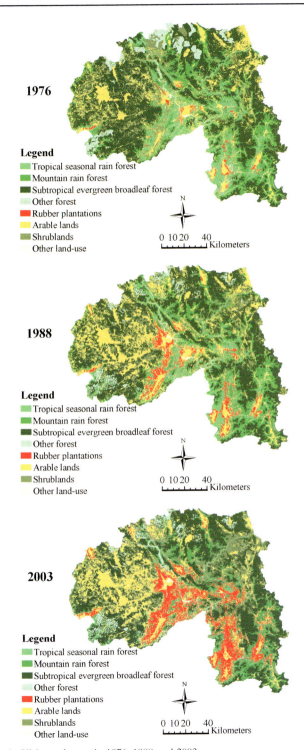

Fig. 3 Land use in Xishuangbanna in 1976, 1988 and 2003

example, in 2003 the area of rubber plantations above 1000 m was seven times greater than in 1988.

An increase in arable land and especially shrublands between 800 and 1300 m also reduced forest cover. From 1976 to 1988, arable lands area expanded at most elevation, but this trend changed between 1988 and 2003 when arable land below 800 m was converted to rubber plantations. There was little change in the elevational distribution of shrublands between 1976 and 1988, but by 2003 there was a large increase in shrubland area between 700 and 1300 m.

Land-cover change in the natural reserves

In 1980, five natural reserves were established in Xishuangbanna, and together they cover approximately 12% of total area. The major land-cover classes within the reserves were: subtropical evergreen broadleaf forest (48%), mountain rain forest (31%), and tropical seasonal rain forest (8%) (Fig. 6). Since the reserves were established there has been a decrease in the cover of mountain rain forest, tropical seasonal rain forest, and arable lands. In contrast, there has been a slight increase in subtropical evergreen broadleaf forest, shrublands and rubber plan-

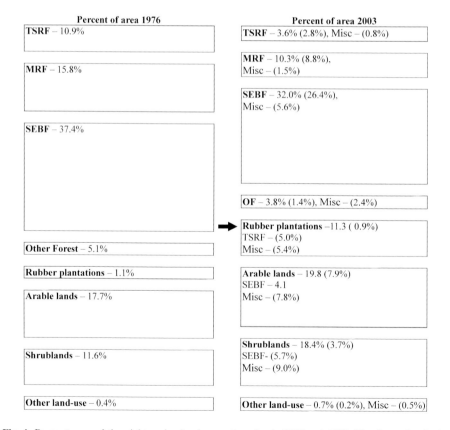

Fig. 4 Percent area of the eight major land-use categories in 1976 and 2003. The first value is the total percent of area for each land use in 1976 and 2003. The values in brackets are the source of the area in 1976. See Table 1 for abbreviations

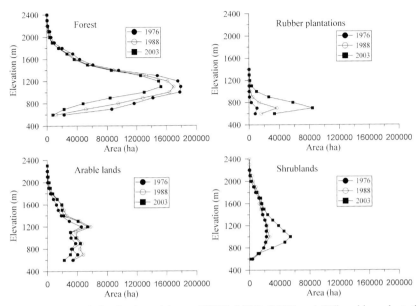

Fig. 5 The elevational distribution of forest (TSRF, MRF, SEBF and OF), rubber plantations, arable lands and shrublands in 1976, 1988, and 2003

tations. Since 1976, the area in shrublands has increased by approximately 15,000 ha, in part reflecting the decrease in arable lands. In 1976, there were no rubber plantations in these areas, but by 2003 they covered 4,310 ha within the reserves.

Discussion

Implications of land-use/land-cover change

Patterns of land-use change varied with elevation due to interactions between environmental limitations and changing economic and land-use policy. The most dramatic change in Xishuangbanna between 1976 and 2003 was the conversion of high diversity tropical seasonal rain forest (below 800 m) to monospecific rubber

Fig. 6 Land-cover change in natural reserve in 1976, 1988, and 2003. See Table 1 for abbreviations

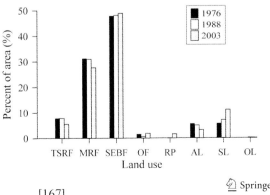

plantations. This transformation has greatly reduced the local plant species richness, and changed plant species composition (Zhu et al. 2004). For example, *Barringtonia macrostachya*, the dominant species in tropical seasonal rain forest, has disappeared from most fragmented forests (Zhu et al. 2004). This transformation also eliminated habitat for many animal species including the Asian elephant (*Elephas maximus*) and tiger (*Panthera tigris tigris*), whose distribution in China is limited to these forests in Xishuangbanna (Xu 2004). Bird species richness has also decreased with the conversion of the tropical rain forest into rubber plantations (Yang et al. 1985).

At higher elevations (>800 m), the area covered by forests (i.e. mountain rain forest and subtropical evergreen broadleaf forest) also was reduced by the conversion to arable lands, shrublands, and more recently to rubber plantations. A series of socioeconomic factors and national policies have lead to the complex land use pattern in these areas. First, the combination of population growth and the expansion of rubber plantations at lower elevations resulted in the expansion of agriculture into higher elevations. This increase in arable lands was most obvious during the 1976–1988 period. During this period, the Household Responsibility Policy provided land to individual households, and much more land was cleared than could actually be worked. Most crops were used for local subsistence, and included many varieties of upland rice, maize, beans, cotton, and many minor crops (Gao 1999). During the period 1988 to 2003, the Sloping Land Conversion Program enacted in 1999 redirected land-use patterns. The goal of this nation-wide program was to increase forest cover on steep slopes to reduce soil erosion and sedimentation. In Xishuangbanna, the policy mainly led to the conversion of arable lands to shrublands, but there was also a shift from local food crops to export crops, such as sugar cane, sun tea, and rubber plantations. Because rubber plantations were classified as "forests" large areas of arable land were converted to plantations, and this explains the increase in plantations at higher elevations.

While increasing forest cover usually implies an improvement in watershed protection, our analysis indicates that this needs further assessment. By dedicating the best agricultural lands for rubber plantations, other areas, usually at higher elevations and on steep slopes, are being deforested for local food production. These areas are often the most vulnerable to soil erosion, and, in general agricultural cover provide little protection for soil erosion. These agricultural activities on marginal lands, plus the open understory of the rubber plantations have increased levels of soil erosion in the region (Wang et al. 1982; Zhang et al. 1997).

Furthermore, the conversion of tropical rain forest into rubber plantations appears to be affecting the local climate, which could influence the long-term future of tropical seasonal rain forest in Xishuangbanna. Tropical seasonal rain forest occurs at its northern limit in Xishuangbanna, and the conditions are drier and cooler than in the rest of the distribution of this forest type. Zhu (1997) has argued that the foggy conditions are the main reason that the forest occurs in this region. Unfortunately, in Jinghong County, an area with a high density of rubber plantations, the number of foggy days has decreased from 166 per year in the 1950s to less than 60 in the 1990s (Gong and Ling 1996). This change has been attributed to the difference in water relations between rubber plantations and natural forest (Huang et al. 2000; Liu et al. 2003). If changes in fog conditions continue, this could result in the decline of many species (Burgess and Dawson 2004; Liu et al. 2004).

Although the government has set aside approximately 12% of the province in reserves, forest cover within the reserves has declined over the last 20 years, and the reserves are virtual islands in a matrix of different land-use practices. In a 5-km buffer around the reserves, most of the area has been converted to rubber plantations or arable lands (Li, unpublished data). Furthermore, because the reserves are isolated and there is little or no appropriate habitat connecting them, the system is not suitable for the migration of large animals such as elephants. In addition, the presence of villages within and around the reserves, has led to conflicts with elephants. The relatively small areas of the reserves result in elephants frequently leaving in search for food, such as bamboo and wild banana, and this often results in damage to crops and even human injury (Xu 2004).

Conservation and management recommendations

It is no surprise that current land use changes involving replacement of tropical forest with rubber plantations will negatively impact the flora, fauna, and some ecosystem services, but economic development, particularly in the developing world, is necessary to help millions of people out of extreme poverty. The challenge is to try to balance the economic and social needs of a growing population, with the habitat requirements of a highly diverse flora and fauna, and the maintenance of ecosystem services. An important first step to assist in determining appropriate strategies is to provide citizens, conservationists, and decision-makers with the best scientific information on land-use dynamics and the implications for conservation. In addition to providing the information, we believe that it is essential to provide recommendations to help decision-makers in understanding and interpreting these results.

In Xishuangbanna, rubber plantations are a critical component of the local economy, and given the increasing demand for natural rubber we can only assume that there will be a continued pressure to establish more plantations. However, we believe that there are strategies that can improve the conservation efforts of the region, without creating barriers to economic development. The area of high-diversity tropical seasonal rain forest has almost been eliminated, and the areas that remain are highly fragmented. Future expansion of rubber plantations below 800 m should be stopped, and restoration efforts that expand bamboo forest along rivers should be promoted to create corridors among the existing stands of tropical seasonal rain forest. In addition, resources should be invested in promoting plantation management strategies, which could increase rubber production. These may include technological advances in breeding, fertilizer applications, or tapping methods.

Areas that are presently dominated by shrublands between 800 and 1200 m may be the most appropriate areas for future rubber plantations expansion. To reduce the negative effects of erosion in the monospecific plantations, we would encourage the establishment of multispecies plantations. One option is to plant tea trees in the understory. Shade tea has become very profitable, it provides habitat for many bird species (Wang and Young 2003), and rubber plantations with a tea understory have been shown to be very productive (Xie 1989). A second approach would be to encourage secondary regeneration in the understory. In Brazil, rubber plantations with understory vegetation had no negative effect on rubber production and actually prolonged latex flow (Schroth et al. 2003). Furthermore, these additional species provided other forest products and reduce erosion.

🕮 Springer

The growing economies of many developing countries are presenting new challenges for conservation. If we are to balance the need for economic development and preserve the biodiversity of these regions, scientists must produce reliable information on land-use dynamics and the implications of these dynamics. But, most importantly, we must work directly with national and local decision-makers to insure that the most appropriate strategies are enacted.

Acknowledgements   This project was funded by the National Natural Science Foundation of China (30570321 and 30570308). TMA received funds from the University of Puerto Rico through the FIPI program. We thank Z. F. Guo and Z. W. Cao who assisted with early stages of geographic information system analysis and fieldwork. Y. H. Liu provided invaluable assistance with data collection. We offer special thanks to R. Grau, J. Thomlinson, X. Velez, and W. J. Liu for their helpful comments on the manuscript.

References

Burgess SSO, Dawson TE (2004) The contribution of fog to the water relations of Sequoia sempervirens (D. Don): foliar uptake and prevention of dehydration. Plant, Cell Environ 27:1023–1034

Cao M, Zhang JH (1997) Tree species diversity of tropical forest vegetation in Xishuangbanna, SW China. Biodiv Conserv 6:995–1006

Cao M, Zhang JH, Feng ZL, Deng JW, Deng XB (1996) Tree species composition of a seasonal rain forest in Xishuangbanna, Southwest China. Trop Ecol 37:183–192

Casson A (2003) Oil palm, soybeans & critical habitat loss. A review prepared for the WWF forest conversion initiative. Data set, Available on-line [http://www.questions.panda.org/downloads/forests/oilpalmsoybeanscriticalhabitatloss25august03.pdf]from the Coordination Office Hohlstrases 110 Hohlstrasse 110 CH–8010 Switzerland.

Chandler W, Gwin H (2004) China's energy and emissions: a turning point? Data set, Available on-line [http://www.pnl.gov/aisu/pubs/chandgwin.pdf] from Pacific Northwest National Laboratory, Richland, Washington

Edit committee of Xishuangbanna Dai Autonomous Prefecture Difangzhi (2002) Xishuangbanna Dai Autonomous Prefecture zhouzhi. Xinhua Publishing House, Beijing

Food and Agricultural Organization of the United Nations (FAO) (2000) Global forest resources assessment 2000. Data set, Available on-line [http://www.fao.org/documents/show_c-dr.asp?url_file=/DOCREP/004/Y1997E/y1997e06.htm]from FAO Forestry Paper 140, United Nations Food and Agriculture Organization, Rome, Italy

Gao LS (1999) On the Dais's traditional irrigation system and environmental protection in Xishuangbannna.Yunnan Nationality Press, Kunming

Gong SX, Ling SH (1996) Fog decreasing in Xishuangbanna region. Meteorology 22:10–14

Grau HR, Aide TM, Gasparri NI (2005a) Globalization and soybean expansion into semiarid ecosystems of Argentina. Ambio 34:265–266

Grau HR, Gaspari NI, Aide TM (2005b) Agriculture expansion and deforestation in seasonally dry forests of north-west Argentina. Environ Conserv 32:140–148

Guo HJ, Padoch PC, Coffey K, Chen AG, Fu YN (2002) Economic development, land use and biodiversity change in the tropical mountains of Xishuangbanna, Yunnan, Southwest China. Environ Sci Policy 5:471–479

Guo YQ, Yang YM, Tang JS, Chen SW, Lei FG, Wang JH, Yang ZH, Yang GZ (1987). The vegetation in the natural reserves of Xishuangbanna. In: Xu YC, Jiang HQ, Quan F (eds) Proceedings of synthetical investigation of Xishuangbanna nature research. Yunnan Science and Technology Press, Kunming, pp. 88–169

Huang YR, Huang YS, Li ZH, Cheng BJ (2000). The influence of ecoenvironmental variation on fog. Sci Meteorol Sin 20:129–135

Li JL, Zhang YH (2004) Forecast for natural rubber consumption in China. Technoecon Manage Res 2:55–56

Liu WJ, Zhang YP, Liu YH, Li HM, Duang WP (2003) Comparison of fog interception at a tropical seasonal rain forest and a rubber plantations in Xishuangbanna, Southwest China. Acta Ecol Sin 23:2379–2386

Liu WJ, Meng FR, Zhang YP, Liu YH, Li HM (2004) Water input from fog drip in the tropical seasonal rain forest of Xishaungbanna, South-West China. J Trop Ecol 20:517–524

Myers N, Mittermeier RA, Mittermeier CG, da Fonseca GAB, Kent J (2000) Biodiversity hotspots for conservation priorities. Nature 403:853–858

Schroth G, Coutinho P, Moraes VHF, Albernaz AL (2003) Rubber agroforests at the Tapajos river, Brazilian Amazon—environmentally benign land use systems in old forest frontier region. Agricult, Ecosyst Environ 97:151–165

Sinton JE, Stern RE, Aden NT, Levine MD, Dillavou TJ, Fridley DG, Huang J, Lewis JI, Lin J, McKane AT, Price LK, Wiser RH, Zhou N, Ku J (2005) Evaluation of China's energy strategy options. Data set, Available on-line [http://www.china.lbl.gov/publications/nesp.pdf] from Prepared for and with the support of the China Sustainable Energy Program

Tuan FC, Fang C, Cao Z (2004) China's soybean imports expected to grow despite short-term disruptions. Data set, Available on-line [http://www.ers.usda.gov/publications/OCS/Oct04/OCS04J01/ocs04j01.pdf] from Report OCS-04J-01, U.S. Department of Agriculture, Economic Research Service

Wang HH, Ma WJ, Deng ZZ (1982) The relationship between exploitation of tropical rain forest and water, soil conservation. Sci Silvae Sin 18:245–257

Wang ZJ, Young SS (2003) Differences in bird diversity between two swidden agricultural sites in mountainous terrain, Xishuangbanna, Yunnan, China. Biol Conserv 110:231–243

Wong J, Chan S (2003) Why China's economy can sustain high performance: an analysis of its sources of growth. Data set, Available on-line [http://www.chathamhouse.org.uk/pdf/research/asia/economy%20can%20sustain%20chan.pdf] from Asia Programme Working Paper, No.6. Asia programme Royal Institute of International Affairs

Wu ZY, Zhu Y, Jiang H (1987) The vegetation of Yunnan. Science Press, Beijing

Xie JW (1989) Study on the productivity of an artificial rubber-tea community in tropical China. Ms. Thesis, Kunming Ecological Institute, Chinese Academy of Sciences

Xu ZF (2004) Approach on ecological effects of Asian elephants isolated from the distribution of bamboo-wild banana and its conservation strategy. Chinese J Ecol 23:131–134

Yang L, Pan LR, Wang SZ (1985) Inverstation of bird in tea garden and rubble plantations in Xishaungbanna. Zool Res 6:353–360

Yin XQ (2001) Summary for forum of eco-environment protection and control & sustainable development. Commun Ecol Econ 125:10–16

Zhang JH, Cao M (1995) Tropical forest vegetation of Xishuangbanna, SW China and its secondary changes, with special reference to some problems in local nature conservation. Biol Conserv 73:229–238

Zhang KY (1988) The climatic dividing line between SW and SE monsoons and their differences in climatology and ecology in Yunnan Province of China. Climatol Notes 38:197–207

Zhang YP, Zhang KY, Ma YX, Liu YH, Liu WJ (1997) Runoff characteristics of different vegetation covers in tropical region of Xishuangbanna, Yunnan. J Soil Erosion Soil Water Conserv 3:25–30

Zhu H (1997) Ecological and biogeographical studies on the tropical rain forest of south Yunnan, SW China with a special reference to its relation with rain forests of tropical Asia. J Biogeogr 24:647–662

Zhu H, Xu ZF, Wang H, Li BG (2004) Tropical rain forest fragmentation and its ecological and species diversity change in Southern Yunnan. Biodiv Conserv 13:1355–1372

Biodivers Conserv (2007) 16:1747–1759
DOI 10.1007/s10531-006-9057-2

ORIGINAL PAPER

Comparing conservation priorities for useful plants among botanists and Tibetan doctors

Wayne Law · Jan Salick

Received: 14 June 2005 / Accepted: 24 April 2006 / Published online: 9 July 2006
© Springer Science+Business Media B.V. 2006

Abstract Perspectives of diverse constituencies need to be incorporated when developing conservation strategies. In *Menri* (Medicine Mountains) of the Eastern Himalayas, Tibetan doctors and professional botanists were interviewed about conservation of useful plants. We compare these two perspectives and find they differ significantly in conservation priorities (Wilcoxon Signed Ranks $P < 0.05$), both in how they prioritized, as well as the priorities themselves. Tibetan doctors first consider which plants are most important to their medical practice and, then secondarily, the conservation status of these plants. Additionally, perceptions of threatened medicinal plants differ among Tibetan doctors who received medical training in Lhasa, who were local trained, and who were self-taught. In contrast, professional botanists came to a consensus among themselves by first considering the conservation status of plants and then considering use. We conclude that, in order to effect community based conservation, opinions from both Tibetan doctors and professional botanists should be considered in establishing conservation priorities and sustainable conservation programs. Furthermore, we set our own research agenda based on combined perspectives.

Keywords Conservation · Tibetan medicine · Threatened plants · Useful plants · Tibetan doctors

Introduction

In response to inadequate, exclusionary conservation policies, the World Conservation Union (1980) urged a shift in the planning and management of natural

W. Law
Department of Biology, Washington University, Campus Box 1137, St. Louis, Missouri 63130, USA

J. Salick · W. Law (✉)
Missouri Botanical Garden, P.O. Box 299, St. Louis, Missouri 63166, USA
e-mail: wlaw@biology2.wustl.edu

resources to include local communities. Since then, there has been growing recognition that local communities need to play an active role if biodiversity is to be preserved (Kellert 1985; Fletcher 1990; Gadgil 1992). However, the success of community based conservation (CBC) projects has been mixed, with performance falling short of expectations (Barett et al. 2001). But many of these programs have been unsuccessful because of two main factors: the mixed objectives of conservation and development (Redford and Sanderson 2000) and improper implementation (Songorwa 1999; Murphree 2002). These factors often lead to unfavorable perceptions by local communities often discouraging participation (Mehta and Kellert 1998).

The goals of many CBC programs have been to socially and economically develop rural areas along with biodiversity conservation. Unfortunately, too often, programs have primarily focused on development for economic growth, and assumed that environmental solutions would arise on their own since communities could invest in more resource efficient ways of life (Adams et al. 2004). Alternatively, some development strategies have prioritized high-productivity agriculture over sustainable management of existing resources. Because varying local priorities are often not taken into consideration, receptivity of these development strategies is low or support is not maintained (Marcus 2001). In fact, in planning and implementing CBC projects, programs have often not included local opinions, knowledge, or priorities, and in the worst cases, have primarily used local people as laborers (Songorwa 1999). These types of projects do not earn support from local people since their needs, which vary from community to community, are often not met. Berkes (2004) states that we must utilize the research within interdisciplinary fields that links human and nature and provides a more sophisticated understanding of social–ecological knowledge and a better insight into CBC.

Central to these emerging conservation fields is traditional knowledge (TK).

"Traditional knowledge is a cumulative body of knowledge, know-how, practices and representations maintained and developed by peoples with extended histories of interaction with the natural environment. These sophisticated sets of understandings, interpretations and meanings are part and parcel of a cultural complex that encompasses language, naming and classification systems, resource use practices, ritual, spirituality and worldview." International Council for Science (2002).

TK of the resource users themselves has been advocated for ecosystem management (Johannes 1978, 1998; Olsson and Folke 2001; Salick et al. 2005). Numerous studies show why and how indigenous knowledge and people can be part of sustainable conservation (e.g., Bennett 1992; Brosius 1997; Salick et al. 2004). In contrast, *not* taking into account TK can hinder formal conservation efforts (Etkin 2002; Chapin 2004).

Local peoples' extensive knowledge of a local area can often surpass some aspects of scientific knowledge if scientists do not reside in that area. Traditional knowledge, passed on from generation to generation, is derived from a close relationship with an environment upon which people depend for their livelihood (Ohmagari and Berkes 1997). TK is frequently very different from scientific knowledge, but when considered together, can be complementary (Berkes et al. 2000).

This is not to downplay the importance or necessity of scientific knowledge. Perspectives of outside experts are also valuable for conservation, since scientists can provide detailed information about plants and animals, communities, and ecosystems from a larger, regional perspective, often synthesizing information from many sources and providing an overview for the local situation (Sheil and Lawrence 2004).

When community based conservation is informed by TK and properly integrated with scientific knowledge, conservation can be effective (Campbell and Vainio-Mattila 2003).

We examine this relationship between local knowledge and that of outside experts at Mt. Khawa Karpo, NW Yunnan, China. We work with academically trained botanists and with local Tibetan doctors from different cultural settings to determine which medicinal plants and fungi are both most valued and threatened. We analyze the differences among Tibetan doctors with different backgrounds in what plants and fungi they deem valuable. Then we evaluate how views of threatened useful plants and fungi differ between local Tibetan doctors and outside expert botanists.

Study area and peoples

Khawa Karpo, the earthly image of the Tibetan warrior god, is one of the eight sacred mountains in Tibet. It lies on the border of NW Yunnan and Tibet in the easternmost Himalayas (Fig. 1). This area is within a biodiversity "hotspot" defined by both WWF/IUCN (Mackinnon et al. 1996) and Conservation International (Mittermeier et al. 1998). It is also home to more than 10 ethnic groups composed of

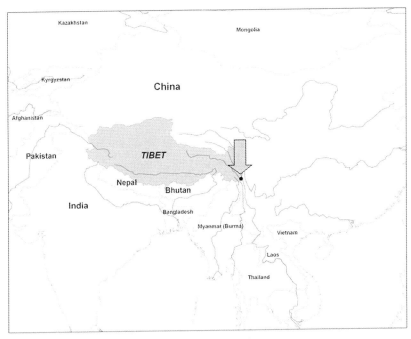

Fig. 1 Khawa Karpo area where this study was conducted

Springer

over five million people, with Tibetans in this area comprising approximately 80% of the population (Xu and Wilkes 2004).

Tibetan doctors (Fig. 2) in this area receive training depending on their background. Tibetan monks are often trained in a formal medical system emanating from Lhasa, as well as in traditional Tibetan Buddhism. Of the seven Tibetan doctors practicing medicine around Khawa Karpo, one is a monk and learned medicinal plants and fungi from monastic training. Four of the doctors in our study, which we will refer to as locally trained, were educated locally by medicine men from around Khawa Karpo whose knowledge includes both training in Lhasa and local knowledge. Finally, there were two local village doctors who started as market collectors, who we will call self-trained.

Methods

After receiving prior informed consent, we conducted a semi-structured, open-ended interview in English, with seven local Tibetan doctors practicing in the Khawa Karpo

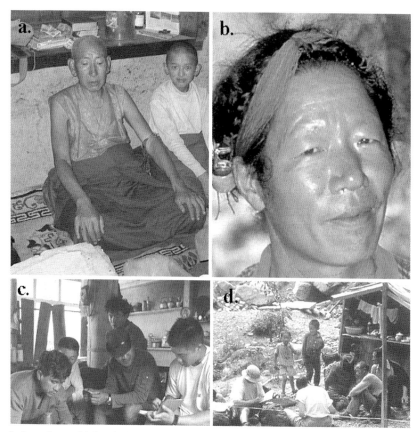

Fig. 2 Interviews with local Tibetan doctors including (**a**) Lhasa trained monk (left), here shown mentoring living Buddha (right), (**b**) locally trained doctor, (**c**) locally trained doctor (left), here with first author (right), and (**d**) self-trained doctors (2 people on right)

⩗ Springer

area (Fig. 2) and three academically trained Chinese botanists most familiar with the plants and fungi and people of NW Yunnan province. Both groups were requested to list 20 useful plants with populations that are threatened in the Khawa Karpo area. They informed us of (1) the name of the plant, if possible, both in Tibetan and Chinese, (2) plant use, (3) what part of the plant is used, and (4) the elevation and habitat where the plant can be found.

Interviews differed somewhat due to literacy and language. Tibetans were interviewed in Tibetan using a translator, while botanists were interviewed in English. With doctors, after listing 20 plants, pile sorting was used to distinguish degrees of usefulness and threat. First they separated the medicinal plants and fungi into three categories (1, less useful; 2, useful; 3, most useful). Then they separated these plants and fungi into three categories of threat (1, not threatened; 2, threatened; 3, very threatened). Finally, we asked the doctors to rate the list of plants and fungi given by the botanist into three threat categories (1, not threatened; 2, threatened; 3, very threatened). Scientific plant names were confirmed using (a) medicinal books used by the doctors, (b) *Diqing Zang Yao* (Yang and Chuchengjiancuo 1989), and/or (c) *The Wildflowers in Hengduan Mountains in Yunnan China* (Fang 1993).

We compared the list of plants and fungi and conservation threats identified by Tibetan doctors and by expert botanists. Because data where non-normally distributed, we used the non-parametric, Wilcoxon Rank Sum (SPSS, v.11.0.1 Inc. 2001). In order to look at the similarity of Tibetan medicinal training, we used Nonmetric Multidimensional Scaling with Jaccard Distance Measure (NMS; PC-ORD v.4.20; (McCune and Medford 1999) to group doctors by similarities of response by which plants were considered most useful. We statistically compared the parts of plants used, ailments cured, and elevation using likelihood ratio tests in JMP (SAS Institute, version 5.1 2003).

Results

The first result we observed was the differing process by which the plant lists were compiled by the two groups, which in itself is indicative of differing orientations. Botanists worked together to form a single list of threatened useful plants by consensus, considering threat status first and use secondarily, and in the end limited their list to only 17 species upon which they all agreed. Working independently, Tibetan doctors did the reverse, listing useful plants first (having trouble limiting themselves to 20) and then ranking threat secondarily. Botanists named plants and fungi with different uses including medicines and ornamentals (see appendix). In contrast, Tibetan doctors named 20 plants and fungi each that were used medicinally (see appendix). Comparing, we see that only 4 of the plant species occurred on both lists, *Pegaephyton scapilflorum*, *Fritillaria delavayi*, *Fritillaria cirrhosa*, and *Saussurea laniceps*, of which the latter three are popularly marketed species.

The botanists presented a list of useful plants and fungi, *all* of which they thought were most threatened (threat level 3). The doctors' evaluation of the botanists' list of threatened plants and fungi indicated that there were significant differences in opinions of threat status between doctors and botanists (Fig. 3; $Z = -5.03$, $P < 0.0001$).

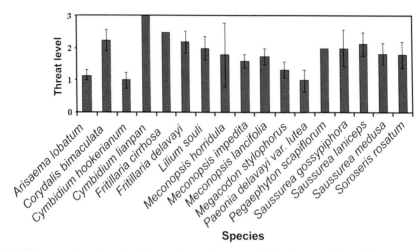

Fig. 3 Threat ratings given by Tibetan doctors (mean ± SE) for botanists' list of threatened and useful plants and fungi. Botanists rated all these plants and fungi as very threatened (value of 3). Tibetan doctors' rankings are significantly different from those of the botanists ($P = 0.001$, Wilcoxon Signed Ranks test)

Fig. 4 Tibetan doctors' most useful medicinal plants and fungi are ordered by Non-metric Multidimensional Scaling (NMS). Doctors with similar training are represented with the same symbol: local doctors with *triangles*, self-trained locals with *diamonds*, and the monk with a *cross*. Only axis one was significant, suggesting the coherent nature of local Tibetan medicine

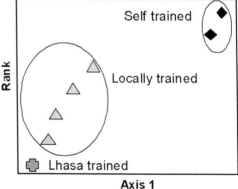

Non-metric Multidimensional Scaling (NMS) of Tibetan doctors' plant lists reveal which doctors were more similar in their opinions (Fig. 4; only Axis 1 was significant ($P = 0.0196$), with a final stress of 18.475). Doctors who were trained similarly held similar opinions on most useful plants and were thus close to each other on the ordination. Whole plants and roots are the plant parts most commonly used by these doctors (Fig. 5a; $\chi^2 = 195.8$, df $= 9$, $P < 0.0001$). Infection, digestion, and circulatory problems are the most commonly treated ailments with these plants (Fig. 5b; $\chi^2 = 189.3$, df $= 16$, $P < 0.0001$). These medicinal plants are most often found between 2,900 m and 3,500 m (Fig. 6; $\chi^2 = 28.5$, df $= 5$, $P < 0.0001$).

Discussion

Botanists and Tibetan doctors have very different perspectives, priorities, and ratings for conservation of threatened useful plants. When asked to list

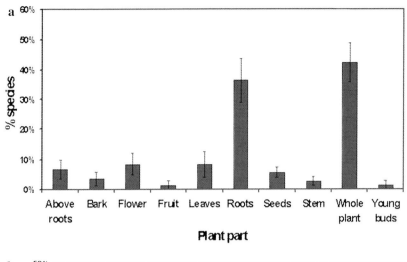

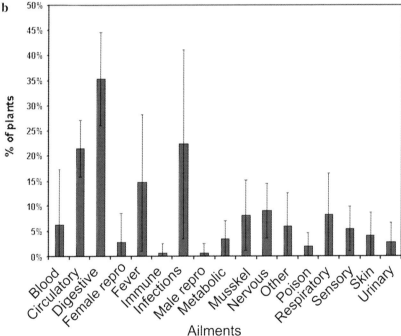

Fig. 5 Characteristics of Tibetan doctors' 20 most useful plants and fungi. (**a**) Plant parts used by the Tibetan doctors (mean frequencies ± SE). Whole plants and roots are most commonly used for medicines (χ^2 = 195.8, df = 9, P < 0.0001). (**b**) Ailments treated (mean frequencies ± SE). Infections and digestive and circulatory problems are the most commonly treated ailments (χ^2 = 189.3, df = 16, P < 0.0001)

threatened useful plants, botanists consider plant threat first and foremost, and medicinal and horticultural uses second; for Tibetan doctors, medicinal use is primary, followed by a more moderate view of plant threat. These results are

�（ Springer

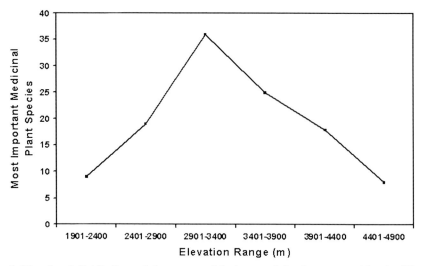

Fig. 6 Elevational distributions of the most useful plants and fungi as reported by the Tibetan doctors. Most of these plants and fungi are found at approximately 3,000 m (χ^2=28.5, df = 5, $P < 0.0001$), the elevation that the doctors inhabit

most likely explained if we consider that botanists want to make sure that plants and fungi are conserved above all, while doctors primarily want to assure that plants and fungi are available for use. Doctors are in favor of regulating commercial collection but want to avoid restricting their own careful harvest of a limited amount of their valuable medicines. This is in response to the growing popularity of traditional medicines, which has caused a shift of sustainable, low level harvesting to a more intense, commercially driven style of collection by plant harvesters. The extreme differences in species identified for conservation (only 4 plants in common) between botanists and Tibetan doctors reflect these different perspectives.

Additionally, there is a difference in geographic scale of evaluation: the botanists assessed plants and fungi threatened at a regional scale, while Tibetan doctors evaluated at a local scale. Some plants and fungi, threatened on a regional scale, may have a refuge near Khawa Karpo, since the *Menri*, Medicine Mountains, have been recognized for centuries for their wealth of medicinal plants.

Most of the threatened medicinal plants that were agreed upon by doctors and botanists are commonly collected for commercial sale. This highlights the pressure imposed by the booming trade in Tibetan medicines both within China and internationally (Olsen and Larsen 2003; Olsen and Bhattarai 2005; Law and Salick 2005). Traditional Tibetan doctors, like those interviewed in this study, observe traditional techniques for harvesting a small amount of herbs needed for their medical practices; however, lack of traditional constraints on rampant collectors serving export markets threaten populations of valuable medicinal herbs (Xu and Wilkes 2004). Conservation of medicinal plants is of the highest priority.

Among Tibetan doctors, views on medicinal plants and conservation varied; doctors similarly trained tended to share the same opinions on which plants and fungi they consider the most valuable. Nonetheless, it is interesting to note that these education systems do not differ greatly from each other, suggesting that local Tibetan medicine has a coherent base. However, medicinal plants vary significantly on a broader scale (Salick et al. 2006), for example only 16 of the 78 plants from this study appear in Kletter and Kriechbaum (2001), which concentrates on Lhasa based Tibetan medicines.

While the opinions of botanists and Tibetan doctors are different, they are both real assessments and both sets of opinions are extremely important in determining conservation needs in this area. These differences indicate the need to integrate both TK and scientific knowledge in conservation of rare plants and fungi. Tibetan doctors may be told they must protect a certain plant species that the botanist have deemed as threatened in other areas but that locally is abundant. If doctors feel these claims are invalid or are unjustified, without consultation and education about the situation, they may disregard the botanists' concerns, warnings, and recommendations. Reciprocally, botanists unfamiliar with the basis of local perceptions may not appreciate local opinions which come from doctors who are familiar with the ecosystems they inhabit. Unified strategies for the management of resources like medicinal plants must be sought. The lack of attention to the overexploitation of plants and animals in several Tibetan regions will threaten the ecology and livelihood of communities (Cardi 2005). Without careful integration into local community perceptions and practices, conservation guidelines may be resisted if they are not in the interest of the people that use the land. Conservation priorities need to take into account perspectives of both TK and science. Perspectives of TK should be carefully considered on a local basis, since priorities can vary depending on areas. Collaboration can provides awareness of the regional scale for local people, and awareness of local conditions for conservationists and botanists. Collaborative efforts such as these can result in exceptionally informative accounts (e.g., Lama et al. 2001).

Furthermore comparisons of TK and science can help identify species that are especially in need of more detailed research. We are concentrating research on *Saussurea laniceps* (snow lotus) based on the joint recommendations of Tibetan doctors and botanists. Snow lotus is over-collected and under great evolutionary pressure (Law and Salick 2005). Our population ecology studies are investigating sustainable harvesting strategies (Law et al. in preparation). With cooperation to identify threatened species like snow lotus, not only do we understand which organisms need to be studied, but we also generate local and conservation support for our efforts. As recommended by the World Conservation Union (1980), we promote joint efforts between conservation and traditional knowledge for Khawa Karpo—the warrior god, sacred mountain, and biodiversity hotspot.

Acknowledgements Support for this study was provided by NSF #408123, the Mellon Foundation, and The Nature Conservancy. We would also like to thank Jessica Woo, Norbu Cili, Luke Harmon, Bob Moseley, Denise Glover, Anja Byg, and all our esteemed informants for their help and guidance with this study.

Appendix

Appendix 1 List of threatened, useful plants and fungi produced by botanists and Tibetan doctors with descriptions of use categories and threat values. Mode was used for threat values of plants given by more than one Tibetan doctor

Botanists	Plant	Use categories	Threat value (Mode)
1	*Arisaema lobatum* Engl.	M-11, 13	3
2	*Corydalis bimaculata* C.Y. Wu & T.Y. Shu	M-3	3
3	*Cymbidium hookerianum* Rchb.f	O	3
4	*Cymbidium lianpan* T. Tang & F.T. Wang ex Y.S. Wu	O	3
5	*Fritillaria cirrhosa* D. Don	M-8, 9	3
6	*Fritillaria delavayi* Franch.	M-8, 9	3
7	*Lilium souliei* Franch.	M-8, 9	3
8	*Meconopsis horridula* Hook.f. & Thomson	M-7	3
9	*Meconopsis impedita* Prain	M-7	3
10	*Meconopsis lancifolia* Franch. ex Prain	M-7	3
11	*Megacodon stylophorus* (C.B. Clarke) Harry Sm.	M-5	3
12	*Paeonia delavayi* Franch. *var. lutea* (Delavay ex Franch.) S.G. Haw	M-15	3
13	*Pegaeophyton scapiflorum* (Hook.f. & Thomson) C. Marquand & Airy Shaw	M-9	3
14	*Saussurea gossypiphora* Wall.	M-1, 4, 9	3
15	*Saussurea laniceps* Hand.-Mazz.	M-1, 4, 9	3
16	*Saussurea medusa* Maxim.	M-1, 4, 9	3
17	*Soroseris rosularis* (Diels) Stebbins	M-1, 4, 9	3
Tibtean doctors			
1	*Cordyceps sinensis*	M-12	3
2	*Fritillaria cirrhosa* D. Don	M-12	3
3	*Aconitum tanguticum* (Maxim.) Stapf	M-11, 12	3
4	*Lagotis alutacea* W.W. Sm.	M-1, 9	3
5	*Pegaeophyton scapiflorum* (Hook.f. & Thomson) C. Marquand & Airy Shaw	M-7	3
6	*Pedicularis longiflora* Rudolph	M-3, 7	2
7	*Gentiana* sp. 1	M-3, 9	2
8	*Dracocephalum bullatum* Forrest ex Diels	M-10	2
9	*Gentiana urnula* Harry Sm.	M-1, 3, 6, 13	3
10	*Aconitum richardsonianum* Lauener	M-8	3
11	*Pedicularis przewalskii* Maxim.	M-3, 14	2
12	*Dracocephalum tanguticum* Maxim.	M-7, 10	2
13	*Rheum officinale* Baill.	M-1, 3	1
14	*Gymnadenia orchidis* Lindl.	M-7	1
15	*Corydalis* sp. 1	M-3, 10, 11	1
16	*Plantago depressa* Willd.	M-2, 14	1
17	*Swertia* sp. 1	M-3	1
18	*Acorus calamus* L.	M-10	3
19	*Lagotis* sp. 1	M-3, 6, 10	3
20	*Punica granatum* L.	M-3, 7	1
21	*Rhododendron* sp. 1	M-1, 4	1
22	*Sinolimprichtia* sp. 1	M-4	3
23	*Pterocephalus hookeri* (C.B. Clarke) L. Diels	M-3	2

Appendix 1 continued

Botanists	Plant	Use categories	Threat value (Mode)
24	*Phlomis* sp. 1	M-9, 13	1
25	*Inula racemosa* Hook.f.	M-3	3
26	*Crocus sativus* L.	M-1, 7	3
27	*Oxytropis reniformis* P.C. Li	M-8	1
28	*Pyrus pashia* Buch.-Ham. Ex D. Don	M-4	1
29	*Herpetospermum* sp. 1	M-3	1
30	*Gentiana* sp. 2	M-9	2
31	*Gentiana straminea* Maxim.	M-11, 13	3
32	*Saxifraga* sp. 1	M-3, 8	1
33	*Vladimiria souliei* (Franch.) Ling	M-3, 12	3
34	*Incarvillea compacta* Maxim.	M-8, 10	3
35	*Corydalis* sp. 2	M-3, 10	2
36	*Chrysosplenium carnosum* Hook.f. & Thomson	M-3, 10	2
37	*Phlomis younghushandii* Mukerjee	M-4	2
38	*Rhodiola crenulata* (Hook.f. & Thomson) H. Ohba	M-1	2
39	*Primula secundiflora* Franch.	M-1	1
40	*Meconopsis horridula* Hook.f. & Thomson	M-3, 14	2
41	*Gastrodia elata* Blume	M-1	3
42	*Gynura japonica* (Thunb.) Juel	M-10	3
43	*Panax japonicus* var. *major* (Burkill) C.Y. Wu & Feng	M-11, 12	3
44	*Coptis* sp.	M-3, 6	1
45	*Plantago asiatica* L.	M-5	1
46	*Verbena officinalis* L.	M-5	1
47	*Stephania delavayi* Diels	M-8	1
48	*Ainsliaea pertyoides* Franch.	M-7, 14	1
49	*Fritillaria delavayi* Franch.	M-1, 10	3
50	*Mahonia mairei* Takeda	M-3	1
51	*Aconitum vilmorinianum* Kom.	M-10	2
52	*Elaeagnus viridis* Serv.	M-3	2
53	*Dioscorea cirrhosa* Lour.	M-3	1
54	*Arisaema consanguineum* Schott	M-10	2
55	*Angelica sinensis* (Oliv.) Diels	M-10	3
56	*Saussurea laniceps* Hand.-Mazz.	M-10	3
57	*Corydalis yanhusuo* W.T. Wang	M-3, 9	2
58	*Hypecoum leptocarpum* Hook.f. & Thomson	M-3, 9	2
59	*Saxifraga* sp. 2	M-3	2
60	*Aconitum* sp.	M-7, 14	1
61	*Lagotis yunnanensis* W.W. Sm.	M-1	2
62	*Meconopsis torquata* Prain	M-7, 9, 14	3
63	*Galium* sp.	M-1, 9	2
64	*Aristolochia griffithii* Hook.f. & Thomson ex Duch.	M-3	1
65	*Delphinium* sp.	M-7	1
66	*Primula* sp.	M-10	1
67	*Pedicularis trichoglossa* Hook.f.	M-3	2
68	*Adhatoda vasica* Nees	M-1	2
69	*Sisymbrium heteromallum* C.A. Mey	M-3	1
70	*Halenia elliptica* D. Don/*Gentianopsis grandis* (Harry Sm.) Ma	M-3, 6	1

🖄 Springer

Appendix 1 continued

Botanists	Plant	Use categories	Threat value (Mode)
71	*Rhodiola* sp. 1	M-1, 3	1
72	*Lancea tibetica* Hook.f. & Thomson	M-3, 8	1
73	*Fragaria orientalis* Losinsk.	M-1, 14	2
74	*Nardostachys grandiflora* DC.	M-6, 9	3
75	*Polygonum* sp.	M-1	1
76	*Gentiana stipitata* Edgew. subsp.*tizuensis* (Franch.) T.N. Ho	M-1, 3, 6	2
77	*Taraxacum tibetanum* Hand.-Mazz.	M-4	2
78	*Thlaspi arvense* L.	M-7	1

Use categories include: O, Ornamental; M, Medicine. Subcategories for medicine: 1, Blood & circulatory system; 2, Dental; 3, Digestive system; 4, Reproduction & sexual health; 5, Urinary System; 6, Immune system; 7, Muscular–skeletal system; 8, Nervous system & mental health; 9, Respiratory system; 10, Sensory system; 11, Skin & related tissue; 12, Infections; 13, Poison; 14, Non-Western medical systems; 15, Belief systems

References

Adams WM, Aveling R, Brockington D, Dickson B, Elliott J, Hutton J, Roe D, Vira B, Wolmer W (2004) Biodiversity conservation and the eradication of poverty. Science 306:1146–1149

Barett CB,, Brandon K, Gibson C, Ghertsen H (2001) Conserving tropical biodiversity amid weak institutions. Bioscience 51(497–502):497–502

Bennett BC (1992) Plants and people of the Amazonian rainforests: the role of ethnobotany in sustainable development. Bioscience 42:599–607

Berkes F, Colding J, Folke C (2000) Rediscovery of traditional ecological knowledge as adaptive management. Ecol Appl 10(5):1251–1262

Berkes F (2004) Rethinking community-based conservation. Conserv Biol 18(3):621–630

Brosius JP (1997) Endangered forest, endangered people: environmentalist representations of indigenous knowledge. Human Ecol 25(1):47–69

Campbell LM, Vainio-Mattila A (2003) Participatory development and community-based conservation: opportunities missed for lesson learned? Human Ecol 31(3):417–437

Cardi F (2005) Evolution of Tibetan medical knowledge in the socio-economic context: the exploitation of medicinal substances among traditional doctors. Milan, Societa Italiana di Scienze Naturali

Chapin M (2004) Challenge to conservationists. World Watch Magazine November/December 2004

Etkin NL (2002) Local knowledge of biotic diversity and its conservation in rural Hausaland, Northern Nigeria. Econ Bot 56(1):73–88

Fang ZD (1993) The wildflowers in Hengduan Mountains in Yunnan China. Yunnan People's Publishing House, Kunming

Fletcher SA (1990) Parks, protected areas and local populations: new international issues and imperatives. Landscape Urban Plan 19:197–201

Gadgil M (1992) Conserving biodiversity as if people matter: a case study from India. Ambio 21(3):266–270

International Council for Science (Fensted JE, Hoyningen-Huene P, Hu Q, Kokwaro J, Nakashima D, Salick J, Shrum W, Subbarayappa BV) (2002) Science, traditional knowledge and sustainable development. ICSU Series on Sustainable Development, Paris

The World Conservation Union (1980) The world conservation strategy. The World Conservation Union, Gland, Switzerland

Johannes RE (1978) Traditional marine conservation methods in Oceania and their demise. Annu Rev Ecol Syst 9:349–364

Johannes RE (1998) The case of data-less marine resource management: examples from tropical near-shore finfisheries. Trends Ecol Evol 13:243–246

Kellert SR (1985) Social and perceptual factors in endangered species management. J Wildlife Manage 49(2):528–536

[184]

Kletter C, Kriechbaum M (2001) Tibetan medicinal plants. CRC Press, London
Lama YC, Ghimire SK, Aumeeruddy-Thomas Y (2001) Medicinal plants of Dolpo. WWF Nepal Program, Kathmandu, Nepal
Law W, Salick J (2005) Human induced dwarfing of Himalayan snow lotus (*Saussurea laniceps* (Asteraceae)). Proc Natl Acad Sci 102:10218–10220
Law W, Salick J, Knight TM (In preparation) Comparative population ecologies and sustainable harvest of Tibetan medicinal snow lotus (Saussurea laniceps and S. medusa (Asteraceae)).
Mackinnon J, Sha M, Cheung C, Carey G, Zhu X, Melville D (1996) A biodiversity review of China. WWF International, Hong Kong
Marcus RR (2001) Seeing the forest for the trees: integrated conservation and development projects and local perceptions of conservation in Madagascar. Human Ecol 29(4):381–397
McCune B, Mefford MJ (1999) PC-ORD. Gleneden Beach, Oregon, MjM software design: multivariate analysis of ecological data
Mehta JN, Kellert SR (1998) Local attitudes toward community-based conservation policy and programmes in Nepal: a case study in the Makalu-Barun conservation area. Environ Conserv 25(4):320–333
Mittermeier RA, Myers N, Thomsen JB, Da Fonseca GAB, Olivieri S (1998) Biodiversity hotspots and major tropical wilderness areas: approaches to setting conservation priorities. Conserv Biol 12:516–520
Murphree MW (2002) Protected areas and the commons. Common Prop Resour Digest 60:1–3
Olsen CS, Larsen HO (2003) Alpine medicinal plant trade and Himalayan mountain livelihood strategies. Geogr J 169:243–254
Olsen CS, Bhattarai N (2005) A typology of economic agents in the Himalayan plant trade. Mountain Res Develop 25:37–43
Olsson P, Folke C (2001) Local ecological knowledge and institutional dynamics for ecosystem management: a study of Lake Racken Watershed, Sweden. Ecosystems 4:85–104
Ohmagari K, Berkes F (1997) Transmission of indigenous knowledge and bush skills among the Western James Bay Cree women of subarctic Canada. Human Ecol 25(2):197–222
Redford KH, Sanderson SE (2000) Extracting humans from nature. Conserv Biol 14:1362–1364
Salick J, Anderson D, Woo J, Sherman R, Norbu C, Na A, Dorje S (2004) Tibetan ethnobotany and gradient analyses, Menri (Medicine Mountains), Eastern Himalayas millennium ecosystem assessment
Salick J, Yang YP, Gunn BF (2005) In situ capacity building: traditional ecological knowledge for conservation and sustainable development. Saint Louis, MO
Salick J, Amend A, Gunn B, Law W, Schmidt H, Byg A (2006) Tibetan medicine plurality. Econ Bot 60(2)
SAS Inc. (2003) JMP. Cary, NC
Sheil D, Lawrence A (2004) Tropical biologists, local people and conservation: new opportunities for collaboration. Trends Ecol Evol 19(12):634–638
SPSS Inc. (2001) SPSS for windows. Chicago, Illinois
Songorwa AN (1999) Community-based wildlife management (CWM) in Tanzania: are the communities interested? World Develop 27:2061–2079
Xu J, Wilkes A (2004) Biodiversity impact analysis in northwest Yunnan, southwest China. Biodivers Conserv 13:955–983
Yang JS, Chuchengjiancuo (1989) Diqing Zang Yao. Kunming Shi, Yunnan Min Zu Chu Ban She

🖄 Springer

Biodivers Conserv (2007) 16:1761–1784
DOI 10.1007/s10531-006-9063-4

ORIGINAL PAPER

Environmental heterogeneity and disturbance by humans control much of the tree species diversity of Atlantic montane forest fragments in SE Brazil

José Aldo A. Pereira · Ary T. Oliveira-Filho ·
José P. Lemos-Filho

Received: 18 July 2005 / Accepted: 13 April 2006 / Published online: 27 October 2006
© Springer Science+Business Media B.V. 2006

Abstract The effects of human impact and environmental heterogeneity on the tree species diversity were assessed in 20 fragments of tropical montane seasonal forest in southeastern Brazil. Previous surveys of the tree community, soils and topography of the fragments provided the bulk of the data. The diversity parameters used were the means of species richness, Shannon diversity (H'), and Pielou evenness (J') obtained from "bootstrap" sub-samplings of 1,000 trees. Morphometric variables obtained for the fragments included total, edge, and inner areas. Investigation forms were used to survey the history of human interventions and prepare an impacts matrix containing scores assigned to assess the extent, severity and duration of selected impacts. Scores for overall environmental impacts were obtained from the ordination scores produced by a multivariate analysis of the impacts matrix. A multivariate analysis of the standard deviations of soil variables was used to identify the variable which contributed most to soil heterogeneity. The same procedure was repeated for the variables related to topography and ground-water regime. The three species diversity parameters were related to the proportions of edges, the overall impacts scores, and the standard deviations of two selected soil and topographic variables. The species diversity in the fragments increased with increasing heterogeneity of both soil chemical properties and topographic features, and decreased with increasing proportion of forest edges. The evenness component of species diversity also increased with increasing severity of overall environmental impacts. This probably occurred because the 20 fragments did not include highly disturbed forests in the range and the intermediate disturbance effect on species diversity was therefore detected.

J. A. A. Pereira (✉) · A. T. Oliveira-Filho
Departamento de Ciências Florestais, Universidade Federal de Lavras, 37200-000 Lavras, MG, Brazil
e-mail: j.aldo@ufla.br

J. P. Lemos-Filho
Departamento de Botânica, ICB, Universidade Federal de Minas Gerais, 30161-970 Belo Horizonte, MG, Brazil

🖄 Springer

Keywords   Disturbance · Environmental heterogeneity · Environmental impacts · Forest fragments · Man-made impacts · Tree species diversity · Tropical seasonal forests

Introduction

Brazil is one of the two richest countries in biodiversity in the world and shelters two of the 25 global biodiversity hotspots recognized by Myers et al. (2000). One of these, the Atlantic Forest, is ranked among the five most important because of its enormous biodiversity, but is also one of the most seriously threatened in the world (PROBIO 1999). From its original area of 1,227,600 km^2, only 91,930 km^2 (7.5%) are presently left, most of which are scattered fragments with a past history of disturbance by man (Mittermeir et al. 1999; MMA 2000; Fundação SOS Mata Atlântica 2002; Galindo-Leal and Câmara 2003). As a result, the Atlantic Forest is certainly (and sadly) among the world biomes with the highest rates of loss of its endemic species (Whitmore 1997).

The Atlantic Forest remnants in the landlocked state of Minas Gerais, Southeastern Brazil, are composed mostly of seasonal semideciduous forests areas concentrated in the east and south of the state. The present study was carried out in 20 forest fragments of the Alto Rio Grande region, in southern Minas Gerais, where the landscape is characterized by a hilly relief covered by vegetation mosaics formed by the contact between Atlantic seasonal forests, *cerrado* (woody savanna) and montane grasslands. As has occurred in other parts of Brazil where European occupation goes back to colonial times, all these plant formations were seriously plundered, and the seasonal forests, in particular, have been reduced to scattered fragments, most of which are small (<10 ha) and disturbed by selective logging, cattle-raising, and fire (Oliveira-Filho et al. 1994a). Among the 182 areas of the Atlantic Forest Domain elected by the federal government as of high conservation priority (MMA 2002) three are found in the Alto Rio Grande region and are the subject of the present study: Tiradentes, Carrancas e Ibitipoca.

The present contribution is part of the vast Brazilian government Project for Conservation and Sustainable Use of Brazilian Biological Diversity (PROBIO) supported by the Global Environmental Facility and International Bank for Reconstruction and Development. Conservation initiatives in a region like the Alto Rio Grande must deal with the reality of scattered and disturbed vegetation remnants. Therefore, it is important to describe the patterns of biodiversity distribution in the relic areas because this is essential information on which to base the preparation of both conservation projects and state environmental policy. Despite their great important for conservation sciences, studies of the kind are still scarce for Atlantic Forest fragments (Metzger 1997; Metzger et al. 1998; Tabarelli et al. 1998; Tabarelli and Mantovani 1999; Nunes et al. 2003).

Current literature contains a plethora of models and theories about the mechanisms that determine and maintain species diversity, particularly the high tree species diversity of tropical forests (Tilman 1999; Chesson 2000; Ashton and LaFrankie 2000; Leigh et al. 2004). For some time, spatial and temporal heterogeneity in the physical environment were strong candidate mechanisms. In fact, variations of the environment in both horizontal (particularly soil chemical and textural properties, and ground-water regime) and vertical/temporal (gap cycle disturbance, canopy

🕮 Springer

layering and rooting zones) dimensions do affect species distribution in tropical forests with potential effects on the local or α-diversity (e.g. Fowler 1988; Terborgh 1992; Clark et al. 1998; Pinto et al. 2006). Disturbance, in particular, is an important force capable of molding plant communities as it produces spatial and temporal heterogeneity therefore determining, to a considerable extent, tree community composition and diversity (Picket and White 1985; Caswell and Cohen 1991; van der Maarel 1993; Phillips and Gentry 1994; Whitmore and Burslem 1998). The progress of the gap-phase dynamics theory for tropical forests (see Denslow 1987 for a review) enhanced that temporal heterogeneity of the environment plays an additional and important role in determining tree species distribution and hence diversity. The resulting model of tropical forests as species mosaics with asynchronous pieces determined by diversity-promoting disturbance factors (Oldeman 1990) combined spatial and temporal dimensions of environmental heterogeneity.

However, it soon became clear that environmental heterogeneity and disturbance regime could not explain on their own the high tree species diversity of tropical forests as compared to the extra-tropical ones. Now that the frenzied noise of those bandwagons belongs to the past, new views arise and most of these consider interacting factors and the role of stochastic events (Tilman 1999; Chesson 2000; Hubbell 2001; Leigh et al. 2004). It is now widely believed that disturbance and microhabitat specialization do underlie the mechanisms maintaining the diversity of tropical trees but are insufficient to explain the whole picture in the tropics where long-term stability and high pressure of pests may play a crucial role in promoting a much higher diversity than in temperate forests. The present contribution provides additional evidence that, although they cannot completely explain the high tree species diversity of tropical forests, disturbance and environmental heterogeneity can never be totally disregarded as factors promoting both their α- and β-diversity (intra- and inter-habitat).

A study of diversity patterns in a fragmented forest landscape such as that of the Atlantic Forest requires giving high priority to the fragmentation itself as a potential factor underlying variations in tree species diversity (Metzger 1997; Metzger et al. 1998; Tabareli and Mantovani 1999). Fragment edges become a new habitat that usually shows a forest structure and composition different from that of the fragment core, and this may add to the original β-diversity (Laurance and Yensen 1991; Laurance and Bierregaard 1997; Ranta et al. 1998; Nunes et al. 2003). Therefore the present study addressed four variables as candidate factors related to the variations in tree species diversity among the 20 forest fragments: a soil heterogeneity variable, a topographic heterogeneity variable, a disturbance severity variable and an edge-related variable. The hypothesis was that tree species diversity in the 20 forest fragments was chiefly determined by the heterogeneity of the physical environment, including edge-effects, and by disturbance impacts, as sources of differential chances for species establishment, growth and reproduction.

Materials and methods

The study region and the 20 forest fragments

Surveys of the tree community, soils and topography of 20 fragments of tropical montane seasonal semideciduous forest (sensu Veloso et al. 1991) of the Alto Rio Grande region, southeastern Brazil, provided the largest part of the information we used in the present study. The name, identification code, total area, geographical co-ordinates and altitudinal range of each forest fragment are given in Table 1 together with the latest literature source where more detailed information can be obtained. The location of the fragments is shown in Fig. 1.

The Alto Rio Grande Region is situated between 21°00′–21°43′S and 43°50′–45°05′W (Fig. 1) and comprises the inland section of the Serra da Mantiqueira that shelters the upper course of the Rio Grande. This river, after merging with the Rio Paranaíba, becomes the Rio Paraná which is the main watercourse of the second largest river system in the whole of South America. The topography of the region is mostly hilly, with altitudes ranging between 800 and 1,000 m. Altitudes of up to 1,600 m, however, occur along the mountain ridges of the Serra de São José, Serra da Bocaina, Chapada das Perdizes and Serra de Ibitipoca. Predominant parent materials are quartzites, granitic gneisses, mica-slates, and calcareous gneisses while most soils are classified as Latosols, Cambisols, Argisols, and Alluvial and Litholic Neosols (Curi et al. 1990). The climate is classified as Köppen's Cw type, i.e. tem-perate/mesothermal with a rainy summer and a dry winter; ca. 66% of total rainfall occurs between November and February (Vilela and Ramalho 1979). Cwa climates (warmer) predominate in the region, while Cwb climates (cooler) are restricted to mountain tops. The mean annual rainfall and temperature of Cwa climates range between 19.3–20.1°C and 1,514–1,588 mm, respectively, while those of Cwb climates range between 14.8–18.6°C and 1,536–1,605 mm.

Methodological procedure

As the methodology of our study was relatively complex and extensive, a flow chart is given in Fig. 2 in order to facilitate the comprehension of how we devised and developed the procedures. The flow chart is horizontally stratified into three levels. The target analyses were linear correlations between three species diversity parameters in the 20 fragments and four environmental variables obtained to appraise several aspects of habitat heterogeneity and man-made impacts in the fragments. We describe below how the diversity parameters and environmental variables used in the correlation analyses were produced following the elements of the flow chart.

Tree community surveys and species diversity parameters

We calculated species diversity parameters for each of the 20 fragments using data from surveys of the tree community and environmental variables carried out in the last 18 years by a research team from the Federal University of Lavras (see Table 1 for references). All those surveys were carried out in sample plots distributed in the fragments following sampling designs that aimed at capturing supposed main species and environmental gradients. The purpose was therefore the same in all surveys and the sampling design was very similar only varying to adapt to the presumed envi-ronmental gradients of each fragment. Total sample areas varied from 1 to 5 ha, and sample plots were distributed throughout the area in most fragments (16), but only on chosen sectors of four of the largest ones (MD, TD, IB and PB). Most surveys (18) used systematic distribution of sample plots; one used random distribution (PI) and one was actually a census (LV). Plot dimensions varied to conform to supposed gradients so that nine surveys used 20 × 20 m plots (LV, SE, ML, IU, IN, LU, CP,

[190]

Table 1 Identification codes, names, total area, geographical co-ordinates, altitudinal range, and latest reference source of the 20 fragments of tropical montane seasonal semideciduous forest surveyed in the Alto Rio Grande Region, southeastern Brazil. Fragments are listed in decreasing order of longitude

Code	Name	Area (ha)	No individuals	Latitude (S)	Longitude (WGW)	Altitude (m)	Latest reference
ML	Mata da Lagoa	3.97	1,743	21°13'00"	44°58'49"	855–902	Machado et al. (2004)
PB	Poço Bonito	88.56	4,153	21°19'45"–21°20'48"	44°58'18"–44°59'24"	925–1,210	Dalanesi et al. (2004)
LV	Lavras, UFLA	5.72	9,363	21°14'42"	44°58'10"	918–937	Nunes et al. (2003)
SE	Subestação	8.73	5,919	21°13'17"	44°57'47"	910–940	Espírito-Santo et al. (2002)
PN	Pedra Negra	72.34	2,091	21°05'25"	44°56'42"	835–885	Carvalho (2002)
LU	Luminárias	77.91	2,343	21°29'11"	44°55'20"	880–1,000	Rodrigues et al. (2003)
BS	Bom Sucesso	83.45	3,064	21°09'27"	44°54'10"	806–832	Appolinário et al. (2005)
IN	Ingaí	16.14	2,683	21°24'26"	44°53'32"	860–890	Botrel et al. (2002)
CP	Capivari	9.78	1,881	21°16'23"	44°52'53"	825–875	Souza et al. (2003)
IU	Ibituruna	59.75	1,573	21°10'00"	44°50'25"	820–980	Silva et al. (2005)
LA	Lafite	25.95	1,694	21°12'58"	44°48'08"	825–863	Carvalho (2002)
MI	Mata da Curva	54.97	1,678	21°13'31"	44°47'31"	815–910	Carvalho (2002)
RM	Rio das Mortes	14.31	1,465	21°06'49"	44°45'00"	845–980	Carvalho (2002)
CM	Camargos	10.36	2,393	21°21'18"	44°36'49"	913–960	van den Berg and Oliveira-Filho (2000)
CR	Carrancas	35.98	2,746	21°36'29"	44°36'38"	1,440–1,513	Oliveira-Filho et al. (2004)
IT	Itutinga	3.77	3,842	21°21'05"	44°36'29"	913–945	Oliveira-Filho et al. (1994c)
MD	Madre de Deus	20.66	2,311	21°29'03"	44°22'32"	915–980	Guilherme et al. (2004)
PI	Piedade	24.95	1,929	21°29'16"	44°06'02"	1,040–1,150	Carvalho (2002)
TD	Tiradentes	271.47	2,183	21°03'55"–21°05'48"	44°05'58"–44°10'17"	920–1,340	Machado and Hargreaves (2000)
IB	Ibitipoca	95.05	1,928	21°42'13"–21°43'24"	43°52'43"–43°53'31"	1,150–1,510	Fontes (1997)

Springer

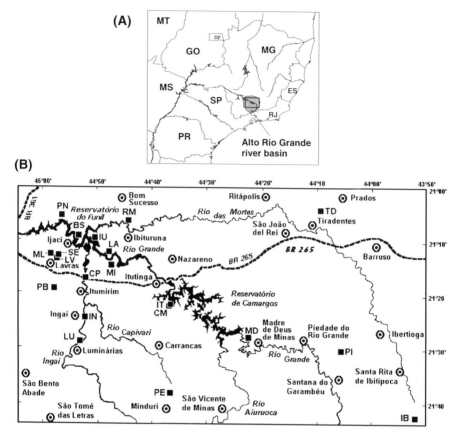

Fig. 1 (**A**) Geographic situation of the Alto Rio Grande river basin in southeastern Brazil, and (**B**) distribution of the 20 forest fragments surveyed in the region (■) identified by their codes in Table 1. Map (**B**) appears in (**A**) as a rectangle overlapping the river basin gray polygon

CR and PI), four used 5× 50 m plots (PN, RM, LA and MI), three used 15 × 15 m plots (BS, IT and MD), two used 10 × 30 m plots (PB and CM), one used 10 × 50 m plots (TD) and one used 10 × 20 m plots (IB). All live trees with diameter at breast height (dbh, i.e. 1.30 m) ≥5 cm found in the plots were identified to the species level and measured (dbh and total height). Multi-stemmed trees were recorded when the square root of the sum of squares of individual diameters matched the criteria. Voucher specimens are lodged at the ESAL Herbarium of the Federal University of Lavras.

We produced a species–area curve where the mean cumulative number of species in *n* fragments is obtained from all possible combinations of *n* fragments (McCune and Mefford 1999). We also calculated the first- and second-order jackknife estimators of the number of species projected by the total sample of the 20 forest fragments (Heltsche and Forrester 1983; Palmer 1991).

We calculated three species diversity parameters for each forest fragment: species richness (*S*), Shannon diversity index (*H'*) and Pielou evenness (*J'*) (Brower and Zar 1984). We obtained means and confidence intervals (95%) of the three

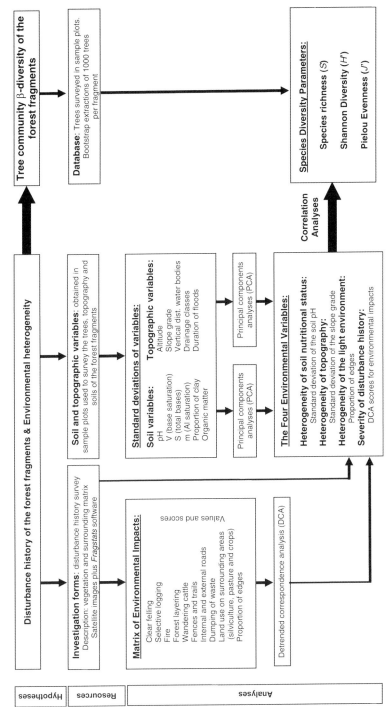

Fig. 2 Flow chart representing the stages of the methodological procedure and the connections among variables. The top level contains the hypotheses, the second the resources used and the third the data analyses

diversity parameters from n "bootstrap" random sub-samplings of 1,000 trees of the original databases; n values were proportional to the sample size and are given in Table 2. Figure 3 gives an example of the stabilization of means and confidence intervals with the progression of "bootstrap" sub-samplings in fragment LV. These three values were used in the correlation analyses with the environmental variables.

Cartographic base and morphometric indices

We obtained the outline of the fragments by the analysis of Landsat 7 satellite imagery—sensor ETM+, using bands 3, 4, 5 (spatial resolution of 25 m) and band 8 (spatial resolution of 12.5 m), as recorded on March 25, 2001. We digitized the outline images using the SPRING 3.4 and ENVI 3.1 programs for further processing by FRAGSTATS, version 2.0 (McGarigal and Marks 1994), and treated the output as morphometric indices. To process the outlines, FRAGSTATS requires the input of edge-effect width for each fragment. According information from the literature (Kapos 1989; Laurance and Yensen 1991; Murcia 1995) and our own experience with edge effects in the region (Oliveira-Filho et al. 1997; Espírito-Santo et al. 2002; Souza et al. 2003), edge-effect widths vary substantially in response to various features of the fragments themselves and of their surrounding landscapes. We therefore defined three edge widths—50, 75, and 100 m—and ascribed one of them to each fragment after analyzing and scoring an aggregation matrix containing the following

Table 2 Means ± confidence intervals (95%) of species diversity parameters calculated for n "bootstrap" sub-samples of 1,000 trees randomly extracted from the tree totals (N) of each of the 20 forest fragments surveyed in the Alto Rio Grande Region, southeastern Brazil. S = number of species, H' = Shannon diversity and J' = Pielou evenness. Fragments with the same P values do not differ significantly among themselves in comparisons of their confidence intervals. Fragments in descending order of H'

Fragment	n	N	S	P	H'	P	J'	P
PB	10	4,137	151.2 ± 3.0	a	4.473 ± 0.021	a	0.891 ± 0.003	a
IU	5	1,128	149.3 ± 1.2	a	4.383 ± 0.004	b	0.876 ± 0.001	b
PI	5	1,776	145.0 ± 7.4	ab	4.299 ± 0.083	bc	0.864 ± 0.008	cd
BS	5	1,114	136.6 ± 2.3	cd	4.275 ± 0.018	c	0.869 ± 0.002	c
TD	5	2,148	116.0 ± 5.3	ef	4.236 ± 0.020	d	0.891 ± 0.008	a
CP	5	1,666	124.6 ± 2.1	de	4.214 ± 0.034	de	0.873 ± 0.008	bc
LA	5	1,015	136.8 ± 1.6	c	4.204 ± 0.008	e	0.855 ± 0.002	d
LU	8	2,343	133.1 ± 2.0	d	4.186 ± 0.026	ef	0.856 ± 0.006	d
RM	5	1,262	139.0 ± 2.2	c	4.155 ± 0.015	f	0.842 ± 0.005	e
SE	8	3,120	128.9 ± 4.3	de	4.125 ± 0.028	fg	0.849 ± 0.005	de
IB	5	1,919	116.8 ± 2.2	ef	4.104 ± 0.037	fg	0.862 ± 0.008	cd
PN	5	1,215	144.8 ± 1.6	b	4.099 ± 0.020	g	0.824 ± 0.005	f
MI	5	1,064	113.8 ± 2.2	f	3.924 ± 0.025	h	0.829 ± 0.006	f
CR	8	2,565	105.0 ± 3.0	g	3.893 ± 0.026	h	0.837 ± 0.005	ef
ML	5	1,294	132.8 ± 4.9	cde	3.846 ± 0.016	i	0.787 ± 0.007	h
IT	5	2,091	123.0 ± 5.3	e	3.753 ± 0.075	jk	0.780 ± 0.010	h
CM	5	1,697	112.4 ± 3.9	f	3.737 ± 0.027	k	0.792 ± 0.006	h
MD	5	1,349	97.6 ± 1.1	h	3.694 ± 0.024	k	0.806 ± 0.005	g
IN	8	2,683	103.4 ± 2.2	g	3.686 ± 0.038	k	0.795 ± 0.007	gh
LV	10	6,507	95.1 ± 1.6	h	3.617 ± 0.022	l	0.794 ± 0.004	h

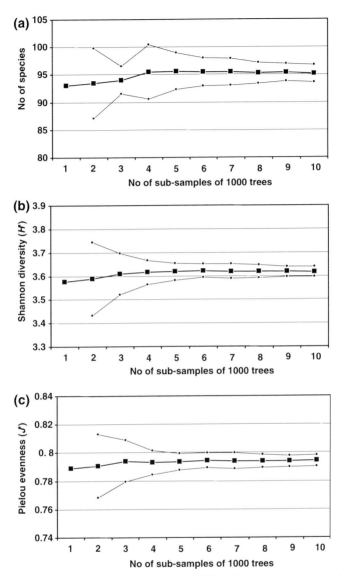

Fig. 3 Progression of means ± confidence intervals (95%) of the species richness (**a**), Shannon diversity index (**b**) and Pielou evenness (**c**) with increasing number of "bootstrap" sub-sampling of 1,000 trees from the total of 6,507 trees surveyed in the forest fragment LV

information: edge type (natural or man-made), forest physiognomy (proportion of old-growth and of early and late regeneration phases), surrounding landscape (proportion of pastures, crops, forest plantations), location (valleys, slopes or hilltops). We used the sum of scores to distribute the fragments and four fell into the 100 m category, nine in the 75 m and seven in the 50 m. Of the various indices coming from an analysis using FRAGSTATS, we only used the total, inner and edge areas to obtain the proportion of edges.

Survey of disturbance history and assessment of environmental impacts

We adopted the technique of Rapid Participatory Appraisal, RPA (Chambers 1987, 1992) to survey the history of (a) human intervention in the fragments, (b) land use in the surrounding areas, and (c) the composition of the following groups of verte-brate fauna: mammal carnivores, primates, large rodents and game birds. The RPA technique was first conceived to obtain information quickly and interactively for sustainable development projects and is characterized by a systemic approach, the triangular gathering of information (interviewer-interviewed-information) and the interaction between collected information and analyses (Beebe 1995). We applied RPA question forms during interviews seeking to obtain the perceptions of owners, neighbors, or older employees with a good knowledge of each fragment. We also consulted existing documents, such as management plans (in the case of protected areas), academic monographs and dissertations, and the publications cited in Table 1.

We assigned scores to forest layering as a possible result of disturbance history but also expressing penetration of light. The scores were assigned as follows: 1—discontinuous canopy 10–15 m high with rare emergent trees; 2—continuous canopy (with some gaps) 10–15 m high with emergent trees of 15–25 m; 3—dis-continuous canopy 15–20 m high with rare emergent trees; 4—continuous canopy (with some gaps) 15–20 m high with emergent trees of 20–35 m; 5—continuous canopy (with some gaps) 20–25 m high with emergent trees of 20–40 m.

We used the Interaction Matrices Method (Leopold et al. 1971) to describe and assess the past and ongoing environmental impacts of each fragment. Past impacts were obtained from the RPAs while the ongoing impacts were obtained during field visits accompanied by the authors of the individual studies (Table 1). These authors also helped us in ascribing scores to account for the severity of both past and ongoing impacts so that the whole assessment process results from the sensibility of a whole team of experienced researchers.

Each matrix intersection between line (fragments) and column (impacts) con-tained four smaller cells where we ascribed scores (ranging from 0 to 4) to assess three aspects of the environmental impact on the fragment—severity, extent and dura-tion—and obtained a final score from the sum of the three previous scores. The last column contained the scores ascribed to forest layering. We eventually multiplied the final score contained in each matrix cell by weights given to impacts by the research team: weight 1 was given to (a) dumping of waste, (b) forestry on surrounding areas and (c) fences along the edges; weight 2 was given to (d) cultivation on surrounding areas (e) trails; weight 3 was given to (f) external roads and (g) pastures on sur-rounding areas; weight 4 was given to (h) internal roads, (i) wandering cattle and (j) proportion of edges (provided by FRAGSTATS); weight 5 was given to (k) selective logging, (l) forest layering and (m) fire; and weight 6 was given to (n) clear felling. Ascribed scores were obviously produced by subjective appraisal based on our field experience and on the literature (Fiszon and Marchioro 2002).

The four environmental variables

We eventually produced the four environmental variables that were individually used in correlation analyses with the three species diversity parameters. These were meant to express the following environmental aspects of the forest fragments: (a) heterogeneity of the soil nutritional status, (b) heterogeneity of the surface topography and related soil moisture availability, (c) heterogeneity of the light environment caused by edge-effects, and (d) appraisal of disturbance history severity (Fig. 2).

(a)   Soil variables were provided by the analyses of soil samples 0–20 cm in depth collected at each sample plot during the forest surveys. Laboratory procedures followed the EMBRAPA (1997) protocol and yielded the following variables: pH in water, base saturation (V%), aluminum saturation (m%), total exchangeable bases (S), and proportion of clay and organic matter. We obtained the standard deviation of each soil variable in each forest fragment to produce a 6×20 matrix that we processed in a Detrended Correspondence Analysis (DCA) (Hill and Gauch 1980) after standardizing all variables to zero mean and unit variance. We then identified the soil variable that most strongly contributed to the first ordination axis and chose it for the correlation analyses.

(b)   The variables related to topography and soil moisture provided by the forest surveys for each sample plot were (i) soil drainage class according to EMBRAPA (1999) criteria and expressed as ranks, (ii) elevation at the center of the plot, (iii) elevation above water-bodies, (iv) slope grade, and (v) duration of floods. We repeated here the same procedure described in (a) to identify the variable that most strongly contributed to the major heterogeneity gradient.

(c)   We used the variable proportion of edges to account for the heterogeneity of the light environment caused by edge-effects.

(d)   We performed a second DCA of the 13×20 matrix of synthetic scores of environmental impacts on each fragment. We then used the ordination scores yielded by the first DCA axis to represent an overall assessment of the severity of the past and ongoing environmental impacts.

We also performed an additional round of correlations to investigate the possibility of interference of fragment size on the target correlations. We therefore correlated the four environmental variables and three diversity parameters to the area of the fragments. In the particular case of the number of species we used the model proposed by McArthur and Wilson (1967) for species–area relationships in ecological islands: $S = C*A^Z$ (S = no of species, A = area, C and Z = constants).

We used the software PC-ORD for Windows version 4.14 (McCune and Mefford 1999) to perform the DCAs, and the software STATISTICA for Windows, release 5.0 A (StatSoft, Inc. 1995), to perform the correlation analyses. We adopted the significance threshold of $\alpha < 0.05/12$ (i.e. $\alpha < 0.004167$) to account for the Bonferroni correction (Miller 1991) for 12 correlations with the same dataset (3 three diversity parameters $\times$ 4 environmental variables). For the correlations with the fragment areas the threshold was $\alpha < 0.05/7$ (i.e. $\alpha < 0.007143$).

Results

Tree community diversity

The surveys of the 20 forest fragments registered a total of 730 tree species distributed into 286 genera and 86 families, following the APG (2003) classification

🕭 Springer

system. The richest families were Myrtaceae (78 species), Fabaceae (74), Melastomataceae (46), Lauraceae (45), Euphorbiaceae (37), Rubiaceae (30), Asteraceae
(25), Solanaceae (19), Annonaceae (16), and Malvaceae (15). These 10 families
included 52.7% of the species' total. On the other hand, 23 families were represented
by only one species (8.0%).

The curve of cumulative number of tree species with increasing number of forest
fragments (Fig. 4) shows no clear indication of stabilization as it assumes a quasi-
linear configuration above 10 surveyed fragments. This strongly suggests that the
species richness of the region is much higher than the 730 species registered in the 20
surveys. In fact, the first- and second-order jackknife estimators were 901.8 and 991.5
species, respectively. Another indication of the high species richness of the region
are given by the large proportion of species found in few fragments: 184 in one
(25.2%), 89 in two (12.2%), and 72 in three (9.9%). In contrast, only five were found
in all 20 fragments: *Tapirira obtusa* (Benth.) J.D.Mitch., *T. guianensis* Aubl. (Ana-
cardiaceae), *Ocotea corymbosa* (Meisn.) Mez, (Lauraceae), *Pera glabrata* (Schott)
Poepp. ex Baill. (Euphorbiaceae) and *Machaerium villosum* Vogel (Fabaceae).

The species diversity parameters of each forest fragment are given in Table 2 as
mean and confidence interval of values obtained for repeated "bootstrap" sub-
samples of 1,000 trees. The species richness (S) ranged from 95.1 ± 1.6 species (LV)
to 151.2 ± 3.0 species (PB). Those two fragments also showed the lowest and highest
Shannon diversity (H'): 3.617 ± 0.022 and 4.473 ± 0.021, respectively. PB also
showed the highest Pielou evenness (J'), 0.891, but the lowest value was registered
for IT: 0.780. H' was significantly correlated with both S ($r = 0.82$, $P = 0.017$) and J'
($r = 0.92$, $P = 0.015$), but S and J were not significantly correlated ($r = 0.53$,
$P = 0.289$).

The comparisons of diversity parameters among fragments using their confidence
intervals showed significant differences among eight groups for S and J' and 12
groups for H', though with many cases of overlapping significance among groups
(Table 2). This demonstrates that the variations in species diversity among the
fragments are meaningful and therefore add strength to analyses of their relationship
to other variables.

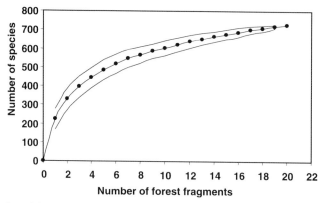

Fig. 4 Progression of the cumulative number of species with increasing number of forest fragments
surveyed in the Alto Rio Grande Region, southeastern Brazil. Curves area means ± standard
deviations obtained from all possible combinations of the 20 fragments at each number of forest
fragments

Table 3 Environmental variables used in the correlation analyses with species diversity parameters in the 20 forest fragments in the Alto Rio Grande Region, southeastern Brazil. Means ± standard deviations are given for soil pH and elevation above water-bodies obtained for N sample plots used in surveys. DCA scores are ordination scores in the first axis yielded by Detrended Correspondence Analysis (DCA) of environmental impacts in the 20 forest fragments. Fragments are ranked alphabetically

Frag.	N	pH	Elevation above water-bodies (m)	Proportion of edges (%)	DCA scores
BS	24	5.30 ± 0.39	9.3 ± 3.6	59.5	23
CM	28	4.14 ± 0.24	8.9 ± 5.6	82.3	23
CP	28	4.60 ± 0.35	27.4 ± 14.2	95.8	114
CR	30	4.10 ± 0.25	27.1 ± 16.2	48.4	105
IB	48	4.25 ± 0.46	38.5 ± 34.2	22.1	12
IN	25	5.00 ± 0.22	7.3 ± 4.1	63.2	9
IT	42	4.68 ± 0.33	9.4 ± 6.7	98.4	33
IU	26	5.30 ± 0.79	28.9 ± 29.5	38.9	143
LA	15	4.32 ± 0.08	32.2 ± 10.3	62.0	46
LU	32	5.03 ± 0.65	50.5 ± 32.4	33.0	129
LV	126	4.18 ± 0.13	55.6 ± 4.0	99.7	71
MD	43	4.96 ± 0.13	7.3 ± 3.1	71.9	0
MI	15	4.66 ± 0.27	17.8 ± 7.6	64.9	128
ML	29	5.09 ± 0.41	33.6 ± 11.4	100.0	86
PB	80	4.39 ± 0.26	24.6 ± 24.3	46.2	127
PI	30	4.69 ± 0.55	214.3 ± 31.0	41.8	45
PN	15	4.76 ± 0.22	9.9 ± 5.7	47.2	33
RM	15	4.52 ± 0.32	33.8 ± 15.0	59.0	84
SE	52	4.63 ± 0.53	36.4 ± 8.5	72.1	119
TD	18	4.56 ± 0.17	36.4 ± 8.5	27.7	128

Environmental variables and tree species diversity

The DCA of soil variables indicated soil pH as the most strongly related to the overall gradient of soil heterogeneity summarized by axis 1. In the case of the variables related to topography and ground-water regime the elevation above water-bodies was pointed out. These two variables plus the proportion of edges and DCA scores for overall environmental impacts made up the set of four environmental variables (Table 3) used in the correlation analyses with the three diversity parameters. Eight out of the 12 paired correlations yielded significant results and are shown in Figs. 5–8.

The species diversity in the forest fragments, expressed by number of species and Shannon diversity (H'), were both significantly correlated with soil pH heterogeneity and both increased with increasing pH heterogeneity (Fig. 5). The diversity expressed by the Pielou evenness (J'), however, was not significantly correlated with pH heterogeneity ($r = 0.339$, $P = 0.144$).

In the case of topographic heterogeneity, the species diversity in the forest fragments, expressed by all three parameters (number of species, H' and J'), yielded significant correlations and increased with increasing heterogeneity of the elevation above water-bodies (Fig. 6).

The proportion of edges in the forest fragments was significantly correlated with both J' and H', but not with the number of species ($r = 0.152$, $P = 0.090$). Both diversity parameters decreased significantly with increasing proportion of edges (Fig. 7).

The overall impacts expressed by DCA scores were significantly correlated with J' only and failed to yield significant correlations with both number of species ($r = 0.088$, $P = 0.203$) and H' ($r = 0.47$, $P = 0.036$). The evenness component of species (J') diversity increased significantly with increasing severity of the records of environmental impacts to the fragments (Fig. 8).

There was no significant correlation between fragment area and the standard deviations of both soil pH ($r = -0.039$, $P = 0.872$) and elevation above water-bodies ($r = 0.154$, $P = 0.517$), and between fragment area and the DCA scores for environmental impacts ($r = 0.284$, $P = 0.224$). The proportion of edges, however, was significantly and negatively correlated with the fragment area ($r = -0.663$, $P = 0.0014$). The species–area relationship ($S = -6.806*A^{2.091}$) was not significant ($r = 0.249$, $P = 0.291$), as was the correlation between H' and fragment area ($r = 0.442$, $P = 0.051$). The evenness index J', however, was significantly and positively correlated with the fragment area ($r = 0.591$, $P = 0.0061$).

Discussion

Regional diversity and species composition

The present study deals with two scales of species diversity. That of any single fragment corresponds to β-diversity because all fragments include at least two habitats, edge and interior, and most also show additional environmental gradients connecting different habitats caused by variations of the substrate and/or disturbance history. The second scale is that encompassing all 20 fragments, which clearly fits regional or γ-diversity because it includes an array of habitats of a geographic area without dispersion barriers to the organisms (Ricklefs 1987; Simberlofff and Dayan 1991; Chesson 2000).

The tree species γ-diversity of the Alto Rio Grande region estimated by the 730 recorded species may be considered high in the context of similar regional surveys in central and southeastern Brazil. For instance, figures extracted from the database used by Oliveira-Filho et al. (2006) totaled 378 tree species in 21 gallery forests of the Federal District, 338 in six fragments of rain forest in the São Paulo city plateau, and 335 in eight fragments of semideciduous forests in the Belo Horizonte region. In fact, as suggested by both the species–fragment curve and jackknife estimators (901.8 and 991.5 species), the total number of forest tree species in the region is probably at least 20% higher than the registered figure.

To a great extent the high tree species richness of the Alto Rio Grande region is probably explained by two main factors: the high environmental heterogeneity and the complex contact zone between vegetation formations of forest and savanna biomes. The region is characterized by a transition between Atlantic forests sensu lato and Cerrados (Central Brazilian woody savannas) taking place at higher altitudes compared to other transitional areas. This adds the contribution of montane rocky savannas ("campos de altitude") that also have a peculiar flora (Harley 1995). Both savanna formations contribute to the species diversity of the regional forests because their typical tree species are commonly found at marginal forest areas and, less frequently and in small numbers, also in the forest interior (Botrel et al. 2002; Oliveira-Filho et al. 2004).

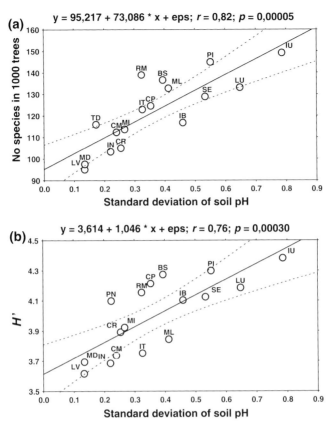

Fig. 5 Relationships between the standard deviations of soil pH and species diversity parameters in 20 forest fragments surveyed in the Alto Rio Grande Region, southeastern Brazil

Variations in altitude (800–1,500 m) are an important component of the environmental heterogeneity of the region that certainly contributes to boost species diversity. Altitude is known to be associated to remarkable variations in both physiognomy and floristic composition of Atlantic forests sensu lato (Oliveira-Filho and Fontes 2000; Scudeller et al. 2001; Ferraz et al. 2004). In fact, fragments situated at higher altitudes (e.g., IB, CR, PB and PI) show higher abundance of epiphytes (particularly bromeliads, orchids and mosses) than those situated at lower altitudes (e.g., LV, IM and IN). The former also contain species typical of upper montane forests; such as *Hyptidendron asperrimum* (Epling) Harley, *Clethra scabra* Pers., *Croton organensis* Baill., *Drimys brasiliensis* Miers, *Weinmannia paulliniifolia* Pohl, *Meliosma sellowii* Urb., *Jacaranda subalpina* W. Morawetz,*Podocarpus sellowii* Klotzsch and *Araucaria angustifolia* (Bert.) O. Kuntze; while the latter contain species typical of lower montane forests, such as *Protium widgrenii* Engl., *Trichilia emarginata* (Kurtz) C.DC., *Vismia brasiliensis* Choisy, *Piptocarpha macropoda* Baker and *Calyptranthes clusiifolia* O. Berg (Oliveira-Filho and Fontes 2000). Typical coastal rain forest species; such as *Coussapoa microcarpa* (Schott) Rizz., *Sloanea guianensis* (Aubl.) Benth., *Vochysia schwackeana* Warm., *Virola bicuhyba* (Schott) Warb and *Marlierea excoriata* Mart.; also have their innermost distribution

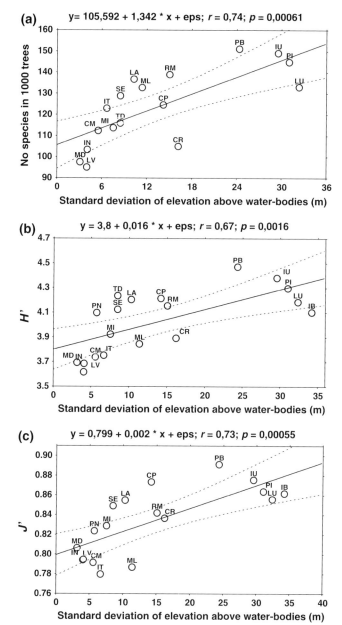

Fig. 6 Relationships between the standard deviations of elevation above water-bodies and species diversity parameters in 20 forest fragments surveyed in the Alto Rio Grande Region, southeastern Brazil

in the upper montane semideciduous forests of the region therefore contributing to increase their species diversity.

A second important factor contributing to augment the environmental heterogeneity and therefore the tree species γ-diversity of the region is that associated with

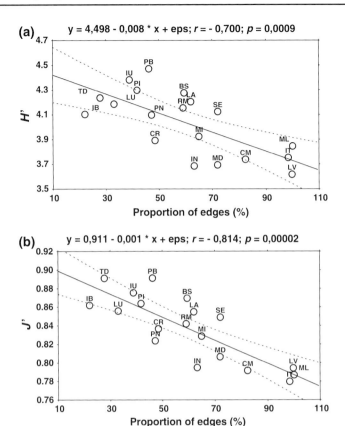

Fig. 7 Relationships between the proportion of edges and species diversity parameters in 20 forest fragments surveyed in the Alto Rio Grande Region, southeastern Brazil

its complex drainage system. Among the 20 fragments there are riparian forests that also contain well drained slopes and also some with no contact with water-bodies (e.g., LV and LA). Among the former, some are associated with large rivers and contain alluvial terraces liable to seasonal floods (e.g., BS, MD, IN, LU and CP) and others are associated with rivulets of both steep mountains (e.g., IB, TD, PB and CR) and lower hills (CM, IT, IU, PN, SE, and PI). Variations in ground-water regime caused by both topography and water-bodies are known to play a strong role in habitat differentiation among forest tree species (Oliveira-Filho et al. 1994b; van den Berg and Oliveira Filho 1999; Rodrigues and Nave 2000; Botrel et al. 2002).

It must be emphasized that the number of tree species registered in any single fragment is considerably lower than the total species richness recorded for the region, as has been found for other fragmented forests (Tabarelli et al. 1998). As a consequence, no forest fragment on its own is able to represent an adequate sample of the region's total diversity. The smaller fragments, in particular, are actually rather incomplete species assemblages. Therefore, the best possible directive aiming at the conservation of the remaining tree biodiversity of the region would require the strict protection of the largest fragments as well as of a large number of smaller fragments representing the whole array of forest habitats.

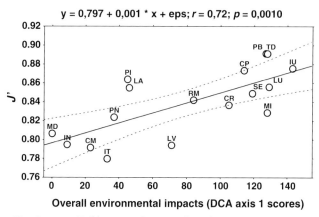

Fig. 8 Relationships between DCA scores for overall environmental impacts and the evenness component of species diversity in 20 forest fragments surveyed in the Alto Rio Grande Region, southeastern Brazil

Substrate heterogeneity effects on tree species diversity

The species–area relationship proposed by the island biogeography theory (McArthur and Wilson 1967) was not confirmed in the present case and only the evenness component of diversity (J') was sensitive to fragment size. Nevertheless, McArthur and Wilson (1967) also stated that the environmental heterogeneity of the islands could interfere strongly in the species–area relationship and their theory also predicts that island area effects on species diversity are boosted when increasing island areas also correspond to increasing habitat diversity. In the present case, the soil and topographic heterogeneity (substrate heterogeneity) of the fragments (islands) was independent of fragment size, and so were the number of species and Shannon diversity (H'). This strongly suggests that the role played by substrate heterogeneity in determining the tree species diversity of the forest fragments is probably much more important than fragment size and that various levels of environmental heterogeneity may be found in fragments of any size within the considered range. The J'–area relationship is discussed further ahead under "edge-effects".

The greater the variation of topography and soil properties in the 20 forest fragments the higher their tree species diversity expressed by both number of species and H'. Evenness, J', was only sensitive to topography. The heterogeneity of the physical environment in the Alto Rio Grande region is highly influenced by its rough topography and associated patchy distribution of different soil types. The resulting mosaic of habitats related to the substrate was found within the majority of the 20 forest fragments. On a local scale, topography has been regarded as the most important substrate-related variable causing spatial variation in the structure of tropical forests since it commonly corresponds to changes in soil properties, particularly ground-water regime and natural soil fertility (Bourgeron 1983). The correlation between tree species distribution and topographic and soil variables has been successfully demonstrated in numerous studies of tropical forests worldwide (e.g., ter Steege et al. 1993; Duivervoorden 1996; Silva-Júnior et al. 1996; Clark et al. 1998) and also in the present forest fragments (all references in Table 1). Nevertheless, it was quite surprising that the degree of differentiation among species

[204]

promoted by the variation of topography and soils did have a clear effect on the tree species diversity of the 20 fragments themselves.

Forest-edge effects on tree species diversity

Tree species diversity decreased with increasing proportional areas of edge habitats in the fragments, although this was only detected for H' and J'. Nevertheless, the proportion of edges in the fragments was the only environmental variable significantly correlated to their total area and this is understandable because the proportions were obtained by dividing the edge area by the total area of each fragment. In addition, J' was also significantly correlated to fragment area. It is therefore difficult to separate edge-effects from total area-effects on the species diversity because they are interconnected. Nevertheless, the proportion of edges produced higher correlation coefficients than did the fragment area and, unlike the latter, also correlated significantly with H', therefore suggesting that it is a better descriptor of habitat differentiation with respect to edge-effects.

Forest-edges are a rather differentiated habitat compared to forest interiors (Murcia 1995) and this commonly corresponds to a peculiar species composition in both natural (van den Berg and Oliveira Filho 1999) and man-made edges (Oliveira-Filho et al. 1997; Dalanesi et al. 2004). As a consequence, small fragments with high proportion of edge areas would have lower species diversity because this habitat is over-represented while fragments with balanced proportions of edge and inner areas would have both habitats represented and therefore higher species diversity. One might reasonably expect that larger fragments would have a similar effect of that of the small ones because edge habitats would be proportionally underrepresented. Notwithstanding, our 20-fragment sample showed a steady increase of species diversity towards the smaller proportions of edge habitats. One probable reason for this lies in the lack of very large fragments in the sample. Our largest, TD (271.5 ha), was so irregularly shaped that the proportion of edges was fairly high (27.7%).

One other reason may be derived from the fact that edge-effects on diversity were only detectable for J' and H', but not for the number of species. The above-cited studies of edge-effects on the fragments demonstrated that the edge tree communities are differentiated basically on their very high density of light-demanding and pioneer species. These species also occur in the forest interior though in much lower density and associated with gap-disturbance events. This explains why the number of species was unaffected by the proportion of edges, since the edge habitat did not add a relevant number of alien species to the forest. The evenness component of species diversity was affected instead because it was probably the dominance of the gap- and edge-species that declined steadily with increasing fragment size therefore increasing species evenness in the whole community.

Disturbance history effects on tree species diversity

The tree species diversity of the fragments also increased with increasing severity of environmental impacts though only the evenness component of diversity, J', was significantly affected. These findings may be related to the theory of intermediate disturbance (Connell 1971) that states that species diversity of tropical forests would be highest at intermediate levels of disturbance regimes. Light habitats would be more homogeneous at both high and low extremes of disturbance levels and the stronger dominance of a few highly adapted species would reduce diversity through competition, while under a moderate disturbance regime the heterogeneity of light habitats would flatten out species dominance (Roberts and Gilliam 1995; Brokaw and Busing 2000). If the theory really fits in the present findings, one must conclude that there is a missing piece among the 20 fragments: highly disturbed forests that should have lower species diversity. In fact, when the 20 forest fragments were chosen for study the research team targeted the less disturbed ones because the focus was on the relationship between tree species distribution and variations of the substrate.

Conclusions

The present contribution brings additional evidence that, although they cannot explain the high species diversity of tropical forests, disturbance and environmental heterogeneity can never be totally disregarded as diversity promoting factors. These findings demonstrate that the conservation of forest biodiversity in a fragmented landscape is a complex task, particularly in mountain systems where habitats are comparatively more diversified and usually have a patchy distribution. The conservation of forest diversity in landscapes with these peculiarities must address habitat heterogeneity and consider edge effects because they reduce the representation of the primitive habitats of the forest interiors. Another important aspect to be considered is that, to some extent, disturbance by humans may boost species diversity in forest fragments. Nevertheless this holds true only up to the level of disturbances of intermediate severity and that this increased diversity only refers to the expansion of disturbance-related species and consequent retreat of species typical of the undisturbed native habitats.

References

APG (2003) An update of the Angiosperm Phylogeny Group classification for the orders and families of flowering plants: APG II. Bot J Linn Soc 141:399–436

Appolinário V, Oliveira-Filho AT, Guilherme FAG (2005) Tree population and community dynamics in a Brazilian tropical semideciduous forest. Rev Bras Bot 28:347–360

Ashton PS, LaFrankie JV (2000) Patterns of tree species diversity among tropical rain forests. In: Kato M (ed) The biology of biodiversity. Springer, Berlin, Germany

Beebe J (1995) Basic concepts and techniques of rapid appraisal. Hum Organ 54:42–51

Botrel RT, Oliveira-Filho AT, Rodrigues LA, Curi N (2002) Influência do solo e topografia sobre as variações da composição florística e estrutura da comunidade arbóreo-arbustiva de uma floresta estacional semidecidual em Ingaí. Rev Bras Bot 25:195–213

Bourgeron PS (1983) Spatial aspects of vegetation structure. In: Golley FB (ed) Ecosystems of the world 14A – tropical rain forest ecosystems, structure and function. Elsevier, Amsterdam Holland, pp 29–47

Brokaw N, Busing RT (2000) Niche versus chance and tree diversity in forest gaps. Tree 15:183–188

Brower JE, Zar JH (1984) Field and laboratory methods for general ecology. Wm C Brown Pub., Dubuque, Iowa

Carvalho WAC (2002) Variações da composição e estrutura do comportamento arbóreo da vegetação de oito fragmentos de floresta semidecídua do Vale do Alto Rio Grande. MG. Masters thesis, Universidade Federal de Lavras, Lavras, Brazil

Caswell H, Cohen JE (1991) Communities in patchy environments: a model of disturbance, competition and heterogeneity. In: Kolasa J, Pickett STA (eds) Ecological heterogeneity. Springer-Verlag, Heidelberg, Germany, pp 97–122

Chambers R (1987) Rural development: putting the last first. Longman Scientific and Technical, New York

Chambers R (1992) Rural appraisal: rapid, relaxed and participatory (IDS Discussion Papers 311). Institute of development Studies, London

Chesson P (2000). Mechanisms of maintenance of species diversity. Annu Rev Ecol Syst 31:343–366

Clark DB, Clark DA, Read JM (1998) Edaphic variation and the mesoscale distribution of tree species in a neotropical rain forest. J Ecol 86:101–112

Connell JH (1971) On the role of natural enemies in preventing competitive exclusion in some marine animals and rainforest trees. In: Boer PJ, Gradwell GR (eds) Dynamics of numbers and populations. Advanced Studies Institute, Wageningen, Holland, pp 298–312

Curi N, Lima JM, Andrade H, Gualberto V (1990) Geomorfologia, física, química e mineralogia dos principais solos da região de Lavras (MG). Ciênc Prát 14:297–307

Dalanesi PE, Oliveira-Filho AT, Fontes MAL (2004) Flora e estrutura do componente arbóreo da floresta do Parque Ecológico Quedas do Rio Bonito, Lavras - MG, e correlações entre a distribuição das espécies e variáveis ambientais. Acta Bot Bras 18:737–757

Denslow JS (1987) Tropical rainforests gaps and tree species diversity. Annu Rev Ecol Syst 18: 431–451

Duivervoorden JF (1996) Patterns of tree species richness in rain forest of the Middle Caquetá area, Colombia, MW Amazonia. Biotropica 28:142–158

EMBRAPA (1997) Manual de métodos de análises de solo. 2. Empresa Brasileira de Pesquisa Agropecuária, Centro Nacional de Pesquisa de Solos Rio de Janeiro, Brazil

EMBRAPA (1999) Sistema brasileiro de classificação de solos. Empresa Brasileira de Pesquisa Agropecuária, Centro Nacional de Pesquisa de Solos, Rio de Janeiro, Brazil

Espírito-Santo FDB, Oliveira-Filho AT, Machado ELM, Souza JS, Fontes MAL, Marques JJGSM (2002) Variáveis ambientais e a distribuição de espécies arbóreas em um remanescente de floresta estacional semidecídua montana no campus da Universidade Federal de Lavras, MG. Acta Bot Bras 16:331–356

Ferraz EMN, Araújo EL, Silva SI (2004) Floristic similarities between lowland and montane areas of Atlantic Coastal Forest in Northeastern Brazil. Plant Ecol 174:59–70

Fiszon JT, Marchioro NPX (2002) Atividades antrópicas e fatores de impacto nos fragmentos. In: Efeitos da fragmentação de habitats: Recomendações de políticas públicas. MMA/Secretaria de Biodiversidade e Florestas/ PROBIO, Brasília, Brazil

Fontes MAL (1997) Análise da composição florística das florestas nebulares do Parque Estadual do Ibitipoca, Minas Gerais. Masters thesis, Universidade Federal de Lavras, Lavras, Brazil

Fowler N (1988) The effects of environmental heterogeneity in space and time on the regulation of populations and communities. In: Davy AJ, Hutchings MJ, Watkinson AR (eds) Plant population ecology. Blackwell, Oxford, UK, pp 249–269

Fundação SOS Mata Atlântica (2002) Atlas da evolução dos remanescentes florestais e ecossistemas associados da Mata Atlântica no período 1995–2000. http://www.sosmatatlantica.org.br

Galindo-Leal C, Câmara IG (2003) Atlantic Forest hotspot status: an overview. In: Galindo-Leal C, Câmara IG (eds) The Atlantic Forest of South America. Center for Applied Biodiversity Science, Washington DC, pp 3–11

Guilherme FAG, Oliveira-Filho AT, Appolinário V, Bearzoti E (2004). Effects of flooding regimes and woody bamboos on tree community dynamics in a section of tropical semideciduous forest in South-Eastern Brazil. Plant Ecol 174:19–36

Harley RM (1995) Introduction. In: Stannard B (ed), Flora of the Pico das Almas, Chapada Diamantina, Bahia, Brazil. Royal Botanic Gardens, Kew, pp 1–78

Heltshe JF, Forrester NE (1983) Estimating species richness using the jackknife procedure. Biometrics 39:1–12

Hill MO, Gauch HG (1980) Detrended correspondence analysis, an improved ordination technique. Vegetatio 42:47–58

Hubbell SP (2001) The unified neutral theory of biodiversity and biogeography. Princeton University Press, Princeton, New Jersey

Kapos V (1989) Effects of isolation on the water status of forest patches in the Brazilian Amazon. J Trop Ecol 5:173–185

Laurance WF, Yensen E (1991) Edge effects in fragmented habitats. Biol Conserv 55:77–92

 Springer

Laurance WF, Bierregaard RO Jr (1997) Tropical Forest remnants: ecology, management, and conservation of fragmented communities. University of Chicago Press, Chicago

Leigh EG Jr, Davidar P, Dick CW, Puyravaud JP, Terborgh J, ter Steege H, Wright SJ (2004) Why do some tropical forests have so many species of trees?. Biotropica 36:447–473

Leopold LB, Clarcke FE, Hanshaw BB, Balsley JR (1971) A procedure for evaluating environmental impact. Geological Survey Circular 645. Government Printing Office, Washington, DC

Machado JNM, Hargreaves P (2000) Ecologia da serra de São José, Minas Gerais – Composição florística lenhosa e estudos fitossociológicos. Relatório técnico, IBAMA Floresta Nacional de Ritápolis Ritápolis, Brazil

Machado ELM, Oliveira-Filho AT, Carvalho WAC, Souza JS, Borém RAT, Botezelli L (2004). Análise comparativa da estrutura e flora do compartimento arbóreo-arbustivo de um remanescente florestal na Fazenda Beira Lago, Lavras, MG. Rev Árvore 28:493–510

McArthur RH, Wilson EO (1967) The theory of island biogeography. Princeton University Press, Princeton, New Jersey

McCune B, Mefford MJ (1999) PC-ORD version 4.0, multivariate analysis of ecological data, Users guide. MjM Software Desing, Glaneden Beach Oregon

McGarigal K, Marks B (1994) Fragstats version 2.0: spatial pattern analysis program for quantifying landscape structure. Oregon State Department, Conallis, Oregon

Metzger JP (1997) Relationships between landscape structure and tree species diversity in tropical forests of South-East Brazil. Landsc Urban Plan 37:29–35

Metzger JP, Goldenberg R, Bernacci LC (1998) Diversidade e estrutura de fragmentos de mata de várzea e de mata mesófila semidecídua submontana do rio Jacaré-Pepira (SP). Rev Bras Bot 21:321–330

Miller RG Jr (1991) Simultaneous statistical inference. Springer-Verlag, New York

Mittermeir RA, Myers N, Gil PR, Mittermeir CG (1999) Hotspots: Earth's biologically richest and most endangered terrestrial ecoregions. CEMEX/Conservation International, Washington, DC

MMA (2000) Avaliação e ações prioritárias para a conservação da biodiversidade da Mata Atlântica e Campos Sulinos Conservation International do Brasil, Fundação, SOS Mata Atlântica, Fundação Biodiversitas, Instituto de Pesquisas Ecológicas, Secretaria do Meio Ambiente do Estado de São Paulo. Instituto Estadual de Florestas-MG; Ministério do Meio Ambiente, Secretaria de Biodiversidade e Florestas, Brasília, Brazil

MMA (2002) Biodiversidade brasileira, Avaliação e identificação de áreas e ações prioritárias para conservação, utilização sustentável e repartição dos benefícios da biodiversidade nos biomas brasileiros. Ministério do Meio Ambiente, Secretaria de Biodiversidade e Florestas, Brasília, Brazil

Murcia C (1995) Edge effects in fragmented forests: implication for conservation. Tree 10:58–62

Myers N, Mittermeir RA, Mittermeir CG, Fonseca GAB, Kents J (2000) Biodiversity hotspots for conservation priorities. Nature 407:853–858

Nunes YRF, Mendonça AVR, Oliveira-Filho AT, Botezelli L, Machado ELM (2003) Variações da fisionomia, diversidade e composição de guildas da comunidade arbórea em um fragmento de floresta semidecidual em Lavras, MG. Acta Bot Bras 17:213–229

Oldeman RAA (1990) Dynamics in tropical rain forests. In: Holm-Nielsen LB, Nielsen IC, Balslev H (eds) Tropical forests, Botanical dynamics, speciation and diversity. Academic Press, London, UK, pp 3–21

Oliveira-Filho AT, Fontes MAL (2000) Patterns of floristic differentiation among Atlantic Forest in South-eastern Brazil, and the influence of climate. Biotropica 32:139–158

Oliveira-Filho AT, Vilela EA, Gavilanes ML, Carvalho DA (1994a) Comparison of the woody flora and soils of six areas of montane semideciduous forest in southern Minas Gerais, Brazil. Edinburgh. J Bot 51:355–389

Oliveira-Filho AT, Vilela EA, Gavilanes ML, Carvalho DA (1994b) Effect of flooding regime and understory bamboos on the physiognomy and tree species composition of a tropical semideciduos forest in Southeastern Brazil. Vegetatio 113:99–124

Oliveira-Filho AT, Vilela EA, Carvalho DA, Gavilanes ML (1994c) Differentiation of streamside and upland vegetation in an area of montane semideciduous forest in southeastern Brazil. Flora 189:287–305

Oliveira-Filho AT, Mello JM, Scolforo JRS (1997) Effects of past disturbance and edges on tree community structure and dynamics within a fragment of tropical semideciduos forest in southeastern Brazil over a five-year period (1987–1992). Plant Ecol 131:45–66

Oliveira-Filho AT, Carvalho DA, Vilela EA, Curi N, Fontes MAL (2004) Diversity and structure of the tree community of a patch of tropical secondary forest of the Brazilian Atlantic Forest Domain 15 and 40 years after logging. Rev Bras Bot 27:685–701

[208]

Oliveira-Filho AT, Jarenkow JA, Rodal MJN (2006) Floristic relationships of seasonally dry forests of eastern South America based on tree species distribution patterns. In: Pennington RT, Ratter JA, Lewis GP (eds) Neotropical savannas and dry forests: Plant diversity, biogeography and conservation. CRC Press, Boca Raton, pp 151–184

Palmer MW (1991) Estimating species richness: the second-order jackknife reconsidered. Ecology 72:1512–1513

Phillips OL, Gentry AH (1994) Increasing turnover through time in tropical forests. Science 263:954–958

Pickett STA, White PS (1985) The ecology of natural disturbance and patch dynamics. Academic Press, San Diego, California

Pinto JRR, Oliveira-Filho AT, Hay JDV (2006) Influence of soil and topography variables on the composition of the tree community of a Central Brazilian valley forest. Edinb J Bot 62:1–22

PROBIO (1999) Estratégias para conservação e manejo da biodiversidade em fragmentos de florestas semidecíduas. Ministério da Ciência e Tecnologia, Brasília, Brazil

Ranta E, Kaitala V, Lindström J (1998) Spatial dynamics of populations. In: Bascompte J, Solé RV (eds) Modelling spatiotemporal dynamics in ecology. Springer-Verlag, Berlin, pp 45–60

Ricklefs RE (1987) Community diversity: relative roles of local and regional processes. Science 235:167–171

Roberts MR, Gilliam FS (1995) Patterns and mechanisms of plant diversity in forested ecosystems: implications for forest management. Ecol Appl 5:969–977

Rodrigues RR, Nave AG (2000) Heterogeneidade florística das matas ciliares. In: Rodrigues RR, Leitã-Filho HF (eds) Matas ciliares: conservação e recuperação. EDUSP São Paulo, Brazil, pp 45–71

Rodrigues LA, Carvalho DA, Oliveira-Filho AT, Botrel RT, Silva EA (2003) Florística e estrutura da comunidade arbórea de um fragmento florestal em Luminárias, MG. Acta Bot Bras 17:71–87

Scudeller VV, Martins FR, Shepherd GJ (2001) Distribution and abundance of arboreal species in the Atlantic ombrophilous dense forest in Southeastern Brazil. Plant Ecol 152:185–199

Silva VF, Oliveira-Filho AT, Venturin N, Carvalho WAC, Gomes JBV (2005) Impacto do fogo no componente arbóreo de uma floresta estacional semidecídua no município de Ibituruna, MG. Acta Bot Bras 19:701–716

Silva-Júnior MC, Furley PA, Ratter JA (1996) Variation in the tree communities and soils with slope in gallery forest Federal District Brazil. In: Anderson MG, Brooks SM (eds) Advances in hillslope processes. John Wiley & Sons, London, pp 451–469

Simberloff D, Dayan T (1991) The guild concept and the structure of ecological communities. Annu Rev Ecol Syst 22:115–143

Souza JS, Espírito-Santo FDB, Fontes MAL, Oliveira-Filho AT, Botezelli L (2003) Análise das variações florísticas e estruturais da comunidade arbórea de um fragmento de floresta semidecídua às margens do rio Capivari, Lavras-MG. Rev Árvore 27:185–206

StatSoft Inc. (1995) STATISTICA for Windows, Release 5.0 A (Computer program manual). Statsoft Inc., Tulsa, Oklahoma

Tabarelli M, Mantovani W (1999) A riqueza de espécies arbóreas na floresta atlântica de encosta no estado de São Paulo (Brasil). Rev Bras Bot 22:217–223

Tabarelli M, Baider C, Mantovani W (1998) Efeitos da fragmentação na Floresta Atlântica da bacia de São Paulo. Hoehnea 2:168–186

Terborgh J (1992) Diversity and the tropical rain forest. Scientific American Library, New York

ter Steege H, Jetten VG, Polak AM, Werger MJA (1993) Tropical rain forest types and soil factors in a watershed area in Guyana. J Veg Sci 4:705–716

Tilman D (1999) Diversity by default. Science 283:495–496

van den Berg E, Oliveira Filho AT (1999) Spatial partiticioning among tree species within an area of tropical montane gallery forest in south-eastern Brazil. Flora 194:249–266

van den Berg E, Oliveira Filho AT (2000) Composição florística e estrutura fitossociológica de uma floresta ripária em Itutinga, MG, e comparação com outras áreas. São Paulo: Rev Bras Bot 23:231–253

van der Maarel E (1993) Some remarks on disturbance and its relation to diversity and stability. J Veg Sci 4:733–736

Veloso HP, Rangel-Filho ALR, Lima JCA (1991) Classificação da vegetação brasileira adaptada a um sistema universal. FIBGE, Rio de Janeiro, Brazil

Vilela EA, Ramalho MAP (1979) Análise das Temperaturas e Precipitações Pluviométricas de Lavras - MG. Ciênc Prátic 3:71–79

Whitmore TC (1997) Tropical forest disturbance, disappearance, and species loss. In: Laurance WF, Bierregaard RO (eds) Tropical forest remnants: ecology, management, and conservation of fragmented communities. University of Chicago Press, Chicago, pp 3–12

Whitmore TC, Burslem DFRP (1998) Major disturbances in tropical rainforests. In: Newbery DM, Prins HHT, Brown N (eds) Dynamics of tropical communities. Blackwell, Oxford, UK, pp 549–565

Biodivers Conserv (2007) 16:1785–1801
DOI 10.1007/s10531-006-9071-4

ORIGINAL PAPER

Correspondence between scientific and traditional ecological knowledge: rain forest classification by the non-indigenous ribereños in Peruvian Amazonia

K. J. Halme · R. E. Bodmer

Received: 11 July 2005 / Accepted: 19 May 2006 / Published online: 12 July 2006
© Springer Science+Business Media B.V. 2006

Abstract Traditional ecological knowledge (TEK) is a potential source of ecological information. Typically TEK has been documented at the species level, but habitat data would be equally valuable for conservation applications. We compared the TEK forest type classification of ribereños, the non-indigenous rural peasantry of Peruvian Amazonia, to a floristic classification produced using systematically collected botanical data. Indicator species analysis of pteridophytes in 300 plots detected two forest types on non-flooded tierra firme, each associated with distinct soil texture and fertility, and one forest type in areas subject to flooding. Nine TEK forest types were represented in the same set of plots. Each TEK forest type was consistently (>82%) associated with one of the three floristic classes and there were also clear parallels in the ecological characterizations of the forest types. Ribereños demonstrated clear preferences for certain forest types when selecting sites for slash-and-burn agriculture and hunting. Our results indicate that the non-tribal inhabitants of Amazonia possess valuable TEK that could be used in biodiversity inventories and wildlife management and conservation for characterizing primary rain forest habitats in Amazonia.

Keywords Amazonia · Beta-diversity · Traditional ecological knowledge · Tropical rain forest · Vegetation classification · Wildlife habitat

Abbreviation
TEK   Traditional ecological knowledge

K. J. Halme (✉)
Department of Biology, University of Turku, FIN-20014 Turku, Finland
e-mail: kati.halme@gmail.com

R. E. Bodmer
Department of Anthropology, Durrell Institute of Conservation and Ecology, University of Kent, Canterbury, Kent CT2 7NS, UK

[211]

 Springer

Introduction

Understanding the local and regional diversity patterns is critical for efficient conservation and management planning. Vegetation inventory and forest classification systems can be extremely valuable for these purposes, but there have been few attempts to develop appropriate methods for the mega-diverse Amazonian rain forests (Condit 1996). Complex vegetation classification schemes were not deemed necessary in the past due to the longstanding but incorrect view of Amazonian forests having a high number of species but relatively little spatial heterogeneity (Posey 1983). Another likely reason is the high diversity of tree species found in Amazonian forests (Gentry 1988; Valencia et al. 1994; ter Steege et al. 2000). Species identification is a daunting task and few tree species can be identified in the field, so it is often impracticable to typify forest types based on characteristic tree species. Consequently, published forest classifications in Amazonia have been primarily based on clearly noticeable geomorphological and physiognomical traits that may not completely reflect the ecological characteristics of the forests.

The two major forest types in Amazonia are inundated forests and upland tierra firme forests (e.g. Pires and Prance 1985). More detailed vegetation classifications for Peruvian Amazonia include the systems of Malleux (1975) and Encarnación (1985). Both systems use a combination of topography, drainage conditions and vegetation structure as classification criteria, but put little emphasis on the species composition. This superficiality is especially prominent in the classification of tierra firme forests that are divided into a small number of forest types, although these forests cover the vast majority of Amazonia (Salo et al. 1986; Hess et al. 2003) and consist of a mosaic of floristic communities that differ in soil characteristics and species composition (Gentry 1981; Tuomisto et al. 1995; Ruokolainen et al. 1997; Ruokolainen and Tuomisto 1998; Phillips et al. 2003; Tuomisto et al. 2003a, b).

These floristic patterns can be documented by analyzing the species composition of two indicator groups, the pteridophytes (ferns and fern allies) and the family Melastomataceae. The method is based on extensive studies on the distribution and edaphic preferences of these indicator taxa (Tuomisto and Ruokolainen 1994, 1998; Ruokolainen et al. 1997; Ruokolainen and Tuomisto 1998; Tuomisto et al. 2003a, b). The floristic patterns of these indicator groups are strongly correlated with the floristic patterns of trees and palms across sites (Ruokolainen et al. 1997; Ruokolainen and Tuomisto 1998; Vormisto et al. 2000) and these groups can apparently be used to characterize the differences in flora and the soil fertility and drainage across inventoried sites. The resulting habitat patterns are also detectable in the spectral reflectance of Landsat TM satellite images (Tuomisto et al. 2003a, b; Salovaara et al. 2005).

We used pteridophytes as an indicator taxon to produce a floristic forest type classification for over 60 km of line transects in Peruvian Amazonia in order to separate primary rain forest types (Salovaara et al. 2004). During the field work we noticed that the local people also recognized numerous ecologically different forest types in the study area and decided to compare their forest classification with the one we produced by quantitative scientific methods. Such comparisons are rare (Shepard et al. 2001), even though there are many earlier accounts on the sophisticated ecological knowledge possessed by indigenous peoples (Balick 1996) and in some cases vegetation scientists have adopted local forest type nomenclature (Encarnación 1985; Pires and Prance 1985).

The term traditional ecological knowledge (TEK) is used here to refer to a body of ecological knowledge and insights that have been accumulated empirically over generations. Many studies on TEK have focused on folk taxonomic classifications of plants and animals or listing of useful species and characterization of their use by different cultures (e.g. Berlin 1973; Balick 1996; Jinxiu et al. 2004) and have repeatedly revealed close parallels between the local and scientific taxonomies (Holman 2002). TEK has also many practical applications in conservation, management and research (Huntington 2000; Sheil and Lawrence 2004). It has been successfully utilized in fisheries research (Poizat and Baran 1997; Mackinson 2001; Huntington et al. 2004), soil classification and mapping (Messing and Hoang Fagerström 2001; Oudwater and Martin 2003), biodiversity inventories (Hellier et al. 1999; Huntington 2000; Jinxiu et al. 2004) and in natural resources management (Phuthego and Chanda 2004).

Previous ethnobiological studies in Amazonian forests have primarily brought together information on just one level of biological diversity, the species diversity (alpha-diversity), whereas habitat diversity (beta-diversity) has received less attention (Shepard et al. 2004). Fleck and Harder (2000) documented the habitat classification system of the Matses indians in northeastern Peru. The system is based on vegetative and geomorphological habitat classes, which can be applied concurrently to recognize 178 rain forest habitat types. Altogether 47 rain forest habitat types had specific names. Sampling of 16 habitat types showed that they differed significantly in vegetation structure, palm species composition and occurrence of small mammals. Shepard et al. (2001) studied the classification system of the Matsigenka indians in southeastern Peru. The Matsigenka distinguish 69 vegetational and 29 abiotic types of forest habitats and many of these were visible in Landsat TM satellite images (Shepard et al. 2004). Thus, at least some of the Matses and the Matsigenka forest classes were distinct also in scientific terms, although they recognized more habitat classes at a much finer spatial resolution than any scientific classification system.

Several other studies have documented less detailed habitat classifications by indigenous peoples and non-tribal dwellers in Amazonia (Carneiro 1978; Parker et al. 1983; Posey 1983; Frechione et al. 1989). All these classifications are based on similar biotic (e.g. indicator species, forest structure, successional stage) and abiotic (e.g. flooding and disturbance regimes, topography and soil type) characteristics that have practical importance for people utilizing the forests for agriculture, hunting and gathering. Thus they are likely to reflect ecologically significant differences between forest types and could provide valuable information about beta-diversity of Amazonian forests. However, these classifications have not been validated by quantitative methods.

This study had three objectives. First, we cross-referenced a forest type classification that was based on a systematic use of indicator plant species (scientific classification) to the classification used by the local ribereños (TEK classification). Peruvian ribereños are riverine dwellers (called *mestizos* or *caboclos* in other parts of Amazonia), whose culture is a mixture of indigenous and more recent immigrant influence (Parker 1989; Kvist and Nebel 2001). The ribereño classification has not been validated by scientific methods earlier, although Encarnación (1985) adopted the ribereño nomenclature in his forest type classification. Second, we explored how the different forest types were perceived and valued by the local people who depend on forest resources for their livelihood. Third, we discuss the potential management and conservation applications of this type of habitat information.

Materials and methods

Study area

The study was conducted in 2001 in the Yavarí-Mirím river basin in northeastern Peruvian Amazonia (Fig. 1). The area is ca. 100–180 m above sea level and the landscape mainly consists of gently undulating hills that are separated by a network of flat stream flood plains. Mean monthly temperature varies between 25 and 27°C and mean annual precipitation is 3,100 mm (Marengo 1998).

In parallel publications we have classified the vegetation along the Yavarí-Mirím (Salovaara et al. 2004, 2005). Pitman et al. (2003) conducted a vegetation inventory along the adjacent Yavarí river. These studies concluded that the forests in the Yavarí basin are rather similar to the forests near the city of Iquitos (e.g. Ruokolainen et al. 1997; Ruokolainen and Tuomisto 1998). Although extremely nutrient poor white sand forests are rare in the Yavarí area, there is considerable heterogeneity in the soil fertility and drainage among sites within the tierra firme forests.

The interviews were conducted in June 2002 in the ribereño village of Nueva Esperanza, which is located in the lower part of the Yavarí-Mirím river. The village had 28 households and approximately 180 inhabitants at the time. There were another three settlements within the river basin, each with less than 20 inhabitants, but they were not included in the study.

The resource use and life style of ribereños closely resemble that of their indigenous ancestors, especially in remote areas away from markets and employment opportunities. In Nueva Esperanza many adults could name at least one indigenous group that their parent or grandparent belonged to. The villagers live in subsistence

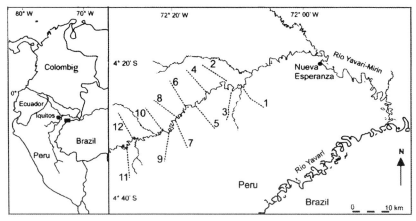

Fig. 1 Location of the study area. The map insert shows the location of the Yavarí-Mirím river basin as a dark square next to the Brazilian border in NE Peruvian Amazonia. Interviews were done in the village of Nueva Esperanza in the lower part of the Yavarí-Mirím basin. TEK vegetation classification was conducted in the middle part of the basin along the line transects 7, 8 and 11 (inventoried completely) and 5 (inventoried partly)

economy that is founded on shifting cultivation, fishing and hunting. Small swiddens (*chacras*) are planted with manioc, plantain, fruit trees, sugar cane and other crops and cultivated actively for a few years before the swidden is gradually abandoned. Main supply of protein is fish and wild meat, which are also the only locally available source of cash income. However, the trade is controlled by a handful of villagers who have the necessary capital to transport the products to market places along the Amazon river. The village participates in a community-based wildlife management program and the inhabitants are used to collaborating with scientists (Bodmer and Puertas 2000).

Data collection

Floristic forest type classification was produced for eight line transects (Fig. 1) based on distribution and abundance of Pteridophytes (ferns and fern allies), which were used as the indicator taxon (Salovaara et al. 2004). Each line transect was 8 km long and the floristic transects were divided into 100 m long and 2 m wide vegetation sampling units. The results of Salovaara et al. (2004) are used in the present paper as scientific reference data, which are compared with the TEK forest classification.

TEK forest types were recorded with the help of experienced field assistants for 100 m long vegetation sampling units along four transects. Three transects (numbered 7, 8 and 11 in Fig. 1) were surveyed completely and transect five partly. This work was completed before conducting the floristic field survey and thus neither the informants nor the investigators had prior knowledge about the floristic classification when the TEK classification was conducted. Transects were walked with local field assistant, who was asked to name the forest type for each 100 m long transect unit and to explain how he recognized it. The TEK classification had somewhat finer resolution than the sampling unit size (100 m) used in the floristic classification. As a result some sampling units were recorded as transitional (representing two forest types) in the TEK classification.

After the transect surveys were completed, the TEK forest types found along the transects were cross-checked with the nomenclature used by the inhabitants in Nueva Esperanza in order to confirm that the forest type names were in general use. Five hunters were informally asked to tell what forest types were found in the area and to explain what their typical characteristics (such as indicator species, structural features and soil characteristics) were.

Based on these discussions six forest types that were common in the area and widely known by the locals were selected to be used in the subsequent interviews. Two of them were selected even though they were not present along the transects: *varillal* was reported to be widespread near the village and *arenal* (white sand forest) represents the nutrient poorest extreme of the soil fertility gradient in Peruvian Amazonia. In conclusion, the questionnaires included two inundated forest types and four tierra firme forest types. Two of the tierra firme forest types were reported to grow on relatively fertile soil and two on relatively infertile soil.

All 26 hunters in the community of Nueva Esperanza were interviewed individually using semi-structured interviews to study their perception of the six selected forest types. All respondents were already familiar with the interviewer. Before starting the interview they were explained the purpose of the interviews, how the results would be used and that participation in the study was voluntary. Each respondent was also given some food supplies as a compensation for the time they spent doing the interview.

Each respondent was asked to rank-order the six TEK forest types from the most suitable to the least suitable for agriculture using cards that each had one forest type written on it. The respondent was asked to consider a situation where he could select any of the forest types for clearing *a chacra*. He was prompted to select the best forest type and then to repeat the selection among the remaining forest types, until all cards were ranked. After the task was completed, the respondent was asked to explain briefly why he ranked the forest types the way he did.

The suitability of different forest types for hunting was studied using triad comparisons, where each triad represented three out of the six forest types. Randomized questionnaires were produced for each respondent in Anthropac software (http://www.analytictech.com/). Each questionnaire had 20 triads, i.e. each forest type triplet appeared only once. All triads were presented in a similar manner, but the order of forest types in the triads and the order of triads were randomized for each respondent.

The objective was to have the respondent to select among three forest types the one that he deemed to offer the best chances of finding game and killing it. Each triad was presented as a simplified map (Fig. 2). The respondent was asked to consider a situation where he is hunting in an area he knows well and he is in the crossing of three trails that each lead to a different forest type. The respondent was then prompted to select the trail (and the forest type) which in his opinion offered the highest probability of a successful hunt. This question could not be completed with one illiterate respondent who had difficulties in understanding the triads.

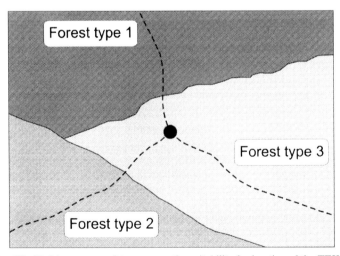

Fig. 2 A simplified habitat map used to compare the suitability for hunting of the TEK forest types in triad comparisons. Respondents were asked to select one of the three marked trails and, consequently, one of the forest types during a hypothetical hunting trip in a situation where they were in the crossing of the trails. The objective was to maximize the probability of a successful hunt

Data analysis

In order to quantify the correspondence between the TEK and scientific forest classification systems, the 100-m transect segments were allocated to a TEK forest type on the basis of field interviews and to a floristic forest type based on Salovaara et al. (2004). As a result, pairs of corresponding forest types in the two classification systems could be formed.

Preference scores were calculated for the TEK forest types based on their perceived suitability for agriculture. Each forest type received from six (highest rank) to one (lowest rank) points from each respondent, and the points were summed across all respondents for each forest type. Results were also summarized by computing the number of times each forest type received a particular ranking.

The triad comparisons of the TEK forest types based on their suitability for hunting were summarized using a simple preference scoring, where each forest type received one point each time it was selected from the triad. Points were then summed across all informants for each forest type. Since there were 25 respondents that each made 20 choices, the sum of preference scores was 500. If no forest type preferences were present, these 500 points would be allocated randomly among the six forest types. Thus each forest type would receive approximately $500/6 = 83$ points. Any significant deviations from this mean value would indicate either preference or avoidance.

A permutation test was used to check whether deviations from the expected random allocation were statistically significant. This was done by simulating 25 questionnaires that had randomly selected responses. After each simulation a preference score was summed for each forest type. This was repeated 1,000 times to test significance at the P 0.001 level. Permutation thus allowed testing whether the observed deviations from the expected mean score of 83 could have been produced by chance alone. The permutation procedure was programmed in Resampling Stats version 5.0.2. (1999, http://www.resample.com).

Results

Parallels between TEK and floristic forest types

Altogether 300 vegetation sampling units were classified using both the floristic and the TEK forest classifications. In total, 10 TEK forest types were found along the four transects. The floristic classification included four main forest types; terrace forests, Pebas formation forests, intermediate tierra firme forest and inundated forests (including palm swamps) (Salovaara et al. 2004, 2005). For the purposes of the present paper Pebas formation and intermediate tierra firme forests were combined to one class, because they were floristically rather similar (Salovaara et al. 2004, 2005). This combined class is called Pebas formation/intermediate tierra firme.

Table 1 presents the allocation of the 300 transect sampling units to different forest types by the two classification systems. Each TEK forest type corresponded closely with one of the floristic forest types, which is indicated by the fact that any given TEK forest type had at least 82% of its sampling units allocated to a single floristic forest type. However, there were several TEK forest types that corresponded with each of the floristic forest types.

Table 1 Comparison of how TEK and floristic classifications allocated 300 transect sampling units to forest types. Shaded cells indicate which forest types most often corresponded with each other in the two classification systems. Characteristics of the forest types are described in Table 2

TEK classification	Floristic classification			Total
	Flooded	Terrace	Pebas/intermediate	
Aguajal	13 (87%)	2	0	15
Bajial	14 (82%)	0	3	17
Restinga	6 (100%)	0	0	6
Chamizal	0	1 (100%)	0	1
Irapayal/Shamposal	1	74 (85%)	12	87
"Monte bueno"	0	0	9 (100%)	9
Shapajal	0	0	110 (100%)	110
Yarinal	0	0	30 (100%)	30
Supaichacra	0	0	1 (100%)	1
Transitional	5	1	18	24
Total	39	78	183	n = 300

Table 2 characterizes the 10 TEK forest types and the matching floristic forest types. Inundated TEK forest types are differentiated on the basis of flooding regime, which varies from seasonal flooding (*bajial* and *restinga*) to permanent waterlogging in palm swamp (*aguajal*). *Aguajal* is furthermore characterized by the dominance of the *aguaje* palm (*Mauritia flexuosa* L. f.). *Restinga* is found on somewhat higher ground and is less frequently inundated than *bajial*.

Chamizal forest is generally found on extremely nutrient poor white sand soil (Anderson 1981) and is poorly drained, but it is never inundated by rivers. Its physiognomy differs drastically from the other inundated forests due to the low stature and openness of the canopy. Only one sampling unit belonged to this TEK forest type. During all our visits the plot was flooded by rain water, but in the floristic analysis it was assigned to the infertile soil terrace forest class. *Irapayal* forest of the TEK classification clearly corresponded with the floristic class terrace forest. *Irapayal* is characterized by the dominance of a small understory palm *Lepidocaryum tenue* Mart., which often forms dense thickets in the understorey. Another forest type, *shamposal*, was recognized within the *irapayal* forests. *Shamposal* typically has a thick layer of litter and humus on the ground, indicating slow rate of decomposition.

Four TEK forest types were matched with the floristic forest type Pebas formation/intermediate tierra firme, which represents relatively fertile tierra firme soils. The Pebas formation is a geological formation with clayey soils of (semi)marine origin (Hoorn 1993). The corresponding TEK forest types were mainly differentiated by characteristic tree species. *Yarinal* and *shapajal* forest classes were both named after the dominant palm species found in these forests (*yarina*, *Phytelephas* spp. and *shapaja*, *Attalea* spp.). In some cases these palm species were not especially abundant, but the informants considered the forest to be in other respects similar to these fertile soil forest types. In these cases they used the general term "*monte bueno*", or "good forest". This term apparently referred to the fertile and productive soils, but also to the fact that the understorey of the "good forest" is relatively open. These three forest types were found in extensive areas, whereas *supaichacra* was only found in patches of 100 m diameter. This peculiar forest type is characterized by a strong dominance of two myrmecophilous tree species (*Duroia hirsuta* (Poepp.)

Table 2 Corresponding forest types in the TEK and floristic classifications. Local forest type terms are described based on their flooding regime, indicator species and soil characteristics. Two forest types that were not present along the line transects but were described by the locals (*arenal* and *varillal*) were added here. The forest types included in the questionnaires are shaded

Floristic classification	TEK classification
Flooded forest Seasonally flooded forest along rivers and streams and permanently waterlogged palm swamps	*Aguajal* Permanently or seasonally flooded swamp forest with high occurrence of the palm *Mauritia flexuosa* L. f. (aguaje) *Bajial* Seasonally flooded forest along rivers and large streams *Restinga* Seasonally flooded forest along rivers and large streams, somewhat higher ground and not as regularly inundated as *bajial*
Terrace forest Tierra firme forest growing on relatively nutrient poor loamy/sandy soil	*Chamizal* Swamp forest inundated by rain water, growing on poorly drained, nutrient poor white sand soil *Irapayal* Tierra firme forest with high dominance of the palm *irapay* (*Lepidocaryum tenue* Mart.) *Shamposal* Tierra firme forest with soils covered by deep litter and humus layer, generally high dominance of *Lepidocaryum tenue* Mart., subtype of *irapayal* forest *Arenal* Low canopy forest growing on relatively infertile sandy soil, generally reffered to as *varillal* in the Iquitos region. Not present along the transects, but according to the respondents is found patchily in the area
Pebas formation/intermediate tierra firme Tierra firme forests on relatively nutrient rich soils (either on Pebas formation clays or soils of intermediate fertility)	*"Monte bueno"* General term for tierra firme forest with relatively fertile soils and open understory, but no particular dominant palm species *Shapajal* Relatively fertile soil tierra firme forest, high frequency of *Attalea* spp. (*shapaja*) palms *Yarinal* Relatively fertile soil tierra firme forest with dominance of *Phytelephas* spp. (*yarina*) palms *Supaichacra* Tierra firme forest where myrmecophilous trees *Duroia hirsuta* (Poepp.) K. Schum. (Rubiaceae) and *Cordia nodosa* Lam. (Boraginaceae) are dominant. Found in small patches throughout tierra firme on relatively nutrient rich clayey soil *Varillal* Relatively fertile soil tierra firme forest not found along the transects but documented in the vicinity of Nueva Esperanza. Based on the fern species composition this forest type belongs to Pebas formation forests

K. Schum., Rubiaceae and *Cordia nodosa* Lam., Boraginaceae) and is thus also physiognomically clearly different from the other tierra firme forests.

Generally the TEK nomenclature and forest type classification followed a system that is widely used in Peruvian Amazonia and has also been adapted by Encarnación (1985). However, there were some important differences. Based on the preliminary interviews with selected hunters, the TEK classification included an infertile soil tierra firme forest type called *arenal* or *varillal arenoso*. This forest type was clearly identical with the *varillal* forest of Encarnación (1985). It grows on extremely infertile white sand which represents one extreme of the soil fertility gradient in Amazonia (Anderson 1981; Encarnación 1985; Ruokolainen and Tuomisto 1998; Ruokolainen et al. 1997), and for that reason it was included in the questionnaires. Although *varillal arenoso* was not present along the classified transects, the hunters reported that it is found patchily in the Yavarí-Mirím basin.

However, in Nueva Esperanza the word *"varillal"* was also used for a clearly different, fertile soil forest that was described to be common in the vicinity of the village. Several sites near the village were visited to check which pteridophyte species were found in this forest type. All the species were typical of Pebas formation/intermediate forest. Thus the questionnaires included both *varillal*, a relatively fertile soil tierra firme forest type, and *arenal* (or *varillal arenoso*), a tierra firme forest type on extremely nutrient poor white sand.

Suitability of TEK forest types for agriculture

Table 3 shows the rankings of forest types according to their suitability for agriculture (i.e. clearing a *chacra*). *Shapajal* and *varillal* represented fertile soil tierra firme forests and *arenal* and *irapayal* infertile soil tierra firme. *Bajial* and *aguajal* are both inundated forest types.

Shapajal and *varillal* were ranked highest. These forest types were declared to be well suited for cultivation because of productive soils; all the species that are commonly cultivated in *chacra* grow well on these soils. In *arenal* production was reported to be lower. For example, one respondent explained that plantain and manioc can not be grown in *arenal*, although they grow well in *shapajal* or *varillal*. *Bajial* received almost similar rankings as *arenal*. Some respondents mentioned that there are several different types of *bajial*, and some are suitable for cultivation (mainly corn, rice and manioc), whereas others are not.

Table 3 The number of times each forest type received certain ranking according to its suitability for chacra cultivation. For each forest type the ranking that received most responses is shaded. Preference score was calculated by allocating scores from six (highest rank) to one (lowest rank) for each forest type and summing the scores across all respondents for each forest type

Forest type	Preference score	Respondent's rankings					
		1st	2nd	3rd	4th	5th	6th
Shapajal	140	14	8	4			
Varillal	139	11	14		1		
Arenal	95	1	3	10	10	2	
Bajial	90		1	12	11	2	
Irapayal	55				4	21	1
Aguajal	27					1	25

Irapayal was ranked to be the least suitable among the tierra firme forests. Many respondents considered it to be completely inapt for cultivation due to a thick layer of organic material on top of the mineral soil. Some respondents mentioned that if such a layer is not present, some fruit trees could be cultivated even in *irapayal*, whereas others said that production is always extremely low or null and thus it is not worth clearing *a chacra* in *irapayal*. As expected, *aguajal* (permanently waterlogged palm swamp) was ranked as the least suitable forest type. The respondents unanimously agreed that *aguajal* cannot be used for agriculture.

Suitability of TEK forest types for hunting

Figure 3 shows the results of the triad comparison of forest types according to their suitability for hunting. *Bajial* was the only forest type that received a score that could be expected by random allocation. All the other forest types' scores deviated considerably from random and these deviations were statistically significant at the $P < 0.001$ level. Fertile soil tierra firme forests, *shapajal* and *varillal*, were the most preferred forest types and they were clearly preferred over the other forest types. The remaining three forest types received noticeably lower scores, indicating clear avoidance by hunters; *irapayal* had the lowest score, while *arenal* and *aguajal* were ranked slightly higher.

More detailed analysis of the questionnaires confirmed that the respondents clearly preferred the fertile soil tierra firme forests over the infertile soil tierra firme for hunting. For example, in all except one out of the 50 cases where the respondents were given a triad that included two fertile soil tierra firme forests *(varillal* and *shapajal)* and one of the infertile soil tierra firme forests *(arenal* or *irapayal)*, respondents preferred one of the two fertile soil forests. In the 50 cases where two

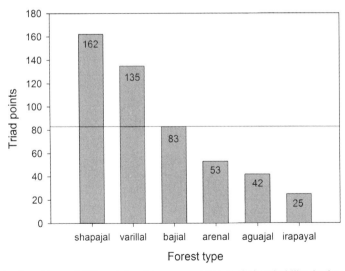

Fig. 3 Hunters' rankings of different forest types according to their suitability for hunting. Totally random allocation would give a score of ca. 83 to all forest types, but based on a permutation test all the other forest types except *bajial* deviated statistically significantly from the expected value ($P < 0.001$)

🖄 Springer

infertile soil forest types were compared against one fertile soil tierra firme type, *arenal* was selected only six times and *irapayal* three times.

Most hunters stated that the abundance of game mammals differs between the forest types. *Shapajal* and *varillal* were considered to have more game mammals than the other tierra firme forests. Some hunters pointed out that especially the red brocket deer *(Manama americana* Erxleben) is more common in these forest types, but also the spider monkey (*Ateles chamek* Humboldt) and the woolly monkey (*Lagothrix lagotricha* Humboldt) were mentioned by many hunters. Only few hunters opined that the number of animals is the same in all forest types. Another reason why *shapajal* and *varillal* forests were considered good hunting grounds is the relatively open understory, which makes it easy for the hunters to move around and sight the animals.

Both *arenal* and *irapayal* were considered to have fewer animals than the fertile soil tierra firme forests. The scarcity of fruits consumed by frugivorous mammals was given as one possible explanation for the low mammal density. *Irapayal* forests were ranked lower than the *arenal* forests. The main negative aspect of the *irapayal* forests are the dense palm thickets that hamper movement and make hunting difficult, although tapirs (*Tapirus terrestris* L.) frequently rest in the palm thickets during the day. A further downside is the commonness of ticks (Acari), whose bites cause intense itching.

Many hunters considered *bajial* forests to offer good hunting environment, but only in the dry season when certain fruiting trees bear fruit and attract frugivores from the tierra firme. Likewise, the *aguajal* forests were stated to have plenty of game, but only during the dry season when the aguaje (*Mauritia flexuosa* L. f.) palms bear fruit. Especially the peccaries (*Tayassu* spp.) and the tapir were mentioned to be common in *aguajal*. However, palm swamps are extremely difficult to move in, and thus the hunters mainly hunt from the edge of the swamps.

Discussion

Comparison of the TEK and the floristic forest classifications

The ribereños used indicator species (mainly palms), flooding regime, forest physiognomy and soil characteristics as classification criteria, whereas our floristic classification was based solely on the species composition of one indicator taxon, the pteridophytes. Despite the apparently different criteria, the TEK classification corresponded closely with the classification produced by floristic classification. The parallel result is probably due to the fact that both methods reflect edaphic site conditions. Our indicator species approach separates forest types that differ in soil fertility and drainage (Ruokolainen and Tuomisto 1998) and an edaphic element is practical in the case of TEK forest classification, which is applied in the selection of suitable sites for different land-uses, such as agriculture and hunting.

The floristic forest class Pebas formation/intermediate tierra firme is found on relatively fertile clayey soils (Salovaara et al. 2004). The corresponding TEK forest types (*shapajal* and *varillal*) were judged to have soils well suited for agriculture. Similarly the infertile soil terrace forests corresponded with the TEK classes *irapayal* and *arenal*, which the ribereños considered to have low productivity and to be unsuitable for agriculture. In Nueva Esperanza these preferences were observable also in practice: the villagers preferred to travel 30 min to their chacras in the fertile soil *varillal* forests rather than to clear the chacras in *irapayal* forests closer to the village.

The inundated *bajial* forests in the Yavarí-Mirím were ranked to have quite low suitability for agriculture, whereas flood plain *varzea* forests along Amazonian white-water rivers generally have high agricultural potential due to sediment input (Kvist and Nebel 2001). The Yavarí-Mirím probably resembles the adjacent Yavarí river, which has acidic river sediments with high concentrations of extractable Al and low concentrations of extractable Ca, Mg and K (Kalliola et al. 1993). Also, the flood cycle is generally shorter and more irregular compared to larger Amazonian rivers. Consequently, the sediment deposition may be smaller and cultivation period less predictable.

Fertile soil forests were clearly preferred and infertile soil forests avoided as hunting grounds. Many hunters claimed that the abundance of game mammals differs between the tierra firme forest types. Also a mammal census conducted in different tierra firme forest types in Yavarí-Mirín indicated higher density of large-bodied mammals in fertile soil Pebas formation/intermediate forests than in infertile soil terrace forests (Halme, unpublished data).

Possible explanation for this observation is that the species composition of other plant taxa than just the indicator groups may differ between the forest types. Furthermore, site productivity and the density of important fruit trees (Ruokolainen et al. 1997; Tuomisto and Ruokolainen 1998; Clark et al. 1998, 1999; Vormisto et al. 2000) as well as the quality and quantity of plant secondary compounds (Janzen 1974; Fine et al. 2004) may differ due to variable soil characteristics. These differences could explain why the abundance of certain herbivorous mammals varies between tierra firme forest types (e.g. Emmons 1984; Peres 2000; Fleck and Harder 2000; Halme, unpublished data).

Generally TEK classifications include more habitat types than published scientific classification systems (Shepard et al. 2001). Also in the present study the main difference between the two classification systems was the finer spatial resolution and higher number of forest classes in the TEK classification. This was partly due to the fact that the floristic classification was produced for purposes of wildlife habitat studies and thus the focus was on relatively large habitat patches (Salovaara et al. 2004). Consequently, only three main floristic classes were separated and a part of the heterogeneity within classes was ignored. Apparently the tierra firme forests comprise more heterogeneity than was taken into consideration in our floristic classification.

Four of the TEK forest classes of the present study (*aguajal, irapayal, yarinal* and *shapajal*) were named after characteristic palm species. All these forest types are also present in the TEK classifications of the Matses (Fleck and Harder 2000) and the Matsigenka indians (Shepard et al. 2001), as well as in the classification of Encarnación (1985). Palms are suitable indicators, because they form one of the most diverse and ubiquitous plant families in Amazonian forests (Gentry 1988) and edaphic and topographic heterogeneity cause marked variation in palm community structure at different spatial scales (Clark et al. 1995; Svenning 1999; Vormisto et al. 2000; Vormisto 2002). Palms are also widely utilized and therefore well known by the ribereños (Vormisto 2002).

Springer

Conservation and management applications

Our results lend support to the notion made in previous studies that TEK forest classification systems can offer novel ecological information for ecologists (Posey 1983; Parker et al. 1983; Fleck and Harder 2000; Shepard et al. 2001, 2004; Jinxiu et al. 2004). Furthermore, our results suggest that indigenous cultures are not the only holders of intriguing TEK, but instead significant components of TEK can be maintained in detribalized cultures even after long periods of acculturation. A great majority of rural Amazonian people still live in subsistence economy or rely heavily on extraction of natural resources. It is thus likely that many non-indigenous Amazonians still apply TEK extensively in their daily lives to improve the profitability of natural resource use.

Tropical forests are renowned for their extremely high species diversity and for the fact that biodiversity inventories are difficult to implement in these environments (Higgins and Ruokolainen 2004). On the other hand, rapid and large-scale forest classification methods are urgently required for conservation and land-use planning in Amazonia. For example, ecologically pertinent habitat classifications could be used as surrogate data for modeling species distributions when selecting priority areas for conservation (Ferrier 2002). In many cases TEK classification might be useful for these purposes even without extensive field verification. For example, in biological inventories it could be useful to interview local inhabitants and ask such questions as "What forest types are present in the area?" and "How would you characterize them?". This information could then be applied in designing efficient sampling schemes for field surveys or to support interpretation of remote sensing data (Hernandez-Stefanoni and Ponce-Hernandez 2004; Shepard et al. 2004).

The results of the present study suggest that TEK could be used in identifying major landscape units in Amazonian rain forest areas. There were clear parallels between the floristic and the TEK forest classifications. The TEK forest types seem to differ in soil characteristics and thus the forest classes could be used as rough indicators of soil fertility in the absence of soil data. The suitability of the forest types for different land-use options may vary due to variable soil fertility. Furthermore, plant species respond to soil conditions and plant species composition influences the distribution of herbivorous animals. Consequently, TEK forest classification may also be useful for estimating wildlife habitat quality.

For wildlife management and monitoring it is essential to know how habitat influences the density of wildlife populations. Since there is a strong agreement between the TEK and the scientific classifications, TEK habitat analyses could be used as a valuable tool in the development and implementation of collaborative management plans. The use of TEK nomenclature could increase the participation of the community members and facilitate their collaboration with conservation professionals. For example, hunting registers that are used for long-term monitoring of hunting sustainability could include habitat information to permit more accurate analysis of hunting practices. Management plans could also incorporate non-hunted source areas that are implemented by the communities. The habitat composition of these source areas is essential for their effectiveness and the TEK system could be used for classifying the habitats. TEK habitat classification could also be used in habitat management, since it is widely known by the local people.

However, it should be kept in mind that local knowledge should not be uncritically applied for scientific purposes (Vásquez and Gentry 1987). TEK is regionally and culturally specific and cannot be generalized over large distances. The same terminology may be applied differently in different areas or forest classifications may diverge between locations simply because the environmental gradients differ. For instance, in our study area the term *varillal* was used for two very different habitat types. Despite such shortcomings, especially in data-poor tropical countries, TEK deserves more attention in ecological studies and conservation projects.

Acknowledgements   We are indebted to Gilberto Asipali; Jorge Pacaya and the inhabitants of Nueva Esperanza for sharing their knowledge with us. Hanna Tuomisto and Kalle Ruokolainen made valuable comments to the manuscript. K.H. is grateful to Helsingin Sanomain 100-vuotissäätiö, Ella and Georg Ehrnrooth's Foundation and the Academy of Finland (through grants to H. Tuomisto and K. Ruokolainen) for financial support. Field work was conducted as part of R.B.'s wildlife conservation program funded by the Wildlife Conservation Society.

References

Anderson AB (1981) White-sand vegetation of Brazilian Amazonia. Biotropica 13:199–210
Balick MJ (1996) Transforming ethnobotany for the new millennium. Ann Mo Bot Gard 83:58–66
Berlin B (1973) Folk systematics in relation to biological classification and nomenclature. Annu Rev Ecol Syst 4:259–271
Bodmer RE, Puertas PE (2000) Community-based comanagement of wildlife in the Peruvian Amazon. In: Robinson JG, Bennett EL (eds) Hunting for sustainability in tropical forests. Columbia University Press, New York, pp 395–409
Carneiro RL (1978) The knowledge and use of rain forest trees by the Kuikuru Indians of central Brazil. In: Ford RI (ed) The nature and status of ethnobotany. Anthropological Papers no. 67, Museum of Anthropology. University of Michigan, Ann Arbor, pp 201–216
Clark DA, Clark DB, Sandoval R, Vinicio Castro M (1995) Edaphic and human effects on landscape-scale distributions of tropical rain forest palms. Ecology 76:2581–2594
Clark DB, Clark DA, Read JM (1998) Edaphic variation and the mesoscale distribution of tree species in a neotropical rain forest. J Ecol 86:101–112
Clark DB, Palmer MW, Clark DA (1999) Edaphic factors and the landscape-scale distributions of tropical rain forest trees. Ecology 80:2662–2675
Condit R (1996) Defining and mapping vegetation types in mega-diverse tropical forests. Trends Ecol Evol 11:4–5
Emmons LH (1984) Geographic variation in densities and diversities of non-flying mammals in Amazonia. Biotropica 16:210–222
Encarnación F (1985) Introducción a la flora y vegetación de la Amazonia peruana: estado actual de los estudios, medio natural y ensayo de una clave de determinación de las formaciones vegetales en la llanura amazónica. Candollea 40:237–252
Ferrier S (2002) Mapping spatial pattern in biodiversity for regional conservation planning: where to from here? Syst Biol 31:331–363
Fine P, Mesones I, Coley PD (2004) Herbivores promote specialization by trees in Amazonian forests. Science 305:663–665
Fleck DW, Harder JD (2000) Matses Indian rainforest habitat classification and mammalian diversity in Amazonian Peru. J Ethnobiol 20:1–36
Frechione J, Posey DA, da Silva LF (1989) The perception of ecological zones and natural resources in the Brazilian Amazon: an ethnoecology of Lake Coari. Adv Econ Bot 7:260–282
Gentry AH (1981) Distributional patterns and an additional species of the Passiflora vitifolia complex: Amazonian species diversity due to edaphically differentiated communities. Plant Syst Evol 137:95–105
Gentry AH (1988) Changes in plant community diversity and floristic composition on environmental and geographical gradients. Ann Mo Bot Gard 75:1–34
Hellier A, Newton AC, Ochoa Gaona S (1999) Use of indigenous knowledge for rapidly assessing trends in biodiversity: a case study of Chiapas, Mexico. Biodivers Conserv 8:869–889

 Springer

Hernandez-Stefanoni JL, Ponce-Hernandez R (2004) Mapping the spatial distribution of plant diversity indices in tropical rain forest using multi-spectral satellite image classification and field measurements. Biodivers Conserv 13:2599–2621

Hess LL, Melack JM, Novo EMLM, Barbosa CCF, Gastil M (2003) Dual-season mapping of wetland inundation and vegetation for the central Amazon basin. Remote Sens Environ 87:404–428

Higgins MA, Ruokolainen K (2004) Rapid tropical forest inventory: a comparison of techniques based on inventory data from western Amazonia. Conserv Biol 18:799–811

Holman EW (2002) The relation between folk and scientific classifications of plants and animals. J Classif 19:131–159

Hoorn C (1993) Marine incursion and the influence of Andean tectonics on the Miocene depositional history of northwestern Amazonia: results of a palynostratigraphic study. Palaeogeogr Palaeoclimatol Palaeoecol 105:267–309

Huntington HP (2000) Using traditional ecological knowledge in science: methods and applications. Ecol Appl 10:1270–1274

Huntington HP, Suydam RS, Rosenberg DH (2004) Traditional knowledge and satellite tracking as complementary approaches to ecological understanding. Environ Conserv 31:177–180

Janzen DH (1974) Tropical blackwater rivers, animals, and mast fruiting by the dipterocarpaceae. Biotropica 6:69–103

Jinxiu W, Hongmao L, Huabin H, Lei G (2004) Participatory approach for rapid assessment of plant diversity through folk classification system in a tropical rain forest: case study in Xishuangbanna, China. Conserv Biol 18:1139–1142

Kalliola R, Linna A, Puhakka M, Salo J, Räsänen M (1993) Mineral nutrients from fluvial sediments in the Peruvian Amazonia. Catena 20:333–349

Kvist LP, Nebel G (2001) A review of Peruvian flood plain forests: ecosystems, inhabitants and resource use. For Ecol Manage 150:3–26

Mackinson S (2001) Integrating local and scientific knowledge: an example in fisheries science. Environ Manage 27:533–545

Malleux J (1975) Mapa forestal del Peru (Memoria explicativa). Univ. Nacional Agraria, Lima, p 184

Marengo JA (1998) Climatología de la zona de Iquítos, Perú. In: Kalliola R, Flores Paitan S (eds) Geoecología y desarrollo amazónico – Estudio integrado en la zona de Iquítos, Perú, vol 114. Annales Universitatis Turkuensis Ser A II, pp 35–57

Messing I, Hoang Fagerström MH (2001) Using farmer's knowledge for defining criteria for land qualities in biophysical land evaluation. Land Degrad Dev 12:541–553

Oudwater N, Martin A (2003) Methods and issues in exploring local knowledge of soils. Geoderma 111:387–401

Parker EP (1989) A neglected human resource in Amazonia: the Amazon caboclo. In: Posey DA, Balée W (eds) Resource management in Amazonia: indigenous and folk strategies. Advances in Economic Botany, vol 7. The New York Botanical Garden, Bronx, pp 249–259

Parker E, Posey D, Frechione J, da Silva LF (1983) Resource exploitation in Amazonia: ethnoecological examples from four populations. Ann Carnegie Mus 52:163–203

Peres CA (2000) Evaluating impact and sustainability of subsistence hunting at multiple Amazonian forest sites. In: Robinson JG, Bennett EL (eds) Hunting for sustainability in tropical forests. Columbia University Press, New York, pp 31–56

Phillips OL, Nuñes Vargas P, Lorenzo Monteagudo A, Peña Cruz A, Chuspe Zans M-L, Galiano Sanchez W, Yli-Halla M, Rose S (2003) Habitat association among Amazonian tree species: a landscape scale approach. J Ecol 91:757–775

Phuthego TC, Chanda R (2004) Traditional ecological knowledge and community-based natural resource management: lessons from a Botswana wildlife management area. Appl Geogr 24:57–76

Pires JM, Prance GT (1985) Notes on the vegetation types of the Brazilian Amazon. In: Prance GT, Lovejoy TE (eds) Key environments: Amazonia. Pergamon Press, Oxford, pp 109–145

Pitman N, Vriesendorp C, Moskovits D (eds) (2003) Perú: Yavarí. Rapid Biological Inventories Report 11. The Field Museum, Chicago, p 282

Poizat G, Baran E (1997) Fishermen's knowledge as background information in tropical fish ecology: a quantitative comparison with fish sampling results. Environ Biol Fishes 50:435–449

Posey DA (1983) Indigenous ecological knowledge and development of the Amazon. In: Moran E (ed) The dilemma of Amazonian development. Westview Press, Boulder, Colorado, pp 225–257

Ruokolainen K, Tuomisto H (1998) Vegetación natural de la zona de Iquitos. In: Kalliola R, Flores Paitan S (eds) Geoecología y desarrollo amazónico – Estudio integrado en la zona de Iquítos, Perú, vol 114. Annales Universitatis Turkuensis Ser A II, pp 253–365

[226]

Ruokolainen K, Linna A, Tuomisto H (1997) Use of Melastomataceae and pteridophytes for revealing phytogeographic patterns in Amazonian rain forests. J Trop Ecol 13:243–256

Salo J, Kalliola R, Häkkinen I, Mäkinen Y, Niemelä P, Puhakka M, Coley PD (1986) River dynamics and the diversity of Amazon lowland forest. Nature 322:254–258

Salovaara KJ, Cárdenas GG, Tuomisto H (2004) Forest classification in an Amazonian rainforest landscape using pteridophytes as indicator species. Ecography 27:689–700

Salovaara K, Thessler S, Malik RN, Tuomisto H (2005) Classification of Amazonian primary rain forest vegetation using Landsat ETM+ satellite imagery. Remote Sens Environ 97:39–51

Sheil D, Lawrence A (2004) Tropical biologists, local people and conservation: new opportunities for collaboration. Trends Ecol Evol 19:634–638

Shepard GH Jr, Yu DW, Nelson BW (2004) Ethnobotanical ground-truthing and forest diversity in the western Amazon. Adv Econ Bot 15:133–171

Shepard GH Jr, Yu DW, Lizarralde M, Italiano M (2001) Rain forest habitat classification among the Matsigenka of the Peruvian Amazon. J Ethnobiol 21:1–38

Svenning JC (1999) Microhabitat specialization in a species-rich palm community in Amazonian Ecuador. J Ecol 87:55–65

ter Steege H, Sabatier D, Castellanos H, van Andel T, Duivenvoorden J, de Oliveira AA, Ek R, Lilwah R, Maas P, Mori S (2000) An analysis of the floristic composition and diversity of Amazonian forests including those of the Guiana Shield. J Trop Ecol 16:801–828

Tuomisto H, Ruokolainen K (1994) Distribution of Pteridophyta and Melastomataceae along an edaphic gradient in an Amazonian rain forest. J Veg Sci 5:25–34

Tuomisto H, Ruokolainen K (1998) Uso de especies indicadoras para determinar características del bosque y de la tierra. In: Kalliola R, Flores Paitan S (eds) Geoecología y desarrollo amazónico – Estudio integrado en la zona de Iquítos, Perú, vol 114. Annales Universitatis Turkuensis Ser A II, pp 481–491

Tuomisto H, Ruokolainen K, Kalliola R, Linna A, Danjoy W, Rodriguez Z (1995) Dissecting Amazonian biodiversity. Science 269:63–66

Tuomisto H, Ruokolainen K, Aguilar M, Sarmiento A (2003a) Floristic patterns along a 43-km long transect in an Amazonian rain forest. J Ecol 91:743–756

Tuomisto H, Poulsen AC, Ruokolainen K, Moran RC, Quintana C, Celi J, Cañas G (2003b) Linking floristic patterns with soil heterogeneity and satellite imagery in Ecuadorian Amazonia. Ecol Appl 13:352–371

Valencia R, Balslev H, Mino GPY (1994) High tree alpha-diversity in Amazonian Ecuador. Biodivers Conserv 3:21–28

Vásquez R, Gentry AH (1987) Limitaciones del use de nombres vernaculares en los inventarios forestales de la Amazonia Peruana. Rev Fores Peru 14:109–120

Vormisto J (2002) Palms as rainforest resources: how evenly are they distributed in Peruvian Amazonia? Biodivers Conserv 11:1025–1045

Vormisto J, Phillips O, Ruokolainen K, Tuomisto H, Vásquez R (2000) A comparison of fine-scale distribution patterns of four plant groups in an Amazonian rainforet. Ecography 23:349–359

Springer

Biodivers Conserv (2007) 16:1803–1821
DOI 10.1007/s10531-006-9072-3

ORIGINAL PAPER

Species richness, endemism and conservation of Mexican gymnosperms

Raúl Contreras-Medina · Isolda Luna-Vega

Received: 12 July 2005 / Accepted: 19 May 2006 / Published online: 16 August 2006
© Springer Science+Business Media B.V. 2006

Abstract An analysis of the distribution patterns of 124 Mexican gymnosperm species was undertaken, in order to detect the Mexican areas with high species richness and endemism, and with this information to propose areas for conservation. Our study includes an analysis of species richness, endemism and distributional patterns of Mexican species of gymnosperms based on three different area units (states, biogeographic provinces and grid-cells of 1° × 1° latitude/longitude). The richest areas in species and endemism do not coincide; in this way, the Sierra Madre Oriental province, the state of Veracruz and a grid-cell located in the state of Oaxaca were the areas with the highest number of species, whereas the Golfo de México province, the state of Chiapas and a grid-cell located in this state were the richest areas in endemic species. A weighted endemism and corrected weighted endemism indices were calculated, and those grid-cells with high values in both indices and with high species richness were considered as hotspots; these grid-cells are mainly located in Southern and Central Mexico.

Keywords Areography · Conservation · Endemism · Gymnosperms · Mexico · Species richness

Introduction

Since the first proposals of biogeographical regionalizations of the world (Sclater 1858; Wallace 1876), Mexico has been considered a transitional zone between the Nearctic and Neotropical biogeographic regions (Halffter 1987). Recently, based on panbiogeographic studies, Contreras-Medina and Eliosa-León (2001) and Morrone and Márquez (2001) proposed that the Mexican biota shows different biogeographic relationships as suggested by two North American tracks, one at the east and other

R. Contreras-Medina · I. Luna-Vega (✉)
Departamento de Biología Evolutiva, Facultad de Ciencias, Univ. Nacional Autónoma de México, Apartado Postal 70-399, C.P. 04510 Ciudad Universitaria, México
e-mail: ilv@hp.fciencias.unam.mx

at the west, and a Gondwanic track, that relate Mexico to the rest of the Neo-tropical region. The geographic distribution of the elements that constitute the Mexican biota has been the result of vicariance, dispersal events and local extinction, as well as climatic changes and speciation processes in situ (Salinas-Moreno et al. 2004), in a complex plate tectonic scenario (Ferrusquía-Villafranca 1993; Ortega et al. 2000).

At the end of the Cretaceous period, the Laramidian orogeny started and determined the main physiographic features of the mountains in Mexico and northern Central America (Halffter 1987; Salinas-Moreno et al. 2004), with the exception of the Transmexican Volcanic Belt that started in the Mid-Tertiary (Ferrusquía-Villafranca 1993; Ortega et al. 2000). Climatic changes during the Pleistocene (Toledo 1982) and orogenic processes contributed to the diversification of the genus *Pinus* in Mexico (Eguiluz 1985; Farjon and Styles 1997). In this scenario, barriers and corridors, as Pleistocenic refugia, played an important role in the spatial evolution of the Mexican gymnosperms (e.g. Perry et al. 1998; Contreras-Medina et al. 2001b; González and Vovides 2002).

Gymnosperms are seed plants that mainly inhabit temperate zones of both hemispheres and have been important elements in fossil and extant plant communities; their appearance in the late Paleozoic represents one of the most important evolutionary phases among the patterns of vascular plant diversification (Niklas et al. 1983). Due to their antiquity, they represent an interesting group for distributional analysis from an historical biogeography approach. In several Mexican floristic studies, gymnosperms represent approximately 2% of species diversity, in contrast with angiosperms and pteridophytes (Contreras-Medina 2004). Notwithstanding, Mexico represents the country with more species of some genera, such as *Ceratozamia*, *Dioon* and *Pinus* (Contreras-Medina 2004) and plays an important role in gymnosperm diversity at a world-wide level (see Takhtajan 1986; Osborne 1995).

Mexican gymnosperms are distributed mainly in temperate forests and arid scrubs. Studies about the geographic distribution of gymnosperms in Mexico are imperative not only theoretically but practically, especially for those groups with great economic value such as *Abies* and *Pinus*, and those threatened taxa included in some risk categories, like cycads. Floristic richness of Mexican gymnosperms is represented by nearly 130 species, included in 14 genera and six families, such diversity represents 15% in a world-wide level; species endemism is frequent, even at state level, mainly in Zamiaceae.

Areography (also named chorology) is defined as the study of distributional areas of taxa (Rapoport 1975; Rapoport and Monjeau 2001). This type of studies can offer information about areas of richness and endemicity of faunistic and floristic groups in a country or continent, and may also contribute to the delimitation of biogeographic regions. Some previous works that have applied this approach to Mexican plant taxa were carried out by Kohlmann and Sánchez (1984) with *Bursera*, Valdés and Cabral (1993) with grasses, and García-Mendoza (1995) with Agavaceae.

Our aim is to detect Mexican areas of richness and endemism of gymnosperms based on their presence on states, biogeographic provinces, and grid-cells and to compare the results obtained. With this task we will be able to generate useful information to carry out several aspects on the geographical distribution of these seed plants in the country, and to detect some important areas for conservation based on this group of plants.

Materials and methods

Distributional data of gymnosperm species were obtained from the revision of 1465 herbarium specimens deposited in the following collections: National Herbarium of the Instituto de Biología, UNAM (MEXU); Herbarium of the Escuela Nacional de Ciencias Biológicas, IPN (ENCB); Herbarium of the Missouri Botanical Garden (MO); Herbaria of the Instituto de Ecología A.C. in Xalapa City (XCAL) and Pátzcuaro City (IEB); Herbarium of the Facultad de Ciencias, UNAM (FCME); Herbarium of the Universidad de Guadalajara (IBUG); Herbarium of the Departamento de Bosques, Universidad Autónoma Chapingo (CHAP); Herbario Nacional Forestal (INIF); Herbarium of the Universidad Veracruzana (XALU); and Herbarium of the Universidad de Sonora (USON). In addition, some floristic and revisionary studies were reviewed (Zanoni and Adams 1979; Wiggins 1980; Zanoni 1982; Vovides et al. 1983; Stevenson et al. 1986; Patterson 1988; Espinosa 1991; McVaugh 1992; Zamudio 1992, 2002; Moretti et al. 1993; Zamudio and Carranza 1994; Fonseca 1994; Farjon and Styles 1997; Medina and Dávila 1997; Narave and Taylor 1997; Vovides 1999; Aguirre-Planter et al. 2000; Felger 2000; Contreras-Medina et al. 2001a, 2003). Finally, in order to obtain field data and to make field observations of natural populations of some gymnosperm species, field exploration was carried out in the Mexican states of Hidalgo, Querétaro, Estado de México, Puebla, and Oaxaca.

In order to perform the biogeographic analysis, Mexican states, biogeographic provinces and $1° × 1°$ latitude/longitude squares were used as units of study. We included a state level in the analysis because in Mexico, as well as in other countries, conservation decisions are generally undertaken considering political boundaries, rather than natural criteria (Dávila-Aranda et al. 2004), and because in megadiverse countries distributional data tend to be organized on the basis of geopolitical units (Gaston and Williams 1996). In order to recognise some patterns at the state level (Fig. 1a), we followed the criteria suggested by Dávila-Aranda et al. (2004), grouping the species in four sets based on their patterns of distribution: (1) scarcely distributed (species recorded only in one state); (2) narrowly distributed (2–4 states); (3) normally distributed (5–9 states); and (4) widely distributed (10 or more states). We also used the Mexican biogeographic provinces proposed by the Comisión Nacional Para el Uso y Conocimiento de la Biodiversidad (CONABIO) (Arriaga et al. 1997), which represent a regionalization of the country based on four different sources (vascular plants, herpetofauna, mammals and morphotectonics, Fig. 1b). In each province, the contribution of endemic gymnosperm species was evaluated and those biogeographic provinces with more species were remarked. Despite more recent regionalizations of Mexico have been proposed (e.g. Morrone 2001), the scheme of CONABIO has the advantage of having been generated in digital format, so the distributional data of gymnosperms can be analysed with a Geographic Information System (GIS).

In many cases the size of the Mexican states studied herein (i.e. Chihuahua versus Tlaxcala) and the biogeographic provinces (i.e. Altiplano Norte versus Soconusco) are extremely different. Thus, it was necessary to carry out an alternative biogeographic analysis with standard units of the same size. For this reason, we chose squares of one geographical degree per side, partially to facilitate the data manipulation and to reduce the effect of sampling artifacts, such as mapping errors and

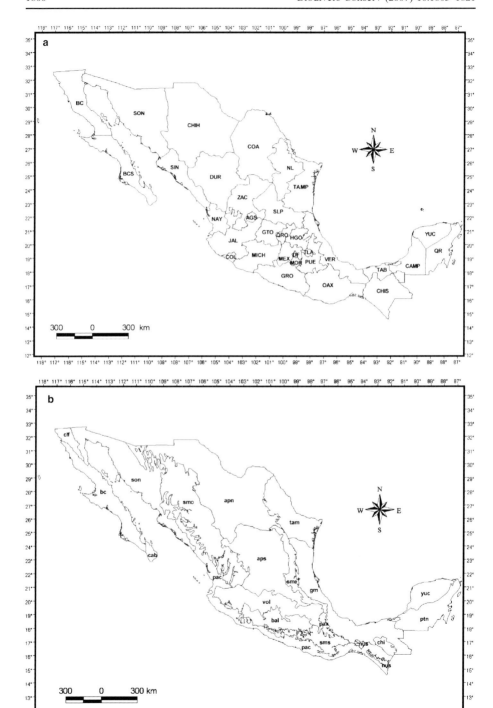

◄**Fig. 1** (**a**) The 32 Mexican states. Abbreviations are: AGS = Aguascalientes, BC = Baja California, BCS = Baja California Sur, CAMP = Campeche, CHIS = Chiapas, CHIH = Chihuahua, COA = Coahuila, COL = Colima, DF = Distrito Federal, DUR = Durango, GTO = Guanajuato, GRO = Guerrero, HGO = Hidalgo, JAL = Jalisco, MEX = México, MICH = Michoacán, MOR = Morelos, NAY = Nayarit, NL = Nuevo. León, OAX = Oaxaca, PUE = Puebla, QR = Quintana Roo, QRO = Querétaro, SLP = San Luis Potosí, SIN = Sinaloa, SON = Sonora, TAB = Tabasco, TAMP = Tamaulipas, TLA = Tlaxcala, VER = Veracruz, YUC = Yucatán, ZAC = Zacatecas; (**b**) the 19 biogeographic provinces of Mexico according to Arriaga et al. (1997). Abbreviations are: apn = Altiplano Norte, aps = Altiplano Sur, bal = Depresión del Balsas, bc = Baja California, clf = California, cab = Del Cabo, chi = Los Altos de Chiapas, gm = Golfo de México, nus = Soconusco, oax = Oaxaca, pac = Costa del Pacífico, ptn = Petén, sme = Sierra Madre Oriental, smo = Sierra Madre Occidental, sms = Sierra Madre del Sur, son = Sonorense, tam = Tamaulipeca, vol = Eje Volcánico, yuc = Yucatán

unsampled grids in sparsely inhabited areas (Crisp et al. 2001). This scale size was chosen because it was tested in previous works on areography and diversity of different groups of Mexican flora (e.g. Kohlmann and Sánchez 1984; García-Mendoza 1995; Dávila-Aranda et al. 2004). We applied richness and endemism indices proposed by Crisp et al. (2001) and Linder (2001) to these grid-cells, which were previously applied to Australian and African floras, respectively, dividing the study areas in squares of one or two geographical degrees per side, in order to detect centres of species richness and endemism of vascular plants. The use of equal-area grids has also been considered as an important tool for studying biogeographic patterns in biological diversity (McAllister et al. 1994).

Species richness was measured simply as the total count of species within each grid-cell and is also known as unweighted species richness (Linder 2001). A first index termed 'weighted endemism' was related to species richness (Crisp et al. 2001). The first step consisted in dividing each grid-occurrence by the total number of grids in which one species occurs. Thus, a species restricted to a single grid was scored as '1' for that grid, and '0' for all other grids, and a species found in four grids, was scored as '0.25' for each of the four grids, and '0' for all remaining grids; then the sum of all score species values for each grid was obtained. A second index named 'corrected weighted endemism' (Crisp et al. 2001), consisted in dividing the weighted endemism index by the total count of species in each grid cell. Those grid-cells with the highest scores in the first index were considered as centres of richness and for the second index as centres of endemism (see Crisp et al. 2001 and Linder 2001 for further details). Grid-cells with none or one species recorded were deleted from the analysis of corrected weighted endemism and are not shown in the resultant map. Grid-cells values obtained for weighted endemism and corrected weighted endemism indices were ranged from 1 to 10. Each species was scored as present in a grid-cell independently of the number of times recorded in it (Linder 2001).

In our work, we do not deal with patterns of similarity among states, biogeographic provinces or grid-cells. We considered as endemic species those with ranges limited to a state or biogeographic province. Species restricted to a single cell were considered endemics with a small distribution range (narrow endemics).

Geographic distribution maps of each gymnosperm species were obtained using ArcView GIS (ESRI 1999). The map of known distribution of each species was first projected on a map of Mexico with state divisions, second on a map of biogeographic provinces produced by CONABIO (Arriaga et al. 1997), and third on a grid map of Mexico divided in cells of 1° per side, in order to detect those richest areas in endemism and species of gymnosperms in Mexico.

[233]

Finally, we selected those grid-cells with more endemic species to Mexico and/or more species richness and compared them with the Mexican priority regions for conservation of CONABIO (Arriaga et al. 2000), which represent areas with high biodiversity, formulated by an expertise set of national researchers coordinated by CONABIO; also we compare our results with others based in other groups of plants.

Results and discussion

State analysis

Sixty-eight species of gymnosperms are endemic to Mexico, representing 56% of the total number of taxa recorded in the country. Notwithstanding that these endemic species are found at different states, many of them are concentrated in southern Mexico, in the states of Chiapas, Oaxaca and Veracruz, especially those of Zamiaceae (Table 1). Many states do not include any endemic species, mainly those

Table 1 Families, genera, species richness, and endemism of wild gymnosperms in the 32 Mexican states

States	Number of species	Number of genera	Number of families	Number of endemic species
Veracruz	**39**	**12**	6	5
Oaxaca	**38**	**12**	6	6
Nuevo León	33	11	5	1
Chihuahua	33	7	3	0
Coahuila	32	8	3	0
Chiapas	31	10	5	7
Hidalgo	31	**12**	6	0
Jalisco	31	8	4	1
Durango	30	9	4	0
San Luis Potosí	29	11	6	1
Puebla	28	**12**	6	1
Michoacán	25	8	4	1
Querétaro	25	**12**	6	0
Tamaulipas	25	11	6	1
Guerrero	24	8	4	0
Zacatecas	24	5	3	1
Sonora	22	8	4	0
México	20	5	2	0
Distrito Federal	16	5	2	0
Nayarit	16	6	3	0
Baja California	15	5	3	0
Morelos	15	4	2	0
Tlaxcala	14	4	2	0
Sinaloa	13	4	2	0
Guanajuato	12	5	3	0
Aguascalientes	9	4	3	0
Colima	5	4	3	0
Tabasco	4	3	2	1
Quintana Roo	2	2	2	0
Baja California Sur	1	1	1	0
Campeche	1	1	1	0
Yucatán	1	1	1	0

Numbers in bold represent the higher in each category

located in northeastern Mexico and central and northern portions of the Pacific coast. States richest in species are: Veracruz, Oaxaca, Nuevo León, Chihuahua, Chiapas, Coahuila, Hidalgo, Jalisco, Durango, San Luis Potosí and Puebla (Fig. 2a). These states are located in southern, central, and northern Mexico, which suggests that they do not obey a latitudinal gradient, following the distribution of the main Mexican mountain chains (i.e. Sierra Madre Oriental, Sierra Madre Occidental, Transmexican Volcanic Belt, and Serranías Transístmicas). States poor in species are Baja California Sur, Campeche, and Yucatán, located in the Baja California and Yucatán peninsulae.

Results obtained show that nine out of the 14 genera represented in Mexico have at least one endemic species, which is more evident in the cases of *Ceratozamia* (13 species) and *Dioon* (10 species); most of the species of these genera are restricted to the Mexican territory.

Several species are shared with adjacent parts of Central America (i.e. *Abies guatemalensis* Rehder, *Pinus teocote* Schltdl. et Cham., *Zamia herrerae* Calderón et Standl.) and with the United States of America (e.g. *Abies concolor* (Gordon et Glendinning) Hildebrand, *Calocedrus decurrens* (Torr.) Florin, *Ephedra nevadensis* S. Watson, and *Pinus coulteri* D. Don). Thus, the percentage of endemism may increase if a broader geographical approach is undertaken. The inclusion of some parts of Central America and southern United States of America generate a more natural geographic regionalization, as suggested by several authors (Rzedowski 1991; Morrone 2001).

There are 47 species that are scarcely distributed in Mexico and 26 of them are represented in only one state (Table 1); also 28 of them are endemic to Mexico. Many of these species are only known from one or few localities, such as *Ceratozamia kuesteriana* Regel, *Dioon califanoi* De Luca et Sabato, and *Pinus maximartinezii* Rzedowski, and the remaining species are distributed also in adjacent countries. Twenty-seven species are narrowly distributed (2–4 states), 12 of them are only found in two states and 15 species of Pinaceae and Zamiaceae are endemic to Mexico. A third group (distributed in 5–9 states) includes 25 species; nine of them are endemic to the Mexican territory. Twenty-one species are distributed in 10 or more states, and eight are endemic to Mexico. *Pinus teocote* Schltdl. et Cham. and *Taxodium mucronatum* Ten. (the National Mexican tree) seem to be the species of gymnosperms most widely distributed (23 states), and represent the dominant trees in some Mexican temperate and riparian forests, respectively.

Biogeographic provinces analysis

Mexican gymnosperms are distributed mainly in the Mesoamerican Mountain region sensu Rzedowski (1978), in which are concentrated more than a half (near 70 species); this region includes the following biogeographic provinces: Sierra Madre Occidental, Sierra Madre Oriental, Eje Volcánico, Sierra Madre del Sur and Altos de Chiapas (Fig. 2b).

The Sierra Madre Oriental province harbours the high number of species (50); this province has been previously considered as an important richness area of Mexican gymnosperms (Contreras-Medina 2004); other rich provinces are Eje Volcánico and Sierra Madre Occidental (35 species each), Sierra Madre del Sur (27), Altiplano Norte (25), Golfo de México (23), Soconusco (22) and Altos de Chiapas (21) (Table 2). Provinces with fewer numbers of species are Depresión del Balsas,

🍂 Springer

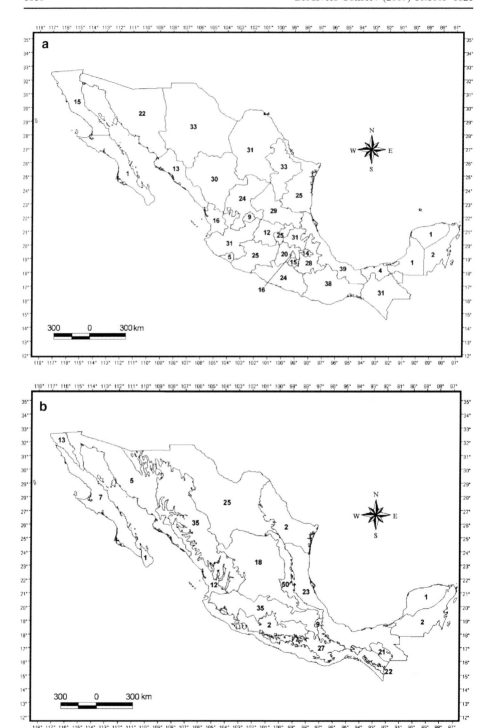

Fig. 2 (**a**) Species richness recorded in each Mexican states; (**b**) species richness recorded in each Mexican biogeographic provinces (sensu Arriaga et al. 1997). See Fig. 1b for province names

Table 2 Families, genera, species richness, and endemic species of wild gymnosperms in the 19 Mexican biogeographic provinces

Biogeographic provinces	Number of species	Number of genera	Number of families	Number of endemic species
Sierra Madre Oriental	**50**	**13**	6	10
Eje Volcánico	**35**	8	4	2
Sierra Madre Occidental	**35**	7	2	2
Sierra Madre del Sur	**27**	9	**5**	1
Altiplano Norte	**25**	5	3	0
Golfo de México	23	7	4	**12**
Soconusco	22	9	4	**5**
Altos de Chiapas	21	9	**5**	0
Altiplano Sur	18	6	3	0
California	13	5	3	0
Costa del pacífico	12	6	4	**3**
Oaxaca	9	5	3	**4**
Baja California	7	3	3	0
Sonorense	5	4	4	0
Depresión del Balsas	2	2	2	0
Petén	2	2	2	0
Tamaulipeca	2	2	2	0
Del Cabo	1	1	1	0
Yucatán	1	1	1	0

Numbers in bold represent the higher in each category

Petén and Tamaulipeca with two species each, and finally Yucatán and Del Cabo provinces with one species each.

Nearly one third of the Mexican species of gymnosperms are restricted to one province, mainly in the case of the following provinces: Golfo de México (12), Sierra Madre Oriental (10), Soconusco (5), Oaxaca (4), and Costa del Pacífico (3) (Table 2). The geographic distribution of some species agree and has been useful to define the province in which they inhabit, i.e. *Pinus coulteri* D. Don is diagnostic to the Californian province (Espinosa et al. 2000) and *Pinus greggii* Parl. and *Ceratozamia kuesteriana* Regel are both endemic to the Sierra Madre Oriental province (Contreras-Medina 2004). Species that have been recorded in a large number of provinces are *Taxodium mucronatum* (12 provinces) and *Pinus teocote* (8 provinces), followed by *Cupressus lusitanica* Mill., *Juniperus flaccida* Schltdl., and *Pinus oocarpa* Schltdl. (7 provinces).

Grid-cells analysis

Mexico was divided in 240 grid-cells and from these, 164 cells include at least one record; for purposes of this work, we included 1155 occurrence records of 124 species of gymnosperms from six families: Cupressaceae, Ephedraceae, Pinaceae, Podocarpaceae, Taxaceae, and Zamiaceae. Statistics revealed that the mean range is 9 grid-cells, the median is 5 cells, and the mode is a single cell (24 species); this last result indicates that nearly one fifth of gymnosperms are distributed in small ranges of Mexico. Similar statistical parameters with this methodology were obtained by Crisp et al. (2001) for the Australian flora. Taxa represented in more grid-cells were *Pinus cembroides* Zucc. and *P. teocote* Schltdl. et Cham., recorded in 44 grid-cells each.

Grid-cells richest in species were concentrated in different areas (Fig. 3a), mainly located in the following biogeographic provinces, all of them represented by

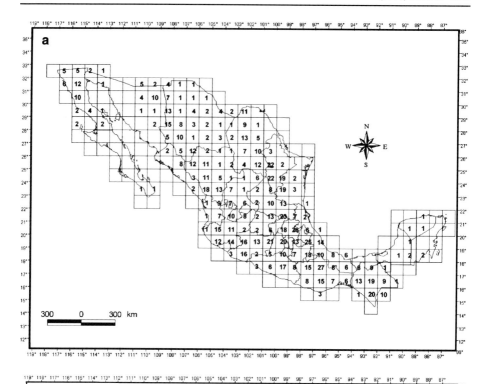

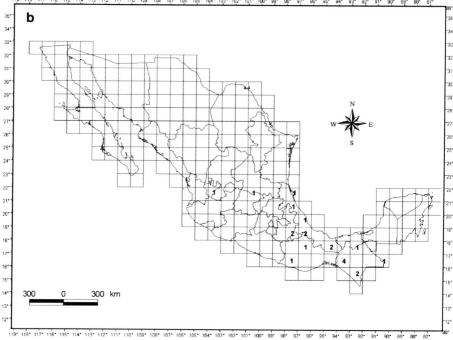

Fig. 3 (**a**) Species richness mapped in $1° \times 1°$ grid-cells; (**b**) species of Mexican gymnosperms restricted to one grid-cell mapped in $1° \times 1°$ grid-cells

mountain chains: Sierra Madre Oriental, Eje Volcánico, Sierra Madre del Sur, Los Altos de Chiapas, and Sierra Madre Occidental. The Sierra Madre Oriental, located in eastern Mexico, comprises most of the richest grid-cells; six of the 12 grid-cells scoring highest for species diversity are located in this province. This result confirms that this area is the richest in species of gymnosperms in Mexico, as suggested earlier by Contreras-Medina (2004). The Transmexican Volcanic Belt (Eje Volcánico province) is other mountain chain that contains several grid-cells with more than 19 species, three of them high-scored. The state of Chiapas comprises four provinces, two of them almost restricted to this state (Soconusco and Los Altos de Chiapas); in these two provinces are located two of the most richest cells (20 and 19 species, respectively) of the country. The richest cell (containing 27 species) is located at the Sierra Madre del Sur province, in the state of Oaxaca (Fig. 3a).

Results obtained in the grid-cells analysis support the state and province analyses done above. They coincide that the Yucatán Peninsula and the southern part of Baja California are the poorest regions in gymnosperm species.

Gymnosperm species occurring in a single grid-cell are shown in Fig. 3b. These species are distributed mainly in southeastern and central Mexico and in the northern portion of the Baja California Peninsula. All the species in the continental plate are endemic to the country, whereas those located in Baja California are shared with the United States of America.

Ranges values of grid-cells obtained for weighted endemism and corrected weighted endemism indices of Mexican gymnosperms are shown in Table 3. With the values of weighted endemism (which counts all the species in an inverse proportion to their range), we produced a map that resembles the pattern of species richness (Fig. 4a). This is an expected result, because Crisp et al. (2001) suggested that there is a high correlation between weighted endemism and species richness.

The map representing the values of corrected weighted endemism (Fig. 4b) showed a remarkable correspondence with several biogeographic provinces. This index emphasizes such areas that are not necessarily high in species richness, but have a high proportion of species with restricted distributions. This is the case of the grid-cells located near the Gulf of Mexico, all of them not considered as richest areas in species (compare Figs. 3a and 4b).

The northern portion of the Baja California Peninsula includes many grid-cells with high values of corrected weighted endemism (Fig. 4b). This is because several species shared with the United States inhabit in this area, with a distribution restricted to the western portion of North America, especially to the Californian Province proposed by Takhtajan (1986) in a floristic regionalization of the world.

Table 3 Range values of grid-cells obtained for weighted endemism and corrected weighted endemism indices of Mexican gymnosperms, ranged from 1 to 10	Ranges	Weighted endemism	Corrected endemism
	1	0–0.504	0–0.053
	2	0.505–1.009	0.054–0.107
	3	1.010–1.514	0.108–0.161
	4	1.515–2.019	0.162–0.215
	5	2.020–2.524	0.216–0.269
	6	2.525–3.029	0.270–0.323
	7	3.030–3.534	0.324–0.377
	8	3.535–4.039	0.378–0.431
	9	4.040–4.544	0.432–0.485
	10	4.545–5.049	0.486–0.539

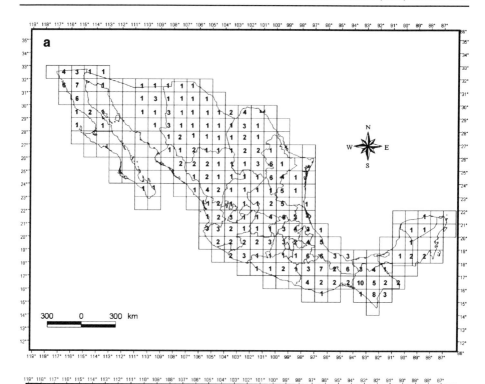

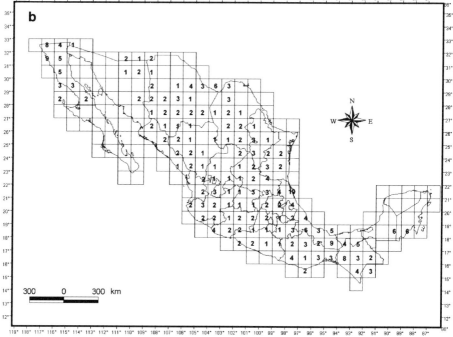

Fig. 4 (**a**) Weighted endemism of Mexican gymnosperms. The value in each grid-cell represents the sum of weights for all species occurring in each grid-cell, and ranged from 1 to 10; (**b**) corrected weighted endemism of Mexican gymnosperms. The value in each grid-cell is the weighted endemism for that grid-cell, divided by the grid-diversity of the grid-cell, and ranged from 1 to 10

✡ Springer

We noted that several of the endemic species restricted to a single grid-cell are located in the Golfo de México province, comprising grid-cells located mainly at the states of Veracruz and Oaxaca; in fact, the state of Veracruz (located almost completely in this province) is one of the states including more endemic species (Table 1). This province appears as an area of endemism (Fig. 4b) but not as an area of high species richness (Fig. 3a). In fact, this province contains the highest number of gymnosperm endemic species (Table 2).

Another apparently important area is located in the southern portion of the Yucatán Peninsula (Fig. 4b), where *Pinus caribaea* Morelet and *Zamia polymorpha* D. W. Stevenson inhabit. This fact may be the result of the delimitation of the study area, because *P. caribaea* is also distributed in other countries of Central America and the Caribbean, but in Mexico occurs only in two grid-cells.

The term 'hotspot' has been used to refer to areas where high levels of richness, threat and endemism coincide (Myers 1988). It has been proposed that biogeographic methods can contribute to the recognition of gymnosperm 'hotspots' based on the coincidence of panbiogeographic nodes, pleistocenic refugia and areas of endemism (Contreras-Medina et al. 2001b). In this study, some grid-cells coincide in their high values in both indices and thus were considered as hotspots (Fig. 4). Since human impact in these grid-cells is not evaluated herein, our hotspots are based only in the data produced by the two indices. We also considered another meaning of hotspot that has been used to refer to those areas with extreme taxonomic richness (Gaston and Williams 1996). Those areas detected by two indices are located in southeastern Mexico (three) and one in the northern portion of the Baja California Peninsula, and contain almost two gymnosperm species each restricted to one grid-cell; only the grid-cell located in the state of Chiapas contains four restricted species; the grid-cells with more species are located in Sierra Madre del Sur, Eje Volcánico and Sierra Madre Oriental (Fig. 3a).

With the previous analyses, we detected seven grid-cells richest in species and/or endemic species for these seed plants (Fig. 3). Some grid-cells found in southern Mexico are congruent in location with the Mesoamerican hotspot of Myers et al. (2000), with three richest in species and/or endemic grid-cells detected for Mexican Ternstroemiaceae (Fig. 5 in Luna et al. 2004) and with some rich in characteristic species grid-cells to Mexican cloud forest conditions (Fig. 5 in Luna et al. 2006).

Finally, when we compared those grid-cells richest in species and /or endemics with the Mexican priority regions for conservation (RTP's) of CONABIO (Arriaga et al. 2000), we detected that they coincide with eight RTP's: Punta Banda, Sierra de Juárez and San Telmo (Baja California state), and Bosques Mesófilos de Montaña de la Sierra Madre Oriental, Cuetzalan, Pico de Orizaba, Sierras del Norte de Oaxaca and Selva Zoque-La Sepultura (southeastern Mexico). The grid-cell located in Baja California coincide with portions of three RTP's.

Conclusions

Veracruz and Oaxaca are the states with the most species of gymnosperms in Mexico; this richness pattern is congruent with other groups of plants as Asteraceae, Cucurbitaceae, Fabaceae, and Poaceae (Dávila-Aranda et al. 2004). Both Mexican states have been earlier ranked in the first places of diversity for these angiosperm families (Dávila-Aranda et al. 2004). Our results coincides with the work of Mittermeier and

Mittermeier (1992),which suggested that these two states, as well as Chiapas, Guerrero, and Michoacán are the Mexican states with the most biodiversity.

Mexican gymnosperm diversity as well as species endemicity are concentrated in some states; many species that inhabit the northern states are also represented in the United States of America, and in general belong to genera with Nearctic affinities, as *Abies* and *Picea*. In the case of the southern states, some species share their distributions with the adjacent countries of Central America, as *Ceratozamia* and *Zamia*.

Pattern of species diversity did not followed a latitudinal gradient; most of the richest states are located in southeastern and eastern Mexico, but also in the north are located some of the richest states (Chihuahua, Coahuila, Durango, and Nuevo León). This fact can be explained if we consider that most of the Nearctic gymnosperm genera, such as *Abies, Cupressus, Juniperus, Pinus*, and *Picea* are more diverse in the Holarctic kingdom, and that those states located in northern Mexico are not the exception and therefore influenced by this distributional pattern.

The so-called 'peninsula effect', which implies the reduction in diversity towards the end of a peninsula (Gaston and Williams 1996) is evident in the geographic distribution of Mexican gymnosperms. In the distal portions of the Yucatán and Baja California peninsula only one species inhabits, *Zamia polymorpha* and *Pinus cembroides*, respectively. However, it has proven that this effect is occasional, rather than a quite general phenomenon (Gaston and Williams 1996).

If we compare the three levels of analysis done herein, we can observe that the state and grid-cell analyses include more artificial geographic units than the province analysis. However, a determined set of grid-cells may produce a larger unit and may show a partial correspondence with a particular biogeographic subprovince. In this work we support the idea suggested previously by Luna et al. (2004), that it is more informative and operative to use small geographic units instead of using the Mexican states, only in the case when we want to detect areas with high values of richness and endemicity. We need to have in mind that it is important to do this in order to detect areas with conservation priorities, but it is also important to protect the non-living environment (Bonn and Gaston 2005), that is to protect biodiversity in all its manifestations, where priority areas for nature conservation are needed to be recognised and networks of protected areas established and maintained (Bonn and Gaston 2005). Analyses of these types are fundamental to undertake other biogeographic studies applying other methods such as track analysis and cladistic biogeography.

Coincidence between richest in species and/or endemism grid-cells with some Mexican priority regions for conservation (RTP's) of CONABIO (Arriaga et al. 2000), suggest that these grid-cells are important for conservation, because these areas harbour high biodiversity. Comparatively, these areas have high values of ecosystem and species richness in relation to other areas of Mexico, as well as a functional ecologic integrity where real opportunities for conservation exist (Arriaga et al. 2000).

Some problems that have been detected with the grid-cell methodology (Crisp et al. 2001) and which we were not completely able to avoid are: (1) the existence of cells without distributional information, (2) the topographic variation found in each grid-cell (each grid-cell comprises an area of approximately 12,100 km^2) that may include different types of abiotic factors (climate, soil, vegetation, etc.) as well as altitudinal parameters, and (3) absence of updated distribution data, at least for rare, threatened, and new species.

Results obtained in this work support the identity of several biogeographic provinces based on high-scored values of several grid-cells obtained from the corrected weighted endemism for the Mexican gymnosperms. It appears that range restricted species are not distributed randomly over the landscape (Crisp et al. 2001), and in the case of Mexican gymnosperms they are aggregated in some areas of endemism that correspond and are useful to define and corroborate the naturalness of the Mexican biogeographic provinces. Grid-cells with high values in both indices and high richness (considered as hotspots herein) are important for conservation, especially those recognised by the corrected weighted endemism, because they have a high proportion of unique species; these grid-cells deserve special attention in Mexican future conservation plans. Those endemic taxa occurring in a single grid-cell are at high risk of human impact and could lead to extinction (McAllister et al. 1994). Gaston (1994) mentioned that most of the species that have small range sizes have more probability of extinction than others with wide range sizes. This is especially true for some gymnosperm species such as *Ceratozamia euryphyllidia* Vázquez-Torres et al., *C. hildae* G. Landry et M. Wilson, *C. norstogii* D. W. Stevenson, *C. zaragozae* Medellín-Leal, *Dioon califanoi* De Luca et Sabato, *D. caputoi* De Luca, Sabato et Vázquez-Torres, *D. holmgrenii* De Luca, Sabato et Vázquez-Torres, *D. rzedowskii* De Luca, Sabato et Vázquez-Torres, *Pinus maximartinezii* Rzedowski, *P. rzedowskii* Madrigal et Caballero, *Zamia inermis* Vovides, Rees et Vazquez-Torres, *Z. purpurea* Vovides, Rees et Vázquez-Torres, and *Z. soconuscensis* Schutzman et al. All of these taxa are examples of species with small ranges that are mostly represented by relatively few individuals within those ranges. In relation to cycads has been estimated that these species include less than 2,500 adult individuals in wild conditions (Osborne 1995). Some species of pines included in this study were considered by Farjon and Styles (1997) of urgent concern for conservation, namely *Pinus culminicola*, *P. rzedowskii*, *P. maximartinezii*, *P. pinceana*, *P. jaliscana*, *P. nelsonii*, and *P. strobus*. Approximately 71 species (57%) of Mexican gymnosperms have been included in some risk category in the latest version of the Mexican official publication named 'Norma Oficial Mexicana 059' (NOM-059-ECOL, Secretaria del Medio Ambiente y Recursos Naturales 2002), which includes native and introduced threatened taxa. In this document are included all the restricted-distribution species cited above in the categories of threatened, endangered or with special protection. For some Mexican gymnosperms several conservation strategies have been proposed (i.e. Styles 1993; Vovides and Iglesias 1994; Farjon and Styles 1997; Sosa et al. 1998; Luna et al. 2006), but it is important to continue with this task.

In general, areas of high species richness coincide with those areas of endemism generated by the corrected weighted endemism. Two exceptions are the Golfo de México province and California province which are not areas of high species richness, although they are confirmed as areas of endemism; in fact, two out of the five grid-cells scoring highest for this index are found in the Golfo de México province, whereas in the California province are found some of the high-scored grid-cells for this same index.

Several areas of high species richness agree with those proposed by Eguiluz (1985) and Styles (1993) for Mexican pines, especially those related to mountain chains, as Sierra Madre Occidental, Sierra Madre Oriental, Transmexican Volcanic Belt, Sierra Madre del Sur, Sierra de San Cristóbal, and Sierra Madre de Chiapas. This resemblance may be due to the influence of the number of species of pines (41), which represents one third of the total of gymnosperm species used in the present study. In relation to an altitudinal range, Mexican species of pines are mainly classified in the categories of montane (1,000–2,600 m) and high montane (2,500–4,000 m), showing a close relationship to montane habitats (Farjon and Styles 1997). Many other gymnosperm species belonging to different genera, such as *Abies*, *Ceratozamia*, *Cupressus*, *Ephedra*, *Juniperus*, *Picea*, and *Taxus*, are also mainly classified as montane species (Contreras-Medina 2004).

Repeatability of the grid-cell method applying the two indices explained above must be tested using other data sources from other well-documented groups, such as non-vascular plants, angiosperms, birds, butterflies and mammals, especially for those distributed in the Mexican montane chains, in order to compare the distributional patterns suggested herein for gymnosperms.

This study represents an example of the value of specimen-based data, such as are held in museums and herbaria of the world. Most of the distributional data of the species of gymnosperms used in this work were obtained from an exhaustive analysis of hundred of specimens of Mexican and North American herbaria. The information from herbaria is of special value because permanently preserved specimens can be physically examined, reexamined on subsequent occasions, and any reservations about identification noted (Hall 1994).

The present study contrasts with those mainly based only on literature, which did not corroborate distributional data and may contain identification and distribution mistakes; i.e. the biogeographic regionalization of Mexico by Espinosa et al. (2000) includes distributional incongruences in the case of the species of pines. Also, we have to consider that the distribution map of any plant species or taxon based strictly on herbarium specimens is in practical terms unrealistic, and assumptions that such maps may be error-free are unjustified, because locating and examining all herbarium specimens of a widely distributed taxon is a process that is not feasible (Hall 1994); this fact is especially evident in the case of the genus *Pinus*, because several species are widely distributed in the country. Revision of herbarium specimens and scientific literature citing voucher specimens and geographical localities should be considered as a major source of data for mapping (Hall 1994), and not those publications, which contain only distributional maps.

Distributional data from scientific collections are only useful if they are available (Crisp et al. 2001); this availability depends on the coordination of Mexican herbaria, and in this kind of studies serious problems are present, shortly commented above; despite this, the present analysis should be considered as a first biogeographic approximation of the areography of Mexican gymnosperms. However, resultant numbers of species richness and endemics per state, biogeographic province or grid-cell as presented above, are relevant to make conservation plans (McAllister et al. 1994). The identification of areas of high taxonomic diversity at more moderate scales than geopolitical and biogeographic regions, such as grid-cells used herein, has been a topic of some concern to conservationists (Gaston and Williams 1996).

The 'Red Mexicana Sobre la Biodiversidad' (REMIB) of the Comisión Nacional Para el Uso y Conocimiento de la Biodiversidad (CONABIO), located in Mexico City, has achieved accession to scientific collections. It represents a web-based flora and fauna information system developed by the cooperation of several American and Mexican scientific institutions. Unfortunately, it only provides direct access to some of the main specimen-based data. This information net has poor distributional information of threatened and rare species, as well as several errors in the

[244]

determination of the specimens. Other problems are that data on the web do not correspond with specimen labels, and it is not continuously updated; however, it represents a first attempt to make accessible information of Mexican scientific collections.

Acknowledgements We thank Juan J. Morrone, Susana Magallón, Oswaldo Téllez, Othón Alcántara and an anonymous referee for useful comments on the manuscript. We are also indebted to the staff of the herbaria cited in the text for their courtesy during our review of their specimens. Assistance in the field provided by Sandra Córdoba, Gimena Pérez, Ana Quintos, Othón Alcántara, Hamlet Santa Anna, Armando Ponce, Jorge Escutia, and Rogelio Aguilar is gratefully appreciated. Figures were done by Othón Alcántara. Support from projects PAPIIT IN206202, SEMARNAT-2004-01-311, and CONABIO W025 is gratefully acknowledged. The first author was supported by a Research Doctoral Fellowship number 169858 from the Consejo Nacional de Ciencia y Tecnología (CONACyT), Mexico.

References

Aguirre-Planter E, Furnier GR, Eguiarte LE (2000) Low levels of genetic variation within and high levels of genetic differentiation among populations of species of *Abies* from southern Mexico and Guatemala. Am J Bot 87:362–371

Arriaga L, Aguilar C, Espinosa D, Jiménez R (eds) (1997) Regionalización ecológica y biogeográfica de México. Workshop developed in the Comisión Nacional para el Conocimiento y Uso de la Biodiversidad (CONABIO), November 1997

Arriaga L, Espinoza JM, Aguilar C, Martínez E, Gómez L, Loa E (coord) (2000) Regiones terrestres prioritarias de México. Comisión Nacional para el Conocimiento y Uso de la Biodiversidad (CONABIO), Mexico

Bonn A, Gaston KJ (2005) Capturing biodiversity: selecting priority areas for conservation using different criteria. Biodivers Conserv 14:1083–1100

Contreras-Medina R (2004) Gimnospermas. In: Luna I, Morrone JJ, Espinosa D (eds) Biodiversidad de la Sierra Madre Oriental: un enfoque multidisciplinario. CONABIO-UNAM, Mexico City, pp 137–148

Contreras-Medina R, Eliosa-León H (2001) Una visión panbiogeográfica preliminar de México. In: Llorente J, Morrone JJ (eds) Introducción a la biogeografía en Latinoamérica: teorías, conceptos, métodos y aplicaciones. UNAM, Mexico City, pp 197–211

Contreras-Medina R, Luna I, Alcántara O (2001a) Las gimnospermas de los bosques mesófilos de montaña de la Huasteca Hidalguense, México. Boletín de la Sociedad Botánica de México 68:69–81

Contreras-Medina R, Morrone JJ, Luna I (2001b) Biogeographic methods identify gymnosperm biodiversity hotspots. Naturwissenschaften 88: 427–430

Contreras-Medina R, Luna I, Alcántara O (2003) Zamiaceae en Hidalgo, México. Anales del Instituto de Biología,Universidad Nacional Autónoma de México, Serie Botánica 74:289–301

Crisp MD, Laffan S, Linder HP, Monro A (2001) Endemism in the Australian flora. J Biogeogr 28:183–198

Dávila-Aranda P, Lira R, Valdés-Reyna J (2004) Endemic species of grasses in Mexico: a phytographic approach. Biodivers Conserv 13:1101–1121

Eguiluz T (1985) Origen y evolución del género *Pinus* (con referencia especial a los pinos mexicanos). Dasonomia Mexicana 3:5–31

ESRI (Environmental Systems Research Institute) (1999) Arc View GIS ver. 3.2. Environmental Systems Research Institute Inc., Redlands, USA

Espinosa D, Morrone JJ, Aguilar C, Llorente J (2000) Regionalización biogeográfica de México: provincias bióticas. In: Llorente J, González E, Papavero N (eds) Biodiversidad taxonomía y biogeografía de artrópodos de México: hacia una síntesis de su conocimiento. UNAM-CONABIO, Mexico City, pp 61–94

Espinosa J (1991) Gymnospermae. In: Rzedowski J, Calderón G (eds) Flora fanerogámica del Valle de México. Instituto Politécnico Nacional, Mexico City, pp 63–76

Springer

Farjon A, Styles BT (1997) *Pinus* (Pinaceae). Flora Neotropica Monograph 75. The New York Botanical Garden, New York

Felger RS (2000) Flora of the Gran Desierto and Rio Colorado of Northwestern Mexico. The University of Arizona Press, Tucson

Ferrusquía-Villafranca I (1993) Geology of Mexico: a synopsis. In: Ramamoorthy TP, Bye R, Lot A, Fa J (eds) Biological diversity of Mexico: origins and distribution. Oxford University Press, New York, pp 3–107

Fonseca RM (1994) Cupressaceae y Taxodiaceae. In: Diego-Pérez N, Rzedowski J, Fonseca RM (eds) Flora de Guerrero. Fascicle 2. Facultad de Ciencias, UNAM, Mexico City, 16 pp

García-Mendoza A (1995) Riqueza y endemismos de la familia Agavaceae en México. In: Linares E, Dávila P, Chiang F, Bye R, Elías T (eds) Conservación de plantas en peligro de extinción: diferentes enfoques. Instituto de Biología, UNAM, Mexico City, pp 51–75

Gaston KJ (1994) Rarity. Chapman & Hall, London

Gaston KJ, Williams PH (1996) Spatial patterns in taxonomic diversity. In: Gaston KJ (ed) Biodiversity, a biology of numbers and difference. Blackwell Science Ltd, Cambridge, pp 202–229

González D, Vovides AP (2002) Low intralineage divergence in *Ceratozamia* (Zamiaceae) detected with nuclear ribosomal DNA ITS and chloroplast DNA trnL-F non coding region. Syst Bot 27:654–661

Halffter G (1987) Biogeography of the montane entomofauna of Mexico and Central America. Ann Rev Entomol 32:95–114

Hall JB (1994) Mapping for monographs: baselines for resource development. In: Miller RI (ed) Mapping the diversity of nature. Chapman & Hall, London, pp 21–35

Kohlmann B, Sánchez S (1984) Estudio areográfico del género *Bursera* Jacq. ex L. (Burseraceae) en México: una síntesis de métodos. In: Métodos cuantitativos en la biogeografía. Instituto de Ecología A. C., Mexico City, pp 45–120

Linder HP (2001) Plant diversity and endemism in sub-Saharan tropical Africa. J Biogeogr 28:169–182

Luna I, Alcántara O, Contreras-Medina R (2004) Patterns of diversity, endemism and conservation: an example with Mexican species of Ternstroemiaceae Mirb. ex DC. (Tricolpates: Ericales). Biodivers Conserv 13:2723–2739

Luna I, Alcántara O, Contreras-Medina R, Ponce A (2006) Biogeography, current knowledge and conservation of threatened vascular plants characteristic of Mexican temperate forests. Biodiversity and Conservation DOI: 10.1007/s10531-5401-9

McAllister DE, Schueler FW, Roberts CM, Hawkins JP (1994) Mapping and GIS analysis of the global distribution of coral reef fishes on an equal-area grid. In: Miller RI (ed) Mapping the diversity of nature. Chapman & Hall, London, pp 155–175

McVaugh R (1992) Gymnosperms. In: Anderson WR (ed) Flora Novo-Galiciana, vol 17. The University of Michigan Herbarium Michigan, pp 4–119

Medina R, Dávila P (1997) Gymnospermae. In: Flora del Valle de Tehuacán-Cuicatlán. Fascicle 12. Instituto de Biología, UNAM, Mexico City, 29 pp

Mittermeier RA, Mittermier CG (1992) La importancia de la diversidad biológica de México. In: Sarukhán J, Dirzo R (ed) México ante los retos de la biodiversidad. CONABIO, Mexico City, pp 63–73

Moretti A, Caputo P, Cozzolino S, De Luca P, Gaudio L, Siniscalco G, Stevenson DW (1993) A phylogenetic analysis of *Dioon* (Zamiaceae). Am J Bot 80:204–214

Morrone JJ (2001) Biogeografía de América Latina y el Caribe. Manuales & Tesis SEA, Zaragoza, Spain

Morrone JJ, Márquez J (2001) Halffter's Mexican transition zone, beetle generalized tracks, and geographical homology. J Biogeogr 28:635–650

Myers N (1988) Threatened biotas: 'hot spots' in tropical forests. Environmentalist 8:187–208

Myers N, Mittermeier RA, Mittermeier CG, da Fonseca GAB, Kent J (2000) Biodiversity hotspots for conservation priorities. Nature 403:853–858

Narave H, Taylor K (1997) Pinaceae. In: Flora de Veracruz. Fascicle 98. Instituto de Ecología A. C. and University of California. Xalapa, Veracruz, Mexico, 50 pp

Niklas KJ, Tiffney BH, Knoll AH (1983) Patterns in vascular land plant diversification. Nature 303:614–616

Ortega F, Sedlock RL, Speed RC (2000) Evolución tectónica de México durante el Fanerozoico. In: Llorente J, González E, Papavero N (eds) Biodiversidad taxonomía y biogeografía de artrópodos de México: hacia una síntesis de su conocimiento. UNAM-CONABIO, Mexico City, pp 3–59

[246]

Osborne R (1995) The 1991–1992 world cycad census and a proposed revision of the threatened species status for cycads. In: Vorster P (ed) Proceedings of the Third International Conference on Cycad Biology. Cycad Society of South Africa, Stellensboch, pp 65–83

Patterson TE (1988) A new species of *Picea* (Pinaceae) from Nuevo León, México. SIDA 13:131–135

Perry JP, Graham A, Richardson DM (1998) The history of pines in Mexico and Central America. In: Richardson DM (ed) Ecology and biogeography of *Pinus*. Cambridge University Press, Cambridge, pp 137–149

Rapoport EH (1975) Areografía. Estrategias geográficas de las especies. Fondo de Cultura Económica, Mexico City

Rapoport EH, Monjeau A (2001) Areografía. In: Llorente J, Morrone JJ (eds) Introducción a la biogeografía en Latinoamérica: teorías, conceptos, métodos y aplicaciones. UNAM, Mexico City, pp 23–30

Rzedowski J (1978) Vegetación de México. Limusa, Mexico City

Rzedowski J (1991) Diversidad y orígenes de la flora fanerogámica de México. Acta Bot Mex 14:3–21

Salinas-Moreno Y, Mendoza MG, Barrios MA, Cisneros R, Macías-Sámano J, Zúñiga G (2004) Areography of the genus *Dendroctonus* (Coleoptera: Curculionidae: Scolytinae) in Mexico. J Biogeogr 31:1163–1177

Sclater PL (1858) On general geographical distribution of the members of class Aves. J Linn Soc Zool 2:130–145

Secretaría del Medio Ambiente y Recursos Naturales (SEMARNAT) (2002) Norma Oficial Mexicana NOM-059-ECOL-2001, Protección ambiental-Especies nativas de México y de flora y fauna silvestres-Categorías de riesgo y especificaciones para su inclusión, exclusión o cambio-lista de especies en riesgo. Diario Oficial de la Federación, México, 6 de marzo, pp 1–80

Sosa V, Vovides AP, Castillo-Campos G (1998) Monitoring endemic plant extinction in Veracruz, Mexico. Biodivers Conserv 7:1521–1527

Stevenson DW, Sabato S, Vázquez-Torres M (1986) A new species of *Ceratozamia* (Zamiaceae) from Veracruz, Mexico with comments on species relationships, habitats, and vegetative morphology in *Ceratozamia*. Brittonia 38:17–26

Styles BT (1993) Genus *Pinus*: a Mexican purview. In: Ramamoorthy TP, Bye R, Lot A, Fa J (eds) Biological diversity of Mexico: origins and distribution. Oxford University Press, New York, pp. 397–420

Takhtajan A (1986) Floristic regions of the world. University of California Press, Berkeley

Toledo VM (1982) Pleistocene changes in vegetation in tropical Mexico. In: Prance GT (ed) Biological diversification in the Tropics. Columbia University Press, New York, pp 93–111

Valdés J, Cabral I (1993) Chorology of Mexican grasses. In: Ramamoorthy TP, Bye R, Lot A, Fa J (eds) Biological diversity of Mexico: origins and distribution. Oxford University Press, New York, pp 439–446

Vovides AP (1999) Familia Zamiaceae. In: Flora del Bajío y de regiones adyacentes. Fascicle 71. Instituto de Ecología, Pátzcuaro, Michoacán, México, 17 pp

Vovides AP, Iglesias CG (1994) An integrated conservation strategy for the cycad *Dioon edule* Lindl. Biodivers Conserv 3:137–141

Vovides AP, Rees JD, Vázquez-Torres M (1983) Zamiaceae. In: Flora de Veracruz. Fascicle 26. INIREB, Xalapa, Veracruz, México, 31 pp

Wallace AR (1876) The geographical distribution of animals. Hafner Press, New York

Wiggins IL (1980) Flora of Baja California. Stanford University Press, Stanford

Zamudio S (1992) Familia Taxaceae. In: Flora del Bajío y de regiones adyacentes. Fascicle 9. Instituto de Ecología, Pátzcuaro, Michoacán, México, 7 pp

Zamudio S (2002) Familia Podocarpaceae. In: Flora del Bajío y de regiones adyacentes.Fascicle 105. Instituto de Ecología, Pátzcuaro, Michoacán, México, 7 pp

Zamudio S, Carranza E (1994) Familia Cupressaceae. In: Flora del Bajío y de regiones adyacentes. Fascicle 29. Instituto de Ecología, Pátzcuaro, Michoacán, México, 21 pp

Zanoni TA (1982) Cupressaceae. In: Flora de Veracruz. Fascicle 23. INIREB, Xalapa, Veracruz, México, 15 pp

Zanoni TA, Adams RP (1979) The genus *Juniperus* (Cupressaceae) in Mexico and Guatemala: synonymy, key, and distributions of the taxa. Boletín de la Sociedad Botánica de México 38:83–121

Biodivers Conserv (2007) 16:1823–1838
DOI 10.1007/s10531-006-9075-0

ORIGINAL PAPER

Impacts of El Niño related drought and forest fires on sun bear fruit resources in lowland dipterocarp forest of East Borneo

G. M. Fredriksson · L. S. Danielsen · J. E. Swenson

Received: 20 October 2005 / Accepted: 19 May 2006 / Published online: 12 July 2006
© Springer Science+Business Media B.V. 2006

Abstract Droughts and forest fires, induced by the El Niño/Southern Oscillation (ENSO) event, have increased considerably over the last decades affecting millions of hectares of rainforest. We investigated the effects of the 1997–1998 forest fires and drought, associated with an exceptionally severe ENSO event, on fruit species important in the diet of Malayan sun bears (*Helarctos malayanus*) in lowland dipterocarp forest, East Kalimantan, Indonesian Borneo. Densities of sun bear fruit trees ($\geq$10 cm DBH) were reduced by ~80%, from 167$\pm$41 (SD) fruit trees ha^{-1} in unburned forest to 37$\pm$18 fruit trees ha^{-1} in burned forest. Densities of hemi-epiphytic figs, one of the main fallback resources for sun bears during periods of food scarcity, declined by 95% in burned forest. Species diversity of sun bear food trees decreased by 44% in burned forest. Drought also affected sun bear fruit trees in unburned primary forest, with elevated mortality rates for the duration of 2 years, returning to levels reported as normal in region in the third year after the ENSO event. Mortality in unburned forest near the burn-edge was higher (25$\pm$5% of trees $\geq$10 cm DBH dead) than in the forest interior (14$\pm$5% of trees), indicating possible edge effects. Combined effects of fire and drought in burned primary forest resulted in an overall tree mortality of 78$\pm$11% ($\geq$10 cm DBH) 33 months after the fire event. Disturbance due to fires has resulted in a serious decline of fruit resources for sun bears and, due to the scale of fire damage, in a serious decline of prime sun bear habitat. Recovery of sun bear populations in these burned-over forests will depend

G. M. Fredriksson (✉)
Institute for Biodiversity and Ecosystem Dynamics/Zoological Museum, University of Amsterdam, P.O. Box 94766, Amsterdam, 1090 GT, Netherlands
e-mail: gmfred@indo.net.id

L. S. Danielsen · J. E. Swenson
Department of Ecology and Natural Resource Management, Norwegian University of Life Sciences, Box 5003 NO-1432 Ås, Norway

J. E. Swenson
Norwegian Institute for Nature Research, Tungasletta 2, NO-7485 Trondheim, Norway

 Springer

on regeneration of the forest, its future species composition, and efforts to prevent subsequent fire events.

Keywords Disturbance · Drought · ENSO · Figs · Fires · *Helarctos malayanus* · Kalimantan · Species diversity · Tree mortality

Introduction

Severe droughts, associated with increased occurrences of El Niño/Southern Oscillation (ENSO) events, have become more frequent over the past decades (e.g., Timmermann et al. 1999). These droughts have created conditions conducive for uncontrolled fires, which have damaged extensive areas of forest throughout the tropics, with fires in Southeast Asia being particularly severe on the islands of Borneo and Sumatra (Goldammer and Mutch 2001; Tacconi 2003). In recent years, forest fires have caused more deforestation than intentional clearing in some tropical regions (e.g., Cochrane et al. 1999). On Borneo between 3 million ha and 5 million ha of primary forest were affected by fires in 1982–1983 during a severe ENSO related drought (Lennertz and Panzer 1984; Malingreau et al. 1985). Smaller fire events occurred in 1990, 1992 and 1994, all coinciding with ENSO episodes (Salafsky 1998). During the severe 1997–1998 ENSO event primary forest areas burned easily (Siegert et al. 2001) as prior drought stress led to the shedding of leaves by evergreen species and accumulation of dry litter on the forest floor (GMF pers. obs.). In 1997–1998 in the Indonesian province of East Kalimantan alone, 5.2 million ha of land were affected by the fires, 2.6 million ha of which were forest, including several protected lowland reserves (Hoffman et al. 1999; Siegert et al. 2001; Fuller et al. 2004). Areas previously affected by fires have also become susceptible to more intense fires due to higher fuel loads and rapid desiccation, now even during "normal" dry seasons (Cochrane and Schulze 1999; Cochrane et al. 1999; Laurance 2003).

Several studies have investigated the effects of the 1997–1998 drought and fires on tree mortality and forest structure (e.g., Nakagawa et al. 2000; Williamson et al. 2000; van Nieuwstadt 2002; Slik et al. 2002; Potts 2003; Slik and Eichhorn 2003; Slik 2004). Their findings showed that drought significantly increased mortality rates in unburned forest (Nakagawa et al. 2000; van Nieuwstadt 2002; Potts 2003; Slik 2004; van Nieuwstadt and Sheil 2005), that overall tree mortality was extremely high in burned areas (van Nieuwstadt and Sheil 2005), that fire resulted in a strong reduction of climax tree density (Slik and Eichhorn 2003), and that species composition changed after fire damage due to disproportionate mortality of certain tree species groups and tree size classes (Slik et al. 2002; Slik 2004). Shifts in species composition in natural forest occur slowly under "normal" conditions (Swaine et al. 1987), but catastrophic disturbances like repeated fires can reduce structural and biological complexity in forests (Schindele et al. 1989). Fire-return intervals of less than 90 years can eliminate rainforest tree species, whereas intervals of less than 20 years may eradicate tree growth entirely resulting in savanna-like landscapes (Cochrane et al. 1999).

Few studies however have investigated the effects of drought and forest fires on wildlife or their food resources in Indonesia (Doi 1988; Anggraini et al. 2000; O'Brien et al. 2003). Forest-dependent species tend to become less abundant or even locally extinct and other, less-forest-dependent species invade the area or increase in abundance (Doi 1988; Anggraini et al. 2000; see also Barlow et al. 2002; Peres et al. 2003). Some of the larger, long-lived species persist, reluctant or unable to relocate themselves (Suzuki 1992; Anggraini et al. 2000), although effects on their life-history remain largely unknown (but see O'Brien et al. 2003). Furthermore, the proximate causes of why individual species perish or flourish—e.g., decreased or increased food-availability, reduced nest-sites—remain unstudied.

We studied the effects of fires on a variety of food resources important in the diet of one of the largest extant mammals on Borneo, the Malayan sun bear. Sun bears are partly frugivores (McConkey and Galetti 1999; Wong et al. 2002; Fredriksson et al. in press), although during periodic mast-fruiting events fruit makes up almost 100% of the diet (Fredriksson et al. in press). These mast-fruiting events provide the opportunity for sun bears to gorge themselves on large amounts of succulent fruits, probably enabling them to build up, or recover, fat and energy reserves for the prolonged period of fruit lows preceding and following these supra-annual mast-fruiting events (Fredriksson et al. in press). Sun bears are considered important actors in forest dynamics, in part due to their seed dispersal abilities (McConkey and Galetti 1999), especially of large-seeded fruits (Fredriksson, unpubl. data). Preliminary data suggests that sun bears make little use of burned forest areas for several years after a fire event (Doi 1988; Fredriksson, unpubl. data).

The aims of this study were (i) to investigate the effect of forest fires on the density of sun bear fruit resources; (ii) to quantify the effect of forest fires on species diversity of sun bear fruit resources; and (iii) to study the effect of drought on mortality of sun bear food trees in unburned forest and based on the above (iv) discuss the effects of these changes on sun bear populations in fire affected areas.

Methods

Study area

The study was carried out in the lowland dipterocarp forest of the Sungai Wain Protection Forest (SWPF), a reserve near Balikpapan, East Kalimantan, Indonesian Borneo (1°16′ S and 116°54′ E) (Fig. 1). The reserve covers a watercatchment area of ca 100 km². Average annual rainfall was 2740±530 mm (1998–2003). The topography of the reserve consists of gentle to sometimes steep hills, and is intersected by many small rivers. The area varies in altitude from 30 m to 150 m a.s.l. Trees with stems greater than 10 cm DBH (diameter at breast height) are dominated by the families Euphorbiaceae, Dipterocarpaceae, Sapotaceae, and Myrtaceae. The relative dominance of Dipterocarpaceae increases substantially in the larger size classes. Due to an altitudinal gradient, with rivers running in a north–south direction, the southern part of the reserve is moister. Dipterocarpaceae, dominant in the northern part of the reserve (above 10 cm DBH) decrease in abundance towards the moister south where Sapotaceae and Euphorbiaceae become more dominant. The 25 most common tree species form 40% of the total stem density (van Nieuwstadt 2002). Several palm genera (*Borassodendron, Oncosperma, Polydocarpus, Licuala*) are

🖄 Springer

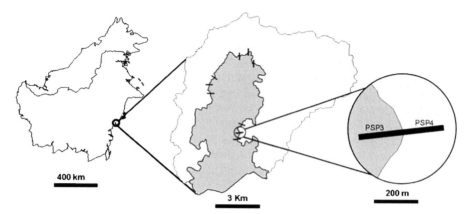

Fig. 1 Map showing the location of Sungai Wain forest on the island of Borneo. The enlargement shows the Sungai Wain forest, indicating the unburned area (dark grey), and burned areas (light grey). The location of pairs of burned and unburned edge plots is shown. The circular inset shows an example of the lay out of one pair burned-unburned edge plots over the fire edge

common in the subcanopy and understory (especially rattans), and ginger species (Costaceae and Zingiberaceae) as well as Marantaceae, Araceae, and Pandanaceae are common in the understory. This paper only deals with trees ≥10 cm DBH. All growth forms of figs (*Ficus* spp.) were included. Only hemi-epiphytic figs with a diameter of ≥3 cm were included as potential food resources, as these were observed to bear fruits and to be fed upon by sun bears (Fredriksson, pers. obs).

History of forest fires at the study site

Most of the reserve was unaffected by fires at the start of the study in 1997, except for a small area near the eastern border that had burned in 1982–1983. The prolonged drought of the 1982–1983 ENSO, and several subsequent shorter ENSO droughts, probably caused elevated levels of mortality among large trees, resulting in an irregular canopy cover. Consequently the primary forest became increasingly vulnerable to desiccation during droughts and more susceptible to fires. Drought in the region started in May 1997 and continued intermittently till late April 1998, with 6 out of 12 months having no rainfall at all, or well below 100 mm. Fires entered the SWPF in March 1998, initially from a neighbouring state-owned logging concession, but subsequently also from surrounding agricultural fields. Fires moved slowly through the leaf litter and remained mainly in the undergrowth. Occasionally the fire reached into the crowns of older trees with hollow trunks, or dead standing trees with resin residues. In burned areas, all leaf litter and surface soil humus was reduced to ash, and mortality of seedlings and saplings was close to 100% (Fredriksson, pers. obs.). Fire breaks were created over a period of 2 months but nevertheless approximately 50% of the reserve was affected by the fires (Fredriksson 2002), leaving an unburned central core of some 4,000 ha of primary forest.

Permanent sampling plots

In the SWPF, 18 permanent sample plots (PSP), of 20 × 200 m (0.4 ha) each, were established in once-burned forest and adjacent unburned forest after the fire event in

🖄 Springer

1998 (Fig. 1) by the Wanariset Research Station (Tropenbos-Kalimantan Project). The set-up of the PSPs was designed around man-made firebreaks of ~1.5 m wide (see van Nieuwstadt 2002). Because the fire-breaks did not correspond to any topographical feature in the places where the PSPs were positioned, this allowed for a random sampling scheme with paired plots of unburned and burned forest at a short distance from each other. The PSPs were laid out in nine pairs, each pair of PSPs adjacent to each other over the firebreak between burned and unburned forest, forming one contiguous transect of 20 × 400 m, half in burned and half in unburned forest. The unburned PSPs lie adjacent to the burn edge and these plots are labelled "unburned edge plots." The PSPs were nested in three groups and spread over a total area of circa 20 km^2 (Fig. 1).

The distance between two pairs of PSPs was more than 500 m. We counted and measured all trees (≥10 cm DBH) in the PSPs 33 months after the fires, and calculated the percentage of live trees in both burned and unburned edge PSPs. Liana's were not sampled, as they only make up a small proportion (1%) of fruits encountered in the diet of sun bears at the study site (Fredriksson et al. in press).

Mortality rates

Annual mortality rates were calculated from ten 0.1-ha phenology plots (total 1 ha, all trees ≥10 cm DBH, $n = 549$ trees at the start of the study) which were monitored on a monthly basis between January 1998 and July 2004. These phenology plots were positioned in unburned forest at least 1 km from the burn edge, and are subsequently called "unburned interior" plots. Annual mortality rates for the interior plots are calculated based on exact 12 month periods (Jan–Dec).

$$\text{Mortality rate was determined as: } m = \{1 - (N_t/N_0)^{1/t}\}^*100$$

where "m" is mortality per year, N_0 is the initial number of live individuals and N_t is the number of live individuals at re-census interval t (e.g., Sheil and May 1996). Percentage of live trees was calculated for the interior plots for the same time interval as for the burned and unburned edge plots.

Densities of tree species important in the diet of the sun bear

A list of 115 fruit species eaten by sun bears in the study area was available from Fredriksson et al. (in press). All trees (≥10 cm DBH) that provided these fruits were subsequently labelled "sun bear fruit trees." Densities of sun bear fruit trees and all *Ficus* spp. were subsequently calculated for the three subsets of plots.

Two common species-rich genera, which occur in the diet of the sun bear (*Syzigium* spp. [Fam. Myrtaceae]; *Diospyros* spp. [Fam. Ebenaceae]) posed a problem for density calculations as identification to species level is difficult. Only certain species from these genera were fed upon by the bears, whereas others were consistently ignored. In order to avoid overestimation of the density of these genera, we first calculated the percentage of trees from these genera in the 1 ha interior plots that belonged to species actually fed upon by sun bears, based on leaf and fruit samples collected during direct feeding observations of sun bears. For both genera this was found to be approximately 50% of the individual trees in the 1 ha interior plots. Therefore, the density of these taxa as potential sun bear food resources was reduced by 50% for further analyses in all plots. Two other taxa, *Madhuca kingiana* [Fam. Sapotaceae] and *Pternandra* spec. [Fam. Melastomataceae], were infrequently encountered in the diet of the bears, but are common in the forest. In order to avoid overestimating the densities of sun bear fruit trees, we reduced the density of these species also by 50% in the analyses.

Statistical analysis

Differences between burnt and unburned forest plots were compared with paired *t*-tests. A one-way ANOVA, with a Bonferroni multiple comparison test, was used to check for differences in mortality rates between sampling regimes and years. Standard deviation is always given when the average is presented (average ± SD). An α-level 0.05 was chosen to indicate statistical significance.

Results

Density of fruit resources

Density of live fruit trees important in the sun bear diet, 33 months after the fire was 167 ± 41 trees ha^{-1} in unburned edge plots and 37 ± 18 trees ha^{-1} in burned plots, indicating a drought and fire-related reduction of nearly 78% (paired *t*-test: $t = 8.7$, df = 8, $P = 0.001$) (Table 1). Densities of sun bear fruit trees in the unburned interior plots was 144 trees ha^{-1} (total area sampled 1 ha) at the start of the study in 1998 and declined to 133 trees ha^{-1} 33 months after the drought.

Densities of all important sun bear fruit genera were lower in burned forest compared to unburned edge plots, although this was only significant for 8 out of 19 genera (Table 1). Several genera, generally occurring at low-medium densities, were not represented by any live trees in burned plots (e.g., *Monocarpia* [Fam. Annonaceae], *Quercus* [Fam. Fagaceae], *Litsea*, *Cryptocarya* [Fam. Lauraceae]). The difference in densities for these genera was not significant due to the large variation in number of trees encountered in the 9 unburned edge plots (Table 1).

Two of the main sun bear fruit genera, *Artocarpus* and *Dacryodes*, which contribute the bulk of fruit eaten by bears during masting events, declined significantly in densities, respectively from 11.9 ± 6.6 trees ha^{-1} to 1.9 ± 2.4 trees ha^{-1} and 11.4 ± 7.8 trees ha^{-1} to 2.5 ± 2.5 trees ha^{-1} (Table 1). The palm *Oncosperma horridum*, whose fruits are favoured by sun bears, declined from 8.1 ± 4.2 trees ha^{-1} in unburned edge plots to zero in burned areas (Table 1).

The most important plant genus for sun bears which provides fruits during intermast periods, *Ficus* spp., declined significantly in burned forest. Most epiphytic, hemi-epiphytic and climber figs were encountered in the unburned edge plots, with only 4 of 74 figs (all sizes combined) observed in burned plots (paired *t*-test: $t = 6.5$, df = 8, $P = 0.001$). Densities of figs important in the sun bear diet declined significantly from 5.6 ± 1.6 figs ha^{-1} in unburned edge plots and 0.3 ± 0.3 figs ha^{-1} in the burned plots (paired *t*-test: $t = 3.7$, df = 8, $P = 0.006$), corresponding to a reduction of ~95% following fire.

No significant differences were found in densities for the variably common genera *Madhuca* and *Diospyros*. *Madhuca kingiana* is a dominant tree species in the permanent sample plots in the south of the reserve, which is the moister part of the

Table 1 Densities (stems ≥10 cm DBH, average ± SD ha^{-1}) calculated from 18 vegetation plots (total 7.2 ha) of the main fruit-bearing genera important in the sun bear diet in unburned edge forest (UBF-edge) and burned forest (BF) 33 months after the 1997–1998 fire event. Genera only include species that have been found to occur in the bear diet. *Densities of *Ficus* mainly comprise of hemi-epiphytic stranglers ≥3 cm diam. *P* indicates significance level for paired *t*-test (df = 8)

Genera	Family	Density				
		UBF-Edge	SD	Burned	SD	*P*
Madhuca	Sapotaceae	16.5	21.3	2.9	4.4	ns
Artocarpus	Moraceae	11.9	6.6	1.9	2.4	0.003
Dacryodes	Burseraceae	11.4	7.7	2.5	2.5	0.017
Syzigium	Myrtaceae	11.4	7	3.1	3.4	0.011
Baccaurea	Euphorbiaceae	8.1	3.7	0.6	1.1	0.001
Oncosperma	Palmae	8.1	10.5	0	–	ns
Diospyros	Ebeneceae	7.8	6.4	3.3	2.8	ns
Santiria	Burseraceae	6.1	4.9	0.6	1.1	0.013
Ficus*	Moraceae	5.6	3.9	0.3	0.8	0.006
Litsea	Lauraceae	5.6	5.3	0	–	0.013
Lithocarpus	Fagaceae	4.7	3.2	0.8	1.8	0.005
Garcinia	Guttiferae	3.6	3.3	0.8	1.8	ns
Polyalthia	Annonaceae	3.3	2.5	1.1	1.8	ns
Durio	Bombacaceae	2.5	3.8	0.6	1.1	ns
Quercus	Fagaceae	1.4	2.8	0	–	ns
Cryptocarya	Lauraceae	1.1	1.8	0	–	ns
Mangifera	Anacardiaceae	0.6	1.7	0.3	0.8	ns
Monocarpia	Annonaceae	0.6	1.1	0	–	ns
Tetramerista	Tetrameristaceae	0	–	0.6	1.1	ns
All sun bear fruit species combined		166.9	40.7	36.8	17.8	0.001

forest due to topography. One unburned edge plot in the southern part of the forest (0.4 ha) contained 22 trees (≥10 cm DBH) of this species compared to several unburned edge plots in the western or northern part of the forest which contained zero trees of this species.

Species diversity of fruit trees in burned and unburned forest

Species diversity of sun bear fruit resources declined significantly following fire (paired *t*-test: *t* = 11.9, df = 8, *P* = 0.001). At least 66 species which feature in the diet of sun bear occurred in unburned edge plots, whereas only 37 of these were observed in burned plots, representing a decrease of ~44% in richness of sun bear fruit species in burned forest areas (Appendix). The family Lauraceae, representing the sun bear fruit genera *Cryptocarya* and *Litsea*, as well as the family Caesalpiniaceae with the sun bear fruit genus *Dialium* were not encountered with any live trees in burned sampling plots, whereas in the unburned edge plots the genus *Litsea* (all species combined) occurred in 8 out of 9 plots and was represented with 20 trees. The genus *Dialium* was found in 6 out of 9 unburned edge plots. A total of 46 sun bear food tree species were encountered in the unburned interior plots (total area sampled 1 ha vs. 3.6 ha sampled of burned and unburned edge plots each).

Tree mortality

Overall tree mortality (both sun bear fruit trees and tree species not fed upon by bears) in burned forest was extremely high, with 77.5±10.8% trees dead 33 months after the fire event (range 58.2–90.0%, n = 9 plots). In unburned edge plots mortality was also high with 24.9±5.4% of trees dead (range 17.7–35.6%, n = 9 plots). Cumulative percentage of dead trees in the unburned interior plots 33 months after the drought was 14.1±5.2%, some 40% lower than encountered in the unburned edge plots.

Mortality rates between years differed significantly (one-way ANOVA F = 6.6 df = 5, P = 0.001) although only mortality in 1999 was significantly different (higher) than all other years (Bonferroni, t > 3.8, P = 0.006 for all comparisons). Mortality remained elevated for 2 years in the interior plots and approached "normal" rates reported for the region in the third year after the drought (Table 2). Mortality rates did not differ significantly between sun bear fruit species and species that do not occur in the diet of bears in the interior plots (t-test, t = 2.3, df = 5, P = 0.067), although the P-value suggested a tendency. No significant difference was found in mortality of sun bear food trees and non-food trees between burned and edge plots 33 months after the fire (paired t-test, t > 0.04, df = 8, P > 0.3).

Discussion

The 1997–1998 fires reduced the density of sun bear fruit trees by nearly 80%, 3 years after the fire. Fruit resources are important in the diet of sun bears, partly to regain energy after periods of fruit scarcity and partly to build up fat reserves to cope with prolonged intermast periods (Fredriksson et al. in press). When few fruit resources are available sun bears subsist primarily on insects although densities of these were also highly reduced in burned forest areas (Fredriksson, unpubl. data). The reduction in fruit trees measured during this study is almost double the 44% decline reported by Leigthon and Wirawan (1986) for fruit species important in the diet of frugivorous primates like Bornean gibbons (*Hylobates muelleri*) and orangutans (*Pongo pygmaeus*) after the 1982–1983 fires in Kutai National Park, East Kalimantan. Possibly the figures presented by Leigthon and Wirawan (1986) are underestimates of the true extent of damage as their sampling was carried out shortly after the fire event. We documented delayed mortality due to the fires and drought, which continued for at least 2 years after the ENSO event. Mortality rates approached "normal" rates reported for the region in the third year after the drought (average 1.7% for 9 study sites in the region *see* Phillips et al. 1994; Wich et al. 1999; Potts 2003).

	Non-bear trees (n = 405)	Bear trees (n = 144)	All trees combined (n = 549)
1998	4.69	3.47	4.37
1999	7.25	4.32	6.48
2000	2.79	0.00	2.04
2001	2.01	0.00	1.46
2002	2.35	2.26	2.32
2003	2.40	3.08	2.59
Average	3.58	2.19	3.21

Table 2 Annual mortality rates (%) of trees (≥10 cm DBH) in unburned interior plots calculated from ten 0.1 ha phenology plots. Mortality rates are presented separately for sun bear food trees and non-food trees, as well as for all trees combined

 Springer

Fire reduced the important sun bear fruit genus *Ficus* spp. by 95% 3 years after the fires. This serious decline might well have negative consequences in terms re-establishment or persistence of frugivore populations, as figs have been found to be one of the main fruit resources during periods of food lows for a variety of wildlife (e.g., Leighton and Leighton 1983). Besides the large reduction of fig densities in burned areas, Harrison (2000) reported on local extinctions of fig wasps after the ENSO drought, which affected fig fruit production, even in forest areas unaffected by fires. Putz and Susilo (1994) however reported that 10 years after the 1982–1983 forest fires regeneration of hemi-epiphytic figs by establishment of new terrestrial connections was close to 60%.

The reduction of almost 44% in species richness of sun bear fruit taxa in once-burned forest could lead to permanent changes in the composition of sun bear fruit resources in these areas, especially as certain sun bear fruit tree genera disappeared altogether from the burned plots. Fires affect dominant tree species more than rare species (Slik et al. 2002; Potts 2003), but the disappearance of rare species is more worrying as they might become locally extinct (Cochrane and Schulze 1999). Regeneration of the forest will largely depend on the crop of trees that sprouts after the fires. The proportion of seedlings that grew up 5 years after the 1982–1983 fires in Sabah showed a close resemblance to the distribution of families in primary forest, although few Dipterocarp saplings were encountered (Woods 1989). Slik et al. (2002) found that lowland dipterocarp forest, 15 years after being affected by fire, did not show an increase in species richness, although stem density increased, but these primarily belonged to a few pioneer tree species (*Macaranga*). The recruitment of *Macaranga*, both in the under- and over-storey, indicated that recovery of species composition in burnt forests takes longer than in selectively logged forest, where after 15 years pioneer species were being replaced by primary forest species (Slik et al. 2002). Whitmore (1985) reported that a lowland dipterocarp forest, extensively damaged by storm and fire in 1880, was still unusually poor in diversity of upper-canopy species when surveyed some 70 years later. Close to 18% of the burned forest in our study area has remained as unburned forest patches in swampy areas or near streams (Fredriksson and Nijman 2004), which might facilitate a more uniform regeneration due to seed dispersal of forest-interior species into the burned areas. But overall, it will take decades, if not centuries, for many slow-growing climax species to begin fruiting and provide food for wildlife (Whitmore 1985).

Fruiting phenologies of remaining fruit trees in burned areas might also deviate from fruiting patterns in unburned forest due to changes in environmental conditions and exposure to higher levels of drought. Kinnaird and O'Brien (1998) reported that occurrence of flowering and fruiting by trees in burned areas was lower than in adjacent unburned areas just after the 1998 fires. The storage of water reserves in trees, which depends on availability of subsoil water, has been found to influence flowering, although water storing capabilities differed greatly between species (Borchert 1994). The possible influences of changed environmental conditions in burned forest on phenological patterns calls for further studies.

Drought by itself had significant effects on sun bear food trees in unburned forest, with elevated mortality rates up to 2 years after the drought. Effects of the drought on trees have been found to be differential, relative to abundance and habitat factors (Potts 2003), with trees in moister areas less affected (Potts 2003;

🍐 Springer

Slik and Eichhorn 2003; Fredriksson and Nijman 2004). The combined effects of fire and drought were far more severe than the effects of drought alone. Heavily burned forest areas have become dominated by pioneer species (Woods 1989; Nykvist 1996; Toma et al. 2000; Slik et al. 2002), have lower species diversity (Matius et al. 2000; Slik et al. 2002; this study), and decreased soil fertility due to a loss of inorganic nutrients (Nykvist et al. 1994; Malmer 1996). Several studies investigated the regeneration potential of burned forest areas after the 1982–1983 fires. Above-ground biomass in a lowland forest in Sabah 8 years after the fire was still only a quarter of adjacent unburned forest (Sim and Nykvist 1991; Nykvist 1996).

The significant decrease in both density as well as species diversity of sun bear fruit resources due to fire could partially explain the reduced usage of burned forest by sun bears (Fredriksson, unpubl. data). Few fresh signs of sun bears were encountered in burned areas up to 5 years after the fire event, and radio-collared bears rarely entered burned forest. Although the extent of the home ranges of our radio-collared bears before the fire was unknown (as they were caught in unburned forest following the fire event), sun bears ranged throughout the reserve prior to fires (Fredriksson 2005; Fredriksson, unpubl. data,). Environmental conditions like an increase in temperature due to lack of canopy cover, exposure to rain, and also the inaccessible nature of burned areas due to a thicket of ferns which blocked the understory for approximately 4 years after the fires, hampering movement for large ground-dwelling mammals, probably discouraged usage by bears. Additionally, densities of various invertebrate food resources, especially termites, declined significantly after the fires (Fredriksson et al., unpubl. data). Doi (1988) reported that sun bears did not recover quickly after the 1982–1983 fires in the largest lowland conservation area in East Kalimantan, the 200,000 ha Kutai National Park, with few bear signs found even in the core of the park 3 years after the fires. Leighton and Wirawan (1986) reported a decrease in vertebrate densities after the 1982–1983 fires, although large-bodied primates reportedly seemed to be the least affected, possibly due to their generalized omnivorous diets and behavioural flexibility to switch food types (van Schaik et al. 1993). On the other hand O'Brien et al. (2003) found a significant decrease in siamang (*Symphalangus syndactylus*) group sizes, as well as infant and juvenile survival, after forest fires in Sumatra to a point where it seems unlikely that groups will survive for more than two generations in burned areas.

Tree mortality figures differed substantially between edge and interior plots in unburned forest. Almost 40% more dead standing trees were encountered in the unburned edge plots compared to interior plots. Although no study as yet has specifically investigated edge effects in Borneo, Laurance et al. (2000, 2001) reported that mortality along forest edges in Amazonia is higher than in the forest interior, with increased tree mortality levels penetrating up to 300 m from the forest edge. The extremely high mortality recorded in burned areas might also be a cumulative effect of several severe ENSO events over the last decades, each of which probably has caused elevated mortality, resulting in a more open canopy and higher water stress during subsequent droughts. The fact that annual mortality rates in unburned forest were elevated for 2 years post-ENSO, could indicate that, with an increased rate of ENSO events, such prolonged elevated mortality rates will also be occurring on a more frequent basis. If forests in these drought prone areas experience mortality rates of 6–7% every 15 years for extended periods of time,

[258]

recruitment might not balance mortality and significant changes in forest structure could occur even without any direct human influence.

Conclusions

Large changes in vegetation structure and environmental conditions in burned forests, coupled with a significant decrease in fruit resources important in the diet of sun bears, have caused a significant reduction of suitable habitat for this bear species in fire-affected areas. The slow regeneration of these burned areas will probably influence re-colonization by bears, even if they have been able to maintain large enough population numbers within the burned forest matrix and adjacent unburned forest areas. Little is known about the ability of sun bears to exploit new pioneer fruit resources which have sprouted since the fires. The most dominant pioneer genus (*Macaranga*) has dehiscent fruits with small arillate seeds primarily attractive to birds (Davies and Ashton 1999; Slik et al. 2000), and has not (yet) been encountered in the sun bear diet.

The damage due to fires in primary forest has far surpassed that encountered in logged-over areas (Woods 1989; Slik et al. 2002). Unburned logged-over areas might have higher potential for biodiversity conservation than primary forest that has burned once, although long-term monitoring of floral regeneration patterns and wildlife diversity and abundance in burned areas needs to be carried out in order to determine the future value of such forest for conservation. The massive spatial scale of these fire disturbances and the relatively short timespan during which they have affected vast areas (usually 2–3 months of fires) is a new and worrying phenomenon. With the increase in the frequency and severity of ENSO (Timmermann et al. 1999), the future of these burned over forests looks grim as regeneration will only take place if no further fires affect these areas, whereas repeated fire damage is common (Cochrane et al. 1999; Siegert et al. 2001). These factors, aggravated by low fire prevention activities and a lack of law enforcement in the region, call for highly increased conservation efforts in these drought prone rainforest areas if productive sun bear habitat is to be retained.

Acknowledgements We thank the Indonesian Institute of Sciences (LIPI) for granting GMF permission to carry out research on sun bears in East Kalimantan. GMF would like to express gratitude to the Forest Research Institute in Samarinda for their assistance. The Tropenbos Foundation and the Conservation Endowment Fund of the American Zoo and Aquarium Association provided financial assistance. We especially thank Dave Garshelis for logistical, financial and technical assistance. We thank the Wanariset Herbarium, especially Ambrianshyah and Arifin for identification of plant samples. The Tropenbos Foundation is thanked for usage of the permanent sample plots. Field assistants Trisno, Lukas Nyagang, and Damy da Costa are thanked for various assistance in the field throughout the years, as well as many people from the Sungai Wain village. LSD received financial and technical assistance from the Norwegian University of Life Sciences. We thank S.B.J. Menken, D. Garshelis, and V. Nijman for useful comments of previous drafts and G. Usher and M. van Nieuwstadt for the map.

Appendix List of sun bear fruit genera/species of trees (≥10 cm DBH) encountered in unburned-edge (UBF) and burned (BF) vegetation plots (total 7.2 ha), 33 months after the 1997–1998 fire. A "√" indicates that the genus/species was encountered in the plots, a "0" indicates it was absent. Number of plots indicates in how many of the 0.4 ha plots (9 plots in burned and 9 plots in unburned edge forest) the genus/species was encountered

Genus	Species	Family	UBF-edge	No. of plots	BF	No. of plots
Aglaia	spec.	Meliaceae	√	7	0	0
Alangium	ridleyi	Alangiaceae	√	3	√	1
Artocarpus	anisophyllus	Moraceae	√	9	√	3
Artocarpus	dadah	Moraceae	√	1	0	0
Artocarpus	integer	Moraceae	√	3	√	1
Artocarpus	lanceifolius	Moraceae	√	2	0	0
Artocarpus	nitidus	Moraceae	√	3	0	0
Artocarpus	spp.	Moraceae	√	4	√	2
Baccaurea	bracteata	Euphorbiaceae	√	3	0	0
Baccaurea	macrocarpa	Euphorbiaceae	√	6	0	0
Baccaurea	spec.	Euphorbiaceae	√	8	√	2
Baccaurea	stipulata	Euphorbiaceae	√	1	0	0
Borassodendron	borneensis	Palmae	√	7	√	8
Canarium	spp.	Burseraceae	√	4	0	0
Crypteronia	spec.	Crypteroniaceae	√	7	√	1
Cryptocarya	spec.	Lauraceae	√	3	0	0
Dacryodes	costata	Burseraceae	√	3	√	1
Dacryodes	rostrata	Burseraceae	√	7	√	1
Dacryodes	rugosa	Burseraceae	√	2	√	1
Dacryodes	spec.	Burseraceae	√	5	√	4
Dialium	indum	Ceasalpiniaceae	√	4	0	0
Dialium	platysepalum	Ceasalpiniaceae	√	1	0	0
Dialium	spec.	Ceasalpiniaceae	√	4	0	0
Diospyros	borneensis	Ebenaceae	√	6	√	4
Diospyros	cf buxifolia	Ebenaceae	0	0	√	1
Diospyros	spec.	Ebenaceae	√	7	√	4
Durio	dulcis	Bombacaceae	√	1	√	1
Durio	graveolens	Bombacaceae	√	1	0	0
Durio	kutejensis	Bombacaceae	√	1	0	0
Durio	oxleyanus	Bombacaceae	√	3	√	1
Glochidion	spec.	Euphorbiaceae	√	2	0	0
Ilex	cymosa	Aquifoliaceae	√	1	√	1
Irvingia	malayana	Simaroubaceae	√	1	√	4
Lansium	domesticum	Meliaceae	√	1	√	1
Lansium	spec.	Meliaceae	√	1	0	0
Lithocarpus	gracilis	Fagaceae	√	1	0	0
Lithocarpus	spp.	Fagaceae	√	8	√	2
Litsea	firma	Lauraceae	√	4	0	0
Litsea	spp.	Lauraceae	√	7	0	0
Madhuca	kingiana	Sapotaceae	√	6	√	4
Magnolia	lasia	Magnoliaceae	√	2	0	0
Mangifera	macrocarpa	Anacardiaceae	√	1	0	0
Mangifera	spp.	Anacardiaceae	√	1	√	1
Monocarpia	kalimantanensis	Annonaceae	√	2	0	0
Nephelium	spec.	Sapindaceae	√	5	√	1
Ochanostachys	amentaceae	Olacaceae	√	9	√	2
Oncosperma	horridum	Palmae	√	6	0	0
Palaquium	spp.	Sapotaceae	√	9	√	3
Polyalthia	lateriflora	Annonaceae	0	2	√	1
Polyalthia	laterifolia	Annonaceae	√	2	0	0
Polyalthia	rumphii	Annonaceae	√	3	√	1
Polydocarpus	spec.	Palmae	√	5	√	2
Prunus	spec.	Rosaceae	√	3	√	2
Pternandra	volgens	Melastomataceae	√	6	0	0
Quercus	spec.	Fagaceae	√	8	√	2
Sandoricum	spp.	Meliaceae	√	2	0	0
Santiria	spp.	Burseraceae	√	3	0	0
Santiria	cf apiculata	Burseraceae	√	1	0	0
Santiria	spp.	Burseraceae	√	6	√	2
Santiria	tomentosa	Burseraceae	√	5	0	0

Appendix continued

Genus	Species	Family	UBF-edge	No. of plots	BF	No. of plots	Genus	Species	Family	UBF-edge	No. of plots	BF	No. of plots
Dysoxylum	spec.	Meliaceae	√	2	√	1	*Syzygium*	spp.	Myrtaceae	√	9	√	7
Eugenia	spec.	Myrtaceae	√	2	0	0	*Tetramerista*	*glabra*	Tetrameristaceae	0	0	√	2
Eugenia	*tawahense*	Myrtaceae	√	5	√	3	*Xerospermum*	spec.	Sapindaeae	√	1	0	0
Ficus	spp.	Moraceae	√	8	√	1	Unburned edge plots: 66 species						
Garcinia	*parvifolia*	Guttiferae	√	4	0	0	Burned plots: 37 species						
Garcinia	spp.	Guttiferae	√	4	√	2							

🍥 Springer

References

Anggraini K, Kinnaird M, O'Brien T (2000) The effects of fruit availability and habitat disturbance on an assemblage of Sumatran hornbills. Bird Conserv Int 10:189–202

Barlow J, Haugaasen T, Peres CA (2002) Effects of ground fires on understorey bird assemblages in Amazonian forests. Biol Conserv 105:157–169

Borchert R (1994) Soil and stem water storage determine phenology and distribution of tropical dry forest trees. J Ecol 75:1437–1449

Cochrane MA, Alencar A, Schulze MD, Souza CMJ, Nepstad DC, Lefebvre P, Davidson EA (1999) Positive feedbacks in the fire dynamic of closed canopy tropical forests. Science 284:1832–1835

Cochrane MA, Schulze MD (1999) Fire as a recurrent event in tropical forests of the Eastern Amazon: effects on forest structure, biomass, and species composition. Biotropica 31:2–16

Davies SJ, Ashton PS (1999) Phenology and fecundity in 11 sympatric pioneer species of Macaranga (Euphorbiaceae) in Borneo. Am J Bot 86:1786–1795

Doi T (1988) Present status of the large mammals in the Kutai National Park, after a large scale fire in East Kalimantan, Indonesia. In: Tagawa H, Wirawan N (eds) A research on the process of earlier recovery of tropical rain forest, pp 82–93

Fredriksson GM (2002) Extinguishing the 1998 forest fires and subsequent coal fires in the Sungai Wain Protection Forest, East Kalimantan, Indonesia. In: Moore P, Ganz D, Tan LC, Enters T, Durst PB (eds) Communities in flames: proceedings of an international conference on community involvement in fire management. FAO and FireFight SE Asia, pp 74–81

Fredriksson GM (2005) Human-sun bear conflicts in East Kalimantan, Indonesian Borneo. Ursus 16:130–137

Fredriksson GM, Nijman V (2004) Habitat-use and conservation of two elusive ground birds (Carpococcyx radiatus and Polyplectron schleiermacheri) in Sungai Wain protection forest, East Kalimantan, Indonesian Borneo. Oryx 38:297–303

Fredriksson GM, Wich SA, Trisno. Frugivory in sun bears (Helarctos malayanus) is linked to El Niño-related fluctuations in fruiting phenology, East Kalimantan, Indonesia. Biol J Linnean Soc (in press)

Fuller DO, Jessup TC, Salim A (2004) Loss of forest cover in Kalimantan, Indonesia, since the 1997–1998 El Nino. Conserv Biol 18(1):249–254

Goldammer JG, Mutch RW (2001) Global forest fire assessment 1990–2000. FAO, pp 1–495

Harrison RD (2000) Repercussions of El Niño: drought causes extinction and the breakdown of mutualism in Borneo. Proc Royal Soc London B-Biol Sci 267:911–915

Hoffmann AA, Hinrichs A, Siegert F (1999) Fire damage in East Kalimantan in 1997/1998 related to land use and vegetation classes: satellite radar inventory results and proposals for further actions. IFFM-SFMP, pp 1–44

Kinnaird MF, O'Brien T (1998) Ecological effects of wildfire on lowland rainforest in Sumatra. Conserv Biol 12:954–956

Laurance WF (2003) Slow burn: the insidious effects of surface fires on tropical forests. Trends Ecol Evol 18:209–212

Laurance WF, Delamonica P, Laurance SG, Vasconcelos HL, Lovejoy TE (2000) Rainforest fragmentation kills big trees. Nature 404:836–836

Laurance WF, Williamson BG, Delamô Nica P, Oliveira A, Lovejoy TE, Gascon C, Pohl L (2001) Effects of a strong drought on Amazonian forest fragments and edges. J Trop Ecol 17:771–785

Leighton M, Leighton DR (1983) Vertebrate responses to fruiting seasonality within a Bornean rain forest. In: Sutton SL, Whitemore TC, Chadwick AC (eds) Tropical rain forest ecology and management. Blackwell Scientific Publications, pp 181–196

Leighton M, Wirawan N (1986) Catastrophic drought and fire in Borneo tropical rain forest associated with the 1982–1983 El Niño southern oscillation event. In: Prance GT (ed) Tropical rain forest and world atmosphere. Westview Press, pp 75–101

Lennertz R, Panzer KF (1984) Preliminary assessment of the drought and forest fire damage in Kalimantan Timur. DFS German Forest Inventory Service for Gesellschaft fur Technische Zusammenarbeit (GTZ), pp 1–45

Malingreau JP, Stephens G, Fellows L (1985) Remote sensing of forest fires: Kalimantan and North Borneo in 1982–1983. Ambio 14:314–321

Malmer A (1996) Hydrological effects and nutrient losses of forest plantation establishment on tropical rainforest land in Sabah, Malaysia. J Hydrol 174:129–148

Matius P, Toma T, Sutisna M (2000) Tree species composition of a burned lowland dipterocarp forest in Bukit Suharto, East Kalimantan. In: Guhardja E, Fatawi M, Sutisna M, Mori T, Ohta S (eds) Rainforest ecosystems of East Kalimantan. El Niño, drought, fire and human impacts. Springer-Verlag, pp 99–106

McConkey K, Galetti M (1999) Seed dispersal by the sun bear (*Helarctos malayanus*) in Central Borneo. J Trop Ecol 15:237–241

Nakagawa M, Tanaka K, Nakashizuka T, Ohkubo T, Kato T, Maeda T, Sato K, Miguchi H, Nagamasu H, Ogino K, Teo S, Hamid AH, Seng LH (2000) Impact of severe drought associated with the 1997–1998 El Niño in a tropical forest in Sarawak. J Trop Ecol 16:355–367

Nykvist N (1996) Regrowth of secondary vegetation after the 'Borneo fire' of 1982–1983. J Trop Ecol 12:307–312

Nykvist N, Grip H, Sim BL, Malmer A, Wong FK (1994) Nutrient losses in forest plantations in Sabah, Malaysia. Ambio 23:210–215

O'Brien TG, Kinnaird MF, Nurcahyo A, Prasetyaningrum M, Iqbal M (2003) Fire, demography and the persistance of siamang (*Symphalangus syndactylus*: Hylobatidae) in a Sumatran rainforest. Anim Conserv 6:115–121

Peres CA, Barlow J, Haugaasen T (2003) Vertebrate responses to surface wildfires in a central Amazonian forest. Oryx 37:97–109

Phillips OL, Hall P, Gentry AH, Sawyer SA, Vasquez R (1994) Dynamics and species richness of tropical rain forests. Proc Natl Acad Sci USA 91:2805–2809

Potts MD (2003) Drought in a Bornean everwet rain forest. J Ecol 91:467–474

Putz FE, Susilo A (1994) Figs and fire. Biotropica 26:468–469

Salafsky N (1998) Drouht in the rainforest, part II. An update based on the 1994 ENSO event. Clim Change 39:601–603

Schaik CPV, Terborgh JW, Wright SJ (1993) The phenology of tropical forests: adaptive significance and consequences for primary consumers. Annu Rev Ecol Syst 24:353–377

Schindele W, Thoma W, Panzer KF (1989) The forest fire 1982/1983 in East Kalimantan. Part 1: the fire, the effects, the damage and technical solutions. German Forest Inventory Service Ltd.

Sheil D, May RM (1996) Mortality and recruitment rate evaluations in heterogenous tropical forests. J Ecol 84:91–100

Siegert F, Ruecker G, Hinrichs A, Hoffmann AA (2001) Increased damage from fires in logged forests during droughts caused by El Niño. Nature 414:437–440

Sim BL, Nykvist N (1991) Impact of forest harvesting and replanting. J Trop Sci 3:251–284

Slik JWF (2004) El Niño droughts and their effects on tree species composition and diversity in tropical rain forests. Oecologia 141:114–120

Slik JWF, Eichhorn KAO (2003) Fire survival of lowland tropical rain forest trees in relation to stem diameter and topographic position. Oecologia 137:446–455

Slik F, Verburg RW, Keßler P (2002) Effects of fire and selective logging on the tree species composition of lowland dipterocarp forest in East Kalimantan, Indonesia. Biodivers Conserv 11:85–98

Slik JWE, Priyono, Welzen PC (2000) Key to the *Macaranga* Thou. and *Mallotus* Lour. species (Euphorbiaceae) of East Kalimantan, Indonesia. Gardens' Bull Singapore 52:11–87

Suzuki A (1992) The population of orangutans and other non-human primates and the forest conditions after the 1982–1983's fires and droughts in Kutai National Park, East Kalimantan. In: Ismail G, Mohamed M, Omar S (eds) Forest biology and conservation in Borneo. Yayasan Sabah, pp 190–205

Swaine MD, Lieberman D, Putz FE (1987) The dynamics of tree populations in tropical forest: a review. J Trop Ecol 3:359–366

Tacconi L (2003) Fires in Indonesia: causes, costs and policy implications. CIFOR, pp 1–34

Timmermann A, Oberhuber J, Bacher A, Esch M, Latif M, Roeckner E (1999) Increased El Niño frequency in a climate model forced by future greenhouse warming. Nature 398:694–696

Toma T, Matius P, Hastaniah, Kiyono Y, Watanabe R, Okimori Y (2000) Tree species composition of a burned lowland dipterocarp forest in Bukit Suharto, East Kalimantan. In: Guhardja E, Fatawi M, Sutisna M, Mori T, Ohta S (eds) Rainforest ecosystems of East Kalimantan. El Niño, drought, fire and human impacts. Springer-Verlag, pp 107–119

van Nieuwstadt MGL (2002) Trial by fire-postfire development of a dipterocarp forest. Ph.D. Thesis, University of Utrecht

van Nieuwstadt MGL, Sheil D (2005) Drought, fire and tree survival in a Borneo rain forest, East Kalimantan, Indonesia. J Ecol 93:191–201

Whitmore TC (1985) Tropical rainforests of the far East. Clarendon Press

✑ Springer

Wich SA, Steenbeek R, Sterck EHM, Palombit RA, Usman S (1999) Tree mortality and recruitment in an Indonesian rain forest. Trop Biodivers 6:189–195

Williamson GB, Laurance WF, Oliviera AA, Delamonica P, Gascon C, Lovejoy TE, Pohl L (2000) Amazonian tree mortality during the 1997 El Niño drought. Conserv Biol 14:1538–1542

Wong ST, Servheen C, Ambu L (2002) Food habits of malayan sun bears in lowland tropical forest of Borneo. Ursus 13:127–136

Woods P (1989) Effects of logging, drought, and fire on structure and composition of tropical forests in Sabah, Malaysia. Biotropica 21:290–298

Biodivers Conserv (2007) 16:1839–1850
DOI 10.1007/s10531-006-9076-z

ORIGINAL PAPER

Pollinator shift and reproductive performance of the Qinghai–Tibetan Plateau endemic and endangered *Swertia przewalskii* (Gentianaceae)

Yuan-Wen Duan · Jian-Quan Liu

Received: 1 November 2005 / Accepted: 19 May 2006 / Published online: 12 July 2006
© Springer Science+Business Media B.V. 2006

Abstract Reproductive failure results in many plant species becoming endangered. However, little is known of how and to what extent pollinator shifts affect reproductive performance of endangered species as a result of the artificial introduction of alien insects. In this study we examined breeding systems, visitor species, visiting frequency and seed set coefficients of *Swertia przewalskii* in two years that had different dominant pollinator species (native vs. alien). Flowers of this species were protandrous and herkogamous and insects were needed for the production of seeds. The stigmatic receptivity of this species was shorter than for other gentians. No significant difference in seed set coefficient was found for hand-pollinated plants between the two years, indicating that pollinator shift only had a minor effect on this plant's breeding system. The commonest pollinators in 2002 were native bumble-bees, alien honeybees and occasional solitary bees, however, only alien honeybees were observed in 2004. The flower visitation rate in both years was relatively high, although the total visit frequency decreased significantly in 2004. The control flowers without any treatment produced significantly fewer seed sets in 2004 than in 2002. In the past decade the seed production of this species may have partly decreased due to pollination by alien honeybees, however, we suggest that they might have acted as alternative pollinators ensuring seed production of *S. przewalskii* when native pollinators were unavailable. The main reason that this plant is endangered is probably the result of habitat destruction, but changes in land use, namely intensified

Y.-W. Duan · J.-Q. Liu (✉)
Key Laboratory of Qinghai-Tibetan Plateau Ecological Adaptation, Northwest Institute of Plateau Biology, The Chinese Academy of Sciences, Xiguan street 59, Xining 810001 Qinghai, P. R. China
e-mail: ljqdxy@public.xn.qh.cn

J.-Q. Liu
Key Laboratory of Arid and Grassland Ecology, Lanzhou University, Lanzhou 730000 Gansu, P. R. China

Y.-W. Duan
Graduate School of the Chinese Academy of Sciences, Beijing 100039, P. R. China

agricultural practice and unfavorable animal husbandry have also contributed to its decline. We recommend that in-situ conservation, including the establishment of a protected area, is the best way to preserve this species effectively.

Keywords *Swertia przewalskii* · Endangered species · Alien pollinators · Pollination · Reproduction · Qinghai–Tibetan Plateau

Introduction

There are diverse reasons why a plant species might become endangered. The most important one is probably poor reproduction during the species' life cycle, which can reduce the recovery time of existing populations. Therefore, understanding the reproductive processes of endangered species will contribute greatly to its successful conservation (e.g. Wiens et al. 1989; Navarro and Guitián 2002). Many factors, including decreased pollen quality (Byers 1995), poor stigma receptivity (He et al. 2000) and the absence of pollinators (e.g. Kwak and Jennersten, 1991; Burd 1994; Evans et al. 2004) can result in decreased sexual reproduction and seed production. Such decreases can result in plant species becoming endangered or extinct if the species is already threatened. This is of particular concern where populations are scattered, small and isolated (see Wagner and Mitterhofer 1998). For rare and outcrossing annuals and perennials, fluctuations in the number of pollinators (species and frequencies) between years create variability in the number of seedlings that establish within a population (Bierzychudek 1981). In addition, the worldwide introduction of honeybees may disturb native plant–pollinator interactions, particularly for those outcrossing species that have adapted to specialized native insect pollinators. Such introduction of alien pollinators can reduce the reproductive success of plant species and may be one of the important factors leading to their endangerment (Paton 1997; Gross and Mackay 1998; Kato et al. 1999; Kato and Kawakita 2004). However, Dick (2002) demonstrated that introduced African honeybees have become important pollinators of the canopy tree, *Dinizia excelsa* (Fabaceae) and may have aided the recovery of this endangered species. These introduced bees have also altered the genetic structure of remnant populations by mediating frequent long-distance gene flow. Pollinators are generally less reliable in alpine habitats (Stenström and Molau 1992), although it remains unknown how the introduction of alien pollinators affects the pollination systems of local species and the role they may play in the reproductive performance of those that are endangered.

Most Gentianaceae species are restricted to highland habitats (Ho and Liu 2001). Previous research has revealed that most gentians are self-compatible, but pollinator visits are vital for their reproduction due to protandry and herkogamy (e.g. Spira and Pollak 1986; Webb and Littleton 1987; Petanidou et al. 1998, 2001; Bynum and Smith 2001; Duan et al. 2005). Moreover, some gentians have evolved diverse strategies in adaptive responses that appear to enhance their reproductive success in arid alpine habitats, including autogamy in annual gentians (Spira and Pollak 1986), longer duration of pollen shedding and stigma receptivity (e.g. Webb and Littleton 1987; Petanidou et al. 2001) and floral closure to protect pollen shedding during thunderstorms and when temperatures decrease (Bynum and Smith 2001; He et al. 2006). In spite of these adaptations, a few perennial gentians are now endangered and are close to extinction. Intensified agricultural practices are one of the important factors that have decreased their reproductive output (Raijmann et al. 1994; Petanidou et al. 1995; Luijten et al. 2000; Lennartsson and Oostermeijer 2001; Lienert et al. 2002).

Swertia przewalskii Pissjauk. (Gentianaceae), a rare alpine perennial, is only distributed in the Qilian and Menyuan counties of the north-eastern Qinghai–Tibetan Plateau in the Qinghai province. Despite being restricted to a small area along the southwest valley of the Qilian Mountains, extending for about 100 km, this species had high abundance only 20 years ago (Yang et al. 1991). However, the number of individuals has decreased greatly in recent years and this species has now become extremely endangered. In the natural distribution range of *S. przewalskii*, alien honeybees have been artificially introduced by beekeepers because a large number of nearby grasslands have been used for planting cole *(Brassica napus)* to produce colza oils which have high commercial value. The abundance of these honeybees fluctuates between years depending on the number of beekeepers and how many hives they can carry. These honeybees, as well as native bumble-bees, visit alpine plants in this region (see, for instance, He and Liu 2004; Duan et al. 2005). In this study, we investigated the reproductive ecology of *S. przewalskii* and discuss possible causes for its endangered status. Specifically, we aimed to address whether there was a difference in the reproductive performance of this species between two years when native bumblebees and alien honeybees were the dominant pollinators, respectively.

Materials and methods

Study site

Twenty years ago, *S. przewalskii* is continuously distributed in the marsh wetlands along the Babao and Datong rivers and forms four large populations in Menyuan county, Qishizui town, the Haibei Alpine Meadow Ecosystem Research Station and Qilian County. At present, according to our recent investigation, most individuals in the sparse populations have disappeared, while the abundance of individuals in the four large sites has decreased to several hundred plants from more than ten thousand. We therefore chose only one large site around the Haibei Alpine Meadow Ecosystem Research Station for our experiments, which contained enough individuals for our controlled experiments and pollinator observations. This station is located on the north-east Qinghai–Tibetan Plateau (Lat. 37°29′–37°45′ N, Long. 101°12′–101°23′ E) at 3200 m above sea level. The average annual air temperature here is –1.7°C with extremes of 27.6°C (maximum) and –37.1°C (minimum), and average annual precipitation ranging from 426 to 860 mm. All of our experiments were carried out from late July to early August, and observations of pollinators were made in mid August in 2002 and 2004. The wetlands where *S. przewalskii* grows are dominated by *Kobresia tibetica* (Poaceae). In 2003, we recorded all similar visiting species as in 2002, however, after two weeks all fenced plants used for experiments and observations were unexpectedly damaged by yaks.

Breeding system

Firstly, we fixed eight flowers in FAA, and then dissected ovules and counted the pollen in the laboratory according to the method described by Dafni (1992). The

Springer

pollen:ovule coefficient was used to crudely estimate the breeding system of *S. przewalskii*.

Ten flowers whose stigmas had just opened were labeled each morning, daily from July 27th to August 2nd in 2002. Each flower was emasculated and bagged and then artificially pollinated with pollen from different individuals on August 2nd. The experiment was designed to test the duration coefficient of stigma receptivity through the production of seed sets. A month later, but before the fruits dehisced, we collected all the fruits from each treatment. To obtain the seed set coefficient we counted the matured seeds, aborted seeds and unfertilized ovules. The seed set coefficient was calculated using the following formula, seed set = no. of matured seeds/(no. of matured seeds + no. of aborted seeds + no. of unfertilized ovules).

Before flowers opened in 2002 and 2004, we randomly selected 60 flowers from different individuals, 30 of which were emasculated and bagged to test apomictic seed production and the others were bagged without emasculation to examine whether *S. przewalskii* can be spontaneously self-pollinated without insect involvement. We then labeled 60 flowers on different plants, 30 of which were emasculated and left to freely pollinate to investigate the degree to which *S. przewalskii* depended on vectors for seed production, while the others were left to pollinate under natural conditions as controls. We further emasculated 80 flowers and bagged them for the following two experiments: 40 of the flowers were pollinated artificially with fresh pollen from the same plant, while the rest were pollinated by hand with pollen from different plants.

Flower visitors

During the full flowering period we recorded how many species visited *S. przewalskii* and their behavior during the two years. We further selected 12 individual plants, on which all the open flowers were used as focal flowers, to record the visiting frequencies of all the pollinators from 10 to 12 in August in 2002 (21 h in total) and four days from 10 to 13 in August in 2004 (26 h in total) at different time intervals (9:00–12:00, 13:00–15:00 and 15:00–17:00). We then calculated the visiting frequencies of each visitor and the total frequency of active visitors for each flower (times/h).

Statistical analysis

We used One-way ANOVA and Post hoc-LSD tests (SPSS Inc., 1997) to compare: (1) seed set among control, emasculated but freely pollinated, selfed within an individual plant and outcrossed flowers in 2002 and 2004, and seed set among flowers that had been artificially outcrossed on different stigma receptivity dates in 2002; (2) seed set of the same treatment between 2002 and 2004; (3) visiting frequencies of the same pollinator species from different time intervals in the two years; and (4) visiting frequencies of all visitors between the two years. All data were tested for normality using 1 Sample K-S in Nonparametric tests before comparison. Nonparametric tests (Kruskal–Wallis Tests for K-Independent Samples) were used to analyze data which were not normally distributed.

Results

Breeding system

The ovule abundance per flower ranged from 38 to 106 (average, ±1.0 SE; 74.5 ± 6.9, $N = 8$), while the number of pollen grains ranged from 12,150 to 31,950 and averaged 18,862.5 ± 2438.3 per flower ($N = 8$). The pollen:ovule ratio varied from 156.7 to 420.4 (average 266.4 ± 36.3), indicating that the breeding system of *S. przewalskii* is facultative xenogamy (Dafni 1992). *S. przewalskii* flowered from late July to late August, with the blue anthers surrounding the pistil shedding pollen first. After plants had finished shedding pollen, the anthers moved to outside with the filaments, and the stigma then opened. Therefore, *S. przewalskii* shows characteristics of dichogamy and herkogamy (Lloyd and Webb 1986; Webb and Lloyd 1986). When temperatures decreased and during thunderstorms, we observed no floral closure or movement, as in other gentians (He et al., 2006).

Flowers labeled to test the duration of stigma receptivity produced seeds successfully, but the seed set produced strongly depended on the age of the flowers. The seed set coefficient was higher on the second day, although not significantly, than on all other days and it was still 0.4 on the seventh day (Fig. 1). Therefore, the stigma can remain receptive for six days when a flower has not been pollinated. However, not all ovules had been fully pollinated with sufficient fresh pollen even at the highest receptivity stage (Fig. 1), indicating that once-pollination is insufficient to ensure the successful pollination of all ovules.

There was no evidence of apomixis or spontaneous self-pollination in *S. przewalskii* as neither the bagged flowers with or without emasculation produced seeds. The seed set coefficient for naturally pollinated emasculated flowers was high, and there was no significant difference between naturally pollinated emasculated and control flowers between the two years (Fig. 2). Seed set coefficients of naturally pollinated emasculated flowers did not differ significantly between the two years ($F = 0.200$, $P = 0.659$) (Fig. 2), but there was a significant difference between 2002

Fig. 1 Seed set coefficients from flowers pollinated at different times after the stigmas had become receptive. Bars with the same letter indicate that the difference was not significant at the 0.05 probability level

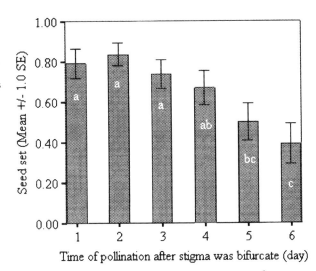

Time of pollination after stigma was bifurcate (day)

Fig. 2 Seed set coefficients from different treatments in 2002 and 2004, denoted by black bars and gray bars, respectively. Bars with the same letters indicate that the difference was not significant at the 0.05 probability level

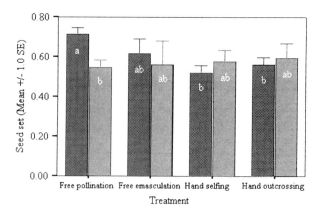

and 2004 in the seed set coefficients of naturally pollinated flowers ($F = 13.145$, $P = 0.001$) (Fig. 2).

All flowers selfed with pollen from the same individual and outcrossed flowers successfully yielded seeds, and no significant difference in the seed set coefficient was found between them (Fig. 2). In addition, the seed set coefficient in the two years did not differ significantly for plants that were hand selfed ($F = 0.738$, $P = 0.394$) and hand outcrossed ($F = 0.187$, $P = 0.667$) (Fig. 2). However, the seed set coefficients from both selfed and outcrossed flowers were lower than that of naturally pollinated flowers in 2002, but not in 2004 (Fig. 2). These comparisons indicate that the control flowers might have received a second pollination, resulting in larger seed set coefficients in 2002. In 2004, however, the control flowers suffered more pollen limitation due to lower pollination from bees and the lack of a second pollination event.

Flower visitors

According to the behavior of flower visitors, bumblebees (*Bumbus keshimirensis* Friese), honeybees (*Apis mellifera* L.) and solitary bees were considered to be the frequent pollinators of *S. przewalskii* in 2002, while other small insects, such as flies, are probably not legitimate pollinators as they only visited flowers for short periods. The frequent flower pollinators landed with the corolla petals either side of their bodies, probing for nectar at the base of the corolla using their proboscises, with the hind part of the body placed at the center of the corolla. Usually they circled the center of the corolla to collect nectar from each petal. During this movement, their hind legs and part of their body continuously touched the anthers or stigmas of the male or female stage flowers. Large quantities of pollen were attached to the hind parts of bumblebees, honeybees and solitary bees.

There were three taxa of the frequent pollinators in 2002 and the visiting frequencies of bumblebees and honeybees were higher than those of solitary bees ($F = 6.514$, df $= 2$, $P = 0.002$), although visitation rates of bumblebees and honeybees varied little ($F = <0.001$, $P = 0.988$). In addition, bumblebees visited *S. przewalskii* more frequently from 9:00 to 12:00 than during other time intervals ($F = 4.686$, df $= 2$, $P = 0.016$), while honeybees visited flowers mainly from 13:00 to 17:00 ($F = 6.162$, df $= 2$, $P = 0.005$). No significant difference was found in the visiting frequency of solitary bees between the different time intervals ($F = 0.912$, df $= 2$, $P = 0.412$) (Fig. 3A).

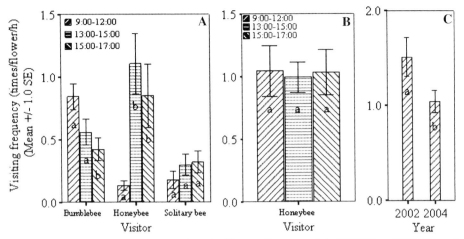

Fig. 3 Visiting frequencies of flower visitors in different time intervals in 2002 (**A**) and 2004 (**B**), and the total visiting frequency averaged across all visitors for each of the two years (**C**). For each comparison, means with the same letter indicate that the difference was not significant at the 0.05 probability level

In 2004, however, no solitary bees were observed to visit flowers, and bumblebees visited only once during 26 hours' of observations, therefore, honeybees were the only major pollinators. Visiting frequencies of honeybees in the three time intervals did not differ significantly ($F = 0.027$, df = 2, $P = 0.973$) (Fig. 3B). Overall, across all pollinators, there was a significant difference between the total visiting frequency between 2002 and 2004 ($F = 4.586$, $P = 0.04$) (Fig. 3C).

Discussion

Breeding systems

Little is known about the floral processes and breeding systems in *Swertia*, which is a large genus containing approximately 150 species from the family Gentianaceae (Ho et al. 1994). Most gentians are protandrous and herkogamous with different floral morphologies during the male and female stages (Lloyd and Webb 1986; Webb and Lloyd 1986; Webb and Littleton 1987; Petanidou et al. 1998, 2001; Bynum and Smith 2001; Duan et al. 2005). Such a combination of dichogamy and herkogamy also exists in *S. przewalskii* and as a consequence of these floral syndromes, spontaneous self-pollination is prevented. Bagged flowers without emasculation produced no seeds, similarly suggesting the dependence on pollinators for seed production. However, hand self-pollinated flowers yielded the same seed set coefficient as the outcrossed flowers (Fig. 2), suggesting that this species is self-compatible and the continuous foraging of pollinators might lead to geitonogamous self-pollination (Duan and Liu 2003). This breeding system remained consistent when the dominant pollinator changed between 2002 and 2004, as there was no difference in the controlled breeding experiments between these two years (Fig. 2).

The infrequent pollinator visits in arid alpine environments are especially disadvantageous to the reproduction of gentians (Spira and Pollak 1986). In circumstances where pollinators are rare, a "sit and wait" strategy of increased floral longevity may be the only means for flowers to successfully reproduce (Ashman and Schoen 1994). Prolonging pollen presentation and duration of stigma receptivity could increase pollination chance and enhance reproductive success. Such a strategy has been reported for most gentians (Webb and Littleton 1987; Petanidou et al. 1998, 2001; Bingham and Orthner 1998; Bynum and Smith 2001; He et al. 2006). However, pollen shedding of *S. przewalskii* lasted on average only 5.2 h in a study by Duan and Liu (2003), which is much shorter than the 3–5 days observed in other gentians with floral closure, such as the North American gentian *Gentiana algida* (Bynum and Smith 2001), the European gentian *G. pneumonanthe* (Petanidou et al. 2001) and *G. straminea* (Duan et al. 2005). The present investigation further suggested that the duration of female receptivity in *S. przewalskii* (Fig. 1) is shorter than that of other gentians (Webb and Littleton 1987; He et al. 2006). Although the seed set coefficient remained at 0.4 on the 6th day of the female stage in *S. przewalskii*, this coefficient is lower than those perennial gentians with floral closure on the same female duration day. For example, Webb and Littleton (1987) reported that the seed set coefficient of two New Zealand alpine *Gentiana* species decreased to zero on the 8th and 10th days, but retained a seed set coefficient of 0.6 on the 6th day of the female phase. All these results suggest that the total floral longevity of *S. przewalskii* is shorter than for other gentians; and this might partly account for the lower seed set coefficients reported for this species when the visiting frequency of native pollinators decreased and then stopped.

Pollinator shift and seed set production

The reproductive success in flowering plants may depend on a number of factors, including pollen limitation, resource limitation, predation and the physical environment (Lee 1988). In alpine species that are dependent on insects for pollination, pollen limitation may be most important, since pollinators are generally less reliable than in other environments (Stenström and Molau 1992). The disruption of the interaction between pollinators and plants may account for the endangered status of some species, especially for those growing in unreliable pollinator environments (e.g. Kunin 1992; Byers 1995; Ärgen 1996; Huang and Guo 2002). Introduced honeybees may have serious effects on plant–pollinator interactions, for example they may cause drastic decreases in native bee populations, changes in patterns of gene flow among plants, or increased reproductive fitness of invasive exotic weeds (Gross and Mackay 1998). Furthermore, introduced honeybees are considered inadequate substitutes for native pollinators (Buchmann and Nabhan 1996) and potential competitors of native bees. They can also reduce the reproductive success of local bumblebees by consuming nectar, which has the potential to cause cascading effects on native communities (Thomson 2004). The native pollinators may become extinct or endangered due to insufficient floral resources when introduced honeybees become established (Kato et al. 1999; Kato and Kawakita 2004). Therefore, the introduced honeybees are thought to have serious effects on local plant–pollinator interactions. However, in some cases the introduction of honeybees may have rescued some endangered plant species by acting as substitute pollinators following declines in native pollinator numbers (Dick 2002).

In the present study, pollinator species changed greatly between the two years. There were three pollinators visiting *S. przewalskii* in 2002, compared with only one (honeybees) in 2004. Bumblebees are important pollinators of alpine ecosystems (Bingham and Orthner 1998). Although the total visiting frequency in both years (Fig. 3) was higher than those reported for other alpine plants (e.g., 0.0029 times/flower/min in the Andes and 0.0017 times/flower/min in North America, Arroyo et al. 1985; Bingham and Orthner 1998), there was a marked decrease in 2004 when honeybees dominated. Such a decrease in visiting rates may account for the reduced seed set coefficients reported in 2004 (Fig. 3). In addition, pollen limitation of this species may be correlated with the second forage of insects, with this assumption being corroborated by hand-pollination experiments. Even though hand-pollinated plants were supplied with abundant pollen, this did not ensure pollination of all ovules even during the highest period of stigma receptivity (at the second day of the female phase). Artificially outcrossed flowers did not have larger seed set coefficients, which were actually lower than those of control flowers that were naturally pollinated by native bumblebees in 2002 (Fig. 2). This pattern has also been observed in another endangered gentian, *Gentianella uliginosa* (Petanidou et al. 1998). Some species in the tribe Delphinieae (Bosch et al. 2001) were reported to have similar pollination limitations on reproductive success, especially *Delphinium nelsonii*, in which seed set coefficients from hand cross-pollination decreased by 30% relative to natural pollination in the field (Bierzychudek 1981). In these species, flowers in the field usually received more than one visit, which ensured all ovules were pollinated (Bierzychudek 1981; Bosch et al. 2001). In 2004, however, seed set coefficients of hand-pollinated flowers were slightly higher than those of flowers naturally pollinated by honeybees. This finding indicates that alien honeybees reinforce the pollen limitation of this species. A possible explanation for this is that honeybees will not make a second visit to the same flower because it has reduced amounts of nectar. In addition, the possibility of competition between native bumblebees and alien honeybees could not be dismissed, although we found that both species pollinated *S. przewalskii* in 2002. These results seem to suggest that alien honeybees have serious affect on the reproductive performance of *S. przewalskii* because their pollination effects are lower than those of bumblebees. However, there are two reasons why the inter-annual fluctuations in native bumblebee abundance in these alpine habitats may not only be caused by the introduction of alien honeybees to this study site. Firstly, oscillation in the numbers of native pollinators has been previously reported in alpine areas (Körner 1999), which may be due to the unpredictable climatic changes of these habitats. Secondly, we monitored the inter-annual changes in bumblebee abundance in the central Qinghai–Tibetan Plateau (lat. 34°21′ N, long. 100°29′ E, alt. 4000 m) where no honeybees have been introduced and found the same fluctuations in bumblebee populations during the past 4 years (authors' unpublished data). In 2004, the dominant honeybees still produced relatively high seed set coefficients (Fig. 2) and we conclude that the introduced honeybees have played only a minor role in endangering *S. przewalskii* in the past decade. Furthermore, we suggest that these alien honeybees might have acted as alternative pollinators when native pollinators' numbers were reduced.

🖄 Springer

Conservation and endangered status

The most important causes of the endangered status of *S. przewalskii* are probably wetland habitat destruction and intensified agricultural practices. The animal husbandry in this area may also contribute to its endangered status. In the past decade, more than 20% of the grasslands have been grubbed to cultivate cole, barley and wheat (according to the local farmers). These changes in land use have not only directly destroyed specimens of *S. przewalskii*, but have also shrunk the wetlands on both sides of the Babao and Datong rivers in the valley bottoms. Some small lakes and streams nearby the rivers have also dried up and disappeared in recent years. Most individuals of *S. przewalskii* disappeared when the wetlands were destroyed. The remaining wetland grasslands were further damaged by the unfavorable animal husbandry. In spite of the reduced grasslands, the numbers of yaks and sheep have doubled in this area since the grassland was partitioned and contracted to private herders approximately 20 years ago (data according to the local government estimates). The wetland grasslands in the valley bottom have been used as winter herding pasture. The livestock generally returns to this area for the winter from the summer pastures at higher altitudes at the end of September when *S. przewalskii* has finished its reproduction. Unfortunately, in recent years the livestock have returned to the winter pastures earlier because of reduced grass in the summer pastures. Furthermore, we found that some livestock did not leave the winter pasture during the summer season between July and September. Some livestock in this area are now also fed with additional artificial fodder made from cole, barley and wheat stems and the wetlands have become exercise sites for those livestock retained during the summer season. The activities of these livestock not only affect the reproduction of *S. przewalskii*, but also directly destroy growing individuals. During this study some individuals used for breeding experiments were destroyed by yaks and sheep. In addition, we found that some local communities collected *S. przewalskii* as medicines, since most species of this genus in the Qinghai–Tibetan Plateau are regarded as effective cures (Yang et al. 1991).

The human activities described above not only destroy existing populations of *S. przewalskii*, but are also responsible for those destroyed in the past. This species is now fragmented into only four separated sites, which might reduce reproductive success, lower the total number of seeds produced and possibly compromise long-term population persistence (Valdivia et al. in press) The creation of a mosaic by the agricultural activities also prevents genetic exchange occurring between populations. Although we observed sufficient pollinators, decreasing plant numbers will undoubtedly decrease their attractiveness to pollinators (Sih and Baltus 1987; Olesen and Jain 1994), while changes in pollinator diversity will further reinforce the endangered status of this species.

We suggest the best way to effectively protect this endangered species is to establish a conservation area along the Dabao and Qilian rivers in the Qinlian valley. In this protected area, further agricultural exploitation and the collection of wild plants should be prohibited, with herding practices being restored as they were 20 years ago. When a continuous wetland habitat forms, the seeds from the four sites can be artificially cultivated in the uninhabited area in order to rapidly restore a continuous distribution. However, ex-situ conservation is more difficult than in-situ conservation. Under this scenario, a suitable site should be carefully selected, and seeds should be collected from the four existing sites to establish an artificial and mixed population. It is important for this site to be near to the species' natural distribution because the artificial population would have to rely on the specific pollinators (bumblebees) for more effective pollination.

Acknowledgements   We are grateful to the anonymous reviewers for their suggested improvements to this manuscript and thank Dr John Blackwell to edit and polish the English of our manuscript. Support for this research was provided by the Chinese Academy of Sciences (Key Innovation Plan grant number KSCX2-SW-106 and the Special Fund for an outstanding Ph.D. dissertation awarded to Jian-Quan Liu) and the National Natural Science Foundation of China (Grant numbers 30370284 and 30270253).

References

Arroyo MTK, Armesto JJ, Primack RB (1985) Community studies in pollination ecology in the high temperate Andes of Central Chile. Effect of temperature on visitation rates and pollination possibilities. Plant Syst Evol 149:187–203

Ashman TL, Schoen DJ (1994) How long should flowers live? Nature 371:788–791

Årgen J (1996) Population size, pollinator limitation, and seed set in the self-incompatible herb *Lythrum salicaria*. Ecology 77:1779–1790

Bierzychudek P (1981) Pollinator limitation of plant reproductive effort. Am Nat 117:838–840

Bingham RA, Orthner AR (1998) Efficient pollination of alpine plants. Nature 391:238–239

Bosch M, Simon J, Molero J, Blanché C (2001) Breeding systems in tribe *Delphinieae* (Ranunculaceae) in the western Mediterranean area. Flora 196:101–113

Buchmann SL, Nabhan GP (1996) The pollination crisis. Science 36(July/August):22–27

Burd M (1994) Bateman's principle and plant reproduction: the role of pollen limitation in fruit and seed set. Bot Rev 60:83–139

Byers DL (1995) Pollen quantity and quality as explanations for low seed set in small populations exemplified by *Eupatorium* (Asteraceae). Am J Bot 82:1000–1006

Bynum MR, Smith WK (2001) Floral movements in response to thunderstorms improve reproductive effort in the alpine species *Gentiana algida* (Gentianaceae). Am J Bot 88:1088–1095

Dafni A (1992) Pollination ecology: a practical approach. Oxford University Press, New York

Dick CW (2002) Genetic rescue of remnant tropical trees by an alien pollinator. Proc Roy Soc Biol Sci (Ser B) 268:2391–2396

Duan YW, Liu JQ (2003) Floral syndrome and insect pollination of the Qinghai-Tibet Plateau endemic *Swertia przewalskii* (Gentianaceae). Acta Phytotaxon Sin 41:465–474

Duan YW, He YP, Liu JQ (2005) Reproductive ecology of the Qinghai-Tibet Plateau endemic *Gentiana straminea* (Gentianaceae), a hermaphrodite characterized by herkogamy and dichogamy. Acta Oecol 27:225–232

Evans MEK, Menges ES, Gordon DR (2004) Mating systems and limits to seed production in two *Dicerandra* mints endemic to Florida scrub. Biodivers Conserv 13:1819–1832

Gross CL, Mackay D (1998) Honeybees reduce fitness in the pioneer shrub *Melastoma affine* (Melastomataceae). Biol Conserv 86:169–178

He YP, Liu JQ (2004) Pollination ecology of *Gentiana straminea* Maxim. (Gentianaceae), an alpine perennial in the Qinghai-Tibet Plateau. Acta Ecol Sin 24:215–220

He TH, Rao GY, You RL (2000) Reproductive biology of *Ophiopogon xylorrhizus* (Liliaceae *s.l.*): an endangered endemic of Yunnan, Southwest China. Aust J Bot 48:101–107

He YP, Duan YW, Liu JQ, Smith WK (2006) Floral closure in response to temperature and pollination in *Gentiana straminea* Maxim. (Gentianaceae), an alpine perennial in the Qinghai-Tibetan Plateau. Plant Syst Evol 256:17–33

Ho TN, Liu SW (2001) A worldwide monograph of *Gentiana*. Science Press, Beijing

Ho TN, Xue CY, Wang W (1994) The origin, dispersal and formation of the distribution patterns of *Swertia* L. Acta Phytotaxon Sin 32:525–537

Huang SQ, Guo YH (2002) Variation of pollination and resource limitation in a low seed-set tree, *Liriodendron chinense* (Magnoliaceae). Bot J Linn Soc 140:31–38

Kato M, Kawakita A (2004) Plant-pollinator interactions in New Caledonia influenced by introduced honey bees. Am J Bot 91:1814–1827

Kato M, Shibata S, Yasui T, Nagamasu H (1999) Impact of introduced honeybees, *Apis mellifera*, upon native bee communities in the Bonin (Ogasawara) Islands. Res Popul Ecol 41:217–228

Körner Ch (1999) Alpine plant life. Springer, Berlin

Kunin WE (1992) Density and reproductive success in wild populations of *Diplotaxis erucoides* (Brassicaceae). Oecologia 91:129–133

Kwak MM, Jennersten O (1991) Bumblebee visitation and seed set in *Melampyrum pratense* and *Viscaria vulgaris*: heterospecific pollen and pollen limitation. Oecologia 86:99–104

Lee TD (1988) Patterns of fruit and seed production. In: Lovett DJ, Lovett DL (eds) Plant reproductive ecology. Oxford University Press, New York

Lennartsson T, Oostermeijer JGB (2001) Demographic variation and population viability in *Gentianella campestris*: effects of grassland management and environmental stochasticity. J Ecol 89:451–463

Lienert J, Fischer M, Diemer M (2002) Local extinction of the wetland specialist *Swertia perennis* L. (Gentianaceae) in Switzerland: a revisitation study based on herbarium records. Biol Conserv 103:65–76

Lloyd DG, Webb CJ (1986) The avoidance of interference between the presentation of pollen and stigmas in angiosperms: dichogamy. N Z J Bot 24:135–162

Luijten SH, Dierick A, Oostermeijer JGB, Raijmann LL, den Nijs HCM (2000) Population size, genetic variation, and reproductive success in a rapidly declining, self-incompatible perennial *(Arnica montana)* in The Netherlands. Conserv Biol 34:1776–1787

Navarro L, Guitián J (2002) The role of floral biology and breeding system on the reproductive success of the narrow endemic *Petrocoptis viscosa* Rothm. (Caryophyllaceae). Biol Conserv 103:125–132

Olesen JM, Jain SK (1994) Fragmented plant populations and their lost interactions. In: Loeschcke V, Tomiuk J, Jain SK (eds) Conservation genetics. Birkhäuser, Basel, Switzerland

Paton DC (1997) Honey bees *Apis mellifera* and the disruption of plant-pollinator systems in Australia, Victoria. Naturalist 114:23–29

Petanidou T, den Nijs JCM, Oostermeijer JGB (1995) Pollination ecology and constraints on seed set of the rare perennial *Gentiana cruciata* L. in The Netherlands. Acta Bot Neerland 44:55–74

Petanidou T, Ellis-Adam AC, den Nijs JCM, Oostermeijer JGB (1998) Pollination ecology of *Gentianella uliginosa*, a rare annual of the Dutch coastal dunes. Nordic J Bot 18:537–548

Petanidou T, Ellis-Adam AC, den Nijs JCM, Oostermeijer JGB (2001) Differential pollination success in the course of individual flower development and flowering time in *Gentiana pneumonanthe* L. (Gentianaceae). Bot J Linn Soc 135:25–33

Raijmann LL, van Leeuwen NC, Oostermeijer JGB, den Nijs HM, Menken SBJ (1994) Genetic variation and outcrossing rate in relation to population size in *Gentiana pneumonanthe* L. Conserv Biol 8:1014–1026

Sih A, Baltus MS (1987) Patch size, pollinator behavior, and pollinator limitation in catnip. Ecology 68:1679–1690

Spira TP, Pollak OD (1986) Comparative reproductive biology of alpine biennial and perennial gentians (*Gentiana*: Gentianaceae) in California. Am J Bot 73:39–47

SPSS Inc. (1997) SPSS 8.0: Statistical analysis software for large databases: SPSS 8.0 for 16 windows. USA

Stenström M, Molau U (1992) Reproductive ecology of *Saxifraga oppositifolia*: phenology, mating system and reproductive success. Arct Antarct Alp Res 24:337–343

Thomson D (2004) Competitive interactions between the invasive European honey bee and native bumble bees. Ecology 85:458–470

Valdivia C, Simonetti JA, Henriquez CA (in press) Depressed pollination of *Lapagertia rosea* Ruiz et Pav. (Philesiaceae) in the fragmented temperate rainforest of southern South America. Biodivers Conserv

Wagner J, Mitterhofer E (1998) Phenology, seed development, and reproductive success of an alpine population of *Gentianella germanica* in climatically varying years. Bot Acta 111:159–166

Webb CJ, Littleton J (1987) Flower longevity and protandry in two species of *Gentiana* (Gentianaceae). Ann Miss Bot Garden 74:51–57

Webb CJ, Lloyd DG (1986) The avoidance of interference between the presentation of pollen and stigmas in angiosperms. N Z J Bot 24:163–178

Wiens D, Nickrent DL, Davern CI, Carlvin CL, Vivrette NJ (1989) Development failure and loss of reproductive capacity in the rare palaeoendemic shrub *Dedeckera eurekensis*. Nature 338:65–67

Yang YC, Ho TN, Lu SL, Huang RF, Wang ZX (1991) Tibetan medicines. Qinghai People's Press, Xining

Biodivers Conserv (2007) 16:1851–1865
DOI 10.1007/s10531-006-9083-0

ORIGINAL PAPER

An analysis of altitudinal behavior of tree species in Subansiri district, Eastern Himalaya

Mukunda Dev Behera ·
Satya Prakash Singh Kushwaha

Received: 3 April 2006 / Accepted: 10 July 2006 / Published online: 16 August 2006
© Springer Science+Business Media B.V. 2006

Abstract Plant species diversity and endemism demonstrate a definite trend along altitude. We analyzed the (i) pattern of tree diversity and its endemic subset (ii) frequency distribution of altitudinal range and (iii) upper & lower distributional limits of each tree species along altitudinal gradients in eastern Himalaya. The study was conducted in Subansiri district of Arunachal Pradesh. Data on the tree species (cbh ≥ 15 cm) were gathered every 200 m steps between 200 m and 2200 m gradients. Tree diversity demonstrated a greater variation along the gradients. A total of 336 species (of which 26 are endemic) were recorded belonging to 185 genera and 78 families. The alpha diversity demonstrated a decreasing pattern with two maxima (i.e., elevational peaks) along the gradients; one in 601–1000 m and the other in 1601–1800 m, corresponding to transition zones between tropical-subtropical and subtropical-temperate forests. Pattern diversity revealed a narrow range along the gradients. Frequency of altitudinal range was distributed between 1 and 41. Only one species (*Altingia excelsa*) showed widest amplitude, occurring over the entire range. Highest level of species turnover was found in 400–600 m step at lower elevational limit whereas for upper elevational limit, the highest turn over was recorded between 800 and 1000 m. Tree diversity decreased and its endemic subset increased along the gradients. Two maximas in tree diversity pattern correspond to forest transition zones with subtropical-temperate transition is narrower than tropical-subtropical. The pattern observed here could be attributed to varied microclimates or environmental heterogeneity. If altitudinal amplitude of a species is considered as an aspect of its niche breadth, it is clear from these results that niche breadth in these organisms is in fact independent of the diversity of the assemblage in which they occur. This analysis calls for detailed floristic studies to determine the

M. D. Behera (✉)
Regional Remote Sensing Service Centre, Indian Space Research Organisation (ISRO),
Kharagpur 721 302, India
e-mail: mukundbehera@gmail.com

S. P. S. Kushwaha
Forestry & Ecology Division, Indian Institute of Remote Sensing, Dehradun 248 001, India

🦌 Springer

breadth of changes between adjacent forest types and details of local species richness in high diversity areas.

Keywords Diversity · Altitudinal gradient · Range · Endemism · Eastern Himalaya · Subansiri

Introduction

The introduction on diversity and endemism is critical to the understanding of overall biological diversity. Large environmental variation within a small geographical area makes altitudinal gradients ideal for investigating several ecological and biogeographical hypothesis (Korner 2000). The majority of elevational studies world over have focused on species richness (e.g., Lieberman et al. 1996; Vazquez and Givnish 1998) and elevational zonation of vegetation types (Fram and Gradstein 1991) both in the tropics and other parts of the world. This relationship of species richness to elevation has been linked to the random elevational association between the extent and the position of elevational ranges of species along the geographical ranges (Cowell and Hurtt 1994). Numerous hypotheses have been proposed to explain this relationship (reviewed by Brown and Lomolino 1998; Brown 2001; Lomolino 2001) of which, area (MacArthur 1972; Rahbek 1997; Odland and Birks 1999), climate (Odland and Birks 1999), and mass effect (Shmida and Wilson 1985; Kessler 2000) are commonly discussed. Areas with high species richness may also have a high number of endemic species, but not necessarily in a coherent pattern (Houston, 1994; Whittaker et al. 2001). Endemism may be expressed as a percentage of all extant taxa present (excluding exotics), or as the absolute number of endemics in the area (Major 1988). In mountain areas, such as the Himalaya, maximum number of endemic species is expected to occur at high elevations due to isolation mechanisms mainly governed by terrain (Shrestha and Joshi 1996). Massive topographic relief in mountains can potentially act as barrier, inhibiting gene flow and facilitating speciation (Brown 2001). The eastern Himalayan forests are among the most floristically diverse regions with high endemism (Rao 1974; Hajra et al. 1996; Behera et al. 2002), but very little is known about the exact elevational distribution of tree diversity and its endemic subset.

The composition of plant communities along tropical mountain transects has rarely been investigated in detail (Vazquez and Givinish 1998), mainly because of difficulty in access (Frahm and Gradstein 1991). Whittaker (1956) in his study on species distribution of the Great Smoky mountains observed a rounded or bell-shaped form in most cases, overlap broadly and have their centres and limits scattered along the gradients. Arroyo et al., (1988) have also observed humped patterns of diversity on elevational gradients with diversity lower near the foot and summit and higher at mid-slope on the western slopes of the Andes in northern Chile. Grytnes (2002) qualitatively observed the altitudinal richness pattern in relation to the influence of the area of the species pool, hard boundaries, temperature and precipitation in northern and southern Norway and observed that the patterns cannot be fully accounted for by any of these factors. The studies on elevational distribution pattern of plant species are relatively few in the Himalayas as evidenced from Gentry's (1988) study (Grytnes and Vetaas 2002; Bhattarai and Vetaas 2003). In the central Himalayan altitudinal gradient, Singh et al. (1994) observed an increase in tree species richness with elevation up to 1500 m in mixed *P. roxburghii* broadleaved forests. Vetaas and Grytnes (2002) used published data on the species distribution along the elevational ranges in Nepal and found a significant monotonically increasing trend in total species richness from 1000 m to 2500 m and gentle decrease from 2500 m to 4000 m a.s.l. They have found that the species endemic to the Himalayas increase monotonically up to 4000 m a.s.l. The overall species endemism was observed to be13 per ha. for Subansiri district of eastern Himalaya (Behera et al. 2002).

Forests of the Himalayan mountains, however, have contributed few data to general summaries on forest characteristics. Himalayan forests grow in an environment dissimilar to that of forests that have been most intensively studied (Singh et al. 1994). In the Himalaya, an elevational transect includes vegetation from tropical monsoon forest to alpine meadow and scrub, constituting an unusually extensive elevational and vegetational gradient (Singh and Singh 1992). The most conspicuous changes in plant community composition in the eastern Himalaya are related to differences in climatic conditions viz. rainfall, temperature, humidity, altitude that affect all these. The eastern Himalayas display an ultra-varied topography, a factor that fosters species diversity and endemism. Many deep and semi-isolated valleys are exceptionally rich in endemic plant species (Myers 1988). Kaul and Haridasan (1987) have tried to relate distribution of various forest types with altitudinal ranges in Arunachal Pradesh. They classified various forest types with respect to six altitudinal zones, viz., (I) tropical (Up to 900 m), (II) subtropical (900–1800 m), (III) temperate-broadleaved (1800–2800 m), (IV) temperate conifers (2800–3500 m), (V) sub-alpine (3500–4000 m) and (VI) alpine (4000–5500 m). Each vegetation zone has an intermixing ecotone extent of about 50 to 200 m (Behera et al. 2001). Behera (2000) have mapped various forest cover types of Subansiri district and found that tropical semi-evergreen, subtropical evergreen and temperate broadleaved forests cover 662 km^2, 3885 km^2 and 6150 km^2 respectively. The floristic altitudinal zonation found along the district corresponds to the physiographic vegetation zones (Behera et al. 2001).

The eastern Himalaya whose mountains exceed the snow line, is well north of the Tropic of Cancer but, since the Tibetan plateau is the motor of the Indian monsoon, hot moist summer winds off the Bay of Bengal support tropical wet forests at elevations up to ca. 2200 m (Ashton 2003). The above defined forest type categories can enable the translation from national classification systems to a global one. The translations of tropical semi-evergeen, subtropical evergreen and temperate broad-leaved are *tropical semi-evergreen moist broadleaf forest, tropical lowland evergreen broadleaf rain forest* and *tropical lower montane forest*. These three formations differ substantially globally in their species composition.

This study was restricted to tropical semi-evergreen, subtropical evergreen and temperate broadleaved forests located between 200 m and 2200 m elevation to understand the altitudinal behavior of trees. Authorities agree in subdividing the vegetation of the tropical zone on a basis of moisture conditions. In India, subtropical temperatures are determined more by altitude than by latitude and are characteristically developed in the hilly/mountain tracts viz., the Himalayas. The subtropical zone is therefore most conveniently considered as transition from the tropical to the temperate zone and is sometimes hardly distinguishable either in form or composition from one or the other. The east Himalayan montane temperate forests seem to be best classified by the rainfall during the season of vegetative

$\textcircled{2}$ Springer

activity, which may be taken as roughly the months with mean temperature over 13°C. We studied the pattern of tree diversity and their endemic subsets along the altitudinal gradients in every 200 m steps covered with tropical semi-evergreen, subtropical evergreen and temperate evergreen forests. The distribution of tree species was analysed w.r.t. available land area (i.e., species pool) for every elevation steps. The general pattern of tree species diversity (point, alpha and pattern) were analyzed along the altitudinal gradients. Simple linear regression analyses were done and trend lines were plotted to depict if there is an elevational trend in species diversity and endemism. The frequency distribution of altitudinal range was observed for each species as a measure of its ecological amplitude or niche breadth. The study also analyses the upper and lower distributional limits of tree species to realize the species turn-over along the gradients.

Study area

Study area, the undivided Subansiri district (26°55'–28°42' N latitude and 92°41'–94° 37' E longitude) falls in the *Himalaya-East Himalaya* biogeographic zone (Rodger and Panwar 1988). It has rich biological diversity owing to its strategic location, at the tri-junction of three biogeographic realms viz. the Afro-tropic, the Indo-Malayan and the Indo-Chinese (Takhtajan 1969). Generally, the district has humid subtropical to temperate climate with a wet summer and dry winter. Major part of the district is covered with primary forests. Numerous streams and rivers dissect the topography of the area and maintain the natural drainage system. The altitude varies from 50 m in tropical foothills to ca. 7000 m in alpine zones having a complex terrain. The district exhibits a mosaic of climatic zones owing to its geographical position and varied topography. The humid conditions have resulted in speciation of several genera, thus adding to high endemicity of the flora. In general, the soils are acidic due to high litter fall and low temperature and are rich in nitrogen. The soils in major part of the area are skeletal due to the hilly terrain and high rainfall. Rainfall is well distributed throughout the year except during April, May and June and increases along elevation. Variation in rainfall and precipitation is relatively small from year to year. The meteorological data available for this region are unreliable and therefore cannot be utilized for any ecological studies. The average annual rainfall data collected for three meteorological stations at different elevations i.e., Yazali (880 m), Ziro (1600 m) and Koloriang (2300 m) are 56.7, 122 and 512 cm respectively for the year 1995. The average annual temperature ranges from 9°C in winter to 38°C in summer at lower elevations, whereas it varies from below freezing point to 25 °C at higher reaches (Anon 1998). The relative humidity varies from 45% to 90% (Anon 1998). The southwest monsoon extends from June to September.

The forests of the district can be broadly categorized into four groups viz., (i) tropical (ii) subtropical (iii) temperate and (iv) sub-alpine and alpine (Rao 1974; Benewal and Haridasan 1992). The tropical semi-evergreen forests are found on the Shiwalik hills and extend up to 900 m above m.s.l. (Benewal and Haridasan 1992; Behera 2000). The top storey consists of a mixture of evergreen and deciduous (towards lowland) trees, whereas lower storey is dominated by evergreen species and thick undergrowths. The dominant tree species in this forest type are; *Altingia excelsa, Anthocephalus chinensis, Artocarpus lakoocha, Chukrasia tabularis, Dillenia indica, Duabanga grandiflora, Gmelina arborea, Phoebe goalparensis, Terminalia myriocarpa, Stereospermum chelonoides etc.* The subtropical broadleaved evergreen forests occur between tropical and temperate forests and are located up to 1800 m above m.s.l. This forest is dense and evergreen in nature. The dominant tree species of this forest are; *Alnus nepalensis, Acer oblongum, Callicarpa arborea, Castanopsis armata, C. indica, Engelhardia spicata, Garcinia acuminata, Gynocardia odorata, Kydia calycina, Magnolia pterocarpa, Michelia oblonga, Ostodes paniculata, Saurauia armata, Schima wallichii* var *khasiana* etc. The temperate broadleaved evergreen forest is located beyond subtropical forests. The characteristic appearance of this forest is open canopy and lax storeyed structure with dominance of oak species, members of Magnoliaceae and Ericaceae, particularly the *Rhododendron* spp. The top storey is represented by tall tree species *viz., Acer oblongum, Acer hookeri, Betula alnoides, Exbucklandia populnea, Castanopsis indica, Magnolia campbelli, Populus ciliata, Symplocos racemosa etc.* Our study is restricted to this three forest types found in Subansiri district of Eastern Himalaya.

Methods

The quadrats of 20 m × 20 m laid for measurement of tree species (≥ 15 cm cbh) during *biodiversity characterization at landscape level* project work were arranged per every 200 m steps between 200 m and 2200 m altitude (Table 1). This resulted in 8-quadrats per step. Plots were placed in similar aspects (i.e., southern and south-eastern) and not in clear-cuts or other areas clearly influenced by humans. Depressions and ridges were avoided when the plots were placed. Judgments of forest types correspond to the lower limit of corresponding elevational step. The total number of individuals, species and endemics encountered at various elevational steps were noted. Data on tree species diversity and endemism were analyzed using MS-Excel. For convenience, each elevational step has been referred to by its mid-value in the analysis. The following analysis were carried out to study if diversity (point, alpha and pattern) of tree species and its endemic subset follow an altitudinal pattern between 200 m and 2200 m in the eastern Himalayan mountain ecosystem. The distributional range and elevation limits of trees are also discussed to realize the species turn-over along the gradients.

Within a step, individual plots contain different amounts of species referred as point diversity. Thus point diversity was estimated as the mean number of species

Table 1 Sampling details

Sl.	Elevation step (m)	Mid value (m)	Forest type
1	200–400	300	Tropical semi-evergreen
2	401–600	500	Tropical semi-evergreen
3	601–800	700	Tropical semi-evergreen
4	801–1000	900	Subtropical evergreen
5	1001–1200	1100	Subtropical evergreen
6	1201–1400	1300	Subtropical evergreen
7	1401–1600	1500	Subtropical evergreen
8	1601–1800	1700	Temperate broadleaved evergreen
9	1801–2000	1900	Temperate broadleaved evergreen
10	2001–2200	2100	Temperate broadleaved evergreen

per plot in each step (Fig. 1). Species-accumulation curves were generated for all sites to see how species numbers vary with elevation (Fig. 2). Alpha diversity, the total number of species per step (for all 8-plots) was calculated and analyzed (Fig. 3). Here, the scale covers the richness within each step considered as a single homogenous community. Endemic subset of the tree species was analyzed step-wise along the gradients (Table 2; Fig. 4). The available area per step was compared with the species diversity to enumerate the relationship (Fig. 5). An approximate estimate of the area for each 200 m interval was made based on SRTM (Shuttle Radar Topographic Mission) data with 90 m contour intervals. Pattern diversity i.e., the ratio of point to alpha diversity was studied. It gives an idea of how well differentiated the communities are at different elevations (Fig. 6). The number of species, genera and families were recorded (Table 3). The ratio of species, genera and families were analyzed along the altitude and the corresponding trend lines were plotted (Fig. 7).

The site-wise frequency distribution of altitudinal range was observed for species along the altitudinal gradients (Fig. 8). Range can be defined as the difference in altitude between the highest and lowest plot in which the species was found. For the analysis, altitudinal range was defined as the difference in altitude between the highest and lowest plot in which a species occurred, regardless of whether the species was found in all intermediate plots. The lower and upper distribution limits of species along the sites were calculated (Fig. 9). For this calculation, species encountered at only one elevational step were excluded, because their altitudinal distribution were judged to be insufficiently known, and to avoid the bias of coupled upper and lower elevational boundaries (Shipley and Keddy 1987). The upper and lowermost sample steps were also excluded from the analysis, since it was impossible to distinguish real elevational limits at this altitude from species occurring beyond the sample range. Elevational distribution gaps were assumed to represent sampling biases and not true disjunctions (Shipley and Keddy 1987; Kessler 2000).

Results

A total of 2645 trees ($\geq$ 15 cm cbh) were recorded along the 2000 m altitudinal gradients in 80-plots of 20 $\times$ 20 m². They have wider taxonomic variation

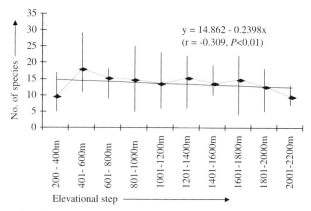

Fig. 1 Step-wise point diversity of tree species per plot along the gradients

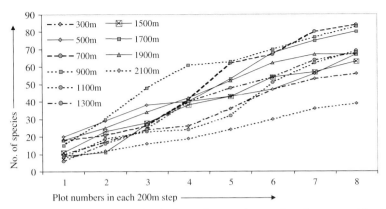

Fig. 2 Species-accumulation curves also indicate that species number decreases with elevation, though sampling seems to be incomplete especially at lower elevations (the mid value of eachelevational steps is shown in figure)

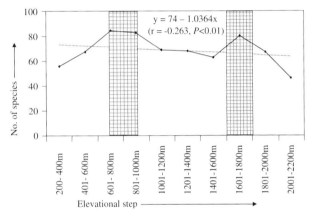

Fig. 3 Alpha diversity of tree species along 2000 m elevation gradients; Two maxima inspecies number can be observed corresponding to the transition zones between two forest types

comprising of 336 species, 185 genera and 78 families. A maximum of 407 and minimum of 206 individuals were recorded in 1401–1600 m and 201–300 m steps respectively. Barring the first step, point diversity demonstrated a decreasing trend along the gradients (Fig. 1). A maximum of 29 species ($\mu = 17.4$) in 401–600 m step and minimum of four species ($\mu = 9.75$) in 2001–2200 m step were recorded. The species accumulation curves also depicted decrease in number of tree species with altitude (Fig. 2). The curves beyond 1600 m steps appeared to turn asymptote with eight quadrats per step. In general, the alpha diversity, the total number of species per step declined along the gradients (Fig. 3). We found two maxima in the alpha diversity pattern, one in 601–1000 m and the other in 1601–1800 m step (Fig. 4). A total of 26 (out of 336) tree species were found endemic to eastern Himalaya (Table 2). *Schima wallichii* and *Beilschmiedia pseodu-microcarpa* to have wider distribution with appearances in five and four steps respectively. *Saurauia grifithii*, *Ixora subsessilis* and *Dysoxylum reticulatum* have appearances in three steps, whereas ten species have appearances in one steps only (Table 2). The

🍦 Springer

Table 2 Distribution of endemic species across 2000 m elevation gradients

Species	A	B	C	D	E	F	G	H	I	J	Total
Schima wallichii ssp. *wallichiana* var *khasiana* (Dyer) Bloembergen Dyer.	1	1				1	1			1	5
Beilschmiedia pseudo-microcarpa (Purk.) Kost.	1				1	1	1				4
Saurauia grifithii (Dyer.) Hk.f.					1	1	1				3
Ixora subsessilis Wall. ex D.Don				1			1	1			3
Dysoxylum reticulatum King	1		1	1							3
Shorea assamica Dyer.		1		1							2
Randia griffithii Hk.f.						1		1			2
Premna bengalensis Cl.					1					1	2
Michelia oblonga Wall.			1	1							2
Magnolia rabianiana (Hk.f. & Th.) Raju & Nayar								1		1	2
Magnolia gustavi King			1	1							2
Lindera latifolia Hk.f.	1	1									2
Lasianthus tubiferus Hk.f.	1			1							2
Ilex khasiana Purk.					1				1		2
Eriobotrya anguistissima Hk.f.							1		1		2
Coffea khasiana Hk.f.		1						1			2
Sapium eugeniaefolium Buch.-Ham.									1		1
Prunus punctata Hk.f. & Th.									1		1
Premna milleflora Cl.				1							1
Phoebe cooperiana U.N.Kanjilal ex A.Das									1		1
Phoebe angustifolia Meissn.									1		1
Miliusa macrocarpa Hk.f. & Th.	1										1
Litsea meissneri Hk.f.								1			1
Lithocarpus milroyi (Purkayastha) Barnett				1							1
Lasianthus hookeri Cl. ex Hk.f.									1		1
Caryota obtusa Griff.						1					1
Total	4	4	6	7	5	4	5	5	7	3	

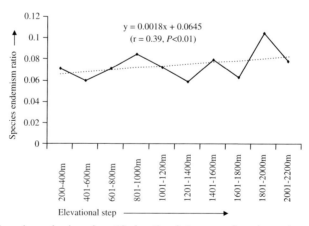

$$y = 0.0018x + 0.0645$$
$$(r = 0.39, P<0.01)$$

Fig. 4 Ratio of species endemism along 10-elevational steps reveals an increasing trend

species endemism ratio has shown an increase ($r = 0.39$, $P < 0.01$) along the gradients with a minima of 0.059 and maxima of 0.104 in 1201–1400 m and 1801–2000 m steps respectively (Fig. 4). We observed that the size of species pool

[284]

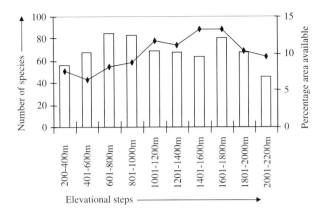

Fig. 5 Species richness (curve) versus area availability (histogram) along 10-elevational steps

increased from 401 m to 1800 m and then decreased beyond (Fig. 5). The ratio of species diversity to species pool was higher in 200–1000 m and in 1801–2000 m steps. The average annual rainfall increased drastically along the altitude. Pattern diversity demonstrated a narrow range between 0.17 and 0.26 along the gradients (Fig. 6). The curve increased and decreased alternatively along the gradients with the regression line denoting a marginal increasing trend ($r = 0.2$, $P < 0.01$). The 401–600 m and 2001–2200 m steps had high pattern diversity (i.e., 0.26 and 0.25). All other steps except 1201–1600 m had pattern diversity values < 0.2. The distributional pattern of genera and families seem to follow the species distribution pattern along gradients (Table 3). A maximum of 65-genera and 44-families were observed in 601–800 m step, whereas minimum of 39-genera and 29-families were observed in 2001–2200 m steps. The ratio between the number of individual trees, species and genera are plotted in Fig. 7. The number of individuals per species varied between 3 and 4, except in 1200–1600 m, where it increased beyond 5 (Fig. 7a). The number of species to genus, showed a decreasing trend along the gradients except at two steps (Fig. 7b). It is also noticed that the ratio of species to genus varied between 1.34 (maximum) and 1.18 (minimum) in 401–6001 m and

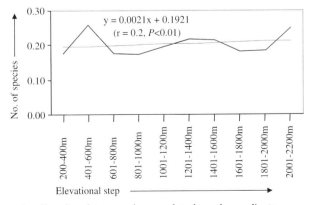

Fig. 6 Step-wise point diversity of tree species per plot along the gradients

Table 3 Number of families, genera, species and trees across each elevational step

Elevation step	No. of families	No. of genera	No. of species	No. of trees
200–400 m	31	47	56	206
401–600 m	30	50	67	243
601–800 m	44	65	84	276
801–1000 m	36	64	83	231
1001–1200 m	33	54	69	221
1201–1400 m	32	55	68	361
1401–1600 m	28	53	63	407
1601–1800 m	43	63	80	227
1801–2000 m	36	52	67	262
2001–2200 m	29	39	46	211

2001–2200 m steps respectively (Fig. 7b). The lowest numbers of species, genera and families (*i.e.*, 46, 39 and 29) occurred in the highest step (Table 3).

The frequency of altitudinal range was distributed between 1 and 41 along the gradients (Fig. 8). Of the total 336 species, *Altingia excelsa* (Hamamelidaeae) was recorded to have the widest amplitude since it occurred over the entire altitudinal range of 2000 m. Hence, the altitudinal range clearly defined the difference in altitude between the highest and lowest plot in which a species occurred regardless of the species presence in all intermediate plots. Further, the lower and upper distribution limits of each species were calculated along the elevational steps and the turnover was adjudged (Fig. 9). The lower and upper distributional limits of tree species along the gradients showed high turn-over rate (Fig. 9). Highest level of species turn-over was found in 401–600 m step at lower elevational limit, whereas for upper elevational limit the highest turn-over was recorded in 801–1000 m. Low level of

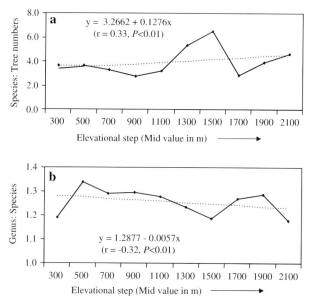

Fig. 7 Ratios of tree numbers, species and genera along the 2000 m gradients. Ratio of (**a**) individuals per species, (**b**) species per genera

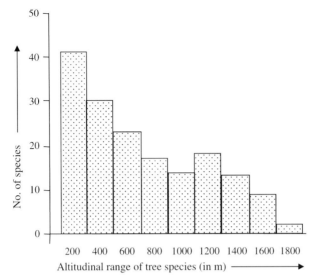

Fig. 8 Frequency distribution of altitudinal range observed for species along 2000 m altitudinal belt.Range is defined as the difference in altitude between the highest and lowest plot in which the species was found. The histogram bar represents species recorded over an elevation step of 200 m or less

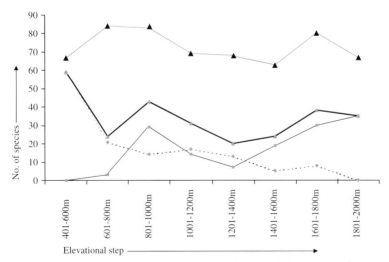

Fig. 9 Total species number (triangles), number of lower (broken line) and upper (thin continuous line) elevational limits, and sum of lower and upper limits (thick continuous line) along 8-elevational steps

species turnover was found between 801 m and 1600 m in both high and low elevation steps. Increase in species turn-over was observed at upper lower elevational limits beyond 1600 m (Fig. 9).

🍃 Springer

Discussion and conclusions

The point diversity demonstrated a greater variation of tree species at any site along the altitudinal gradients, which indicates that the species have their own distributional limits. The large variation of species between the quadrats in each step support the individualistic hypothesis of community organization (Gleason 1926) that posits the distribution of each species is determined by its own ability to survive, compete and produce successfully in different environments, resulting in each species having its own distinctive distribution, and in community composition changing more or less continuously along ecological and altitudinal gradients. Species-accumulation curves though indicate inadequate sampling in lower elevations (Fig. 2), yet they provide good indications of how tree distribution vary with elevation.

In general tree diversity declined and its endemic subset increased progressively along the gradients. The alpha diversity demonstrated a decreasing pattern with two maxima along the gradients; one in 601–1000 m and the other in 1601–1800 m (Fig. 3). The results showed that quadrats falling in the transition zones between two forest types possess significantly more number of species than other steps (Fig. 3). This correspond to the transition zones between tropical-subtropical and subtropical-temperate forests. The former transition seem to be (i) larger/gradual than the later one and/or (ii) a finer analysis at 50/100 m steps might reveal a sharper transition. Sudden increase in tree species at two sites along the gradients corresponds to ecotone. Two maximas in tree diversity pattern correspond to forest transition zones and can be explained by *mass effect* and upper elevation limits (Fig. 9). The most obvious way the mass effect can influence the altitudinal richness pattern is through a feedback among zonal communities. In their reconnaissance of forest vegetation on Volcan Barva, Hartshorn and Peralta (1988) noted that the transition between tropical premontane and lower montane rainforest occurs gradually between 1300 and 1600 m a.s.l.

The species richness pattern observed here is similar to that in the central Himalayan forests but differs from the pattern observed in Nepal. For Nepal, the species richness increased monotonically from 100 m to 2500 m a.s.l. This opposite trend could partly be attributed to data analysis methodology that uses published data on distribution and elevational ranges of the Nepalese flora and not quadrat sampling. Similarly, in the Santa Catalina mountains of Arizona, Whittaker and Niering (1975) observed that vascular plant diversity decreased from high-elevation fir forests to pine forests. The transition zones are not very much noticeable as it is expected in temperate and alpine forests. Although seven per cent (26 of 336 species) trees were found endemic to the eastern Himalaya, they showed an increasing pattern with altitude. However, the ratio of species diversity and endemism demonstrated an increasing pattern along the gradients (Fig. 4). Observation of poor pattern diversity along the gradients could be due to the presence of minor proportion of the total tree species at each steps. This is evident from Figure 8 that 60% tree had less than 200 m distributional range and 12% species had less than 400 m distributional range.

Previously, differences in altitudinal richness patterns have been attributed to differences in sampling methods (McCoy 1990; Rahbek, 1997). In the present study, the observed pattern in species diversity cannot be solely due to differences in sampling intensity. The pattern observed here might be due to the presence of microclimates or environmental heterogeneity, which resulted in different number of individuals at different altitudes (Gotelli and Colwell 2001). In mountains, small streams run almost everywhere, which may enhance heterogeneity in two ways i.e., (i) very different moisture regimes may occur within small distances and (ii) small streams create local variations and different soil conditions (Gentry 1988). The larger explanatory power of the regressions predict the relationship pattern between altitude and richness with confidence. In other words, a decreasing trend in species diversity is evident along altitude with increasing species diversity in transition zones. The different 200 m intervals used in the analysis do not represent equal areas because of the topography of eastern Himalaya (Fig. 5). The fact that species number increases as a function of area was not found suitable here. The pattern diversity values < 0.2 indicates that each plot had minor proportion of the total species (Fig. 6). The two peaks in pattern diversity at 601–1000 m and 1601–1800 m elevational steps reflect *mass effects* (Shemida and Wilson 1985), i.e., the expansion of species ranges outside their centers of importance. The importance of species-pool area (gamma diversity *Sensu* Whittaker 1960) for plant species richness is therefore probably minor. Lomolino (2001) reached a similar conclusion when discussing alternative explanations for altitudinal richness patterns.

Givinish (1999) found that the decline in tree diversity with elevation on mountains may be related to elevational declines in the rates of plant growth and forest turn-over. Highest level of species turn-over was in 401–600 m step at lower elevational limit could be consequent to changes in topography and prevailing microclimates. For upper elevational limit, the highest turn-over was recorded in 801–1000 m indicating the transition zones between tropical and subtropical forests (Fig. 9). Increase in species turnover was observed at upper lower elevational limits beyond 1600 m corresponding to transition zones between subtropical and temperate forests (Fig. 9). An idea proposed by MacArthur (1972) which retains acceptance among many contemporary ecologists is that tropical species, because of the diversity of the assemblages in which they occur, must either have narrower niches, or greater tolerance of ecological overlap than their counterparts living in less diverse assemblages. We found that species occurring in very high diversity assemblages at lower elevations did not differ in their altitudinal breadth from species growing in much lower diversity stands higher on the gradients. At all altitudes, we found some species with narrow distributions and others with wide ranges. If altitudinal amplitude of a species is considered to be an aspect of its niche breadth, it is clear from these results that niche breadth in these organisms is in fact independent of the diversity of the assemblage in which they occur.

The presence of two transition zones as revealed from this study needs to be more clearly defined with intense sampling at narrower elevational belts for the eastern Himalaya. We call for detailed floristic studies to determine the breadth of changes between adjacent forest types and details of local species richness in high diversity areas. The plot size does however appear to give a representative picture of tree diversity pattern, niche breadth and upper and lower elevational limits of species. The endemism rate and pattern is expected to improve when other life forms such as shrubs and herbs would be analyzed. The number of meteorological stations in this region are few and very sparsely distributed and therefore not very reliable for any ecological studies.

Ecological studies on eastern Himalayan forests are relatively less (Singh et al. 1994). Behera et al. (2005) have analyzed *biological richness* as a cumulative property of an ecological habitat and its surrounding environment that has an emerging implications in terms of management and planning. Six biodiversity attributes (i.e., spatial, phytosociological, social, physical, economical and ecological) were linked together based on their relative importance to qualitatively stratify *biological richness* of forest vegetation in Subansiri district of Eastern Himalaya. Subtropical forests claimed highest degree of *biological richness* as well as *disturbance index*. Roy and Behera (2005) attempted to establish the relationship existing between *biological richness* and *disturbance index* along the altitudinal gradients in Arunachal Pradesh and observed that disturbance decreases with increase in altitude, whereas *biological richness* demonstrated a hump-shaped pattern. These studies are the contributions towards understanding the eastern Himalayan ecosystems, that is categorized as one of the important *hotspots* of the World.

Acknowledgements Authors are grateful to Dr PS Roy, Deputy Director, NRSA & Project Director *Biodiversity Characterization at Landscape Level* for his constant support and guidance. We thank Dr A Jeyaram, Head, RRSSC, Kharagpur and Dr VK Dadhwal, Dean, IIRS for their encouragements and Dr PK Hajra, Dr K Haridasan, Dr GD Pal, Mr S Das, Ms S Srivastava, Mr TP Singh and Mr Ashish Kumar for helping with species identification and analysis. We sincerely thank the Reviewer for his critical comments to the earlier versions of this manuscript. This study was undertaken with the financial assistance from Department of Biotechnology and the Department of Space, Government of India in form of a research project on *Biodiversity Characterization at Landscape level* using remote sensing and GIS techniques. The senior author acknowledges Council of Scientific and Industrial Research, New Delhi for award of senior research fellowship (No. –9/735/ UC/99-EMR-I) to him.

References

Anon. (1998) Statistical Abstract of Arunachal Pradesh. Directorate of Economics & Statistics, Government of Arunachal Pradesh, Itanagar, 11–12

Arroyo MTK, Squeo F, Armesto J, Villagran C (1988) Effects of aridity on plant diversity in the northern Chile Andes. Ann Missouri Bot Gard 75:58–78

Ashton PS (2003) Floristic zonation of tree communities on wet tropical mountains revisited. Perspect Plant Ecol Evol Syst 6(1,2):87–104

Behera MD (2000) Biodiversity characterization at landscape level using remote sensing and GIS in Subansiri district of Arunachal Pradesh (Eastern Himalaya). Ph.D. Thesis. Gurukul Kangri University, Hardwar

Behera MD, Kushwaha SPS, Roy PS (2001) Forest vegetation characterization and mapping using IRS-1C satellite images in Eastern Himalayan region. Geocarto Int 16:53–62

Behera MD, Kushwaha SPS, Roy PS (2002) High plant endemism in Indian *Hotspot* – Eastern Himalaya. Biodivers Conserv 11:669–682

Behera MD, Kushwaha SPS, Roy PS (2005) Geo-spatial modeling for rapid biodiversity assessment in Eastern Himalayan region. Forest Ecol Manage 207:363–384

Benewal BS, Haridasan K (1992) Natural distribution and status of Gymnosperms in Arunachal Pradesh. Indian Forester, February, 96–101

Bhattarai RK, Vetaas OR (2003) Variation in plant species richness of different life forms along a subtropical elevation gradient in the Himalayas, east Nepal. Global Ecol Biogeogr 12:327–340

Brown J (2001) Mammals on mountainsides: elevational pattern of diversity, Global Ecol Biogeogr 10:101–109

Brown JH, Lomolino MV (1998) Biogeography 2nd edn. Sinauer

Cowell RK, Hurtt GC (1994) Nonbiological gradients in species richness and a spurious Rapoport effects. Am Nat 144:570–595

Fram JP, Gradstein SR (1991) An altitudinal zonation of tropical rain forests using bryophytes. J Biogeogr 18:669–678

Gentry AH (1988) Changes in plant community diversity and floristic composition on environmental and geographical gradients. Ann Missouri Bot Gard 75:1–34

Givinish TJ (1999) On the causes of gradients in tropical tree diversity. J Ecol 87:193–210

Gleason HA (1926) The individualistic concept of the plant association. Bull Torrey Bot Club 53:7–26

Gotelli NJ, Colwell RK (2001) Quantifying biodiversity: procedures and pitfalls in the measurement and comparison of species richness. Ecol Lett 4:379–391

Grytnes JA (2002) Species richness patterns of vascular plants along seven altitudinal transects in Norway. Ecography 26:291–300

Grytnes JA, Vetaas OR (2002) Species richness and altitude: a comparison between simulation models and interpolated plant species richness along the Himalayan altitude gradient, Nepal. Am Nat 159:294–304

Hajra PK, Verma DM, Giri GS (eds) (1996) Materials for the Flora of Arunachal Pradesh, vol I. Botanical Survey of India, Calcutta

Hartshorn G, Peralta R (1988) Preliminary description of primary forests along the La Selva-Volcan Barva altitudinal transect, Costa Rica. Almeda F, CM Pringle (eds) Tropical rainforests: diversity and conservation. California Academy of Sciences, San Fransisco, CA, pp 281–295

Kaul RN, Haridasan K (1987) Forest types of Arunachal Pradesh – A preliminary study. J Econ Taxon Bot 9:379–389

Kessler M (2000) Altitudinal zonation of Andean cryptogram communities. J Biogeogr 27:275–282

Korner C (2000) Why are there global gradients in species richness? Mountains might hold the answer. Trends Ecol Evol 15:513–514

Lieberman D, Lieberman M, Peralta R, Hartshorn GS (1996) Tropical forest structure and composition on a large scale altitudinal gradient in Costa Rica. J Ecol 84:137–152

Lomolino MV (2001) Elevational gradients of species-density: historical and perspective views. Global Ecol Biogeogr 10:3–13

MacArthur RH (1972) Geographical ecology, patterns in the distribution of species, Harper and Row

Major J (1988) Endemism: a botanical perspective. In: Myers AA, Giller PS (eds) Analytical biogeography. An integrated approach to the study of animal and plant distributions. Chapman and Hall, New York, pp 117–146

McCoy ED (1990) The distribution of insects along elevational gradients. Oikos 58:313–322

Myers N (1988) Threatened biotas: 'hotspots' in tropical forestry. The Environmentalist 8:1–20

Odland A, Birks HJB (1999) The altitudinal gradient of vascular plant species richness in Aurland, western Norway. Ecography 22:548–566

Rahbek C (1997) The relationship among area, elevation and regional species richness in neotropical birds. Am Nat 149:875–902

Rao RR (1974) Vegetation and phytogeography of Assam–Burma. In: Mani MS (ed) Ecology and biogeography of India. Dr. W. Junk B.V. Publishers, The Hague, pp 204–246

Rodgers WA, Panwar SH (1988) Biogeographical classification of India. New Forest, DehraDun

Roy PS, Behera MD (2005) Biological Richness assessment along the eastern Himalayan altitudinal zones in Arunachal Pradesh, India. Curr Sci 88:250–257

Shipley B, Keddy PA (1987) The individualistic and community-unit concepts as falsifiable hypothesis. Vegetatio 69:47–55

Shmida A, Wilson MW (1985) Biological determinants of species diversity. J Biogeogr 12:1–20

Shrestha TB, Joshi RM (1996) Rare, endemic and endangered plants in Nepal. WWF Nepal Program, Kathmandu, Nepal

Singh JS, Singh SP (1992) Forests of Himalaya: structure, functioning and impact of man. Gyanodaya Prakashan, Naini Tal India

Singh SP, Adhakari BS, Zobel DB (1994) Biomass, productivity, leaf longevity and forest structure in the central Himalaya. Ecol Monogr 64:401–421

Takhtajan A (1969) Flowering plants, origin and dispersal. Tr. Jeffrey, Edinburgh

Vazquez JA, Givinish TJ (1998) Altitudinal gradients in tropical forest composition, structure and diversity in the Sierra de Manantlan. J Ecol 86:999–1020

Vetaas OR, Grytnes JA (2002) Distribution of vascular plant species richness and endemic richness along the Himalayan elevational gradient in Nepal. Global Ecol Biogeogr 11:291–301

Whittaker RH (1956) Vegetation of the great smoky mountains. Ecol Monogr 26:1–80

Whittaker RH (1960) Vegetation of the Siskiyou Mountains, Oregon and California. Ecol Monogr 28:453–470

Whittaker RH, Niering WA (1975) Vegetation of the Santa Catalina mountains, Arizona. V. biomass, production and diversity along the elevation gradient. Ecology 56:771–790

Whittaker RJ, Willis KJ, Field R (2001) Scale and species richness: towards a general, hierarchical theory of species diversity J Biogeogr 28:453–470

🖉 Springer

Biodivers Conserv (2007) 16:1867–1884
DOI 10.1007/s10531-006-9086-x

ORIGINAL PAPER

Distribution, diversity and environmental adaptation of highland papayas (*Vasconcellea* spp.) in tropical and subtropical America

X. Scheldeman · L. Willemen · G. Coppens d'Eeckenbrugge
E. Romeijn-Peeters · M. T. Restrepo · J. Romero Motoche
D. Jiménez · M. Lobo · C. I. Medina · C. Reyes
D. Rodríguez · J. A. Ocampo · P. Van Damme ·
P. Goetgebeur

Received: 24 October 2005 / Accepted: 5 July 2006 / Published online: 27 October 2006
© Springer Science+Business Media B.V. 2006

Abstract *Vasconcellea* species, often referred to as highland papayas, consist of a group of fruit species that are closely related to the common papaya (*Carica papaya*). The genus deserves special attention as a number of species show potential as raw material in the tropical fruit industry, fresh or in processed products, or as genetic resources in papaya breeding programs. Some species show a very restricted

X. Scheldeman (✉) · L. Willemen · G. Coppens d'Eeckenbrugge · M. T. Restrepo ·
D. Jiménez · J. A. Ocampo
International Plant Genetic Resources Institute (IPGRI), Regional Office for the Americas,
Apartado Aéreo 6713, Cali, Colombia
e-mail: x.scheldeman@cgiar.org

E. Romeijn-Peeters · P. Goetgebeur
Department of Biology, Faculty of Sciences, Ghent University, K.L. Ledeganckstraat 35,
B-9000 Ghent, Belgium

J. Romero Motoche
Naturaleza & Cultura Internacional, Pío Jaramillo y Venezuela, Loja, Ecuador

M. Lobo · C. I. Medina
CORPOICA, Centro de Investigación "La Selva", A.A. 470, Rionegro, Antioquia, Colombia

C. Reyes
Universidad Nacional de Colombia, Sede Medellín, Medellín, Colombia

D. Rodríguez
Centro Nacional de Conservación de los Recursos Genéticos, Oficina Nacional de Diversidad
Biológica, Ministerio del Ambiente, El Limón, Apartado Postal 4661, Maracay, Venezuela

P. Van Damme
Laboratory of Subtropical and Tropical Agriculture and Ethnobotany, Department Plant
protection, Faculty of Agricultural and Applied Biological Sciences, Ghent University,
Coupure links 653, B-9000 Ghent, Belgium

G. Coppens d'Eeckenbrugge · M. T. Restrepo · J. A. Ocampo
CIRAD/FLHOR, UPR 'Gestion des ressources génétiques et dynamiques sociales', Campus
CNRS/Cefe, 1919 route de Mende, 34 293 Montpellier, France

distribution and are included in the IUCN Red List. This study on *Vasconcellea* distribution and diversity compiled collection data from five *Vasconcellea* projects and retrieved data from 62 herbaria, resulting in a total of 1,553 georeferenced collection sites, in 16 countries, including all 21 currently known *Vasconcellea* species. Spatial analysis of species richness clearly shows that Ecuador, Colombia and Peru are areas of high *Vasconcellea* diversity. Combination of species occurrence data with climatic data delimitates the potential distribution of each species and allows the modeling of potential richness at continent level. Based on these modeled richness maps, Ecuador appears to be the country with the highest potential *Vasconcellea* diversity. Despite differences in sampling densities, its neighboring countries, Peru and Colombia, possess high modeled species richness as well. A combination of observed richness maps and modeled potential richness maps makes it possible to identify important collection gaps. A Principal Component Analysis (PCA) of climate data at the collection sites allows us to define climatic preferences and adaptability of the different *Vasconcellea* species and to compare them with those of the common papaya.

Keywords   Americas · Biodiversity mapping · Caricaceae · Climatic modeling · GIS · Plant genetic resources · Richness · Tropical fruits

Introduction

Nowadays, less than 5% of all edible fruit species native to tropical areas are cultivated and marketed on a commercial basis (Wijeratnam 2000). Five species alone, banana, mango, pineapple, avocado and papaya, account for over 90% of fruit exports. As their demand is even expected to increase by 8% in the period 2000–2010 (FAO 2003) and as in recent years there has been a growing trend to identify and develop new crops for export and domestic markets (Padulosi et al. 1999), it is clear that a rise in demand for new tropical fruit species can be anticipated. A common constraint in answering this potential demand is a lack of knowledge on the nature and potential of many of these fruit species. The latter problem can be addressed by assessing existing diversity of new species and analyzing their climatic requirements. One of the promising tropical fruit families are the Caricaceae.

Caricaceae is a small family of six genera and 35 species, most of which originated in the Americas. The only non-American genus is *Cylicomorpha*, with two West African tree species. *Horovitzia* is a monotypic genus of hairy herbaceous plants that occur around Oaxaca, Mexico. The genus *Jarilla has* three herbaceous species in southern Mexico and Guatemala. *Jacaratia* has seven tree species that are widely spread in tropical climates. *Carica* is monotypic and includes the economically most important representative of the family, the common papaya (*Carica papaya* L.) (Badillo 1971, 1993, 2000, 2001). This species, which gives a large, bland, juicy fruit, is extensively cultivated throughout the tropics. Indeed, with a total annual world production of more than 6.5 Mt, covering nearly 400,000 ha (FAOSTAT 2005) the common papaya is considered the fourth most important tropical fruit crop. *Vasconcellea* was established as a genus by Saint Hilaire in 1837 and later treated as a section within the genus *Carica*. Recently it has been restored as a genus by Badillo (2000), and the genus*Vasconcellea* is now the largest within the family, holding 21 species. The present study focuses on this relatively unknown genus, *Vasconcellea*, using papaya as a reference for climate studies.

Vasconcellea species are often collectively called 'highland papayas' or 'mountain papayas' (National Research Council 1989) because many of them occur at higher altitudes. Compared to their better-known lowland cousin, *Carica papaya*, highland papaya fruits are generally smaller and have distinct texture, taste and aroma. In the Andes, they are consumed fresh, roasted, processed in juices, marmalades, preserves or dairy products or even prepared in sauces, pie fillings and pickles (National - Research Council 1989; CAF 1992; Van den Eynden et al. 1999, 2003). Most *Vasconcellea* fruits and processed products are consumed at household level or, less frequently, sold on local markets. Only the largest mountain papaya, the babaco (*V.* × *heilbornii*), has been commercially developed, albeit on a small scale, outside of its region of origin. It was introduced as a crop in New Zealand in 1973 (Endt 1981; Harman 1983) from where it spread during the eighties to Australia (Cossio 1988), Italy (Cossio and Bassi 1987; Ferrara et al. 1993), Spain (Merino Merino 1989), France (CTIFL 1992), South Africa (Wiid 1994) and even Switzerland (Evéquoz 1990, 1994), Canada (Kempler and Kabaluk 1996) and the Netherlands (Heij 1989) where greenhouse trials have been done. *Vasconcellea cundinamarcensis* is marketed locally in Ecuador, Colombia and Peru and has been successfully introduced in northern Chile (National Research Council 1989) where it has gained some local importance and from where some preserves are exported to Europe and the US. Other *Vasconcellea* species such as *V. candicans*, *V. crassipetala*, *V. goudotiana*, *V. microcarpa*, *V. monoica*, *V. palandensis*, *V. parviflora*, *V. quercifolia*, *V. sphaerocarpa* and *V. stipulata* are consumed locally (Badillo 1993, Van den Eynden et al. 1999; Scheldeman 2002). In addition to their existing use, highland papayas show potential (1) as a source of papain (Baeza et al. 1990; Dhuique Mayer et al. 2001; Scheldeman et al. 2002), a proteolytic enzyme complex used in pharmaceutical and food industries; and (2) as genetic resources for improvement of the common papaya. *Vasconcellea* carry resistance genes, particularly for the most severe and widespread disease, the papaya ringspot virus (Manshardt and Wenslaff 1989a, b; Magdalita et al. 1997; Drew et al. 1998), cold tolerance and organoleptic characteristics (Manshardt and Wenslaff 1989b). Improvement breeding is, however, hampered because interspecific gene flow between *Carica papaya* and *Vasconcellea* species faces considerable postzygotic barriers (Mekako and Nakasone 1975; Drew et al. 1998).

Five of the 21 described *Vasconcellea* species (*V. horovitziana*, *V. omnilingua*, *V. palandensis*, *V. pulchra*, *V. sprucei*) are included in the IUCN Red List of Threatened Species (IUCN 2004a) and require special monitoring attention for their conservation. A better knowledge of these threatened species could therefore contribute to their conservation. For example, *V. palandensis*, was, despite its local importance (Van den Eynden et al. 1999), only described as new to science in 2000 by Badillo et al. (2000). Agricultural extension and intensification causes genetic erosion in all *Vasconcellea* species, including the cultivated and tolerated forms. Their distribution has been declining with a concomitant loss of traditional knowledge (Scheldeman 2002). A detailed study on their distribution, diversity and environmental adaptability will undoubtedly generate valuable information for future conservation actions.

Indeed, in spite of its importance, *Vasconcellea* crop ecology and distribution have been little studied. The present study is looking for general answers to these questions based on spatial information. Spatial analyses based on georeferenced

🍂 Springer

herbarium or collecting data, often in combination with environmental spatial information, have previously generated valuable information in diversity studies (Skov 2000; Guarino et al. 2002; Vargas et al. 2004; Rodríguez et al. 2005). Such studies allow a clearer understanding of distribution of specific taxa and definition of areas of high diversity (Hijmans and Spooner 2001), on evolutionary origin (Jarvis et al. 2002), on defining sampling strategies and collection gaps (Jones et al. 1997; Greene et al. 1999a, b), on sampling biases (Reddy and Davalos 2003), on climatic adaptation (Berger et al. 2003), and on prioritization and definition of conservation areas (Kress et al. 1998; Funk et al. 1999; Bystriakova *et al.*, 2003; Jarvis et al. 2003).

The main objectives of the presented study were: (1) to summarize the distribution of all 21 *Vasconcellea* species using collection data as well as herbarium specimen data; (2) to identify areas of high diversity and (3) collection gaps; and (4) to describe climatic and altitude preferences for *Vasconcellea* species.

Materials and methods

Data collection

Biogeographic data were obtained from collecting trips organized in the framework of five research projects (see Acknowledgements), complemented by herbarium records extracted from label data provided by the following herbaria: A, AAU, AMD, ASU, AWH, B, BIGU, BM, BR, BRIT, CAY, CAUP, CHAPA, CLEMS, COL, CONC, CPUN, CR, CSAT, CTES, CUVC, DAV, DS, F, FAUC, GB, GENT, HUA, HNMN, HULE, HUMO, HUT, JAUM, JEPS, JVR, K, LP, LZ, M, MAD, MEDEL, MFA, MPU, MO, MU, MY, NY, P, PSO, QCA, QPLS, S, SEL, SSUC, TEFH, U, ULM, ULS, UPS, USCG, USJ, VALLE, VT (acronyms based on Index Herbariorum, Holmgren et al. 1990). Species collected were identified using the taxonomic keys developed by Badillo (1993) applying the nomenclature as revised by the same author (Badillo 2000, 2001).

All data points were entered into the Geographic Information System DIVA-GIS 4.1 (Hijmans et al. 2001, 2003), chosen to carry out spatial analysis. Georeferenced data were checked for inconsistencies. Data points without coordinates were assigned coordinates where possible while duplicate or doubtful data were removed. To define the altitude for each collection point, elevation data provided with collection site information were used where possible. In other cases, altitude was extracted from the $2.5 \times 2.5'$ (approximately 4.5×4.5 km at the equator) Worldclim data available at the DIVA-GIS website (http://www.diva-gis.org/).

Diversity and distribution

The measurement of diversity and distribution of *Vasconcellea* species was addressed in several ways. Firstly, the number of observations was tabulated per species and per country. Secondly, the area of occupancy (AOO), defined by IUCN (IUCN 2001, 2004b) as the total area occupied by a specific taxon, was selected as an indicator of abundance or rarity of a particular species. IUCN recommends the use of a 2×2 km (4 km^2) grid to define this parameter (IUCN 2004b). Thus, AOO was calculated by superposing a 2×2 km grid (based on a Lambert Equal Area

Azimuthal projection with central meridian 0 and reference latitude – 80) over the study area, followed by determining the number of grid cells occupied by each species (using the DIVA-GIS histogram tool) and by converting these to an effective area by multiplying the number of occupied cells by 4 km². Thirdly, the extent of the distribution area for each species was also estimated by determining the average distance between all possible pairs of collection points of each species. Finally, observed species richness was mapped using the point-to-grid richness analysis tool in DIVA-GIS, using a $1 \times 1°$ grid (corresponding to 111×111 km at the equator). The circular neighborhood option, with a 2° diameter (Hijmans and Spooner 2001), was applied to eliminate border effects due to the assignation of the grid origin.

Modeled richness and collecting gaps

Potential area of distribution of each species was modeled based on climatic preferences using the Bioclim model (Nix 1986; Busby 1991) in DIVA-GIS, using a $10 \times 10'$ Worldclim climate data set. In the Bioclim model, a site is considered suitable for a particular species if its climate data fall within the range prevalent at the sites where the species occurs (Hijmans et al. 2003). In this study, all 19 available climatic parameters were used (see Table 3). For each climatic parameter, the lower and upper 2.5 percentile of the total range were excluded from the suitability range to avoid the influence of extreme values. In order to prevent modeling outside the natural distribution area, a buffer zone with a radius of 3° (Jarvis et al. 2003) around all collection sites of each species was applied to limit the modeled richness. This climatic modeling resulted in 21 modeled distribution maps, one per *Vasconcellea* species. The sum of these 21 maps resulted in a map that indicates modeled potential species richness at a $10 \times 10'$ grid. Subtracting modeled species richness with the observed richness allowed detection of possible collection gaps.

Climatic requirements and adaptability

For each collection site, values for the 19 climatic parameters were extracted from the $2.5 \times 2.5'$ Worldclim data set and species collection points within the same grid cell were removed. Principal component analysis (PCA) was carried out on all 19 climatic parameters, applying a varimax normalized rotation, with the Statistica® package. To promote visibility, the centroid, i.e., the arithmetic average of the factor scores for each species, was used to represent each species' general climatic preferences. Standard deviation of factor scores was used to represent variation around the centroid to give an indication of the adaptability of each species for each climatic principal component. The well-known *Carica papaya* served as a reference in the *Vasconcellea* climate study.

Results and discussion

Diversity and distribution

The total dataset included 1,702 records representing all 21 *Vasconcellea* species (1,553) and *Carica papaya* (149). Table 1 presents a synthesis for the different species and countries of collection. *Vasconcellea cauliflora, V. cundinamarcensis,*

Table 1 List of all known *Vasconcellea* and *Carica* species and the countries where they were collected

Species (Red list Cat.**)	Country* (with respective number of collections)	Total number of collections	AOO (km²)	Average Distance (km)
V. cundinamarcensis	Col(163), Ecu(75), Ven(14), Per(5), Bol(4), Pan(3), Chl(1), Cri(1).	266	856	705
V. microcarpa	Ecu(110), Per(46), Col(35), Ven(14), Bra(12), Bol(3), Cri(1), Pan(1).	222	792	872
V. stipulata	Ecu(180)	180	424	50
V. × heilbornii	Ecu(164), Per(4)	168	420	116
C. papaya	Col(49), Nic(37), Ecu(18), Bol(8), Blz(6), Cri(6), Per(4), Ven(4), Pan(3), Dma(2), Guf(2), Mex(2), Pry(2), Slv(2), Bra(1), Dom(1), Hnd(1), Pri(1)	149	600	1,584
V. quercifolia	Arg(78), Bol(19), Bra(17), Pry(11), Per(1)	126	464	842
V. cauliflora	Col(54), Cri(25), Ven(11), Nic(8), Pan(7), Gtm(5), Mex(5), Hnd(4), Slv(4)	123	456	985
V. goudotiana	Col(108), Ecu(7), Ven(1)	116	372	318
V. parviflora	Ecu(63), Per(18)	81	288	241
V. chilensis	Chl(62)	62	244	147
V. glandulosa	Bol(17), Arg(17), Per(12), Bra(1)	47	168	1,039
V. monoica	Ecu(17), Bol(6), Per(3), Col(2)	28	100	863
V. crassipetala	Col(21), Ecu(3)	24	68	286
V. candicans	Ecu(13), Per(9)	22	72	618
V. pulchra (Nt)	Ecu(18), Col(1)	19	68	82
V. sphaerocarpa	Col(18)	18	68	318
V. longiflora	Ecu(8), Col(4)	12	44	231
V. weberbaueri	Ecu(11), Per(1)	12	36	102
V. horovitziana (En)	Ecu(8)	8	32	222
V. sprucei (Nt)	Ecu(8)	8	32	109
V. palandensis (Vu)	Ecu(6)	6	8	7
V. omnilingua (En)	Ecu(5)	5	20	23

*Country Codes: Arg: Argentina; Bra: Brazil; Blz: Belize; Bol: Bolivia; Chl: Chile; Col: Colombia; Cri: Costa Rica; Dma: Dominica; Dom: Dominican Republic; Ecu: Ecuador; Gtm: Guatemala; Guf: French Guyana; Hnd: Honduras; Mex: Mexico; Nic: Nicaragua; Pan: Panama; Per: Peru; Pri: Puerto Rico; Pry: Paraguay; Slv: El Salvador; Ven: Venezuela

**Red List Categories: En: endangered; Vu: vulnerable; Nt: Near threatened

V. goudotiana, V. microcarpa, V. quercifolia, V. stipulata and *V. × heilbornii*, are the species that were most commonly collected as evidenced by number of specimens as well as number of countries, while the Red List species *V. horovitziana, V. omni-lingua, V. palandensis, V. pulchra* and *V. sprucei* together with *V. longiflora*, mostly from Ecuador, are the least collected and should therefore be regarded as rare. The AOO confirms this sharp contrast in distribution extent between these two groups of common and rare species. *Vasconcellea cundinamarcensis* is the species with the widest distribution, covering mountainous zones from Costa Rica to Chile. *Vasconcellea microcarpa* has a similar AOO with an area ranging from Colombia to Brazil and Bolivia. *Vasconcellea quercifolia* is widespread in the south of the continent whereas *V. cauliflora* has an ample distribution in the northern part, ranging from Mexico southwards to Colombia and Venezuela. All species mentioned in the IUCN Red List show both a limited AOO and short distances between collection sites, confirming their status. The analyses presented indicate that both *V. longiflora* and *V. weberbaueri* are very rare species as well, and their possible inclusion in the Red List should be considered, after a more detailed analysis. *Vasconcellea palandensis*, which has the lowest values for both AOO and average distance, should be regarded as the rarest species in the genus.

Figure 1 gives a more detailed view of the geographic distribution of each *Vasconcellea* species. Ecuador, where 16 of the 21 *Vasconcellea* species have been recorded, is without any doubt the country with the highest richness for *Vasconcellea*, with its neighbor countries Colombia and Peru coming in second place with nine species. In all other countries, five or less species have been collected. The Andes of northwestern South America clearly constitute the center of diversity of *Vasconcellea* (Fig. 2). Ecuador has the highest diversity, and southern Ecuador (provinces El Oro, Loja and Zamora-Chinchipe) alone accounts for 12 of the 21 species, whereas southern Colombia (Nariño Department), where all nine Colombian species have been recorded, is another area with high *Vasconcellea* diversity.

Despite the size of the dataset, the study faces some methodological difficulties to precisely assess the extent of the distribution area and the rarity of a species, particularly when using a large, continent scale. The use of average distance between collection sites does not give a good indication of local abundance while the AOO does not constitute a direct indicator of abundance either. Considering that the dataset reflects the distribution at continent level, the combination of both indicators allows obtaining a better insight in the abundance of highland papayas in tropical America. Some species, e.g., *V. stipulata* and *V. × heilbornii*, show a rather wide AOO but a small average distance between collection points. They are very common in their area of origin (Loja Province, Ecuador; Badillo, 1993), but are rarely found outside this restricted area. Opposed to this, *V. monoica* and *V. candicans* show a small AOO but have a large distance between collection points, which indicates that they are never very abundant where they occur. The latter situation can however also be caused by undercollection of a species, highlighting the importance of a complete trustworthy dataset in this type of analysis. In fact, only cautious (and iterative) field verification of potential distribution maps, allows one to distinguish rarity from incomplete collecting. Despite these possible methodological shortcomings, our spatial analyses provide useful indications on the rarity of *Vasconcellea* species.

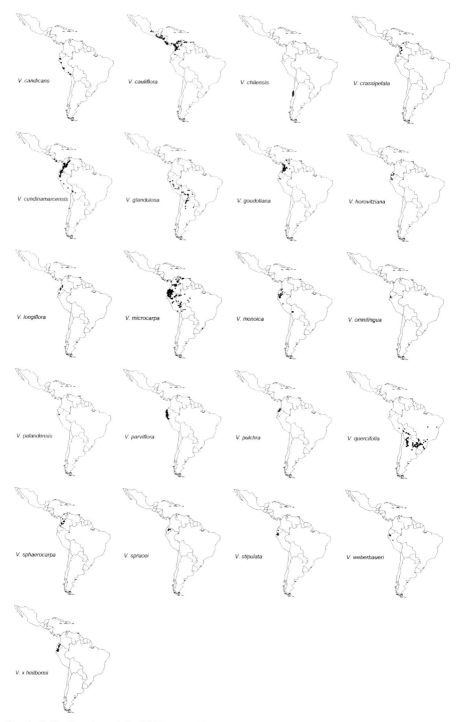

Fig. 1 Collection sites of the 21 *Vasconcellea* species within tropical America

Fig. 2 Observed *Vasconcellea* species richness in Latin America on a $1 \times 1°$ grid using 1,553 *Vasconcellea* observations. The darker areas indicate the highest richness (maximum value: 12 species per grid cell)

Modeled richness and collection gaps

The sum of the modeled potential distribution maps of the 21 *Vasconcellea* species resulted in a map indicating the modeled species richness in tropical America (Fig. 3). This map, which is based on the species climatic preferences, gives a slightly different view on *Vasconcellea* diversity. The Andean region of northern Ecuador and southern Colombia are now identified as those areas that posses the most favorable climate for *Vasconcellea* diversity and that therefore have the highest potential richness. Comparing, at country level, the observed richness with the modeled richness (Table 2), Ecuador, with a potential of 18 species, is confirmed as the most important center of *Vasconcellea* diversity. Its neighboring countries, Peru and Colombia are obviously also countries where a relatively high number of species occur, confirming the importance of the northern Andes for *Vasconcellea* diversity.

The sampled species richness map (Fig. 2), based on collection data, can be compared to the modeled potential species richness map (Fig. 3) to identify collection gaps. Figure 4 shows the areas where the discrepancy between observed and modeled richness is highest for *Vasconcellea*. The most important collection gaps are located in Colombia, where the higher parts of the departments of Nariño, Putumayo, Cauca, Valle de Cauca, Huila, Tolima, Caquetá and Meta should be given top priority for future collecting. Less important, but still significant, gaps are located in

Fig. 3 Modeled *Vasconcellea* species richness in Latin America on a $1 \times 1°$ grid using the sum of the modeled distribution maps of the 21 *Vasconcellea* species. The darker areas indicate the highest potential richness (maximum value: 11 species)

Table 2 Sampled *Vasconcellea* species richness versus modeled species richness in key countries

Country	Sampled species richness	Modeled species richness
Ecuador	16	18
Colombia	9	12
Peru	9	12
Bolivia	5	5
Venezuela	4	5
Brazil	3	3
Costa Rica	3	3
Panama	3	5
Argentina	2	2
Chile	2	1
Guatemala	1	1
Mexico	1	1
Paraguay	1	2
Nicaragua	1	2
Uruguay	0	2

eastern Venezuela (Zulia, Portuguesa, Trujillo, Barinas), in northeastern Colombia (Cundinamarca, Boyacá, Santander, Norte de Santander, Cesar and Bolivar), in northern Ecuador (Imbabura, Pichincha and Esmeraldas), in Central Peru (Cajamarca, San Martin, Amazonas, Huánuco, Pasco, Junín, Ucayali) and in eastern

Fig. 4 *Vasconcellea* collection gaps in northwestern South America based on an overlay of the sampled richness map and the modeled richness map (on a 1 × 1° grid). The darker areas indicate those zones were a high *Vasconcellea* richness is likely to occur, but where up to this moment no or only a limited number of specimens have been collected

Bolivia (La Paz). These are all highly suitable zones where only a limited number of *Vasconcellea* species have been collected so far. Figure 4 also shows that Ecuador, the country with the highest diversity, generally appears well-sampled as few undercollected areas could be identified.

Diversity studies often face the risk of a bias related to uneven collection intensity, particularly oversampling in areas of high species richness and in easily accessible areas (Hijmans et al. 2000; Reddy et al. 2003). This is illustrated clearly in a comparison between Ecuador and Colombia, both located in the center of highland papaya diversity. From the observed richness map (Fig. 2) Ecuador is identified as the country with the highest species number. It is also a country where environmental and political conditions make it easy to sample, as proven by the numerous herbaria records. Colombia on the other hand, has less observed diversity, but the modeled richness (Fig. 3) shows that, especially in the southern part, species richness is equal or higher than in Ecuador. These areas are often difficult to access due to security problems. Extrapolation of species ranges using environmental data, as shown in Fig. 3, is generally acknowledged to counter the sampling bias (Jarvis et al. 2003; Sommer et al. 2003). One could therefore wonder whether the fact that Ecuador is considered more diverse in highland papayas than Colombia could be partly due to a different sampling intensity.

Climatic requirements and adaptability

The extraction of climatic data on the 1,702 collection sites, covering 1,278 grid cells, resulted in a matrix of more than 24,000 values. The PCA allowed the identification of climatic conditions governing the distribution of the 21 *Vasconcellea* species and permitted a comparison of the climatic preferences of all these species, together with those of the better-known *Carica papaya*.

Taking into account only factors with eigenvalues superior or equal to one, four principal components were retained (Table 3). The first one, representing 44% of the observed variance, is a temperature factor differentiating warm climate species from cool climate species. The second factor shows a positive correlation with precipitation in the driest period and a negative correlation with precipitation seasonality. Thus, it differentiates species found in areas with year-round precipitation from those found in areas with a marked precipitation seasonality. Similarly, factor 3 is related to temperature seasonality, separating species found at higher latitudes, where monthly temperatures vary according to seasons, from the species living near the equator. Finally, the fourth factor shows a high correlation with precipitation and separates dry climate species from humid climate species. These four factors together accounted for 88.73% of total variance.

Figure 5 shows the centroids of the PCA factor scores for each *Vasconcellea* species and for papaya. Figure 5a places them in the plan of the two temperature-related factors (1 and 3) whereas Fig. 5b places them in the plan of the precipitation-

Table 3 Factor loadings, eigenvalues and percentages of variance for the first four axes, resulting from the PCA analysis of 19 climatic parameters for the 1,702 collection points (*Vasconcellea* and *Carica*)

	Factor 1	Factor 2	Factor 3	Factor 4
Eigenvalue	8.30	4.86	2.31	1.39
Percentage of variance	43.66	25.59	12.16	7.32
Description bioclim parameter				
Annual mean temperature	**0.98**	0.04	− 0.03	0.19
Mean monthly temperature range	− 0.19	− 0.51	− 0.48	0.26
Isothermality	− 0.14	0.07	**0.85**	0.25
Temperature seasonality	0.06	− 0.05	**− 0.95**	− 0.20
Max. temperature of warmest month	**0.91**	− 0.06	− 0.34	0.17
Min. temperature of coldest month	**0.90**	0.17	0.36	0.16
Temp. annual range	− 0.06	− 0.31	**− 0.93**	0.00
Mean temperature of wettest quarter	**0.93**	0.00	− 0.19	0.21
Mean temperature of driest quarter	**0.95**	0.07	0.18	0.15
Mean temperature of warmest quarter	**0.95**	0.02	− 0.27	0.13
Mean temperature of coldest quarter	**0.93**	0.05	0.24	0.24
Annual precipitation	0.26	0.58	0.15	**0.73**
Precipitation of wettest month	0.34	0.17	0.17	**0.88**
Precipitation of driest month	0.05	**0.90**	0.04	0.30
Precipitation seasonality	0.11	**− 0.82**	− 0.16	− 0.09
Precipitation of wettest quarter	0.33	0.18	0.17	**0.88**
Precipitation of driest quarter	0.05	**0.91**	0.06	0.33
Precipitation of warmest quarter	0.23	0.36	− 0.02	0.66
Precipitation of coldest quarter	0.09	0.60	0.26	0.46

The climatic parameters that have a high (> 0.7) contribution are highlighted in bold

 Springer

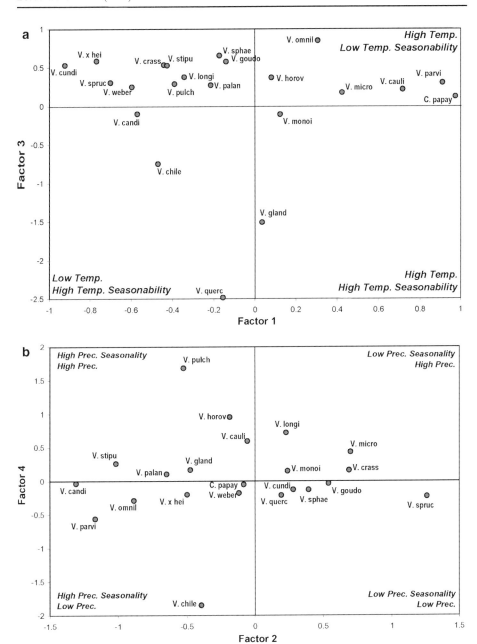

Fig. 5 Distribution of the species centroids for the first four PCA axes, for 1,702 *Vasconcellea* and *Carica* collection sites in Latin America and 19 climatic parameters. **a**: temperature-related axes (1 and 3). **b**: precipitation-related axes (2 and 4)

related factors (2 and 4). As expected, factor 1 allows a clear distinction between typical lowland species, *C. papaya*, *V. parviflora*, *V. cauliflora* and *V. microcarpa*, and highland species as *V. cundinamarcensis*, *V. × heilbornii* and *V. weberbaueri*, whereas factor 3 allows to distinguish the three species found at higher latitudes,

🖄 Springer

i.e., *V. quercifolia*, *V. glandulosa* and *V. chilensis*, from the majority of species most commonly found around the equator. In Fig. 5b, factor 4 clearly separates *V. chilensis*, a species with adaptation to a marked dry season whereas factor 2 ranks species according to their general humidity requirements. Thus, species such as *V. candicans*, *V. parviflora* and *V. stipulata*, from drier regions like southern Ecuador and northern Peru, are logically placed on the left side of the plan, while species adapted to the humid Amazon region, such as *V. monoica V. microcarpa*, and *V. sprucei* appear grouped on the right side, together with species of the humid forests of the Colombian Andes, such as *V. crassipetala*, *V. goudotiana* and *V. sphaerocarpa*.

Table 4 gives the standard deviation around the centroids for each species, and for each of the four main axes. Very low deviations for *V. chilensis* reflect its narrow geographic distribution, probably in relation with very strict climatic adaptations, whereas large values for *V. microcarpa* reveal a very wide adaptability, especially for precipitation. Comparing the deviations for the four factor scores across species, the standard deviation of Factor 3 generally shows the lowest value, reflecting the essentially tropical distribution of *Vasconcellea*. Indeed, only the three species distributed well beyond the equator up to subtropical latitudes, show a high (*V. glandulosa* and *V. quercifolia*) or relatively high (*V. chilensis*) adaptability for temperature seasonality.

Of all the species studied, *Carica papaya* is the one that is found in the warmest zones, with a mean annual temperature of 24.2°C, with little seasonal temperature variations. From a precipitation perspective it nearly falls in the center of the precipitation plan (Fig. 5b) indicating that is occurs at average precipitations (annual

Table 4 Climatic adaptability indicated by the standard deviation of the mean factor scores for each species

Species	Factor 1	Factor 2	Factor 3	Factor 4
V. candicans	1.01	0.54	0.59	1.22
V. cauliflora	0.67	0.86	0.55	1.41
V. chilensis	0.28	0.10	0.39	0.39
V. crassipetala	1.00	0.80	0.16	0.46
V. cundinamarcensis	0.73	0.56	0.37	0.76
V. glandulosa	0.94	0.58	1.11	0.58
V. goudotiana	0.86	0.50	0.21	0.70
V. horovitziana	1.24	0.98	0.21	1.82
V. longiflora	0.68	0.97	0.22	0.68
V. microcarpa	0.90	1.28	0.41	1.05
V. monoica	0.82	1.32	0.54	0.41
V. omnilingua	0.64	0.28	0.11	0.22
V. palandensis	0.53	0.16	0.10	0.08
V. parviflora	0.73	0.39	0.42	0.87
V. pulchra	0.96	0.41	0.21	0.93
V. quercifolia	0.82	1.08	0.79	0.58
V. sphaerocarpa	1.07	0.50	0.24	0.64
V. sprucei	0.97	0.84	0.21	0.74
V. stipulata	0.45	0.56	0.21	0.58
V. weberbaueri	0.75	0.33	0.49	0.55
V. × heilbornii	0.48	0.66	0.27	0.62
C. papaya	0.71	0.86	0.68	0.81

precipitation 1,830 mm) in areas without extreme precipitation seasonality. These values correspond very well with values given in literature, i.e., 24.5°C and 1,920 mm for mean annual temperature and annual precipitation respectively (Duke 1983). Compared to *Vasconcellea* species, *Carica papaya* shows a more constant standard deviation among the four factors (Table 4), which is in general very close to the overall average (0.64) of the four axes for all species, indicating a similar and average adaptability for both temperature and precipitation. Most *Vasconcellea* species show a low standard variation for one of the four factors, which indicates a more limited adaptability for a specific climatic environment.

Altitude

Figure 6 presents an analysis of the relation between elevation and species richness. An average of 14 species can be found at altitudes up to 2,500 m. Above this elevation, species richness gradually decreases, as shown by the second order regression.

The *Vasconcellea* richness map (Fig. 2) illustrates a clear relation between *Vasconcellea* diversity and the Andes. Even lowland species are mostly confined to the Andean foothills. *Vasconcellea cauliflora*, also found in Central America, *V. quercifolia*, also present in southern Brazil, southern Paraguay and northern Argentina and *V. microcarpa*, also collected in the Brazilian Amazon, are only partial exceptions to this link between *Vasconcellea* species and the Andes. Thus, even though several *Vasconcellea* species can be found at sea level, the use of the term "highland papayas" or "mountain papayas" as a collective term for *Vasconcellea* species appears justified.

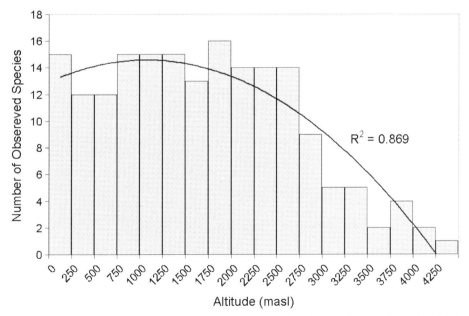

Fig. 6 Relation between altitude and number of observed *Vasconcellea* species using 1,553 *Vasconcellea* observations in Latin America. The solid line represents a second order regression

Conclusions

This study, based on the most complete *Vasconcellea* collection data set compiled so far, provides an overview of distribution, diversity and climatic preferences of 21 *Vasconcellea* species. It is therefore an important addition to the monographs on Caricaceae by Badillo (Badillo 1971, 1993). Despite their wide distribution in tropical America, ranging from Mexico down to Argentina and Chile, the majority of the *Vasconcellea* species are confined to the northern Andean region (Colombia, Ecuador and Peru). This is also reflected in their temperature preferences, generally cooler climates with limited seasonality, as prevalent in the equatorial Andes.

Although some highland papaya species are commonly grown in backyard gardens to be consumed and/or marketed locally, other species such as *V. horovitziana*, *V. omnilingua*, *V. palandensis*, *V. pulchra* and *V. sprucei* are only found in the wild within a very limited distribution range, and are included in the IUCN Red List. Results of the analysis of the distribution results suggest that two more *Vasconcellea* species, *V. longiflora* and *V. weberbaueri*, should be considered for inclusion in the Red List as well.

This study shows that by combining georeferenced species collection data together with available detailed climate data, the use of GIS can add significant value and knowledge to the existing information sources. This study is also a clear illustration of the importance of trustworthy collection data and its sharing, which might allow a better understanding of many other potential neglected and underutilized species to enhance their conservation and use.

Acknowledgements The authors wish to thank the herbaria that provided specimens or collection data. Besides herbarium data, new collection contributed significantly to this study. In Ecuador, the project '*Conocimientos y Prácticas Culturales sobre los Recursos Fitogenéticos Nativos en el Austro Ecuatoriano*' (1996–2000) was funded by the Belgian Development Cooperation (DGDC) through the Flemish Interuniversitary Council (VlIR) while the project 'Taxonomy of wild and semidomesticated *Vasconcellea* species (Caricaceae) in Ecuador' (2000–2003) was made possible through funds from the Flemish Fund for Scientific Research (FWO). The project "*Conservación* ex situ *y obtención de parámetros para el manejo agrotécnico y micropropagación de los géneros* Carica *y* Rubus *en la región sur del Ecuador*" (2001–2003) included also germplasm collecting and was funded by the Ecuadorean Programa de Modernización de los Servicios Agropecuarios (PROMSA). Included collection sites in Ecuador, Colombia and Venezuela were obtained from the project '*Aprovechamiento de los Recursos Genéticos de las Papayas para su Mejoramiento y Promoción*' (1999–2003), funded by the Regional Fund for Agricultural technology (FONTAGRO) while the Colombian Ministry for Environment and the Research Center of the Colombian Coffee Grower Federation provided additional funds for collecting in Colombia through the project '*Estudio de la Diversidad de las Passifloraceae y Caricaceae en la Zona Cafetera*' (2003–2004). The authors express their gratitude to the donors of these projects.

References

Badillo VM (1971) Monografía de la familia Caricaceae. Universidad Central de Venezuela, Maracay
Badillo VM (1993) Caricaceae. Segundo esquema. Revista de la Facultad de Agronomía de la Universidad Central de Venezuela, Alcance 43, Maracay
Badillo VM (2000) *Carica* L. vs. *Vasconcella* St.-Hil. (Caricaceae) con la rehabilitación de este último. Ernstia 10(2):74–79
Badillo VM (2001) Nota correctiva *Vasconcellea* St. Hill. y no *Vasconcella* (Caricaceae). Ernstia 11(1):75–76
Badillo VM, Van den Eynden V, Van Damme P (2000) *Carica palandesis* (Caricaceae), a new species from Ecuador. Novon 10:4–6

Baeza G, Correa D, Salas C (1990) Proteolytic enzymes in *Carica candamarcensis*. J Sci Food Agric 51:1–9

Berger J, Abbo J, Turner NC (2003) Ecogeography of annual wild *Cicer* species: the poor state of the world collection. Crop Sci 43:1076–1090

Busby JR (1991) BIOCLIM: a bioclimate analysis and prediction system. Plant Prot Quart 6:8–9

Bystriakova N, Kapos V, Lysenko I, Stapleton CMA (2003) Distribution and conservation status of forest in the Asia-Pacific Region. Biod Conserv 12(9):1833–1841

CAF (1992) Manual técnico del cultivo de chamburo. Centro Agrícola de Quito, Corporación Andino de Fomento, Quito

Cossio F, Bassi G (1987) Babaco, dopo il boom tiriamo le somme. Terra e Vita (Cagliari) 7:88- 93

Cossio F (1988) Il Babaco. Edizioni Calderini Edagricole, Bologna

CTIFL (1992) Nuevas especies frutales. Ediciones Mundi-Prensa, Madrid

Dhuique Mayer C, Caro Y, Pina M, Ruales J, Dornier M, Graille J (2001). Biocatalytic properties of lipase in crude latex from babaco fruit (*Carica pentagona*). Biotechnol Lett 23(13):1021–1024

Drew RA, O'Brien CM, Magdalita PM (1998) Development of *Carica* interspecific hybrids. Acta Hortic 461:285–291

Duke JA (1983) *Carica papaya* L. In: Handbook of energy crops. Unpublished. Available via Center for New Crops and Plants Products, Purdue University. Department of Horticulture and Landscape Architecture, West Lafayette. http://www.hort.purdue.edu/newcrop/duke_energy/ Carica_papaya.html. Cited 30 Aug 2005

Endt DW (1981) The babaco: a new fruit in New Zealand to reach commercial production. Orchardist New Zeal 54(2):58–59

Evéquoz N (1990) Premiers résultats d'un essai de culture de babaco. Revue Suisse de Viticulture, d'Arboriculture et d'Horticulture 22(2):137–141

Evéquoz N (1994) Culture en serre de babaco (*Carica pentagona*), Résultats d'un essai de 4 ans (2ᵉ partie). Revue Suisse de Viticulture, d'Arboriculture et d'Horticulture 26(5):323–325

FAO (2003) Medium-term prospects for Agricultural Commodities. Projections to the Year 2010. FAO Commodities and Trade Technical Paper 1. Food and Agriculture Organization of the United Nations, Rome

FAOSTAT (2005) Statistical databases of the Food and Agriculture Organization of the United Nations. http://www.apps.fao.org. Cited 30 Aug 2005

Ferrara E, Barone F, Calabrese F, D'Ascanio R, De Michele A, Giorgio V, Martelli S, Monastra F, Nieddu G (1993) Babaco (*Carica pentagona* Heilb.). L'Informatore Agrario (Verona) XLIX(1):41–46

Funk VA, Zermoglio MF and Nasir N (1999) Testing the use of specimen collection data and GIS in biodiversity exploration and conservation decision making in Guyana. Biod Conserv 8:727–751

Guarino L, Jarvis A, Hijmans RJ, Maxted N (2002) Geographic Information Systems (GIS) and the conservation and use of plant genetic resources. In: Engels JMM, Ramantha Rao V, Brown AHD, Jackson MT (eds) Managing plant genetic diversity. CABI publishing, Wallingford, pp 387–404

Greene SL, Hart TC and Afonin A (1999a) Using geographic information to acquire wild crop germplasm for ex situ collections: I. Map development and field use. Crop Sci 39:836–842

Greene SL, Hart TC Afonin A (1999b) Using geographic information to acquire wild crop germplasm for ex situ collections: II. Post collection analysis. Crop Sci 39:843–849

Harman JE (1983) Preliminary studies on the postharvest physiology and storage of babaco fruit (*C. × heilbornii* Badillo nm. *pentagona* (Heilborn) Badillo). New Zeal J Agric Res 26:237–243

Heij G (1989) Exotic glasshouse vegetable crops: Dutch experiences. Acta Hortic 242:269–276

Hijmans RJ, Garrett KA, Huamán Z, Zhang DP, Schreuder M, Bonierbale M (2000) Assessing the geographic representativeness of genebank collections: the case of Bolivian wild potatoes. Conserv Biol 14(6):1755–1765

Hijmans RJ, Guarino L, Cruz M, Rojas E (2001) Computer tools for spatial analysis of plant genetic resources data: 1. DIVA-GIS. Plant Genet Res Newslett 127:15–19

Hijmans RJ, Guarino L, Bussink C, Mathur P, Cruz M, Barrentes I, Rojas E (2003) DIVA-GIS. A geographic information system for the analysis of species distribution data. Manual, version 4.0. http://www.diva-gis.org. Cited 30 Aug 2005

Hijmans RJ, Spooner DM (2001) Geography of wild potato species. Am J Bot 88:2101–2112

Holmgren PK, Holmgren NH, Barnett LC (1990) Index herbariorum. Part I: the herbaria of the world, 8th edn. New York Botanical Garden, New York

IUCN (2001) IUCN Red List categories and criteria: version 3.1. IUCN Species Survival Commission, IUCN, Gland and Cambridge

IUCN (2004a) 2004 IUCN Red List of threatened species. http://www.redlist.org. Cited 30 Aug 2005

IUCN (2004b) Guidelines for using the IUCN Red List categories and criteria. Standards and petitions Subcommittee of the IUCN SSC Red List Programme Committee. IUCN, Glandand Cambridge

Jarvis A, Ferguson ME, Williams DE, Guarino L, Jones PG, Stalker HT, Valls JFM, Pittman RN, Simpson CE, Bramel P (2003) Biogeography of wild *Arachis*: assessing conservation status and setting future priorities. Crop Sci 43(3):1100–1108

Jarvis A, Guarino L, Williams DE, Williams K, Vargas I, Hyman G (2002). Spatial analysis of wild peanut distributions and the implications for plant genetic resources conservation. Plant Genet Res Newslett 131: 29–35

Jones PG, Beebe SE, Tohme J, Galwey NW (1997) The use of geographical information systems in biodiversity exploration and conservation. Biod Conserv 6:947–958

Kempler C, Kabaluk T (1996) Babaco (*Carica pentagona* Heilb.): a possible crop for the greenhouse. Hortscience 31(5):785–788

Kress WJ, Heyer P, Acevedo P, Coddington J, Cole D, Erwin TL, Meggers BJ, Pogue M, Thorington RW, Vari RP, Weitzman MJ, Weitzman SH (1998) Amazonian biodiversity: assessing conservation priorities with taxonomic data. Biod Conserv 7:1577–1587

Magdalita PM, Drew RA, Adkins SW, Godwin ID (1997) Morphological, molecular and cytological analysis of *Carica papaya* × *C. cauliflora* interspecific hybrids. Theor Appl Genet 95:224–229

Manshardt RM, Wenslaff TF (1989a) Zygotic polyembryony in interspecific hybrids of *Carica papaya* and *C. cauliflora*. J Am Soc Hortic Sci 114(4):684–689

Manshardt RM, Wenslaff TF (1989b) Interspecific hybridisation of papaya with other *Carica* species. J Am Soc Hortic Sci 114(4):689–694

Mekako HU, Nakasone HY (1975) Interspecific hybridisation among 6 *Carica* species. J Am Soc Hortic Sci 100(3):237–242

Merino Merino D (1989) El cultivo del babaco. Ediciones Mundi-Prensa, Madrid

National Research Council (1989) Lost crops of the Incas: little-known plants of the Andes with promise for worldwide cultivation. National Academy Press, Washington

Nix HA (1986) A biogeographic analysis of Australian elapid snakes. In: Longmore R (ed) Atlas of Elapid Snakes of Australia, Australian Flora and Fauna Series No. 7, Australian Government Publishing Service, Canberra, Australia, pp 4–15

Padulosi S, Eyzaquirre P, Hodgkin T (1999) Challenges and strategies in promoting conservation and use of neglected and underutilized crop species. In: Janick J (ed), Perspectives on new crops and new uses, ASHS Press, Alexandria, USA, pp 140–145

Reddy S, Davalos LM (2003) Geographical sampling bias and its implications for conservation priorities in Africa. J Biogeogr 30(11):1719–1727

Rodríguez D, Marín C, Quecan H, Ortiz R (2005) Áreas potenciales para colectas del género *Vasconcellea* Badillo en Venezuela. Bioagro 17(1):3–10

Scheldeman X (2002) Distribution and potential of cherimoya (*Annona cherimola* Mill.) and highland papayas (*Vasconcellea* spp.) in Ecuador. PhD Dissertation, Ghent University

Scheldeman X, Van Damme P, Romero Motoche J (2002) Highland papayas in southern Ecuador: need for conservation actions. Acta Hortic 575(1):199–205

Sommer JH, Nowicki C, Rios L, Barthlott W, Ibisch PL (2003) Extrapolating species ranges and biodiversity in data-poor countries: The computerized model BIOM. Revista de la Sociedad Boliviana de Botánica 4(1):171–190

Skov F (2000) Potential plant distribution mapping based on climatic similarity. Taxon 49:503–515

Van den Eynden V, Cueva E and Cabrera O (1999) Plantas silvestres comestibles del sur del Ecuador—Wild edible plants of southern Ecuador. Ediciones Abya-Yala, Quito

Van den Eynden V, Cueva E, Cabrera O (2003) Wild foods from Ecuador. Econ Bot 57(4):576–603

Vargas JH, Consiglio T, Jørgensen PM, Croat TB (2004) Modelling distribution patterns in a species-rich plant genus, *Anthurium* (Araceae), in Ecuador. Divers Distrib 10(3):211–216

Wiid M (1994) Aanpasbaarheid van babako in subtropiese gebiede. Instituut vir Tropiese en Subtropiese Gewasse Inligtingbulletin 15(12):17–19

Wijeratnam RSW (2000) Identification of problems in processing of underutilized fruits of the tropics and their solutions. Acta Hortic 518:237–240

Biodivers Conserv (2007) 16:1885–1900
DOI 10.1007/s10531-006-9091-0

ORIGINAL PAPER

Ecological niche modeling and geographic distribution of the genus *Polianthes* L. (Agavaceae) in Mexico: using niche modeling to improve assessments of risk status

Eloy Solano · T. Patricia Feria

Received: 16 November 2005 / Accepted: 7 July 2006 / Published online: 29 November 2006
© Springer Science+Business Media B.V. 2006

Abstract The genus *Polianthes* (Agavaceae) is endemic to Mexico and is important at the scientific, economical, and cultural level since prehispanic times. Habitat destruction is one of the main factors affecting populations of *Polianthes* species, yet little is known about the geographic distribution of this genus, and thus its vulnerability to habitat change. We compared three different approaches to measure the *Polianthes* species area of distribution to assess the risk of species extinction applying the MER (Method of Evaluation of Risk extinction of wild species for Mexico): area of occupancy, extent of occurrence, and ecological modeling. We also found the richness areas of distribution of this genus. We compared the species distributions with Terrestrial Protected Regions (TPR) and Natural Protected Areas (NPA). Although the three methods used to calculate the species area of distribution agree about the highly restricted nature of *Polianthes* species. The area of occupancy sub-estimate the species distribution, while the extent of occurrence over-estimate it for species with disjoint distribution. Thus, we recommend the use of ecological modeling to improve the assessment of the current species distribution area to apply the MER. Most *Polianthes* species are distributed in the Sierra Madre Occidental and Transvolcanic Belt. Three species do not occur in any of the NPA or TPR, one species has suitable habitat in three TPR but has not been recorded there, and one species, *P. palustris*, is likely extinct.

Keywords Agavaceae · Area of occupancy · Conservation · Distribution · Extent of occurrence · Ecological niche · Endemism · Genetic algorithm · MER · *Polianthes*

E. Solano (✉)
Herbario FEZA, Facultad de Estudios Superiores Zaragoza, UNAM,
AP 9-020, Mexico D. F., Mexico
e-mail: solanoec@correo.unam.mx

T. P. Feria
Department of Biology, University of Missouri-St. Louis, St. Louis, MO 63121, USA

🕿 Springer

Introduction

The genus *Polianthes* (Agavaceae) is endemic to Mexico, and includes 14 species, three varieties, and two cultivars (Solano 2000). Most species in this genus are used for ornamental and ceremonial purposes. The best-known taxon is *Polianthes tuberosa*, which has been cultivated and used for medicinal, ornamental, and ceremonial practices since prehispanic times. Other species (e.g. *P. bicolor*, *P. geminiflora*, *P. longiflora*) are important for their medicinal properties (Solano 2000). To date, however, there have been few attempts to determine geographic distributions at fine scales or to evaluate the conservation status of *Polianthes* species, despite the fact that habitat destruction has been identified as one of the main factors affecting populations of these species in Mexico (García-Mendoza 1995, 2004; IUCN 1997), where the rate of habitat alteration has been high (Mas et al. 2004). Furthermore, since this is an endemic genus, documenting the areas of high species richness of *Polianthes* will help to measure the conservation significance (Brooks et al. 2002) of the protected areas already established in Mexico.

Species of *Polianthes* have been recorded in 18 Mexican states, where most occur in pine forest, oak forest, or pine-oak forest, some in grasslands, and a few in tropical dry or semideciduous forests (Solano 2000). Five species (*Polianthes densiflora*, *P. howardii*, *P. longiflora*, *P. palustris*, and *P. platyphylla*) are listed as rare by the IUCN (IUCN 1997) and are considered to be in the category of special protection according to Mexican law (SEMARNAT 2002). The geographical distribution of a species is one of the main criteria required for the Mexican government to evaluate the risk of species extinction applying the MER (Method of Evaluation of Risk extinction of wild species for Mexico; SEMARNAT 2002). Therefore, understanding the distribution of species, and thus is imperative to establish their species conservation status in Mexico.

Environmental variables are frequently used to estimate species' distributions (Austin 2002) because survival and reproduction can occur only within a certain range of environmental conditions, the species' ecological niche (Brown et al. 1996). Using species' occurrence data and environmental variables, GIS technology can be used to model and visualize the ecological niche using a variety of different algorithms (reviewed by Guisan and Zimmermann 2000; Elith et al. 2006) and to predict species' distributions by projecting a model of the ecological niche onto geographic space GARP (Genetic Algorithm for Rule set Production) has seen particularly strong statistical support for its predictive abilities under diverse conditions (Feria and Peterson 2002; Illoldi-Rangel et al. 2004; Martínez-Meyer et al. 2004; Nakazawa et al. 2004; Peterson and Cohoon 1999; Peterson 2001; Peterson and Vieglais 2001; Peterson et al. 2002a; Raxworthy et al. 2003; Stockwell and Peterson 2002, 2003; but see Elit et al. 2006).

Maps obtained using GARP have been interpreted as maps of the potential distribution of a species (Anderson and Martínez-Meyer 2004). A more realistic distribution is obtained by matching the predicted distribution based on the ecological niche with those geographic features (i.e., basins, drainages, ecoregions) where the species has been recorded (Soberón and Peterson 2005) and overlaying these maps with a map of current land use (Ortega-Huerta and Peterson 2004; Sánchez-Cordero et al. 2005a). The current area of distribution can then be calculated considering the number of grid cells where the species has been predicted to occur, converting those grids to an equal area projection.

The reliability of extinction risk assessments is often compromised by biases in collection data, particularly for species with very few collections or disjoint distributions. As a consequence of the sparseness of collections, calculating the area of occupancy by counting the number of cells occupied by each species tends to underestimate the species' true distribution whereas calculating the extent of occurrence based on the minimum convex polygon defined by all known localities typically overestimates the area occupied. In comparison to these approaches, maps of species' distributions derived from SDM are superior for conservation planning because they allow distinguishing suitable and unsuitable environments for populations within the extent of occurrence. Therefore, we propose to follow the use of ecological niche models to estimate a more realistic assessment of the area of distribution of the species (Anderson and Martínez-Meyer 2004; Ortega-Huerta and Peterson 2004; Peterson et al. 2002b; Sánchez-Cordero et al. 2005a; Soberón and Peterson 2005) and thus more objectively assign risk status based on MER criteria.

Here we have three different objectives based on the following questions: (1) How the different approaches to estimate the area of distribution (area of occupancy, extent of occurrence, and ecological modeling) vary on the assessment of risk status criteria?; (2) are all *Polianthes* taxa included in any of the already protected areas in Mexico?; and (3) what are the richness areas of distribution of *Polianthes*?; thus, our goals in this paper are: (1) to assess the current species area of distribution and to assign the risk status scores to each *Polianthes* species based on three different approaches; (2) to determine whether species' ranges are effectively included in national natural protected areas; and (3) to find the richness areas of *Polianthes* taxa in Mexico. We calculated the area of occupancy, extent of occurrence, and modeled the ecological niches of 12 species and three varieties of species of the genus *Polianthes*. Ecological niches were modeled and projected onto geographic space using the GARP algorithm.

Methods

Distributional data

We compiled a database of 885 locality records of a total of 14 species and three varieties of *Polianthes* based on historical voucher specimens from different herbaria, recent field work conducted by Solano from 1994 to 2004, and on consultation of databases at the Comisión Nacional para el Conocimiento y Uso de la Biodiversidad (Conabio; http://www.conabio.gob.mx) and REMIB (Red Mundial de Información sobre Biodiversidad; http://www.conabio.gob.mx/remib/remib.html). All historical records without geographical coordinates were georeferenced via direct consultation of 1:50,000 topographic maps and evaluated by Solano prior to modeling in order to avoid the use of imprecise distributional information.

Less than ten independent locality records (i.e., records ≥ 1 km distance) exist for 7 of the 14 *Polianthes* taxa and the three varieties (Table 1). *Polianthes geminiflora* var. *geminiflora* was the taxon with the highest number of independent localities (128), while whereas *Polianthes palustris* was recorded only in one locality. Two species were excluded from this analysis, *Polianthes tuberosa* and *P. palustris*. The first is a widely cultivated species and the second has only one historical record.

Table 1 Distributional data for all the species of the genus *Polianthes* analyzed

Species	Records	Localities
Polianthes bicolor E. Solano & García-Mend.	20	6
P. densiflora (B.L. Rob. & Fernald) Shinners	13	5
P. geminiflora var. *clivicola* McVaugh	65	34
P. geminiflora (Lex.) Rose var. *geminiflora*	248	128
P. geminiflora var. *pueblana* E. Solano & García-Mend.	29	10
P. graminifolia Rose	29	12
P. howardii Verh.-Willl.	11	7
P. longiflora Rose	59	33
P. montana Rose	12	6
P. multicolor E. Solano & Dávila	26	18
P. nelsonii Rose	74	25
P. oaxacana García-Mend. & E. Solano	4	3
P. palustris[a] Rose	1	1
P. platyphylla Rose	70	30
P. sessiliflora (Hemsl.) Rose	140	64
P. venustuliflora E. Solano & Castillejos	6	3

[a] *P. palustris* was not considered for the ecological niche modeling analysis

Geographical distribution and risk status assessment

We employed three different approaches to characterize the geographic distribution of species, and thereby assess their risk status: (1) calculation of area of occupancy; (2) calculation of extent of occurrence; and (3) ecological niche modeling of potential distributions. Area of occupancy was calculated for all taxa by superimposing a grid of 1 km², cells onto the distributional species data and counting the number of cells occupied by each species using an avenue script developed by T. Consiglio. The cell size of 1 km × 1 km was chosen to allow comparisons with the results gathered by the ecological modeling procedure (see below). Extent of occurrence was calculated for each taxa (except for *P. palustris* and *P. venustuliflora* due to the small number of distributional data) as the area encompassed by a convex polygon enclosing all occurrence points using an avenue script developed by Willis et al. (2003) which implements the Rapoport (1982) mean propinquity method. Here, all distributional data (points) were joined by the shortest distance with straight lines and distances between points were measured, allowing for the calculation of the arithmetic mean distance between localities.

Ecological niche models were developed for 12 species and three varieties of *Polianthes*, using the following environmental variables: total annual precipitation, absolute maximum temperature, absolute minimum temperature, mean maximum temperature, and mean minimal temperature, aspect (sine and cosine transformations), slope, soils, elevation, and solar radiation. These environmental data were available in digital format from Conabio. The grain resolution of all the environmental variables was 0.01° (approximately 1 × 1 km² grid cells). Models of potential distributions were built based on these environmental variables using the desktop-computer implementation of GARP (http://www.lifemapper.org/desktopgarp). GARP is a machine-learning procedure based on a genetic algorithm (Stockwell and Noble 1992) that divides distributional data into training and test data sets, and works in an iterative process of rule selection, evaluation, testing, and incorporation or rejection. Here a method is chosen from a set of possibilities (e.g., logistic regression, bioclimatic rules), applied to the training data (i.e., distributional data), and a rule is developed or evolved. Testing data are used to internally evaluate the predictive accuracy. Because each run results in a different prediction due to the genetic-based algorithm, final products are based on the best results from 100 runs for each species. From the 100 models, we selected best models from among those with 0% omission error (i.e., no instances of prediction that a species is absent from a location when in fact it is present). Then, we ranked these 0% omission error models by their commission error (this error occurs when the model predicts that the species is present at a location when in fact it is not) and selected five models with less than 20% of commission error. The best output models were exported to ArcGIS as grids, where we summed them to obtain a unique map per species where cell values varied from 0 to 5 (5 indicates that all five models scored the species as present in a particular cell). These predictions were converted to binary predictions assigning values of 1 for grid cells with value of 5 and 0 for grid cells with values from 1 to 4.

To have a more realistic scenario of the geographic distribution of taxa in this genus taking into account biogeographic considerations, we restricted predicted distributions to those Mexican biogeographic provinces and sub-basins (http://www.conabio.gob.mx) in which each species had actually been recorded (Fig. 1; The genus is distributed in eight biogeographic provinces: Balsas Basin, Mexican Pacific Coast, Mexican Plateau, Sierra Madre Oriental, Sierra Madre Occidental, Sierra Madre del Sur, Tamaulipeca, and Transmexican Volcanic Belt.), and calculated the current area of distribution by overlaying the resulting maps on a map of land use and vegetation (http://www.inegi.gob.mx). Biogeographic provinces are reasonable estimates of areas of endemism (Lomolino et al. 2005) and together with sub-basins provide historical information as barriers for *Polianthes* species dispersal. Due to the small number of locality records available for most of the species, we only evaluated

Fig. 1 Mexican biogeographic provinces and sub-basins: apn: North Altiplano, aps: South Altiplano, bal: Balsas basin, bc: Baja California, cab: Del Cabo, chis: Los Altos de Chiapas, clf: California, gm: Gulf of Mexico, nus: Soconusco, oax: Oaxaca, pac: Mexican Pacific Coast, ptn: Peten, sme: Sierra Madre Oriental, smo: Sierra Madre Occidental, sms: Sierra Madre del Sur, son: Sonorense, tam: Tamaulipeca, vol: Transmexican Volcanic belt, yuc: Yucatán

the distributional maps for those species with more than 20 localities (Table 1). Model accuracy was tested originally via a random 50% partition of input data; test points were set aside prior to modeling and overlaid on the resulting prediction. Model significance was evaluated using a chi-square test (df = 1). After converting maps to an equal area projection, the area of distribution was calculated for each species by summing the number of cells where the species was predicted to occur and multiplying this number by the area of the cell. All process were developed on ArcView 3.2 (ESRI 1999).

We followed the first MER criterion (A) to evaluate risk of species extinction. Species that have a distribution range that represents < 5% of the national territory are considered highly restricted and are assigned a risk factor of 4; those between 5% and 15% are considered restricted and have a risk factor of 3; those between 15% and 40% are considered widely distributed or moderately restricted and have risk factor of 2; and those > 40% are considered widely distributed and have a risk factor of 1 (SEMARNAT 2002).

Richness areas and species distribution on the current protected areas on Mexico

Areas of high species richness were assessed by summing of all the resulting distribution maps. Geographic distributions of each species as well as the areas of endemism were compared with Terrestrial Protected Regions (TPR) (Arriaga et al. 2000) and Natural Protected Areas (NPA) (Instituto Nacional de Ecología 1999). The former (TPA) are areas based on different taxa and the knowledge of several experts on every group, and thus it is one of the most formal proposes for areas of conservation, but without a legal decree for their protection. The later are the already protected areas with a national decree for protection.

Results

Geographic distribution and risk status assessment

The three different approaches to calculate the species area of distribution give different results (Table 2) (Fig. 2 shows an example of the area of distribution assessed for the three approaches followed in this paper). In all cases the species with the greater area of distribution was *Polianthes geminiflora* var. *geminiflora* followed for *P. sessiliflora* (Table 2).

Assessments based on area of occupancy consider all taxa as "highly restricted" since all their total current distribution is less than 5% of the National Mexican surface. MER criteria based on extent of occurrence and ecological niche include all species, except *Polianthes geminiflora* var. *geminiflora* (Fig. 3) and *P. sessiliflora* (Fig. 5), as "highly restricted." *Polianthes geminiflora* var. *geminiflora* and *P. sessiliflora* are considered as "restricted" since their distribution is greater that the 5% but less than 15% of the total area of Mexico.

Ecological niche models (Figs. 3–6) were statistically significant (Chi-square; $P < 0.05$) for all taxa. The species that occur at the northern limits of the distribution are *P. nelsonii* (Fig. 3) and *Polianthes densiflora* (Fig. 4); *P. bicolor* (Fig. 5) and *P. oaxacana* (Fig. 3) establish the southern limit of the distribution. Only one species

Table 2 Area of distribution per species and MER risk category based in tree different criteria

Species	AOO[a] (km²)	MS[b] (%)	MER[c]	EOO[d] (km²)	MS (%)	MER	CA[e] (km²)	MS (%)	MER
Polianthes bicolor	6	0.0003	4	1284.56	0.0654	4	1413	0.0719	4
P. densiflora	5	0.0003	4	3616.38	0.1841	4	1462	0.0744	4
P. geminiflora var. clivicola	33	0.0017	4	52349.0	2.6649	4	34843	1.7737	4
P. geminiflora var. geminiflora	118	0.0060	4	282650.	14.3888	3	179414	9.1334	3
P. geminiflora var. pueblana	10	0.0005	4	156.81	0.0080	4	17	0.0009	4
P. graminifolia	12	0.0006	4	11232.0	0.5718	4	1898	0.0966	4
P. howardii	7	0.0004	4	1235.6	0.0629	4	187	0.0095	4
P. longiflora	31	0.0016	4	29138.4	1.4833	4	15303	0.7790	4
P. montana	6	0.0003	4	22355.4	1.1380	4	3444	0.1753	4
P. multicolor	18	0.0009	4	16003.4	0.8147	4	8727	0.4443	4
P. nelsonii	25	0.0013	4	57933.3	2.9492	4	54925	2.7961	4
P. oaxacana	3	0.0002	4	40.246	0.0020	4	262	0.0133	4
P. palustris*	1	0.0001	4	–	–	–		0.0000	–
P. platyphylla	29	0.0015	4	38576.9	1.9638	4	43021	2.1901	4
P. sessiliflora	59	0.0030	4	194662.	9.9096	3	155867	7.9347	3
P. venustuliflora	3	0.0002	4	–	–	4	20	0.0010	4

[a] Area of Occupancy based on 1 km × 1 km grids.

[b] Mexican surface based in Mexico total area = 1 964 375 km² (http://www.inegi.gob.mx)

[c] 4 = Highly restricted and 3 = restricted

[d] Extent of occurrence

[e] Current area of distribution based on ecological niche modeling by GARP, biogeographic provinces, sub-basins, and current land use

is widely distributed occurring in seven provinces (*Polianthes geminiflora* var. *geminiflora*: Fig. 3), while seven species are narrowly distributed and restricted to a single province (Table 3). The most important Mexican Biogeographic Provinces for the distribution of this genus are the Sierra Madre Occidental and the Transmexican Volcanic Belt, in which eight species and the three varieties occur. Three of those species are highly restricted across the Transmexican Volcanic Belt province (*Polianthes geminiflora* var. *pueblana*, Fig. 5; *P. longiflora*, Fig. 6; and *P. venustuliflora* sp. nov. [E. Solano and Castillejos in prep.], Fig. 4). Two species are highly restricted to the Sierra Madre del Sur (*P. bicolor* and *P. oaxacana* sp. nov. [García-Mend. and E. Solano in prep.],), specifically in the state of Oaxaca. Three more species can be considered as endemic to a particular state of Mexico: *P. densiflora* is endemic to Chihuahua, in the northern part of the Sierra Madre Oriental; *P. multicolor* is endemic to Guanajuato, Fig. 4; and *P. venustuliflora* is endemic to Michoacán in the central part of the Transmexican Volcanic Belt.

Richness areas

Areas of species richness can be visualized as the sum of the individual species maps (Fig. 7), which suggests zones of high and low richness across Mexico. The areas with highest predicted richness (six species) occur along the southern part of the Sierra Madre Occidental and the northwest part of the Transmexican Volcanic Belt.

🖄 Springer

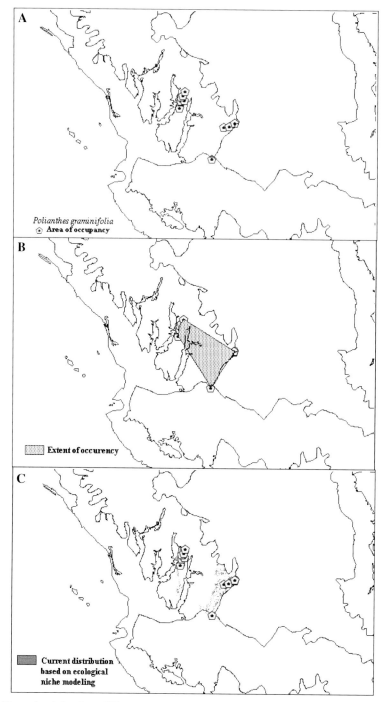

Fig. 2 Example of the three different approaches used to estimate the species area of distribution for *Polianthes graminifolia*: (**A**) area of occupancy, (**B**) extent of occurrency, and (**C**) ecological niche after overly the biogeographic provinces, sub-basins, and current land use and vegetation

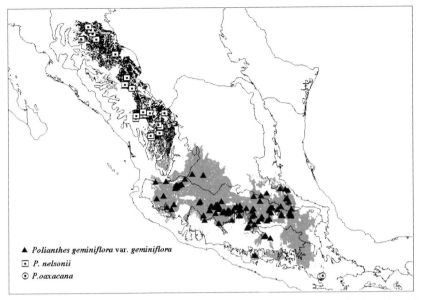

Fig. 3 Modeled ecological niches per species using a genetic algorithm. Distribution of species is ordering to show not overlapping species distributions. Species localities are indicated by different symbols per species. Shadow color represents the potential distribution predicted

The highest richness occurs in the state of Jalisco, where six species and two varieties of *Polianthes* were found (*Polianthes geminiflora* var. *clivicola*, Fig. 5; *P. geminiflora* var. *geminiflora*, Fig. 3; *P. graminifolia*, and *P. howardii*, Fig. 4; *P. longiflora*, *P. montana*, and *P. platyphylla*, Fig. 6; and *P. sessiliflora*, Fig. 5).

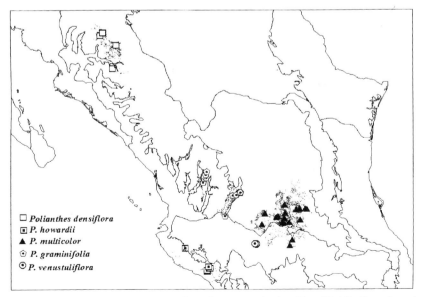

Fig. 4 Modeled ecological niches per species using a genetic algorithm. Distribution of species is ordering to show not overlapping species distributions. Species localities are indicated by different symbols per species. Shadow color represents the potential distribution predicted

⨍ Springer

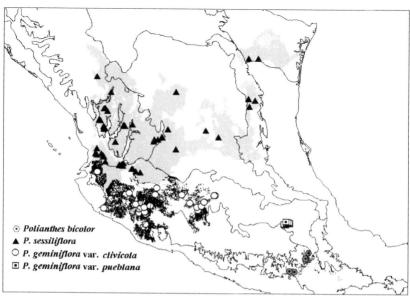

Fig. 5 Modeled ecological niches per species using a genetic algorithm. Distribution of species is ordering to show not overlapping species distributions. Species localities are indicated by different symbols per species. Shadow color represents the potential distribution predicted

Species distribution vs. NPA and TPR

Four species and two varieties occur on both NPA and TPR (*Polianthes geminiflora* var. *geminiflora, P. geminiflora* var. *clivicola, P. longiflora, P. montana, P. nelsonii,*

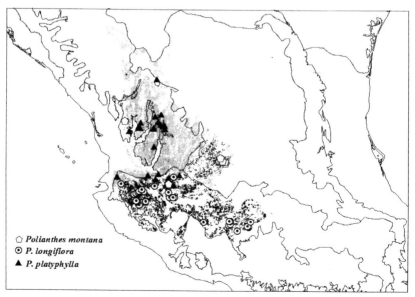

Fig. 6 Modeled ecological niches per species using a genetic algorithm. Distribution of species is ordering to show not overlapping species distributions. Species localities are indicated by different symbols per species. Shadow color represents the potential distribution predicted

Table 3 Distribution of *Polianthes* spp. by Mexican biogeographic provinces

Species	MBP							
	bal	pac	mpl	sme	smo	sms	tam	vol
Polianthes bicolor						X		
P. densiflora					X			
P. geminiflora clivicola	X	X				X		X
P. geminiflora geminiflora	X	X	X	X	X	X		X
P. geminiflora pueblana								X
P. graminifolia					X			X
P. howardii		X						X
P. longiflora								X
P. montana			X		X			X
P. multicolor			X					X
P. nelsonii	X	X			X			
P. oaxacana						X		
P. palustris		X						
P. platyphylla	X	X			X			X
P. sessiliflora	X	X	X		X		X	X
P. venustuliflora								X

MBP: Mexican biogeographic provinces; bal: Balsas Basin; pac: Mexican Pacific Coast; mpl: Mexican Plateau; sme: Sierra Madre Oriental; smo: Sierra Madre Occidental; sms: Sierra Madre del Sur; tam: Tamaulipeca; vol: Transmexican Volcanic Belt (Morrone et al. 2002)

and *P. sessiliflora*). Five species occur only on some TPR (*Polianthes bicolor*, *P. densiflora*, *P. multicolor*, *P. oaxacana*, and *P. platyphylla*). In contrast, three species (*Polianthes geminiflora* var. *pueblana*, *P. howardii*, and *P. venustuliflora*,) are not found in any of the TPR or NPA. Three TPR (Sierra Los Huicholes, Sierra de Morones, and Sierra Fría) include suitable habitat for one species (*P. graminifolia*), but the species has not been recorded there (Fig. 8).

Fig. 7 Endemism pattern summed across 12 species and three varieties of the genus *Polianthes* in Mexico: (*top*) species patterns of endemism (gray scale, 0 species; white, 6 species; black)

Springer

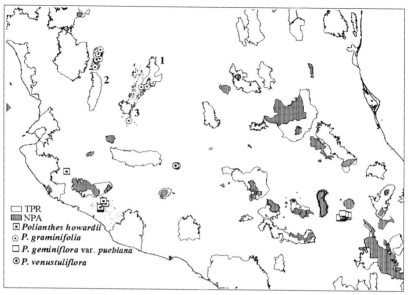

Fig. 8 Distribution of *Polianthes geminiflora* var. *poblana*, *P. howardii*, *P. venustuliflora*, and *P. graminifolia* vs. TPR and NPA. Numbers 1: Sierra Fría, 2: Sierra los Huicholes, and 3: Sierra de Morones

Discussion

Geographic distribution and risk status assignment

MER criterion A (geographical distribution of species) was assessed for 12 species and three varieties of *Polianthes* based on three different approaches to measure the species area of distribution. Area of occupancy sub-estimates the species distribution. No one species has even a distribution of 1% of the Mexican surface based on this approach. Following this criteria all species will have the highest risk score based on MER or other methods of risk status assessment (e.g., IUCN (http://www.iucn.gob) upholds that "only the criteria for the highest category of threat that the taxon qualifies for should be listed"). This result is not realistic since for most species, such as *Polianthes geminiflora* var. *geminiflora*, results clear that its distribution is greater that the estimated, but not all the areas where the species can occur have been sampled, and maybe never will be due to economic and time constrains. Extent of occurrence, on the other hand, frequently over-estimates the species distribution particularly for species, such as *Polianthes graminifolia*, which has a disjoint distribution. This fact is not considered by this method since the estimated distribution is based on a solid convex polygon. SDM give frequently intermediate values of area of distribution between area of occupancy and extent of occurrence. This method includes the current species distribution when ecological niches match geographic features (i.e., biogeographic provinces, sub-basins) where the species has been recorded (Soberón and Peterson 2005) and current land use and vegetation (Ortega-Huerta and Peterson 2004; Sánchez-Cordero et al. 2005a), thus consider both punctual (area of occupancy) and disjoint distribution, and therefore can give a better picture of the current species area of distribution.

Although five species of *Polianthes* (*P. densiflora, P. howardii, P. longiflora, P. palustris*, and *P. platyphylla*) are listed as rare by the IUCN (IUCN 1997) and considered to be in the category of special protection according to the Mexican Official law, most, if not all the species in this genus should be considered as threatened (e.g., vulnerable) due to their highly restricted distribution in Mexico and the fact that their habitat has been severely modified. Currently, *Polianthes palustris*, is under special protection by the Mexican Official Norm, but according to our findings this species is likely extinct as no individuals were found during 10 years of field work by Solano.

Detailed knowledge of a species distribution is needed before management decisions can be made (Lim et al. 2002). In general, less than a quarter of vascular plants have been evaluated for their conservation status, such is the case of the genus *Polianthes*. MER has been required by Mexican law since 2002 for listing organisms for protection. Although MER may be vulnerable to subjectivity or misinterpretation (i.e., criterion B—habitat condition—and criterion D—impact of human activity—are not independent from each other), "MER can meet its intended goals of facilitating timely conservation decisions and generating testable hypotheses" (Olson et al. 2005). Criterion A (geographical distribution of species) of this method of evaluation of risk extinction is generally the most important criterion (i.e., small distribution is a major factor contributing to the status of species for conservation concern) and should be assessed carefully. By just considering the "punctual" (distributional data) distributions were the species have been recorded sub-estimate the species area of distribution, whereas the opposite, over-estimations, can occur by considering the extent of occurrence when species have a disjoint distribution. The approach followed on this paper (Anderson and Martínez-Meyer 2004; Ortega-Huerta and Peterson 2004; Sánchez-Cordero et al. 2005a; Soberón and Peterson 2005) provides a methodology that can be readily applied to most species, and can be more useful for these species poorly sampled or that have a disjoint distribution. More work is already being considered to evaluate all the criteria of the MER (B: habitat conditions, C: The intrinsic biological vulnerability of the species, and D: impact of human activity) for *Polianthes* in order to establish a final conservation status for all the species in this genus (Feria and Solano in prep.).

Richness areas

Due to their restricted ranges, endemic taxa are potentially more sensitive to habitat perturbation, and thereby, are especially vulnerable to extinction; these forms are generally considered critical in indicating areas of importance for conservation action (Peterson and Watson 1998). Endemic taxa with very small ranges are likely to also have small total population sizes, which makes them highly prone to extinction as a consequence of stochastic factors. The fact that most species of *Polianthes* are distributed in the Sierra Madre Occidental and the Transmexican Volcanic Belt has serious conservation implications since natural habitats have been extensively transformed in those regions (Mas et al. 2004; Sánchez-Cordero et al. 2005b). In general, at the national level, between 1976 and 2000 more than 20,000 km^2 of temperate forest, 60,000 km^2 of tropical forest and 45,000 km^2 of scrubland were cleared, which represents an annual average of habitat loss of 90,000,

🕮 Springer

265,000, and 195,000 ha, and rates of deforestation of 0.25%, 0.76% and 0.33% per year, respectively (Mas et al. 2004).

Centers of endemism are identified as areas where the distributions of endemic species overlap (or concentrate), in this case these areas are the areas where the highest *Polianthes* species richness. The degree of endemism in an area is often used as a measure of conservation significance (Brooks et al. 2002), and thus predicting areas with potentially high concentrations of endemic species becomes particularly important in the Neotropic where several areas remain unexplored (Vargas et al. 2004). Although it is well known that the genus *Polianthes* is endemic to Mexico (García-Mendoza 1995; Solano 2000), the present work is the first attempt to understand the current distribution and areas of concentration of *Polianthes* species within Mexico through ecological niche modeling and use this information to inform assessment of risk status for these species. In general, most species have a highly restricted distribution to only one Mexican Biogeographic Province or state of Mexico.

The genus *Polianthes* has the highest concentration of species in the southern Sierra Madre Occidental and the northwest side of the Transmexican Volcanic Belt. This same pattern of species richness has been shown in different groups such as birds (Escalante et al. 1993), mammals (Fa and Morales 1993), grasses (Valdés and Cabral 1993), and Lamiaceae family (especially *Salvia*, Ramamoorthy and Elliott 1993). The patterns of distribution found for this genus agree with the region delimited and named by McVaugh (1989) as Nueva Galicia (see García-Mendoza 1995 and Solano 2000). In particular, the Transvolcanic Belt is considered very important as a biographic barrier for *Polianthes* species dispersal, thus this might be the main factor for which only two species (*Polianthes bicolor* and *P. oaxacana*) occur in Oaxaca (Solano 2000).

Species distribution vs. NPA and TPR

The four species not included in any of the NPA or TPR are highly restricted, implying a high risk of future extinction. The case of *P. howardii* is of particular concern because even though it is considered in the category of special protection by the Mexican official law, it is not included in any NPA or TPR. The results of this study can be used to propose modifications to present protected area systems. In Mexico, NPA are formally protected through a government decree, but the TPR are just proposed regions for conservation without a current legal status. Consequently, without effective management for the TPR, five more species (*Polianthes bicolor*, *P. densiflora*, *P. multicolor*, *P. oaxacana*, and *P. platyphylla*) would not be protected, increasing the risk for the other two already proposed species as in special protections category by the Mexican Official Norm (*P. densiflora* and *P. platyphylla*).

Acknowledgements This paper has been supported by Comision Nacional para el Conocimiento y Uso de la Biodiversidad (Conabio, project FB291/H-230/96), led by the first author. We thank Carlos Correa D., for his kind assistance during field work. We also thank Abisaí García-Mendoza, Bette Loiselle, Daniel Cadena, and Trish Consiglio for their helpful comments on this manuscript. Modeling was developed at GEOMATICA lab at FES Zaragoza, UNAM with the kind assistance of Eliseo Cantellano. Curators at the following herbaria were most helpful with specimen loans: A, CHAPA, ENCB, FEZA, GH, GUADA, HUAA, IEB, MICH, MEXU, NY, RSA, UAMIZ, UAT, US and XAL. Data were also gathered from REMIB.

References

Anderson RP, Martínez-Meyer E (2004) Modeling species' geographic distributions for preliminary conservation assessments: an implementation with the spiny pocket mice (*Heteromys*) of Ecuador. Biol Conserv 116:167–179

Arriaga L, Espinoza-Rodríguez JM, Aguilar-Zúñiga C, Martínez-Romero E, Gómez-Mendoza L, Loa-Loza E, Larson J (2000) Regiones prioritarias terrestres de México. Comisión Nacional para el Conocimiento y Uso de la Biodiversidad, México, DF

Austin MP (2002) Case studies of the use of environmental gradients in vegetation and fauna modelling: theory and practice in Australia and New Zealand. In: Scott JM, Heglund PJ, Morrison ML, Haufler JB, Raphael MG, Wall WA, Samson FB (eds) Predicting species occurrences. Issues of accuracy and scale. Island Press, London, pp 73–82

Brooks TM, Mittermeier RA, Mittermeier CG, da Fonseca GAB, Rylands AB, Konstant WR, Flick P, Pilgrim J, Oldfield S, Magin G, Hilton-Taylor C (2002) Habitat loss and extinction in the hotspots of biodiversity. Conserv Biol 16:909–923

Brown JH, Stevens GC, Kaufman DM (1996) The geographic range: size, shape, boundaries, and internal structure. Annu Rev Ecol Syst 27:597–623

Elith J, Graham C, Anderson RP, Dudik M, Ferrier S, Guisan A, Hijmans RJ, Huettmann F, Leathwick JR, Lehmann A, Li J, Lohmann LG, Loiselle BA, Manin G, Moritz C, Nakamura M, Nakazawa Y, Overton JMcC, Peterson AT, Phillips SJ, Richardson KS, Scachetti-Prereira R, Schapire RE, Soberón J, Williams S, Wisz MS, Zimmermann NE (2006) Novel methods improve prediction of species' distributions from occurrence data. Ecography 29:129–151

Escalante PP, Navarro SA, Peterson AT (1993) A geographic, ecological and historical analysis of land bird diversity in Mexico. In: Ramamoorthy TP, Bye R, Lot A, Fa J (eds) The biological diversity of Mexico: origins and distribution. Oxford Press, New York, USA, pp 281–307

ESRI (1999) ArcView GIS ver 3.2. Environmental System Research Inc., USA

Fa JE, Morales LM (1993) Patterns of mammalian diversity in Mexico. In: Ramamoorthy TP, Bye R, Lot A, Fa J (eds) The biological diversity of Mexico: origins and distribution. Oxford Press, New York, USA, pp 319–361

Feria ATP, Peterson AT (2002) Prediction of bird community composition based on point-occurrence data and inferential algorithms: a valuable tool in biodiversity assessments. Divers Distrib 8:49–56

García-Mendoza A (1995) Riqueza y endemismos de la familia Agavaceae en México. In: Linares E, Chiang F, Bye R, Elias TS (eds) Conservación de Plantas en Peligro de extinción: Diferentes Enfoques. Instituto de Biología, Jardín Botánico, UNAM, México, DF, pp 51–75

García-Mendoza A (2004) Agaváceas. In: García-Mendoza AJ, Ordoñez MJ, Briones-Salas M (eds) Biodiversidad de Oaxaca. Instituto de Biología, UNAM-Fondo Oaxaqueño para la Conservación de la Naturaleza-Worldlife Fund, México, DF, pp 159–169

Guisan A, Zimmermann NE (2000) Predictive habitat distribution models in ecology. Ecol Model 135:147–186

Illoldi-Rangel P, Sánchez-Cordero V, Peterson AT (2004) Predicting distribution of Mexican mammals using ecological niche modeling. J Mammal 85:658–662

Instituto Nacional de Ecología (1999) Red del Sistema Nacional de Áreas Naturales Protegidas. México, DF

IUCN (1997) Cactus and succulent plants. Status survey and conservation plan. Sara Oldfield (comp). Gland, Switzerland and Cambridge, UK

Lim BK, Peterson AT, Engstrom MD (2002) Robustness of ecological niche modeling algorithms for mammals in Guyana. Biodivers Conserv 11:1237–1246

Lomolino MV, Riddle BR, Brown JH (2005) Biogeography. 3rd edn. Sinauer Associates, Inc, Sunderland, MA

Martínez-Meyer E, Peterson AT, Navarro Sigüenza AG (2004) Evolution of seasonal ecological niches in the Passerina buntings (Aves: Cardinalidae). Proc Roy Soc Lond B 271:1151–1157

Mas JF, Velásquez A, Díaz-Gallegos JR, Mayorga-Saucedo R, Alcantara C, Bocco G, Castro R, Fernández T, Pérez-Vega A (2004) Assessing land use/cover changes: a nationwide multidate spatial database for Mexico. Int J Appl Earth Observ Geoinform 5:249–261

McVaugh R (1989) Liliaceae. In: Anderson WR (ed) Flora Novo-Galiciana. A descriptive account of the vascular plants of western Mexico, vol.15. Annals arboretum. The University of Michigan Herbarium, USA, pp 20–293

Morrone JJ, Espinosa-Organista D, Llorente-Bousquets J (2002) Mexican biogeographic provinces: preliminary scheme, general characterizations, and synonymies. Acta Zool Mex 85:83–108

⑴ Springer

Nakazawa Y, Peterson AT, Martínez-Meyer E, Navarro-Sigüenza AG (2004) Seasonal niches of Neartic-Neotropical migratory birds: implications for the evolution of migration. Auk 121:610–618

Olson ME, Lomeli JA, Cacho I (2005) Extinction treat in the *Pedilanthus* clade (*Euphorbia*, Euphorbiaceae), with special reference to the recently rediscovered *E. conzatti* (*P. pulchellus*). Am J Bot 92:634–641

Ortega-Huerta MA, Peterson AT (2004) Modelling spatial patterns of biodiversity for conservation prioritization in North-eastern Mexico. Divers Distrib 10:39–54

Peterson AT (2001) Predicting species' geographic distribution based on ecological niche modelling. Condor 103:599–605

Peterson AT, Cohoon KC (1999) Sensitivity of distributional prediction algorithms to geographical data completeness. Ecol Model 117:159–164

Peterson AT, Vieglais DA (2001) Predicting species invasions using ecological niche modelling. BioScience 51:363–371

Peterson AT, Watson M (1998) Problems with aerial definitions of endemism: the effects of spatial scaling. Divers Distrib 4:189–194

Peterson AT, Ball LG, Cohoon KC (2002a) Predicting distributions of tropical birds. Ibis 144(online):E27–E32

Peterson AT, Ortega-Huerta M, Bartley G, Sánchez-Cordero V, Soberón J, Buddemeier RH, Stockwell D (2002b) Future projections for Mexican faunas under global climate change scenarios. Nature 416:626–629

Ramamoorthy TP, Elliot M (1993) Mexican Lamiaceae. In: Ramamoorthy TP, Bye R, Lot A, Fa J (eds) The biological diversity of México: origins and distribution. Oxford Press, New York, USA, pp 513–539

Rapoport EH (1982) Aerographic-geographic strategies of species. Pergamon Press, New York

Raxworthy CJ, Martínez-Meyer E, Horning N, Nussbaum RA, Schneider GE, Ortega-Huerta MA, Peterson AT (2003) Predicting distributions of known and unknown reptile species in Madagascar. Nature 426:837–841

Sánchez-Cordero V, Cirelli V, Munguia M, Sarkar S (2005a) Place prioritization for biodiversity representation using species' ecological niche modeling. Biodivers Inform 2:11–23

Sánchez -Cordero V, Iloldi-Rangel P, Linaje M, Sarkar S, Peterson AT (2005b) Deforestation and extant distributions of Mexican endemic mammals. Biol Conserv 126:465–473

SEMARNAT (Secretaría de Medio Ambiente y Recursos Naturales) (2002) Norma Oficial Mexicana NOM-059-ECOL-2001, Protección ambiental-Especies nativas de México de flora y fauna silvestres-Categorías de riesgo y especificaciones para su inclusión, exclusión o cambio-Lista de especies en riesgo. Diario Oficial de la Federación. 6 de marzo de 2002, Primera sección, México, DF

Soberón J, Peterson AT (2005) Interpretation of models of fundamental ecological niches and species' distributional areas. Biodivers Inform 2:1–10

Solano CE (2000) Sistemática del género *Polianthes* L. (Agavaceae). Tesis de Doctorado. Facultad de Ciencias, División de Estudios de Posgrado, UNAM, México, DF

Stockwell D, Noble R (1992) Induction of sets of rules from animal distribution data a robust and informative method of data analysis. Math Comput Simul 33:385–390

Stockwell D, Peterson AT (2002) Effects of sample size on accuracy of species distribution models. Ecol Model 148:1–13

Stockwell D, Peterson AT (2003) Comparison of resolution of methods for mapping biodiversity patterns from point-occurrence data. Ecol Ind 3:213–221

Valdés RJ, Cabral CI (1993) Chronology of Mexican grasses. In: Ramamoorthy TP, Bye R, Lot A, Fa J (eds) The biological diversity of Mexico: origins and distribution. Oxford Press, New York, USA, pp 439–446

Vargas JH, Consiglio T, Jørgensen PM, Croat TB (2004) Modelling distribution patterns in a species-rich plant genus, *Anthurium* (Araceae), in Ecuador. Divers Distrib 10:211–216

Willis F, Moat J, Paton A (2003) Defining a role for herbarium data in Red list assessments: a case study of *Plectranthus* from eastern and southern tropical Africa. Biodivers Conserv 12:1537–1552

Biodivers Conserv (2007) 16:1901–1915
DOI 10.1007/s10531-006-9097-7

ORIGINAL PAPER

The uses, local perceptions and ecological status of 16 woody species of Gadumire Sub-county, Uganda

John R. S. Tabuti

Received: 12 August 2005 / Accepted: 5 July 2006 / Published online: 27 October 2006
© Springer Science+Business Media B.V. 2006

Abstract Populations of naturally growing woody species valued for their contribution to human livelihoods are threatened with extinction. Most at risk are those existing in human inhabited areas outside protected areas that are subjected to high population pressure and to a variety of land use demands. The sustainable utilization of these plants requires as a first step knowledge, including, their ecology and an understanding of the peoples attitudes to conservation. This study was conducted to generate data that would contribute to the management for conservation and sustainable use of woody resources. The study objectives were to document local knowledge covering the uses, status, threats, habitats and management solutions of woody species; determine the abundances, distribution and population structure of 16 woody species, and assess the conservation status of the selected woody species. The study was carried out in Gadumire Sub-county, Uganda using both an ethnobotanical approach and quantitative ecological methods. The species are multipurpose and are exploited to satisfy different subsistence needs. They had population densities ranging between 3.6 and 2630 individuals ha^{-1}, and distributions ranging between 0.3 and 39.5%. The species *Acacia hockii*, *Albizia zygia*, *Acacia seyal*, *Markhamia lutea* and *Albizia coriaria* had a good conservation status. The remainder of the species appear threatened either because they had low densities, frequencies or less steep size class distribution (SCD) slopes. *Securidaca longipedunculata* Fres. was not encountered at all in the study plots. Community perceptions collaborated the measured population dynamics. The major threats believed to be impacting the species by the community are the growing human population, expanding crop agriculture, poor harvesting methods and over-exploitation of the species.

Keywords Ethnobotany · Harvesting patterns · Population structure · Savanna woodland

J. R. S. Tabuti (✉)
Department of Botany, Makerere University, P.O. Box 7062, Kampala, Uganda
e-mail: jtabuti@botany.mak.ac.ug

[327]

Introduction

Many households in tropical areas depend on wild plant parts or products gathered from wild management systems. The rural poor are especially dependant on wild plants for their subsistence. They harvest products such as firewood, construction material, food, fodder and browse fodder for their wellbeing (Walter 2001; Tabuti et al. 2003a, b, c; Tabuti et al. 2004; Ticktin 2004). Furthermore, activities related to wild plants contribute to household incomes by providing employment opportunities and cash incomes from the sale of some of the products (Shackleton et al. 1998; Walter 2001). Livelihoods dependant on gathered plants or their parts are threatened by the widely acknowledged on-going loss of plant diversity. This is especially true of wild plants found in human inhabited areas outside protected areas that are subject to high population pressure and various land use demands. These areas experience habitat degradation coming as a consequence of overgrazing, increasing crop agriculture or logging (Walter 2001; Dalle et al. 2002; National Environment Management Authority 2002; Kaimowitz et al. 2004). Furthermore plants outside protected areas are routinely over-exploited. Woody plants are especially vulnerable to human related impacts because they are slow growing (Cunningham 1993; Aumeeruddy 1994; Schippmann et al. 2002).

In order to conserve vulnerable plants for sustainable utilization it is necessary to have information on aspects such as the effect of human activities on the populations of the target plant species, their ecology (Peters 1999; Dalle et al. 2002), and local harvesting methods, existing threats and attitudes to plants conservation. The measurement of ecological status is straightforward (Hall and Bawa 1993; Lykke 1998) however the investigation of human influence on plants is complicated by the need to have long-term monitoring data on population trends. There are few species for which such long-term data exists. In the absence of this information, population dynamics are commonly inferred from population structure data. Population structures are easy to assess from single survey size class frequency distributions (Hall and Bawa 1993; Cunningham 2001; Obiri et al. 2002). From this type of data, preliminary indications of how plant populations may be affected by extractive activities or other land-uses can be attained (Hall and Bawa 1993; Lykke 1998; Peters 1999; Obiri et al. 2002; Dalle et al. 2002).

This study was conducted in Gadumire Sub-county with the specific objectives of: (1) documenting local knowledge covering the uses, status, threats, habitats and management solutions of woody species; (2) measurement of the abundances, distribution and population structure of selected woody species, and (3) assessing the conservation status of selected woody species. Some uses, but not all, of the study species were documented in an earlier study (Tabuti et al. 2003a, b, c; Tabuti et al. 2004); this current study concluded the inventory of uses. It is expected that this information will help clarify decisions for management of woody plant resources for sustainable use.

Study area

Gadumire Sub-county is one of the five sub-counties that together make up Bulamogi County (The County was upgraded to district status in 2005). It is located 200 km north-east of Kampala, the capital city of Uganda, 33°30′ – 33°35′ E and

1°04′ – 1°15′ N at an altitude of 1030–1045 m a.s.l. (Government of Uganda 1963). It is made up of five parishes, namely, Gadumire, Kisinda, Lubulo, Bupyana and Panyolo.

According to Langdale-Brown et al. (1964), *Albizia-Combretum* woodlands and *Cyperus papyrus* swamp vegetation types dominate in the Sub-county. The most extensive soil type of Gadumire is the Mazimasa complex of catenas. This soil type is usually a shallow grey or brown sandy loam on laterite base rock (Ollier and Harrop 1959; Department of Land and Survey 1962). Other types are the mineral and organic hydromorphic soils influenced by permanent or seasonal water logging.

The Sub-county has an estimated population of 22,344 people and a population density of approximately 250 people/km². The population of Bulamogi is growing at a rate of 3% per annum (Uganda Bureau of Statistics 2005), and has more than tripled in the last four decades from an estimated 60–95 people/km² in 1962 to 180 people/km² by 2002 (Department of Land and Survey 1962; Uganda Bureau of Statistics 2005). The people of Gadumire are a rural peasant community whose main source of livelihood is mixed agriculture. They cultivate a variety of annual and perennial crops and keep livestock including cattle, goats and sheep.

Methods

Ethnobotanical study

Field work was carried out between August 2003 and July 2004. The study methods consisted of both an ethnobotanical study and a quantitative ecological study. This study began by holding a focus group discussion with community members using a checklist of questions. Sixteen participants (including 4 women) were selected with the help of the local area leaders to participate in the discussion. At least one participant was selected from each of the parishes of the Sub-county. All participants were over 25 years of age. During the group discussion, consensus was reached on the study species (Table 1) with regard to their uses; their preferred habitats; threats;

Table 1 Study species including their growth habit and management status. All species are indigenous apart from *Mangifera indica* L

Species	Growth habit	Management status
Acacia hockii De Wild.	Tree	Wild
Acacia seyal Del. var. *fistula* (Schweinf.) Oliv.	Tree	Wild
Albizia coriaria Oliv.	Tree	Wild
Albizia zygia (DC.) Macbr.	Tree	Wild
Antiaris toxicaria Lesch.	Tree	Wild
Carissa edulis Vahl	Shrub	Wild
Combretum collinum Fresen. subsp. *elgonense* (Exell) Okafor	Tree	Wild
Ficus natalensis Hochst.	Tree	Cultivated
Mangifera indica L.	Tree	Semi-cultivated
Markhamia lutea (Benth.) K. Schum.	Tree	Semi-cultivated
Milicia excelsa (Welw.) C.Berg	Tree	Wild
Sarcocephalus latifolius (Smith) Bruce	Shrub	Wild
Sclerocarya birrea (A. Rich.) Hochst.	Tree	Wild
Securidaca longipedunculata Fres	Shrub/Tree	Wild
Steganotaenia araliacea Hochst.	Shrub	Wild
Tamarindus indica L.	Tree	Wild

🙂 Springer

which species populations were declining or growing; which species populations were rare or abundant; and possible management solutions to protect them. The focus group discussion data was analysed and responses grouped into classes alongside the major themes of the inquiry here.

Ecological survey

The quantitative ecological study was aimed at determining the population structure, density and distribution of the study species. To achieve these objectives, the following villages were randomly selected from the five parishes that make up Gadumire Sub-county: Mpambwa (64), Bugonya (48), Nawandyo (42), Busiiro (40) Kisinda (38), Bukayale A (38), Bukayale B (37), Nyende-Igembe (35), Nabweyo (34), Nansohera (34), Kavule (31), Kyamba (31), Nandele (30), Kibale (30), Wataka (5). The numbers in parenthesis indicate study plots that were systematically placed within each village. The first plot was randomly placed. Thereafter each suceeding plot was systematically placed 50 m from the preceding one along a transect running in a south to north compass direction. Transects were followed for a maximum distance of 2000 m before a new transect would be laid out. The new transect was then laid 500 m from the preceding one.

Each study plot was nested i.e., divided up into differently sized quadrats. Within the largest quadrat of 20×50 m, large trees (≥ 30 cm dbh) were sampled; small trees (> 15 cm dbh < 30 cm dbh) were sampled in a 20×25 m quadrat placed in the larger quadrat; poles (> 5 cm dbh < 15 cm dbh) in a 10×25 m quadrat; and shrubs/saplings (> 1 cm dbh < 5 cm dbh) in a 10×12 m quadrat. Seedlings (< 1 cm dbh or < 1 m high) were sampled in 1 m sq plot randomly placed in the larger plot.

In each plot individuals of the study species encountered, were enumerated and their diameter at breast height (dbh) measured and recorded. All stumps and coppices of individuals of the study species encountered in the field were treated similarly. The soil types and land use types were also recorded according to the local classification.

Data analysis

Population structure

Data from all the study plots (537) was pooled together for the analysis. Diameter at breast height data were grouped into 27 size classes: 0–0.9, 1–1.9, 2–2.9, 3–4.9, 5–6.9, 7–8.9, 9–10.9, 11–13.9, 14–16.9, 17–19.9, 20–23.9, 24–27.9, 28–31.9, 32–35.9, 36–39.9, 40–43.9, 44–47.9, 48–51.9, 52–55.9, 56–59.9, 60–69.9, 70–79.9, 80–89.9, 90–99.9, 100–109.9 and 110–129.9 cm. The last size class included all individuals ≥ 130 cm. The class widths were made progressively wider as dbh increased from individuals with small dbh to those with larger dbh. This was done to balance samples across size classes because the number of individuals generally declines with size class (Condit et al. 1998).

For each species, the number of individuals per hectare (N_i) in each size class was plotted against class midpoint (d_i). In addition, for each species, a least-squares linear regression slope was calculated for the size class distribution (SCD) using the software SPSS for windows. The size-class midpoint (d_i) was treated as the independent variable and the density (N_i) in each size class as the dependent variable. In order to get straight line plots the N_i was transformed by ln (N_i + 1) because some classes had zero individuals. The regression was calculated between ln (d_i) and ln (N_i + 1) (see Condit et al. 1998). This regression model was chosen over the ln (N_i) vs. d_i used by Lykke (1998) and Obiri et al. (2002) because it explained most of the variation around the regression line for 10 of the 15 species ($r^2 \geq 0.5$).

When computing the slope, only size classes up to the largest size class with individuals were included. Larger size classes without individuals were omitted. The SCD slope was used to summarize in a single number the shape of the SCD for each species (Condit et al. 1998). In this analysis negative slopes (indicative of reverse J-shaped SCD) imply stable populations that are naturally replacing themselves. While weak-negative slopes and flat slopes are taken to mean poor recruitment and declining populations (Hall and Bawa 1993; Lykke 1998; Obiri et al. 2002). Conclusions about the population structures for the different species were made on the basis of the SCD slopes and the graphical SCD plots.

Harvesting index

The influence of a tree species and stem sizes on the likelihood that an individual of a species would be harvested was estimated by a harvesting index (HI), calculated as the proportion of harvested stems (stumps ha^{-1}) to the harvestable stems (stumps and standing trees ha^{-1}) (Obiri et al. 2002). The HI was calculated only for species that had more than 10 stumps.

Species distribution and environmental variables

The relationship between species distribution and the environmental variables of habitats or land use types, and soil types were investigated by the Canonical Correspondence Analysis procedure using the software PC-ORD for windows version 4.2. Canonical ordination techniques help detect patterns of variation in species data that can be explained by environmental variables. The resulting ordination diagrams expresses both patterns of variation in species composition and the main relationships between the species and each of the environmental variables (Jongman et al. 1995).

Results

Community knowledge of uses of woody species, their status, habitats and threats

Uses of the species

The study species have multiple uses; *Ficus natalensis* had the highest number of uses (14) and*Securidaca longipedunculata* the least reported uses (2). Parts of the species are harvested for use in treating and feeding of both humans and livestock, to provide energy for cooking (firewood and charcoal), and for construction materials (Table 2). The species also have spiritual value. Field observations revealed that some plant products such as timber, charcoal, firewood, crafts and mangoes are sold in the community.

Table 2 Reported uses; major uses are denoted with ⊗ (Hm = human medicine, Ev = ethnoveterinary medicine, Fw = firewood, Char = charcoal, Spir = Home for spirits or other spiritual use)

Species	Hm	Ev	Food	Fo	Fw	Char	Timber	Construction	Crafts	Spir.	Minor uses[a]
Carissa edulis	⊗	×	×	×	×	×			×		a
Milicia excelsa	×	×	×	×	×	×	⊗	×	×	×	b, c, d, e, f, g
Acacia hockii	×	×		×	×	×		⊗	×		c
Albizia coriaria	×	×		×	×		⊗	×	×	×	i, j, b, k
Steganotaenia araliacea	⊗	×		×							
Acacia seyal	×	×			⊗	×		×			a, h
Sarcocephalus latifolius	⊗	×								×	l
Securidaca longipedunculata	⊗	×									
Antiaris toxicaria	×	×				×					m, n
Mangifera indica	×	⊗	×	×	×					×	c, w
Tamarindus indica	×	⊗	×	×	×	×				×	b, c
Sclerocarya birrea	×	⊗			×	⊗					
Albizia zygia	×		×	×	×	×		⊗	×	×	k, c
Combretum collinum	×		×	⊗	×						o
Ficus natalensis	×		×	⊗						×	b, c, h, n, p, q, r, s, t, u
Markhamia lutea	×		×	×	×	×		⊗	×	×	c, v

[a] a = Thorn used to extricate jiggers; b = Shade for crop plants e.g., bananas, coffee; c = Shade for people; d = rain formation; e = protect environment; f = fertiliser; g = season indicator (it shades its leaves at onset of dry season); h = Bark burnt to chase snakes; i = *Nkonkolo* (tool hit on anthills to force edible white ants out); j = Bark smoked to chase mosquitoes; k = Bark used to wash clothes; l = Dried flowers are burnt to produce smoke that forces edible white ants out of anthill; m = ignition wood used for lighting cooking fires; n = Bark used to make textile; o = ferment local beer; p = fibre used to make ropes (*ebibohe*); q = demarcate clan land; r = leaves used to make a "spoon" that is used when drinking porridge or to dispense medicine (*kadandi*); s = collect rain water; t = used by 'traditional priests' during traditional religious rituals; u = planted on graves of princes; v = ornamental plant; w = used to make juice and local gin (*waragi*)

Traditional landscape classification and species distribution

The people of Gadumire recognize the following habitats: *Mutala* = characterized by red well-drained soils, *Lusenyi* = characterized by sandy soils, *Kibali/Mui-ga* = characterized by seasonally or permanently water logged soils, and *Mabale* = rocky outcrops. Common landuse types in the area include: *Maka* = homestead, *Nimilo* = cultivated land, *Kisambu* = old abandoned garden field (1–3-year rest), *Kihindila* = fallow land (land of 5 years or more but before it has reverted to scrubland). The known vegetation types are: *Kigoola* = Scrub-land/Shrubland, *Kibila* = closed scrub or tree forest, *Muiga* = swamp, and *Kisaka* = thicket. The people also classify soils into three types: *mutala* (red laterite well drained soils), *lusenye* (sandy loams) and *lumansi* (clayey seasonally water logged soils).

According to community knowledge, the species *Sarcocephalus latifolius* prefers thickets, while *Carissa edulis* prefers ant hills. Others namely, *Acacia hockii*,

Table 3 Local community knowledge of distribution of plants in different land and soil types

Species	Preferred habitat/soil types	Remarks
Acacia hockii	Everywhere	
Mangifera indica	Everywhere	
Albizia zygia	Everywhere	Coppices easily and sprouts readily from roots
Markhamia lutea	Everywhere	Germinates readily but is overexploited.
Tamarindus indica	Everywhere	
Milicia excelsa	Everywhere	
Steganotaenia araliacea	Well drained soils	
Combretum collinum	Well drained soils	Threatened by crop cultivation
Antiaris toxicaria	Well drained soils	
Sclerocarya birrea	Well drained soils	Seeds do not readily germinate
Albizia coriaria	Water logged or sandy loams	Protected by community where found
Acacia seyal	Water logged or well drained soils	Threatened by use as charcoal, does not germinate easily (sets few seeds), overexploited as herbal medicines.
Sarcocephalus latifolius	Well drained soils or brushland	Has slow regeneration
Securidaca longipedunculata	Well drained soils or sandy loams	Over exploited
Carissa edulis	Ant hill (water logged soils [a])	
Ficus natalensis	Homesteads	Cultural use as bark cloth is declining. As a consequence there is declining cultivation of the species.

[a] Personal observation

Mangifera indica, Albizia zygia, Markhamia lutea, Tamarindus indica, Milicia excelsa had no preferred habitats and could grow anywhere (Table 3).

Factors responsible for species population decline

Participants in the Focus Group Discussion stated that some of the study species had declined in their abundance and distributions (Table 4), and attributed such declines to: a growing human population associated with over-exploitation of the species to satisfy growing subsistence needs, and for sale; poor harvesting methods such as the harvesting of large quantities of bark and roots of *Sarcocephalus latifolius* for use as herbal medicine; conversion of woody plant habitats to other land uses such as crop agriculture or construction of homesteads; and use of bush fires to clear land for crop agriculture, affecting species that are poorly adapted to withstand fires like *Steganotaenia araliacea*.

Other factors mentioned were loss of traditional spiritual and cultural values. For instance the traditional use of bark cloth fabricated from the bark of *Ficus natalensis* has waned in the community leading to lowered cultivation of the species. Other factors mentioned were: short resting periods of agricultural land (fallows); the attitude that naturally regenerating species do not require propagating; live-stock grazing; and poor recruitment from seed by *Sclerocarya birrea* and *Acacia seyal*.

 Springer

Table 4 Size class distribution least-squares regression slopes for the study species. Also shown are the community opinions regarding abundances and changes in population sizes of the study species

Species name	Slope	r^2	P value	Community opinions	
				Availability[a]	Changes in population sizes[b]
Group I					
Carissa edulis	− 2.0420	0.98	0.087	R	D
Acacia hockii	− 1.8600	0.83	0.000	VA	I
Sarcocephalus latifolius	− 1.2150	1.00	–	VR	D
Acacia seyal	− 1.0500	0.61	0.000	A	D
Albizia zygia	− 1.0340	0.62	0.000	A	I
Group II					
Markhamia lutea	− 0.8740	0.63	0.000	R	D
Combretum collinum	− 0.8380	0.66	0.001	R	D
Mangifera indica	− 0.7390	0.59	0.000	VA	I
Albizia coriaria	− 0.7180	0.56	0.000	A	I
Group III					
Antiaris toxicaria	− 0.6530	0.51	0.000	R	I
Steganotaenia araliacea	− 0.6050	0.93	0.035	R	D
Tamarindus indica	− 0.4710	0.39	0.001	R	D
Sclerocarya birrea	− 0.4440	0.54	0.010	VR	D
Milicia excelsa	− 0.3700	0.52	0.000	VR	D
Group IV					
Ficus natalensis	− 0.0518	0.31	0.004	R	D

[a] A = abundant; R = rare; VA = very abundant; VR = very rare
[b] D = declining; I = increasing

Ecological measurements

Population density and frequency

Among the studied species, the species with the highest densities were *Acacia hockii* followed by *Albizia zygia, Acacia seyal* and *Albizia coriaria* in that order (Fig. 1a). On the other hand *Sclerocarya birrea, Sarcocephalus latifolius* and *Ficus natalensis* were quite rare. *Securidaca longipedunculata* was not encountered at all in all the study plots.

Frequency for the species ranged from a low of 0.3% to a high of 39.5%. Apart from *Albizia coriaria* and *Milicia excelsa*, the majority of species under study were observed in less than 15% of the study plots (Fig. 1b). The Canonical Correspondence Analysis showed that the study plants avoided seasonally flooded habitats and water logged soils in preference to well drained soils (Fig. 2, 3). However, *Acacia hockii* and *Acacia seyal* appeared to thrive on water logged soils. These results are in agreement with the community perceptions on the species distribution. The cultivated species *Ficus natalensis* was associated with cultivated fields, and *Markhamia lutea* and *Mangifera indica* with homesteads.

Population structure

The study species were divided into four groups basing on their SCD regression slopes. Group I includes *Carissa edulis, Acacia hockii, Sarcocephalus latifolius, Acacia seyal*, and *Albizia zygia*. These species had the most negative SCD slopes

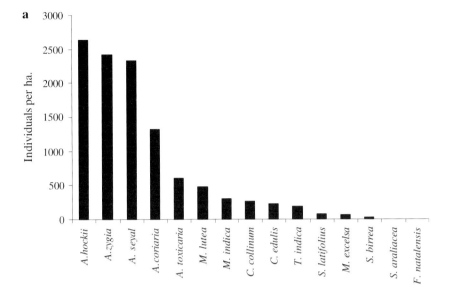

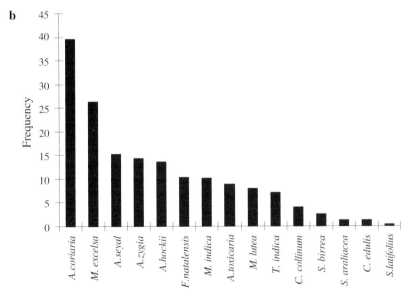

Fig. 1 (**a**) Densities (stems ha^{-1}) of the study species, (**b**) Distribution (frequency) of the study species. The densities of *Steganotaenia araliacea* and *Ficus natalensis* are 3.9 and 3.6 stems ha^{-1} respectively

ranging between − 1.034 and − 2.042 (Table 4). Group II includes *Markhamia lutea*, *Combretum collinum* Fresen. subsp. *Elgonense*, *Mangifera indica* and *Albizia coriaria*. This group exhibited a reverse J-shaped SCD but with weaker negative slopes than species in group I (− 0.718 and − 0.874). Group III includes *Antiaris toxicaria*, *Steganotaenia araliacea*, *Tamarindus indica*, *Sclerocarya birrea*, *Milicia excelsa*. Group IV contains only *Ficus natalensis*. This species had an almost flat slope − 0.0518. The SCD plots (Fig. 4) show that species in groups I and II have

🍁 Springer

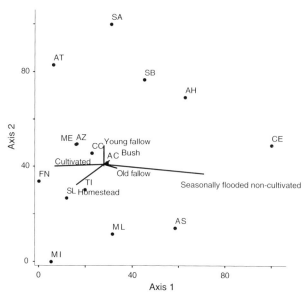

Fig. 2 Relationship between species distribution, and habitat types and land use types (AH = *Acacia hockii*; CE = *Carissa edulis*; AZ = *Albizia zygia*; AS = *Acacia seyal*; ML = *Markhamia lutea*; AC = *Albizia coriaria* Oliv.; AT = *Antiaris toxicaria*; CC = *Combretum collinum*; SL = *Sarcocephalus latifolius*; ME = *Milicia excelsa*; SB = *Sclerocarya birrea*; TI = *Tamarindus indica*; SA = *Steganotaenia araliacea*; MI = *Mangifera indica* L.; FN = *Ficus natalensis*)

Fig. 3 Relationship between species distribution and soil types. Abbreviations as in Fig. 2 (translation of soil types are in the text)

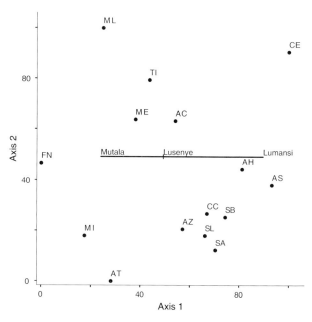

reverse J-shaped plots with relatively more individuals in the small size classes (juveniles) compared to those in the large size classes (mature/reproductive); a trend that is characteristic of species with good and stable regeneration.

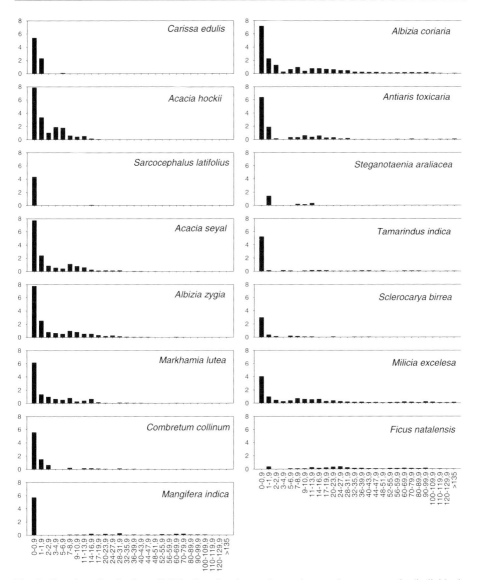

Fig. 4 Size class distributions (SCD) plots for the study species. y-axis represents ln (individuals ha^{-1} + 1) and the x-axis the diameter at breast height (cm) size classes

There was good agreement between local knowledge and results from the quantitative plant study with regard to plant population dynamics. The species for which the analysis indicated a healthy population structure, accompanied with relatively higher densities and wider distributions, were also known to be abundant and/or increasing by the local community (Table 4). Contrariwise those species known by the community to be scarce and declining had relatively weaker SCD slopes, low densities and frequencies.

All species except *Carissa edulis* and *Sarcocephalus latifolius* had coppices (Fig. 2). *Sclerocarya birrea* had an average of five coppices per stump; *Markhamia*

🖄 Springer

lutea had three; while *Acacia hockii*, *Albizia coriaria*, *Albizia zygia*, *Milicia excelsa*, *Antiaris toxicaria* had two. The rest of the study species had one coppice each per stump.

Harvesting patterns

All the species studied here are harvested destructively apart from *Steganotaenia araliacea* for which leaves are used. Stumps of all study species, excepting *Carissa edulis* and *Sarcocephalus latifolius* were found in the study area. Species with the primary use, or which possess a secondary uses, of producing firewood, charcoal or construction materials are felled down to get the required wood. For these end-uses large sized stems are harvested in preference over small sized ones (Fig. 5). Roots of *Carissa edulis* and *Sarcocephalus latifolius* are extensively harvested to produce herbal medicines.

Further to destructive harvesting, plants are also destroyed to clear land for cultivation. *Combretum collinum*, *Albizia zygia* and *Acacia seyal* had many stumps in the small size classes in areas recently cleared for cultivation. Out of the 136 plots with stumps, 113 (83%) were on lands undergoing use by community members viz cultivated fields (67), young fallows (31), cleared fields (7) and homesteads (8) (Fig. 6). This distribution highlights the negative impacts of crop cultivation on species survival.

Discussion

Importance of the study species

The woody species studied here are important to the local community, firstly the products and services acquired from them satisfy subsistence as well as cultural

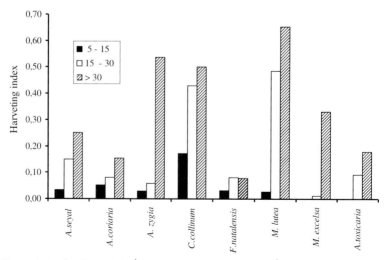

Fig. 5 Harvesting index (stumps ha⁻¹/stumps and standing trees ha⁻¹) calculated for species with 10 or more stumps. Solid bars (Dbh 5–15 cm), open bars (15–30 cm) and > 30 cm (hatched bars)

🙶 Springer

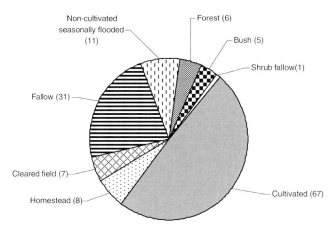

Fig. 6 Distribution of stumps in different land uses and vegetation types

needs. Besides which, their ready availability helps households to not only satisfy their consumptive needs, but to also escape from poverty (Uganda Participatory Poverty Assessment Process 2002). In a community like Gadumire, the availability of firewood, for instance, on which people depend as a source of energy (National Environment Management Authority 2002; Tabuti et al. 2003b) enables households to save money they would otherwise have spent on its purchase. This may make possible for such households to invest in the education of their children, and provide them with opportunities to escape from poverty. Secondly, whenever possible parts or products from these plants are sold to earn cash incomes. People commonly sell timber, firewood, charcoal, crafts or gathered foods like mangoes. Cash incomes from the sale of plant products can supplement subsistence agriculture and are particularly significant for rural women (Konstant et al. 1995; Shackleton et al. 1998). However, the monetary contribution of plant products to the local economy is not known and needs to be studied and quantified. Overall, the continued survival of the species studied here is important to the livelihoods of the local community of Gadumire Sub-county and their loss could result in much hardship for the communities that have traditionally exploited them

Conservation status

The continued survival into the future of the study species was evaluated against their current densities (availability of the species), frequencies (extent of their distribution) and magnitude of SCD regression slope (population structure). The calculated indices were complemented by results from the Focus Group Discussion. Species that are readily available or abundnant, are widely distributed or have relatively more juvenile individuals than mature ones have better chances of regenerating into future generations and vice versa (Cunningham 1993; Primack 1998). Five species *Acacia hockii, Albizia zygia, Acacia seyal, Markhamia lutea* and *Albizia coriaria* appear to have the potential to survive under the prevailing harvest regimes and land use management practices. *Markhamia lutea*, though poorly distributed, is included in the species not immediately threatened because it is actively managed as a cultivated species. The remainder of the species ($n = 11$) appear to be

ℒ Springer

locally threatened because they are either not readily available in the study area (e.g., *Securidaca longipedunculata*), are restricted in their distribution (e.g., *Carissa edulis*) or have a weak population structure (e.g., *Ficus natalensis*).

According to the Focus Group Discussion, the growing human population of Gadumire is seen as the major threat to species survival. It is widely believed that growing human populations are significant drivers of negative environment changes (National Environment Management Authority 2002; Uganda Participatory Poverty Assessment Process 2002; Rosa et al. 2004; Twine 2005). The human population in the study area has been growing at a rate of 3% per annum (Uganda Bureau of Statistics, 2005) and has had the effect of forcing the community to allocate most of the available land to crop agriculture and to settlement. Land devoted to agricultural cropping doubled in the period 1962 to 1997 from 34.2% to 67.4% (Department of Land and Survey 1962; Forest Department 1997). The relationship between a growing human population, decisions to allocate land to crop agriculture and to settlements and its influence on woody species survival is illustrated in part by the relatively large number of small sized stumps of *Combretum collinum* stumps encountered in the two land types. As there is no reported use for juvenile plants of *C. collinum*, it can be assumed that they were destroyed to clear land for crop agriculture.

Besides the growing human population and related land use demands, over-exploitation is another threat thought to impact on plants' survival. Over-exploitation has had an impact on the survival of the prized timber species *Milicia excelsa*. The decline in *M. excelsa* has led to intense harvest pressure on *Antiaris toxicaria* as a substitute. On its part *Sarcocephalus latifolius* has been over-exploited for its bark and root used in herbal medicines.

Participants in the focus group discussion suggested that the conservation of woody species would be enhanced by protecting them where found or by planting. They also suggested that there was need to conduct awareness workshops to sensitize the community on the values of plant diversity and its conservation. And lastly, that there should be some effort to conserve cultural values.

Acknowledgements Permission to carry out this research was granted by the Uganda National Council for Science and Technology (Research No. EC 606). Funding for the study was provided by NORAD. The community of Gadumire Sub-county is acknowledged for collaborating on this study. The following people A-M. Lykke, J.M. Kasenene, P. Mucunguzi, J. Kalema, H. Tushabe and two anonymous referees are thanked for their comments on different aspects of the research. P. Sebulime assisted with field work.

References

Aumeeruddy Y (1994) Local representations and management of agroforests on the periphery of Kerinci Seblat National Park, Sumatra, Indonesia, People and plants working paper 3. Paris, UNESCO

Condit R, Sukumar R, Hubbell SP, Foster RB (1998) Predicting population trends from size distributions: a direct test in a tropical tree community. Am Nat 152:495–509

Cunningham AB (1993) African medicinal plants: setting priorities at the interface between conservation and primary health care. People and plants working paper 1. Paris, UNESCO

Cunningham AB (2001) Applied ethnobotany: people, wild plant use & conservation. People and plants conservation manual. Earthscan Publications Ltd., London

Dalle SP, López H, Díaz D, Legendre P, Potvin C (2002) Spatial distribution and habitats of useful plants: an initial assessment for conservation on an indigenous territory, Panama. Biodivers Conserv 11:637-667

Department of Land and Survey (1962) Atlas of Uganda. Department of Land and Survey, Uganda, Entebbe, pp 22–23

Forest Department (1997) The National Biomass Study. Land Cover Stratification map, Bulamogi County. Forest Department, Uganda, Kampala

Government of Uganda (1963) Map Series Y732: Sheets 52/4, 53/3. Scale 1:50 000 edn. I-DOS. Directorate of Overseas Survey for Uganda Government, Entebbe

Hall P, Bawa K (1993) Methods to assess the impact of extraction of non-timber tropical forest products on plant populations. Econ Bot 47:234–247

Jongman RHG, ter Braak CJF, van Tongeren OFR (eds) (1995) Data analysis in community and landscape ecology. Cambridge University Press, UK, p 137

Kaimowitz D, Mertens B, Wunder S, Pacheco P (2004) Hamburger connection fuels Amazon destruction: Cattle ranching and deforestation in Brazil's Amazon. Bogor, CIFOR

Konstant TL, Sullivan S, Cunningham AB (1995) The effects of utilization by people and livestock on Hyphaene petersiana (Arecaceae) basketry resources in the palm savanna of North-Central Namibia. Econ Bot 49:345–356

Langdale-Brown I, Osmaston HA, Wilson JG (1964) The vegetation of Uganda and its bearing on land-use. Uganda Government, Kampala

Lykke AM (1998) Assessment of species composition change in savanna vegetation by means of woody plants' size class distributions and local information. Biodivers Conserv 7:1261–1275

National Environment Management Authority (2002) State of the environment report for Uganda 2002. National Environment Management Authority, Kampala

Obiri J, Lawes M, Mukolwe M (2002) The dynamics and sustainable use of high-value tree species of the coastal Pondoland forests of the Eastern Cape Province, South Africa. For Ecol Manage 166:131–148

Ollier CD, Harrop JF (1959) The soils of the Eastern Province of Uganda: a reconnaissance survey, memoirs of the research division, series I: soils, number 2. Department of Agriculture, Kawanda Research Station, Kampala

Peters CM (1999) Ecological research for sustainable non-wood forest product exploitation: an overview. In: Sunderland TCH, Clark LE, Vantomme P (eds) Non-wood forest products of central africa: current research issues and prospects for conservation and development. Food and Agriculture Organization, Rome, pp 19–35

Primack RB (1998) Essentails of conservation biology, 2nd edn. Sinauer Associates Inc., Massachusets USA

Rosa EA, York R, Dietz T (2004) Tracking the anthropogenic drivers of ecological impacts. Ambio 33(8):509–512

Schippmann U, Leaman DJ, Cunningham AB (2002) Impact of cultivation and gathering of medicinal plants on biodiversity: global trends and issues. In biodiversity and the ecosystem approach in agriculture, forestry and fisheries: satellite event on the occasion of the ninth regular session of the commission on genetic resources for food and agriculture. Rome, 12–13 October 2002, FAO, Rome

Shackleton SE, Dzerefos CM, Shackleton CM, Mathabela FR (1998) Use and trading of wild edible herbs in the central Lowveld Savanna region, South Africa. Econ Bot 52:251–259

Tabuti JRS, Dhillion SS, Lye KA (2003a) Ethnoveterinary medicines for cattle (Bos indicus) in Bulamogi county, Uganda: plant species and mode of use. J Ethnopharmacol 88:279–286

Tabuti JRS, Dhillion SS, Lye KA (2003b) Fuelwood use in Bulamogi County, Uganda: species harvested and consumption patterns. Biomass Bioenerg 25:581–596

Tabuti JRS, Dhillion SS, Lye KA (2004) The status of wild food plants in Bulamogi County, Uganda. Int J Food Sci Nutr 55:485–498

Tabuti JRS, Lye KA, Dhillion SS (2003c) Traditional herbal drugs of Bulamogi, Uganda: plants, use and administration. J Ethnopharmacol 88:19–44

Ticktin T (2004) The ecological implications of non-timber forest products. J Appl Ecol 41:11–21

Twine WC (2005) Socio-economic transitions influence vegetation change in the communal rangelands of the South African lowveld. Afr J Range For Sci 22:93–99

Uganda Bureau of Statistics (2005) The 2002 Uganda population and housing census—main report. UBoS, Kampala

Uganda Participatory Poverty Assessment Process (2002) Deepening the understanding of poverty: second participatory poverty assessment report. Ministry of Finance Planning and Economic Development, Kampala

Walter S (2001) Non-wood forest products in Africa: A regional an national overview. FAO

Biodivers Conserv (2007) 16:1917–1925
DOI 10.1007/s10531-006-9110-1

ORIGINAL PAPER

Mapping the geographic distribution of *Aglaia bourdillonii* Gamble (Meliaceae), an endemic and threatened plant, using ecological niche modeling

**M. Irfan-Ullah · Giriraj Amarnath ·
M. S. R. Murthy · A. Townsend Peterson**

Received: 16 August 2005 / Accepted: 22 May 2006 / Published online: 27 October 2006
© Springer Science+Business Media B.V. 2006

Abstract *Aglaia bourdillonii* is a plant narrowly endemic to the southern portion of the Western Ghats (WG), in peninsular India. To understand its ecological and geographic distribution, we used ecological niche modeling (ENM) based on detailed distributional information recently gathered, in relation to detailed climatic data sets. The ENMs successfully reconstructed key features of the species' geographic distribution, focusing almost entirely on the southern WG. Much of the species' distributional potential is already under protection, but our analysis allows identification of key zones for additional protection, all of which are adjacent to existing protected areas. ENM provides a useful tool for understanding the natural history of such rare and endangered species.

Keywords *Aglaia bourdillonii* · GARP · Niche modeling · Southern Western Ghats · Species distribution modeling

Introduction

Geographic distributions of species and their relation to environmental variables have been studied for centuries, and have long fascinated scientists and naturalists (Darwin 1859; Fisher 1958; Krishtalka and Humphrey 2000). Association of particular

M. Irfan-Ullah (✉)
Ashoka Trust for Research in Ecology and the Environment (ATREE), #659, 5th A Main
Road, Hebbal, Bangalore, Karnataka 560024, India
e-mails: irfan@atree.org; irfan26@gmail.com

G. Amarnath · M. S. R. Murthy
Forestry and Ecology Division, National Remote Sensing Agency (NRSA), Hyderabad,
Andhra Pradesh 500037, India

A. T. Peterson
Natural History Museum and Biodiversity Research Center, The University of Kansas,
Lawrence, KS 66045, USA

 Springer

species with particular environmental conditions has long been documented (Colding and Folke 1997; Hubbell 2005), but quantitative analyses have been possible only recently (Cutler et al. 2002), with the advent of new tools, as well as broad availability of continuous spatial information about various environmental parameters (Kerr and Ostrovsky 2003).

Inferential procedures that provide robust and reliable predictions of species' geographic and ecological distributions are thus critical to biodiversity analyses. This approach has recently been explored under the rubric of "ecological niche modeling" (ENM; (Soberón and Peterson 2005), and refers to reconstruction of ecological requirements of species that are analogous to the Grinnellian ecological niche (Grinnell 1917). ENM can provide diverse insights into the ecological and geographic extents of species' distributions (Soberón and Peterson 2004).

In this study, we develop a fine-scale distributional understanding of a threatened plant species (*Aglaia bourdillonii*) endemic to the Western Ghats (hereafter "WG") using ecological niche modeling, and explore areas of highest conservation concern, given the current management scenario. Since ENM focuses on the ecological dimensions that are distributed across geographic landscapes, we also explore the range of environmental conditions prevailing within the larger landscape of Southern Western Ghats. The WG forms a homogenous biome running along a 1600 km escarpment parallel to the southwestern coast of the Indian Peninsula, recognized as a global biodiversity hotspot (Myers et al. 2000). WG biodiversity is influenced by gradients in rainfall (east to west and north to south), length of dry season (south to north), temperature (south to north), and topography. WG is home to a wide range of endemic tree species, including wet evergreen, dry deciduous, and scrub forest types. Within evergreen forests, ~63% of tree species are endemic to the region (Ramesh et al. 1991). The Agastyamalai Hills, near the southern end of the region, are known for particularly high species diversity and endemicity, harboring ~2,000 flowering plant species, with 7.5% endemism (Henry et al. 1984).

Aglaia bourdillonii is a microendemic species apparently confined to the Agastyamalai Hills region. IUCN places this species under its "VU B1+2c" category (extent of occurrence < 20,000 km^2 or distributional area <2,000 km^2, with estimates indicating that habitat is severely fragmented or known to exist at ≤10 locations and populations continuing to decline). In this particular case, large areas of the species' distribution have been exposed to fire, grazing, establishment of commercial plantations and cutting for fuelwood. About 1,000 km^2 of forest in the region are protected within sanctuaries (Pannell 1998).

This species, however, apparently requires both specialized habitat conditions and specific plant associations to survive and maintain populations (Pascal 1988). Being an understory species, it requires protection from emergent species (e.g., *Cullenia exarillata* and *Palaquium ellipticum)* for survival (Ganesh et al. 1996; Pascal 1988). *Aglaia bourdillonii* is also sensitive to anthropogenic disturbance (e.g., logging), and is reduced in numbers where selective logging of *Palaquium ellipticum* and *Gluta travancorica* has occurred (Giriraj 2005; Pascal 1988); it is restricted to flatter topographies, seldom occurring in deep valleys (Giriraj 2005). Such restricted-range plants are often the most vulnerable, as they have narrow ecological tolerances.

Accurate distributional information thus becomes crucial for management and conservation efforts. Most often, management decisions for conservation are based on known occurrence sites—sites associated with field observations or specimens in museums and herbaria. Such records are generally used in constructing crude range maps: vaguely extrapolated polygons that enclose known occurrences, based on subjective interpolations. These maps are often highly biased towards accessible or well sampled areas, and rarely can be extended to remote and poorly known locations. Detailed, finescale, validated maps, however, can be developed based on known occurrences and are essential in designing conservation strategies; moreover, specific requirements of species regarding favorable environmental conditions can educate these decisions enormously.

Methods

Input data

We collected, in all, 53 spatially unique point locations for *Aglaia bourdillonii's* current distribution from various primary and secondary sources. Records of the species' current distribution were derived from on-ground surveys using GPS during the regional biodiversity inventory of WG (Roy 2002). To include historical occurrences, herbarium specimen data were gathered from the French Institute of Pondicherry (Ramesh et al. 1997; Ramesh and Pascal 1997), and Botanical Survey of India (BSI; Ahmedullah and Nayar 1987). All records were geocoded via reference to large scale (1:50,000 scale) topographic maps.

We used 19 "bioclimatic" variables derived from globally interpolated datasets of monthly temperature and precipitation available from (Hijmans et al. 2004), including annual and seasonal aspects of temperature and precipitation that are presumed to be maximally relevant to plant survival and reproduction (Hutchinson et al. 2000). We also included elevation, slope, aspect, and compound topographic index from the USGS Hydro-1K dataset (USGS 2001). All analyses were conducted at the native 30″ (~ 1 × 1 km pixels) spatial resolution of the environmental data sets.

Ecological niche modeling

ENM has been used in numerous applications and subjected to various tests, based on diverse analytical approaches (Csuti 1996; Gottfried et al. 1999; Manel et al. 1999a, b; Miller 1994; Tucker et al. 1997). The particular approach to modeling species' ecological niches and predicting geographic distributions used herein (summarized below) is described in detail elsewhere (Peterson et al. 2002a; Stockwell and Peters 1999; Stockwell and Noble 1992). Previous tests of the predictive power of this modeling technique for diverse phenomena in various regions have been published elsewhere (Anderson et al. 2002, 2003; Peterson 2001; Peterson et al. 1999, 2002a, b; Peterson and Vieglais 2001; Stockwell and Peterson 2002a, b).

The ecological niche of a species can be defined as the set of ecological conditions within which it is able to maintain populations without immigration (Grinnell 1917; Holt and Gaines 1992). Several approaches have been used to approximate species' ecological niches (Austin et al. 1990; Nix 1986; Scott et al. 1993, 1996, 2002; Walker and Cocks 1991), of these, one that has seen considerable testing is the Genetic Algorithm for Rule-set Prediction (GARP), which includes several inferential approaches in an iterative, evolutionary computing environment (Stockwell and

Peters 1999). All modeling in this study was carried out on a desktop implementation of GARP now available publicly for download.[1]

For GARP analyses, we initially divided (randomly) available occurrence points as follows: (1) 8 training data points (for rule generation), (2) 8 intrinsic testing data points (for model optimization and refinement), (3) 17 extrinsic testing data points (for choosing best subsets models), and (4) 20 independent validation data points (for final model validation); this procedure was repeated four times, based on different random subsets of available data. After validation trials were completed, to maximize occurrence data available to the algorithm, we eliminated the validation step, and thus provided 13 training points, 13 intrinsic testing points, and 27 extrinsic testing points to the algorithm.

GARP is designed to work based on presence-only data; absence information is included in the modeling via sampling of pseudo-absence points from the set of pixels where the species has not been detected (Stockwell and Peters 1999). GARP works in an iterative process of rule selection, evaluation, testing, and incorporation or rejection: first, a method is chosen from a set of possibilities (e.g., logistic regression, bioclimatic rules), and then is applied to the training data and a rule developed; rules may evolve by a number of means (e.g., truncation, point changes, crossing-over among rules) to maximize predictivity. Predictive accuracy is then evaluated based on 1,250 points resampled with replacement from the intrinsic testing data and 1,250 points sampled randomly from the study region as a whole to represent pseudoabsences. The change in predictive accuracy from one iteration to the next is used to evaluate whether a particular rule should be incorporated into the model, and the algorithm runs either 1,000 iterations or until convergence.

We developed 100 replicate model runs for *Aglaia*, and filtered out suboptimal models based on characteristics in terms of omission (leaving areas of known presence out of predictions) and commission (including areas not actually inhabited) error statistics. Following recent recommendations (Anderson et al. 2003) and also to represent a balance between optimizing model selection and practicalities of computing time required for the analysis, we selected best models in DesktopGARP using a 0% extrinsic hard omission threshold and 50% commission threshold. Experiments with different thresholds indicate that results are quite robust to minor variation in thresholds chosen. Throughout our analysis, we masked analyses to include only the southern WG region.

To permit visualization of patterns of *Aglaia* ecological niche variation, we combined the input environmental grids with the final ENM to create a new grid with a distinct value for each unique combination of environments; we exported the attributes table associated with this grid in ASCII format for exploration in a graphic program.

The initial validation step was repeated four times, in each of which 20 randomly selected points were set aside for testing. The coincidence between the independent points and model prediction was used as a measure of model predictive ability. Binomial tests (based on the proportional area predicted present and the number of independent test points successfully predicted) were used to compare observed predictive success with that expected under random (null) models of no association between predictions and test points. As model results are in the form of a 'ramp' of model agreement from 0 to 10, we repeated binomial tests across all thresholds of model agreement (prediction levels 1 to 10).

---

[1] http://www.lifemapper.org/desktopgarp/.

Results and discussion

Predictions of the distribution of *Aglaia bourdillonii* were good, given current knowledge of the species. To date, this species has been reported only from middle-elevation evergreen pockets of the southern WG, particularly in the Agastyamalai Hills region adjoining the Kalakkad Mundathurai Tiger Reserve (KMTR). The species has never been documented from other sectors of the WG.

In each of four replicate validations of *Aglaia* model predictions, 20 points were available to test predictions. For each replicate, we calculated binomial probabilities for each of the 10 predictive levels, eliminating the need to choose a single threshold for prediction of presence/absence. In all cases (four replicate tests, 10 predictive levels each), agreement between test occurrence points and model predictions was statistically significantly better than random (binomial tests, all $P \ll 0.05$).

Given our successful model validation, we used all data to produce a final model. The distribution of *Aglaia bourdillonii*, as predicted by all 10 best-subsets models in this analysis (Fig. 1) occupies 947 km^2 out of the 20,750 km^2 that constitute the southern WG, or ~5% of the total. Correspondence between known occurrences and the predictions of this final model is very close. The species occupies small areas in four districts, two in each of Tamil Nadu and Kerala states: ~71% (670 km^2) of the species' distribution is predicted in Tamil Nadu State, while the remaining ~29% (277 km^2) falls in Kerala State.

Much of the Agasthyamalai Hills region is included in four protected areas: Neyyar, Peppara, and Shendurni Wildlife Sanctuaries (WLS) and the Kalakkad

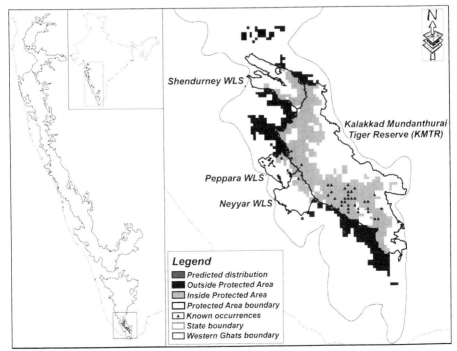

Fig. 1 Map showing known and predicted distribution of *Aglaia bourdillonii*, within Western Ghats

Mundanthurai Tiger Reserve (KMTR); as such, almost 66% (624 km²) of the species' distribution is already under protection. KMTR alone covers >54% (516 km²) of the species' total predicted distribution, and the remaining 3 WLSs cover 11% (108 km²); the remaining 34% (323 km²) that is unprotected is nonetheless close to protected areas. An augmentation of 5 km along the western boundary of the protected areas (Fig. 2) would cover an additional 27% (259 km²) of the species' distribution, bringing the total to 97% (883 km²) of its total distribution.

Further explorations of the implications of our models explored the species' distribution in ecological dimensions. As such, we compared ecological characteristics of areas predicted present for the species with the complete environmental range in the southern WG. For the 19 bioclimatic and physiographic variables, we removed eight highly correlated, redundant variables, and then developed bivariate plots to summarize the species' distribution (Fig. 2). These explorations reveal a very narrow niche (even against the already specialized conditions of the southern

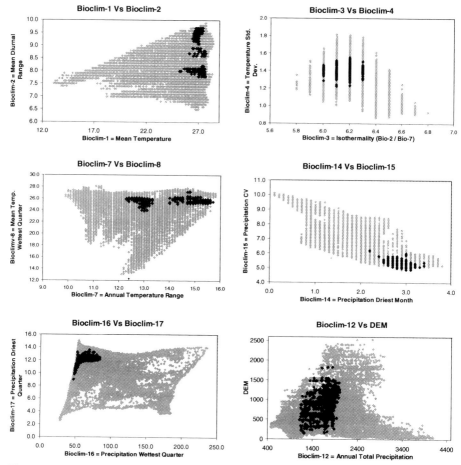

Fig. 2 Exploratory visualization of *Aglaia bourdillonii* niche in environmental space. Gray = Conditions available in the Southern Western Ghats, and Black = modeled potential distribution of species

🕊 Springer

WG) for the species—restricted to areas preserving narrow ranges of annual mean temperatures (25–28°C) in three distinct zones of diurnal temperature variability, and high precipitation (1100–2000 mm).

These results are crucial in addressing conservation issues related to a species that is so narrowly restricted, occurring only in the Agasthyamalai Hills region of the southern WG. Protected area boundaries have rarely been defined in light of real data on biodiversity, more frequently following existing administrative limits and land tenures. In the Agasthyamalai Hills region, protected area boundaries simply follow state boundaries. The results developed here—combined with results of similar ongoing studies that address other restricted range species—suggest clear ways to modify boundaries to include ecologically significant areas based on scientific evidence. This is even crucial to bring about protection from widespread logging of the species and its associates as mentioned in the IUCN Red List of Threatened Species (Pannell 1998).

Conclusions

Aglaia bourdillonii is a low-to-middle elevation evergreen species narrowly endemic to the southern WG. As such, its conservation will depend critically on a few well-placed protected areas. Indeed, much of its small distributional area is already under protection, although this protection could be improved considerably with a quite-small addition to those existing protected areas. ENM is a useful tool in outlining and understanding the distributions—in geographic and ecological spaces—of such species, and may prove useful in a variety of applications to biodiversity conservation, bioprospecting, and any application that requires detailed information about species' geographic distributions.

Acknowledgments We acknowledge the support of the Indo-US Science and Technology Forum, Ministry of Environment and Forest, Government of India. The Ford Foundation and Hewlett Packard. The Authors 2 and 3 thank Director NRSA, for extending his support and encouragement, The Author 2 also acknowledges Rufford Small Grant, Rufford Foundation U.K. for financial support.

References

Ahmedullah M, Nayar MP (1987) Endemic plants of the Indian region. Botanical Survey of India, India
Anderson RP, Peterson AT, Gómez-Laverdem M (2002) Using niche-based GIS modeling to test geographic predictions of competitive exclusion and competitive release in South American pocket mice. Oikos 98:3–16
Anderson RP, Lew D, Peterson AT (2003) Evaluating predictive models of species distributions: criteria for selecting optimal models. Ecol Model 162:211–232
Austin MP, Nicholls AO, Margules CR (1990) Measurement of the realized qualitative niche: environmental niches of five *Eucalyptus* species. Ecol Monogr 60:161–177
Colding J, Folke C (1997) The relations among threatened species, their protection, and taboos. Conserv Ecol [online] (Available from the Internet: http://wwwconsecolorg /vol1/iss1./art6/) 1
Csuti B (1996) Mapping animal distribution areas for gap analysis. In: Scott JM, Tear TH, Davis FW (eds) Gap analysis: a landscape approach to land management issues. American Society of Photogrammetry and Remote Sensing, Bethesda, Maryland, pp 135–145

Springer

Cutler R, Edwards TC, Alegria JJ, McKenzie D (2002) A sample design framework for Survey and Manage species under the Northwest Forest Plan. In: Proceedings of the section on statistics and environment, 2001 joint statistical meeting. American Statistical Association, Alexandria, Virginia, USA

Darwin C (1859) On the origin of species. Murray, London

Fisher RA (1958) The Genetical theory of natural selection, 2nd edn. Dover, New York

Ganesh T, Ganesan R, Devy MS, Davidar P, Bawa KS (1996) Assessment of plant biodiversity at a mid-elevation evergreen forest of Kalakad-Mudanthurai Tiger Reserve, Western Ghats, India. Curr Sci 71:379–392

Giriraj A (2005) Spatial characterization and conservation priortization in tropical evergreen forests of Southern Western Ghats using Geoinformatics. Ph.D. thesis, Barathidasan University, Thirunelveli

Gottfried M, Pauli H, Reiter K, Grabherr G (1999) A Fine-Scaled Predictive model for changes in species distribution patterns of high mountain plants induced by climate warming. Divers Distribut 5:241–251

Grinnell J (1917) Field tests of theories concerning distributional control. Am Nat 51:115–128

Henry AN, Chandrabose M, Swaminathan MS, Nair NC (1984) Agasthyamalai and its environs: a potential area for Biosphere Reserve. J Bombay Nat History Soc 81:282–290

Hijmans RJ, Cameron SE, Parra JL, Jones PG, Jarvis A (2004) The WorldClim interpolated global terrestrial climate surfaces. Version 1.3

Holt RD, Gaines MS (1992) Analysis of adaptation in heterogeneous landscapes: Implications for the evolution of fundamental niches. Evol Ecol 6:433–447

Hubbell SP (2005) Neutral theory in community ecology and the hypothesis of functional equivalence. Funct Ecol 19:166

Hutchinson MF, Nix HA, McMahon J (2000) ANUCLIM User's Guide. Australian National University, Canberra

Kerr JT, Ostrovsky M (2003) From space to species: ecological applications for remote sensing. Trends Ecol Evol 18:299-305

Krishtalka L, Humphrey PS (2000) Can natural history museums capture the future? Bioscience 50:611–617

Manel S, Dias JM, Buckton ST, Ormerod SJ (1999a) Alternative methods for predicting species distribution: an illustration with Himalayan river birds. J Appl Ecol 36:734–747

Manel S, Dias JM, Ormerod SJ (1999b) Comparing discriminant analysis, neural networks and logistic regression for predicting species distributions: a case study with a Himalayan river bird. Ecol Model 120:337–347

Miller RI (eds) (1994) Mapping the diversity of nature. Chapman & Hall, London

Myers N, Mittermeir RA, Mittermeir CG, da Fonseca GAB, Kent J (2000) Biodiversity hotspots for conservation priorities. Nature 403:853–858

Nix HA (1986) A biogeographic analysis of Australian Elapid Snakes. In: Longmore R (ed) Atlas of Elapid Snakes of Australia. Bureau of Flora and Fauna, Canberra, p 415

Pannell CM (1998) *Aglaia bourdillonii*. In: IUCN 2004: Red List of Threatened Species < http://wwwredlistorg/> Downloaded on 05 July 2005

Pascal JP (1988) Wet evergreen forests of the Western Ghats of India, ecology, structure, floristic composition and succession, 1st edn. Institut francais de Pondichery, p 345

Peterson AT (2001) Predicting species' geographic distributions based on ecological niche modeling. The Condor 103:599–605

Peterson AT, Vieglais DA (2001) Predicting species invasions using ecological niche modelling. BioScience 51:363–371

Peterson AT, Soberón J, Sanchez-Cordero V (1999) Conservatism of ecological niches in evolutionary time. Science 285:1265–1267

Peterson AT, Ball LG, Cohoon KC (2002a) Predicting distributions of Mexican birds using ecological niche modelling methods. Ibis 144:e27–e32

Peterson AT, Stockwell DRB, Kluza DA (2002b) Distributional prediction based on ecological niche modeling of primary occurrence data. In: Scott JM, Heglund PJ, Morrison ML (eds) Predicting species occurrences: issues of scale and accuracy. Island Press, Washington, DC, pp 617–623

Ramesh BR, Pascal JP (1997) Atlas of endemics of the Western Ghats (India). French Institute, Pondicherry

Ramesh BR, Pascal JP, De Franceschi D (1991) Distribution of endemic, arborescent evergreen species in the Western Ghats. In: Proceedings of the rare, endangered, and endemic plants of the Western Ghats, Thiruvananthapuram, pp 1–7

Ramesh BR, Menon S, Bawa Kamaljit S (1997) A vegetation based approach to biodiversity gap analysis in the Agastyamalai region, Western Ghats, India. Ambio XXVI:529–536

Roy PS (2002) Biodiversity characterization at landscape level in Western Ghats using satellite remote sensing and geographic information system. A report by: Indian Institute of Remote Sensing (IIRS-NRSA), Department of Space, Dehradun India

Scott JM, Davis F, Csuti B, Noss R, Butterfield B, Groves C, Anderson H, Caicco S, D'Erchia F, Edwards Jr TC, Ullman J, Wright RG (1993) Gap analysis: a geographic approach to protection of biological diversity. Wildlife Monogr 123:1–41

Scott JM, Tear TH, Davis FW (eds) (1996) Gap analysis: a landscape approach to biodiversity planning. American Society for Photogrammetry and Remote Sensing, Bethesda, Maryland

Scott JM, Heglund PJ, Haufler JB, Morrison M, Raphael MG, Wall WB, Samson F (eds) (2002) Predicting species occurrences: issues of accuracy and scale. Island Press, Washington, DC

Soberón J, Peterson AT (2004) Biodiversity informatics: managing and applying primary biodiversity data. Philos Trans, Roy Soc 359:689–698

Soberón J, Peterson AT (2005) Interpretation of models of fundamental ecological niches and species distributional areas. Biodivers Inform 2:1–10

Stockwell DRB, Noble IR (1992) Induction of sets of rules from animal distribution data: a robust and informative method of data analysis. Math Comput Simul 33:385–390

Stockwell DRB, Peters D (1999) The GARP modeling system: problems and solutions to automated spatial prediction. Int J Geograph Inform Sci 13:143–158

Stockwell DRB, Peterson AT (2002a) Controlling bias in biodiversity data. In: Scott JM, Heglund PJ, Morrison ML (eds) Predicting species occurrences: issues of scale and accuracy. Island Press, Washington, DC, pp 537–546

Stockwell DRB, Peterson AT (2002b) Effects of sample size on accuracy of species distribution models. Ecol Model 148:1–13

Tucker K, Rushton SP, Sanderson RA, Martin EB, Blaiklock J (1997) Modeling bird distributions—a combined GIS and Bayesian rule-based approach. Landscape Ecol 12:77–93

USGS (2001) HYDRO1k Elevation Derivative Database. U.S. Geological Survey, http://edcdaac.usgs.gov/gtopo30/hydro/, Washington, DC

Walker PA, Cocks KD (1991) HABITAT: a procedure for modelling a disjoint environmental envelope for a plant or animal species. Global Ecol Biogeogr Lett 1:108–118

Biodivers Conserv (2007) 16:1927–1957
DOI 10.1007/s10531-006-9112-z

ORIGINAL PAPER

Optimizing conservation of forest diversity: a country-wide approach in Mexico

Martin Ricker · Iliana Ramírez-Krauss ·
Guillermo Ibarra-Manríquez · Esteban Martínez ·
Clara H. Ramos · Guadalupe González-Medellín ·
Gabriela Gómez-Rodríguez · José Luis Palacio-Prieto ·
Héctor M. Hernández

Received: 29 August 2005 / Accepted: 22 May 2006 / Published online: 17 November 2006
© Springer Science+Business Media B.V. 2006

Abstract A recent vegetation study [Palacio-Prieto et al. (2000) Bol Inst Geogr UNAM 43:183–203] showed that Mexico's forest area has declined to 33.3%, from originally 52.0% of the country's land area. In order to assess strategies for tree diversity conservation, we compiled a list of 846 tree species native to Mexico, and determined for each the presence or absence in 234 geographical squares of 1° latitude by 1° longitude (approximately 106 × 106 km). On the average, any two squares shared only 6% of their species composition. Using a standard optimization method from engineering and economics [Dantzig (1963) Linear programming and extensions. Princeton University Press, Princeton, NJ, USA, 625 p], we determined the minimally necessary land area in Mexico to conserve the 846 tree species, while securing that each species is found in an area of (approximately) 1,100 km² of currently existing forest vegetation. Furthermore, we took into account 15 existing protected areas with a size of at least 1,100 km² each. With these constraints, the total minimum area needed to conserve all 846 tree species is 45,136 km² of currently existing forest vegetation, or 2.3% of Mexico's surface. While this analysis can be refined with subsequent field work, the proposed reserve network indicates that efficient land use planning on a national

M. Ricker (✉)
Estación de Biología Tropical "Los Tuxtlas", Apartado Postal 94, San Andrés Tuxtla, Veracruz 95701, Mexico
e-mails: mricker@ibiologia.unam.mx; martin_tuxtlas@yahoo.com.mx

I. Ramírez-Krauss · E. Martínez · C. H. Ramos · G. González-Medellín · H. M. Hernández
Herbario Nacional de México, Instituto de Biología, Universidad Nacional Autónoma de México, Apartado Postal 70–367, Delegación Coyoacán, México D.F. 04510, Mexico

G. Ibarra-Manríquez
Centro de Investigaciones en Ecosistemas, Universidad Nacional Autónoma de México, Antigua carretera a Pátzcuaro 8701, Colonia San José de la Huerta, Morelia, Michoacán 58190, Mexico

G. Gómez-Rodríguez · J. L. Palacio-Prieto
Instituto de Geografía, Universidad Nacional Autónoma de México, Apartado Postal 20–850, Delegación Coyoacán, México D.F. 04510, Mexico

Springer

scale may be able to conserve tree species diversity in a relatively small portion of Mexico, even after severe deforestation has taken place.

Keywords   Vegetation analysis · Biogeographical distributions · Linear optimization · Simplex method · Minimum conservation area

Introduction

Mexico is considered to be a biologically megadiverse country, with at least 22,000 species of seed plants (Rzedowski 1991a, 1993). About 52% of its seed plants are endemic (Rzedowski 1991b). The only taxonomic compilation of all Mexican tree and shrub species was elaborated by Paul Carpenter Standley of the U.S. National Herbarium in the Smithsonian Institute, at the beginning of the 20th century: he counted about 5,700 tree and shrub species in the Mexican Flora (Standley 1920–1926, p. 1643), including all woody plants, small shrubs, as well as species exotic to Mexico. After almost 80 years, unfortunately, the taxonomy in Standley's work is largely outdated. A recent taxonomic revision of Mexican native tree species of the legume family (Leguminosae: Caesalpinoideae, Mimosoideae, and Papilionoideae), reaching a height of at least 3 m within its overall distribution in Mexico, resulted in 623 species of this large family. Of these, about 46% are endemic to Mexico (Sousa et al. 2001, 2003). Combining the tree species from the legume family with the tree species from the *Flora Neotropica* series (see below), Pennington and Sarukhán's (1998) book on Mexican tropical trees, two floristic studies (Ibarra and Sinaca 1995, 1996a, 1996b, and Lott 1993) and a doctoral thesis (Ibarra 1996), and complementing species from Standley's work, we got a preliminary new checklist of 2,263 tree species in 127 seed plant families that reach a height of at least 3 m. In comparison, Villaseñor and Ibarra (1998) in an herbarium survey counted 3,639 species of 128 families, that are native Mexican flowering plants with arboreal growth form, but revision of many taxonomic groups are still pending.

If the taxonomy of Mexico's tree species is incompletely known, then their distribution is even less so. The Flora Neotropica series published by the New York Botanical Garden attempts to provide both the taxonomy and distribution maps for plant species (not only trees) for the whole Neotropical region (Mexico to Argentina). The elaboration of these very useful volumes, however, is a slow enterprise (still in progress), having started with Cowan (1967) and to date having worked only through a small proportion of all intended species: we encountered only 183 (8.1%) out of the estimated 2,263 Mexican tree species in the *Flora Neotropica* series.

Given this situation, in Mexico an attempt has been made to develop a "fast track" to a broader data base about species distributions: The Mexican "National Commission for Knowledge and Use of Biodiversity" (*Comisión Nacional para el Conocimiento y Uso de la Biodiversidad* = CONABIO) was created in March 1992. This governmental institution has provided funds for scientific projects that compile georeferrenced data on biodiversity in form of data bases for all taxonomic groups.

The procedure to accumulate within a few years a large database on biodiversity has not always found the approval of taxonomists. These have complained about the lack of taxonomic verification of the specimens in the traditional way, where one specialist compares (and reclassifies) collected specimens within a taxonomically related specimen group (as is done for example in the *Flora Neotropica*). For the purpose of the present study, however, we consider that the distribution data from CONABIO is reasonably accurate to draw conclusions on general trends of tree species diversity, and derive locations that are strategically important for forest diversity conservation.

Any effort to propose a network of forest conservation has to take into account the present vegetation cover. Mexico, as many tropical countries, has experienced extensive deforestation during the last decades (Cairns et al. 1995; Dirzo and García 1992). Mas et al. (2002) report current annual deforestation rates of 0.25% for temperate forests and 0.76% for tropical forests in Mexico. This fact is important because many of the specimens, from which the distributions of the 846 tree species of this study were compiled, were collected decades ago. Therefore, we needed an updated forest cover analysis for Mexico.

In the year 2000, the Mexican Ministry of the Environment (Secretaría del Medio Ambiente, Recursos Naturales y Pesca = SEMARNAP) funded the Geography Institute of the Mexican Autonomous National University to carry out a nationwide vegetation-type and cover analysis, as part of a Mexican forest inventory (Palacio-Prieto et al. 2000). This made it possible to assess the extension and the geographical distribution of the forest cover and forest types as of the year 1999–2000.

The descriptive information on tree species, their distribution, and the vegetation situation together makes it possible to draw interesting conclusions on tree species diversity patterns and the possible strategies for the conservation of their habitat. To derive an efficient conservation strategy for heterogeneously distributed tree species, however, one has to think about a non-trivial mathematical problem: the species are found in different combinations at different sites; which combination of conservation sites minimizes the needed area for conservation?

The literature on biological conservation has addressed the problem of optimizing networks of protected areas. Integer programming models have been applied in a number of studies (Sætersdal et al. 1993; Csuti et al. 1997; Snyder et al. 1999; Church et al. 2000; Rodrigues and Gaston 2002; ReVelle et al. 2002). It is surprising, however, that the original, non-integer model of linear programming from Dantzig (1963) has not been used more intensively to propose practical real-world solutions for biodiversity conservation. Rather, some pseudo-optimization methods have frequently been used, where reserve networks have been determined that are smaller than the whole area of the biodiversity in question, but do not necessarily represent the minimum possible areas. One popular such method has been "complementary analysis", where iterative algorithms augment the network stepwise, using as criteria missing species or endemisms (see Margules et al. 1988; Vane-Wright et al. 1991; Nicholls and Margules 1993; Hernández and Bárcenas 1996; Williams et al. 1996; Gómez-Hinostrosa and Hernández 2000). While there can be advantages of this approach in terms of fullfilling some desired conservation criteria (Bedward et al. 1992), it has been shown that complementary analysis and other "greedy" heuristic methods may provide suboptimal results (Underhill 1994; Church et al. 1996; Csuti et al. 1997). In addition, linear optimization can take into account objectives and constraints that complementary analysis cannot.

The linear programming problem of the present study was to find the minimally necessary land area in Mexico to conserve 846 native tree species, while ensuring that the conservation area was still forested in the year 2000 and that for each species it covered at least 1,100 km^2 (more exactly the protected area had to sum up to 10% of a geographical square of 1° latitude by 1° longitude). Rather than using the (0–1) integer programming model, we use a linear programming model that finds continuous solution parameters, which here are restricted to be proportions between 0 and 1. As a consequence, the algorithm finds not "include" versus "don't include" solutions, but optimized proportions of geographical squares.

Methods

Species selection and determination of their distribution

The detailed information on the included species and their distributions in Mexico is provided in Appendix 1. Only seed plants native of Mexico were considered (Angiosperms and Gymnosperms). A total of 846 species was compiled from the following sources:

(a) 19 *Flora Neotropica* volumes (Berg 1972; Bohs 1994; Cowan 1967; Farjon and Styles 1997; Gentry 1980, 1992; Hekking 1988; Kaastra 1982; Landrum 1986; Maas et al. 1992; Pennington 1990; Pennington et al. 1981; Poppendieck 1981; Prance 1972, 1989; Rohwer 1993; Sleumer 1980, 1984; Todzia 1988);
(b) Ibarra and Sinaca (1995, 1996a, b), a floristic checklist of 940 vascular plant species for the 644-ha reserve belonging to the Mexican Autonomous National University (UNAM) in the Los Tuxtlas region in Veracruz (square #38 in Figs. 1, 2, and 5);
(c) Lott (1993), a floristic checklist of 1,120 vascular plant species of the Chamela Bay region (Jalisco) that includes the 3,200-ha Chamela university reserve, as well as the 13,142-ha Chamela-Cuixmala Biosphere Reserve (square #62 and extending into #61 in Figs. 1, 2, and 5);
(d) Pennington and Sarukhán (1998), a synthesis comprising 190 tree species found in tropical Mexico; and
(e) Ibarra (1996), where the distributions of tree species within the Yucatán Peninsula are analyzed.

The *Flora Neotropica* volumes provided 21.6% of the total number of species in the list, whereas the remaining species (78.4%) were derived from the four additional literature references.

We define the growth form "tree" here as those plants that stand self-sustained (thus excluding lianas) and reach a maximum height of at least 3 m anywhere in its distribution. We do not distinguish monopodial growth ("trees") and polypodial growth ("shrubs"), and also do not exclude monocotyledonous plants such as palms. Heights were either given in the literature references, or were checked in herbarium samples at the National Herbarium of Mexico (MEXU), housed in the Biology Institute of the UNAM. The average height of the 846 species was 17 m, from 3 m of 31 species up to 92 m of Mexico's largest known species, *Ulmus mexicana*.

The selection of the 846 species represents 37% of the 2,263 roughly estimated tree species that exist in Mexico. The emphasis of the species considered here is on tropical tree species, as provided by the mentioned references. Several temperate genera that are characteristic in some mountainous and northern vegetation types in Mexico are missing on our list, such as *Abies* ("fir"), *Cupressus* ("cedar"), *Juniperus* ("juniper"), *Picea* ("spruce"), and many species of *Quercus* ("oak"). Also, not all

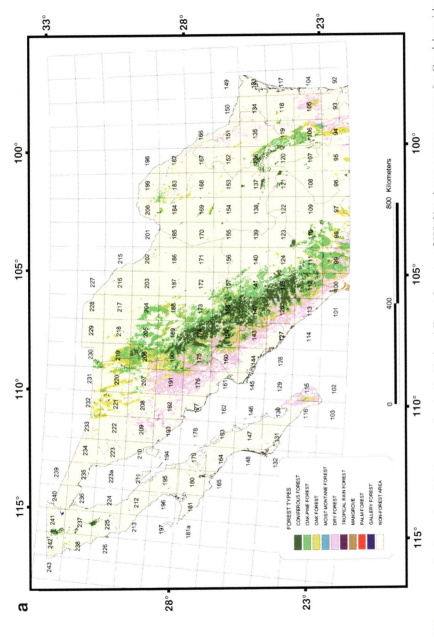

Fig. 1 (**a**) Forest vegetation in northern Mexico according to the National Forest Inventory 2000. Not shown are two squares for Guadalupe island (28–30°N, 118–119°W). The divisions between political states are shown. (**b**) Forest vegetation in southern Mexico according to the National Forest Inventory 2000. The divisions between political states are shown

⚛ Springer

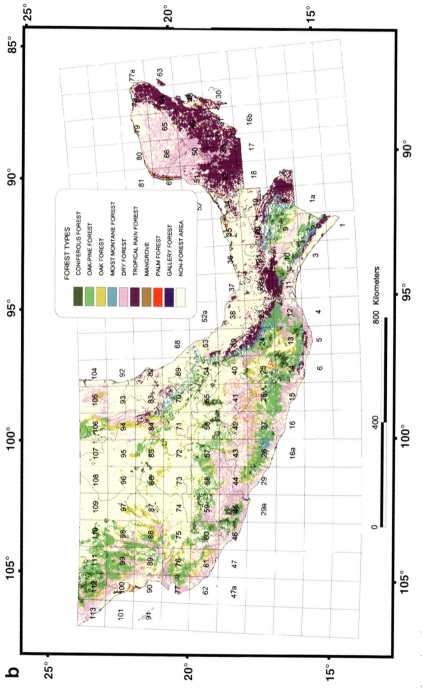

Fig. 1 continued

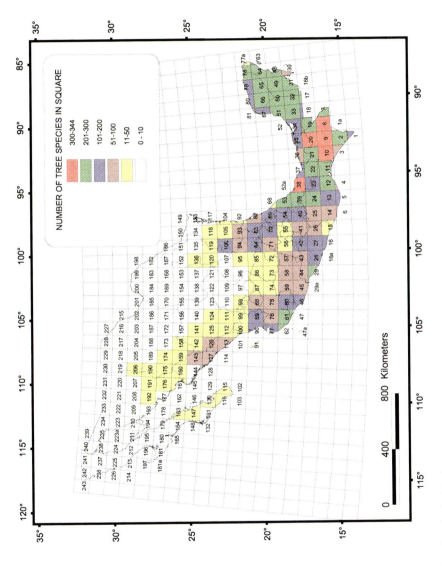

Fig. 2 Tree species diversity in Mexico

tree species found in the selected references were included in the final list. In some cases there were taxonomic uncertainties, and in other cases the distribution data were doubtful or incomplete.

Distributions were determined by assigning the presence or absence in geographical squares of 1° latitude by 1° longitude throughout Mexico. It is noteworthy that the mathematical area of the earth's geographical squares decreases in direction from south to north. This can be seen best on a globe model: when going from the equator northwards, squares become ever smaller, because all longitudinal lines eventually meet in the northern pole (here the "squares" become triangles). For Mexico, we calculated the average surface area of a square to be 11,181 km², or 105.74 km by 105.74 km. The range is from 10,453 to 12,089 km², 6.5% lower and 8.1% higher than the average, respectively. When using proportions of squares, however, no distinction was made between different absolute areas of different squares.

The studied species were found in 234 squares, out of a total of 251, each of which received a code number (see Fig. 1). Seventeen squares (7%) of the 251 squares did not contain any of the 846 tree species in this study. A spreadsheet matrix was elaborated with 234 columns for the geographical squares, and 846 rows for the species, and in each square the presence or absence was indicated with a "1" or "0". When applying this limited resolution, we assume first that a given species occurs in a homogeneous way throughout the forest vegetation of the square. While this assumption will in many cases be unrealistic, the method serves for guiding us efficiently to geographical squares where the attention of forest conservation should be focused.

The information on geographical coordinates stems originally from herbarium collections. In the case of the tree species taken from *Flora Neotropica* volumes, maps with collection points and a geographical coordinate grid are given. When the provided squares of the grid were larger than 1° latitude by 1° longitude, these were subdivided by us. Collection points were also given for the species of the Yucatán Peninsula in Ibarra (1996). For species that were not in the Flora Neotropica (78.5%), and distributions of tree species outside of the Yucatán Pensinsula, geographical coordinates of collection points were provided by the "National Commission for Knowledge and Use of Biodiversity" (CONABIO). For our database, collection coordinates were based on 56 field projects, which are cited in Appendix 1. Distributions were checked and some were modified by two of the coauthors (E. Martínez and C. H. Ramos), who have collected extensively herbarium material of tree species in southern Mexico.

Vegetation-type and cover analysis of Mexico

In the year 2000, the Mexican Ministry of the Environment (Secretaría del Medio Ambiente, Recursos Naturales y Pesca = SEMARNAP) funded the Geography Institute of the UNAM to carry out a nationwide vegetation-type and cover analysis, as part of a Mexican forest inventory. The analysis was based on satellite images, a nationwide grid of aerial photos, previous vegetation maps, and botanical interpretation. The system of vegetation types is derived from the two major sources of Mexican vegetation classification: Miranda and Hernández (1963), and Rzedowski (1986). The result were 121 maps covering all of Mexico on a scale of 1:250,000. The methods used in this geographical–botanical study are described in Palacio-Prieto et al. (2000), Gómez-Rodríguez and Palacio-Prieto (2001), and Mas et al. (2002). The information was processed for the present study in ArcView, for combining vegetation subtypes and calculating their relative area per geographical square.

Optimization of forest conservation

Optimization is mathematically not an easy issue, and the Simplex algorithm of linear programming is probably the most famous method, because in practice it can be used for problems with thousands of variables and constraints, while one can still be sure that the resulting solution is optimal. The objetive function and constraint inequalities, however, have to be linear. The Simplex method of linear optimization (or linear programming) was invented by the U.S. mathematician and economist George B. Dantzig in 1947 as a technique for planning the activities of the U.S. Air Force. According to Dorfman et al. (1958), the fundamental paper was circulated privately for several years, before being published by Dantzig (1951). The main source cited today is Dantzig's book (1963). Linear programming is widely used nowadays in economics, engineering, and production logistics, and contributed fundamentally to the scientific field of operations research (Hillier and Lieberman 1990). The method has been proposed already decades ago for optimizing land allocation (Church and ReVelle 1974; Dykstra 1984; Gilbert et al. 1985).

Here, linear optimization was carried out with the software LINDO (*L*inear *I*nteractive, and *D*iscrete *O*ptimizer) Hyper Version Release 6.0, 1998 by LINDO Systems Inc., Chicago, IL (http://www.lindo.com). A linear optimization problem consists of a to-be-minimized or to-be-maximized objective function and of constraint (in)equalities. The technical formulation of the linear optimization problem for this study is given in Appendix 2. The first entry is the to-be-minimized objective function, which here consists of 234 terms, corresponding to the 234 geographical squares where tree species were found. Squares without any of the 846 tree species are presented as gaps in the objective function.

The objective function is the sum of all $(a_i \cdot x_i)$, where a_i is the proportion of forest vegetation in square number i, and x_i is the optimal proportion of that proportion of forest vegetation that is to be conserved. For example, the term $(0.4702 \cdot x_2)$ can be expressed in words that square #2 contains 47.02% forest vegetation. If subsequently the optimal solution is $x_2 = 0.21267$, then 21.267% of those 47.02% should be set aside for conservation, corresponding to 10% of square #2 $(0.4702 \cdot 0.21267)$. Those squares where the forest inventory (Palacio-Prieto et al. 2000) did not detect any forest cover, are presented with a coefficient of $a_i = 0.0000$.

The objective function is followed by four sets of constraints inequalities:

(1)　The first set are 828 distribution constraints, corresponding to 828 of the 846 tree species, while 18 species are treated separately in point 2. Each constraint polynomial is a subselection of polynomial terms from the objective function, corresponding to those terms where collections of the species have been registered. For example in the case of *Acacia acatlensis*, collections were reported in squares #2, 12, 24, 25, 27, 39, 40, 41, 42, 43, 60, 62, 75, 87, and 112. The sum of those terms is set to be at least 0.1 or 10% of a square. This percentage corresponds to approximately 1,100 km², and is arbitrarily chosen as a reasonable value (approximately, because the absolute size of the squares varies from South to North). Expressed in words: from the overall known geographical distribution in Mexico of each tree species, an existing forest area is to be set aside for conservation of the species that corresponds to at least 10% of a geographical square. The proportion can be set aside from a single square or as a combination of subproportions from a number of squares where the species occurs, depending on what is found by the algorithm to contribute to a lower total conservation area for all species combined.

(2)   The second set of constraints refers to 18 tree species that present a total distribution that has a forest cover smaller than 10% of a geographical square. Requiring the optimization algorithm in these cases to search for conservation areas that represent together at least 10% of the area of a square, would result in an infeasible solution (as we do not consider restoration of the original habitat, thus constraining the optimized coefficients to be maximally 1). In those cases, the maximum possible area was requested as conservation habitat for these tree species (e.g., 9.30% in the case of *Bernardia spongiosa* in Appendix 2).

(3)   The third set of constraints "forces" 15 large existing conservation areas into the solution. These are 12 Biosphere Reserves and 3 Areas of Protection of Flora and Fauna. The conservation areas were selected from Mexico's Forest Atlas (Varela et al. 1999). Only those areas were included that are at least 1,100 km^2 in size, and that contain substantial forest vegetation. In the squares where these 15 protected areas are found, a proportion of 10% was required to be set aside for conservation (the same proportion as for each species, allowing these protected area to represent complete conservation areas for species). Two biosphere reserves, Montes Azules in Chiapas and Los Tuxtlas in Veracruz, expand each substantially into two neighboring geographical squares, so that the neighboring squares were also required to be in the solution. In the cases of three Biosphere Reserves (Los Tuxtlas, Maderas del Carmen, Pantanos de Centla) the currently existing forest cover in the squares turned out to be less than 10%. Consequently, the proportion to be set aside for conservation had to be the maximum possible one (e.g., 3.54% for Maderas del Carmen).

(4)   The fourth and final set of constraints is to assure that the proportions of squares in the optimal solution are not larger than 1 (or 100%). That the proportions would not be smaller than 0 was automatic here, and did not necessarily need to be specified as additional constraints.

Comparative complementary analysis

The selection of squares followed the modified criteria from Margules et al. (1988), Nicholls and Margules (1993), and Villaseñor et al. (2003): the square with the highest number of species is selected. After eliminating the species of that square from the subsequent analysis, the process was repeated. In case of squares being tied, the square was selected that contributed most to increasing the number of sites where species are conserved. If that still led to a tie, the square was selected that was closest to previously selected squares. And if there was another tie, the square was selected that showed up first in the species-site matrix.

Results

Forest vegetation analysis

Mexico is covered by a large variety of vegetation types, ranging from humid tropical rain forest to dry scrub and natural grasslands (Table 1). We considered to be "forest" only those vegetation types that are likely to develop a height over 3 m. Extensive areas in Mexico of various types of natural scrub were excluded (27.5% "matorral" in legend "d" of Table 1). In Fig.1, we distinguish and present nine forest types. Table 1 explains in greater detail the forest types that were recognized by us, together with their percentage of cover of the Mexican territory (1,945,748 km^2 without Guadalupe island). The absolute and relative areas were calculated assuming a plane surface of the landscape, i.e., without taking into

Table 1 Recognized forest types and their cover in Mexico in 2000[a-d]

Dry forest (10.858%):

(a) *Low deciduous and subdeciduous lowland forest* (7.017%): "selva baja caducifolia y subcaducifolia", forest of 4–15 m height, with characteristic genera being *Plumeria, Bursera, Gyrocarpus, Lonchocarpus, Spondias*, and *Pseudosmodingium*; over 50% of the trees loose their leaves during the dry season.

(b) *Intermediate deciduous and subdeciduous lowland forest* (2.007%): "selva mediana caducifolia y subcaducifolia", forest of 15–20 m height, with characteristic genera being *Enterolobium, Hymenaea*, and *Orbignya*; over 50% of the trees loose their leaves during the dry season.

(c) *Mesquite and wattle forest* (1.443%): "mezquital y huizachal", tree vegetation of up to 12 m height dominated by *Prosopis* ("mesquite") and/or *Acacia* ("wattle").

(d) *Low thorny lowland forest* (0.391%): "selva baja espinosa", vegetation of thorny trees of 4–15 m height, with typical genera being *Parkinsonia, Olneya*, and *Pithecellobium*.

Oak–pine forest (6.972%):

(a) *Oak–pine (or pine–oak) forest* (6.353%): "bosque de encino-pino (o pino-encino)", forest of 5–20 m height, with both *Quercus* ("oak") and *Pinus* ("pine") being present, either genus being dominant.

(b) *Low open forest* (0.620%): "bosque bajo abierto", tree community of 5–10 m height, with *Quercus, Pinus, Juniperus*, and a grass layer; a transition zone between forest and natural grassland.

Oak forest (5.135%):
"bosque de encino", forest of 2–30 m height, dominated by *Quercus*.

Tropical rain forest (5.115%):

(a) *High and intermediate semi-evergreen lowland forest* (2.853%): "selva alta y mediana subperennifolia", forest of 20 m height or more, with characteristic genera being *Alseis, Pterocarpus, Carpodiptera*, and *Manilkara*; 25–50% of the trees loose their leaves during the dry season.

(b) *High and intermediate evergreen lowland forest* (1.742%): "selva alta y mediana perennifolia", forest of 20–35 m height, with characteristic genera being *Terminalia, Swietenia*, and *Brosimum*.

(c) *Low semi-evergreen lowland forest* (0.433%): "selva baja subperennifolia", forest of 15–20 m height, with characteristic species being *Haematoxylum campechianum, Metopium brownei*, and *Byrsonima crassifolia*; 25–50% of the trees loose their leaves during the dry season.

(d) *Low evergreen lowland forest* (0.088%): "selva baja perennifolia", forest of 4–15 m height, generally in permanently flooeded areas, with characteristic species being *Pachira aquatica, Chrysobalanus icaco*, and *Calophyllum brasiliense*.

Coniferous forest (3.851%):

(a) *Pine forest* (3.607%): "bosque de pino", forest of 8–25 m height, dominated by *Pinus*.

(b) *Juniper forest* (0.146%): "bosque de táscate", open forests of 2–6 m height, dominated by *Juniperus*.

(c) *Fir forest* (0.097%): "bosque de oyamel", forests of 20–40 m height, dominated by *Abies* ("fir"), *Pseudotsuga, Picea*, and *Cupressus*.

Table 1 continued

Moist montane forest (0.892%):
"bosque mesófilo de montaña", humid montane forest of 15–35 m height, with characteristic genera being *Magnolia, Fagus, Oreomunnea, Clethra, Liquidambar, Podocarpus, Styrax, Symplocos,* and *Chiranthodendron.*

Mangrove (0.445%):
"manglar", forest flooded by (salty) sea water, generally of 3–15 m height, in Mexico dominated by the mangrove tree species *Rhizophora mangle, Avicennia germinans, Laguncularia racemosa,* and *Conocarpus erecta.*

Palm forest (0.059%):
"palmar", mostly tropical vegetation dominated by palms (family Arecaceae), up to 40 m high, with typical genera being *Sabal, Brahea, Orbignya,* and *Attalea.*

Gallery forest (0.005%):
"vegetación de galería", conspicuous tree vegetation along rivers and water bodies, where the soil humidity is higher than otherwise in the area. Typical genera are *Astianthus, Ficus, Salix,* and *Taxodium.*

[a] Forest type categories may include secondary vegetation

[b] Forest type explanations are as defined for the Mexican forest inventory 2000 (Palacio-Prieto et al. 2000 and *Anexo VI.7. Diccionario del Inventario Nacional Forestal 2000–2001*), with some details added from Rzedowski (1986)

[c] 100% = 1,945,748 km^2 area of Mexico without Guadalupe island (including inland water bodies)

[d] Excluded, non-forest coverages: scrub (27.460%), agricultural land (16.944%), cultivated grassland (6.644%), induced grassland (5.006%), natural grassland (4.365%), halophilous and gypsumphilous vegetation (2.545%), sand desert vegetation (1.049%), swamp vegetation (0.592%), water bodies (0.532%), land without apparent vegetation (0.511%), human settlements (0.485%), savanna (0.196%), coast dune vegetation (0.163%), suspended irrigated land (0.151%), tree plantations (0.011%), and high-mountain prairie (0.010%)

account differences of relief. In decreasing order, the largest cover is "dry forest" (10.9%), followed by "oak–pine forest" (7.0%), "oak forest" (5.1%), "tropical rain forest" (5.1%), "coniferous forest" (3.9%), "moist montane forest" (0.9%), "mangrove" (0.4%), "palm forest" (0.06%), and "gallery forest" (0.005%). Four forest types in Table 1 include subtypes: for instance, within the "dry forest" type, four different forest subtypes were recognized. For each forest type or subtype, a short definition is given in Table 1. Also, the original Spanish term for the forest subtypes of the Mexican forest inventory is mentioned.

Table 2 shows a comparison of the vegetation cover as calculated from the Mexican forest inventory, with the original vegetation cover as calculated from the

Table 2 Original vegetation cover and vegetation cover in the year 2000 of Mexico

	Original vegetation cover[a]	Cover in 2000[b]	Loss
Forest	52.0%	33.3%	36%
Scrub	38.5%	27.5%	29%
Natural grassland	8.3%	4.4%	47%
Human-made grassland	0.0%	11.7%	
Agricultural land	0.0%	16.9%	
Other coverages	1.2%	6.2%	

[a] Calculated from Rzedowski and Reyna-Trujillo (1990)

[b] Calculated from the data of the Inventario Nacional Forestal 2000 (see Palacio-Prieto et al. 2000)

map of Rzedowski and Reyna-Trujillo (1990). Rather than talking of "original cover" they refer to "potential vegetation", as they do not define a certain point in time as "original", and do not take into account human influences. The term "potential vegetation", however, can be misleading here, as it could imply what type of vegetation could "potentially" and "artificially" be implemented in Mexico by human management. Therefore, we use the term "original cover".

The two references (Palacio-Prieto 2000, and Rzedowski and Reyna-Trujillo 1990) do not use exactly the same classification system. The forests types of each reference nevertheless can be pooled into comparable forest categories. Doing so, the original forest cover was 52.0% of Mexico's total area, which by the year 2000 had been reduced to 33.3%. The corresponding loss given in the last column of Table 2 is 36% of the original forest cover (33.3· 100/52.0). Similarly, the losses are calculated for Mexico's scrub and natural grassland. It is estimated that originally 38.5% of Mexico was covered with scrub, while today it is approximately 27.5%, reflecting a loss of 29%. For the originally smaller area of natural grassland the loss is even larger (47%). This is less obvious, however, because the area of human-induced grasslands has increased considerably, and for many grasslands it is difficult to tell today if it was originally natural or if it is human-induced.

Figure 1a and b present a map with the distribution of the nine forest types of Table 1. In general terms, forests are found predominantly all over western Mexico along the mountain range of the Sierra Madre Occidental, and in most of southern Mexico. Less predominant is the forest cover in northeastern Mexico along the mountain range of the Sierra Madre Oriental. The highest present-day forest cover proportions per geographical square were found in square #32 with 90.54% (States of Campeche and Quintana Roo), square #126 with 89.44% (Durango and Sinaloa), and square #190 with 87.75% (Chihuahua and Sonora). These percentages are given for all squares as proportions by the coefficients of the objective function in Appendix 2.

Diversity patterns of tree species

Figure 2 shows the species numbers per geographical square that were found for the 846 species of this study. The highest tree species diversity occurs in the southern, primarily tropical portions of the country, and diminishes towards the north. The square with most registered tree species is square #38 in southern Veracruz with 344 species, or 41% of the 846 tree species. Most of these species (239 or 69% of the 344) were reported from the 644-hectare Los Tuxtlas reserve. Together with the Los Tuxtlas area, the State of Chiapas in southern Mexico contains many of the tree species: square #8 contains 332 species, followed by square #10 with 326, square #9 with 318, and square #20 with 314 tree species (red squares in Fig. 2).

When elaborating a frequency distribution of the number of squares in which a given species is found, it has a negative exponential form (Fig. 3): 337 species (40% of the 846 species) were found to be restricted spatially to 10 geographical squares or less. Of these, 32 species were found in one square only (distributed over 17 squares). On the other extreme there is one widely distributed species (*Tecoma stans*) that was found in 155 squares (66.2% of the 234 squares).

Before searching for an optimal combination of tree species conservation areas, it is worthwhile to analyse how similar the geographical squares are in species

🖉 Springer

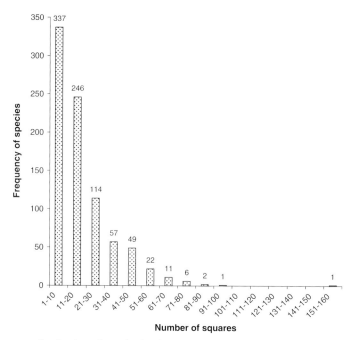

Fig. 3 Frequency distribution of species by the number of squares in which the species is found. Most species are distributed over few squares. Only one species (*Tecoma stans*) was found in 155 squares

composition. Theoretically it could be that groups of geographical squares are similar enough to let a single square of each group represent the diversity of the whole squares-group. To analyze this possibility, we calculated Jaccard similarity coefficients between all 234 squares, based on the tree species compositions. The Jaccard coefficient calculates here between any two squares the number of species that coincide as a proportion of the total number of species found in either of the two squares (Sneath and Sokal 1973). If square-groups existed, then high Jaccard coefficients should result, indicating that several squares share pairwise a large proportion of the same species. The complete procedure results here in $(234^2 - 234)/2 = 27,261$ pairwise comparisons. The 27,261 coefficients are shown in a frequency distribution in Fig. 4. A total of 7,866 pairwise comparisons between squares (28.9%) has no species in common at all (Jaccard coefficient = 0), and most comparisons result in very low coefficients. On the other side, only 49 pairwise comparisons (0.2%) revealed identical species composition. The mean Jaccard coefficient is 0.063, the median 0.021. This means that on the average a comparison of the species between any two geographical squares results in a coincidence of 6.3% common species of all species that are in either one of the two squares. The lack of high Jaccard coefficients makes the search for groups of species-similar squares irrelevant, and causes the optimization of a conservation network to be non-trivial.

Optimization of a reserve network for tree species conservation

Which combination of conservation sites minimizes the needed area for conservation? Linear optimization reveals that a total of 58 squares enter the solution (Fig.5).

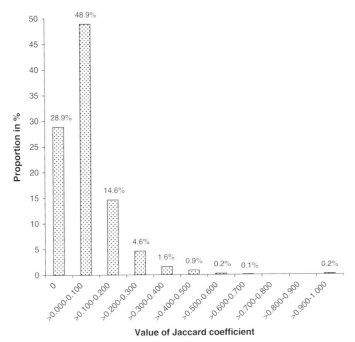

Fig. 4 Frequency distribution of Jaccard similarity coefficients between all 234 squares, based on the tree species compositions. The high frequency of low coefficients indicates that it is non-sensible to look for groups of squares that share the same tree species composition, thus making the optimization of a conservation network non-trivial

The value of the objective function with the optimal solution is 3.9344 squares. In words: the distribution of 846 tree species in 234 geographical squares can be reduced to (rounded) 3.93 squares, while securing that each species is found in a still-forested area corresponding to at least 10% of a square (the latter under the assumption that the species can be found within the suggested forested area).

How sensitive is the optimal solution to a change of the 10%-value (approximately 1,100 km^2) as a lower bound for protection for each species? Using for all species a 5% bound rather than the 10% (a 50% decrease) results in a 50.6% decrease of the optimally protected areas (1.94 instead of 3.93 squares). Using 15% rather than 10% (a 50% increase) results in a 51.5% increase of the optimally protected areas (5.96 instead of 3.93 squares). Consequently, the change of the area in the optimal solution is approximately proportional to the change of the lower bound area for each species. Note that, although not done here, the lower bound area can be chosen individually for each species, according to its distribution and conservation needs.

Another possibility for flexible use of the formulation of the optimization problem lies in varying the maximum possible proportion of the forested area in a given square that could enter a conservation program (here $x_1 \leq 1, x_{1a} \leq 1, x_2 \leq 1,...,$ $x_{242} \leq 1$ in Appendix 2). It could well be that in many (if not most) squares, the idea of protecting all forested area in a given square is politically infeasible, and one has to think only of a proportion that is available. Choosing the forested area of the square that can be set aside for conservation in the optimal solution to be maximally 80% ($x_1 \leq 0.8, x_{1a} \leq 0.8, x_2 \leq 0.8,..., x_{242} \leq 0.8$), the to-be-conserved area increases

🖄 Springer

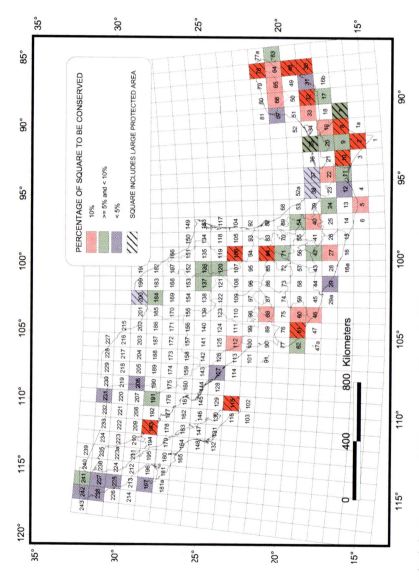

Fig. 5 One optimal solution to conserve all 846 tree species. Maximally 10% of a square is to be conserved. Squares with large protected areas (of at least 1,100 km²) are shaded. The divisions between political states are shown. Note that alternative solutions do exist

slightly by 0.57%. The number of squares in the optimal solution increases from 58 to 60, of which 56 squares are the same as the ones presented in Fig. 5. Note in this approach that again the upper bound x_1, x_{1a}, x_2,..., x_{242} could be chosen individually for each square.

The exact area of the suggested optimal network of protected area in Fig. 5 is calculated in Table 3. The columns provide the square number, the exact area of the square in km^2, the forested proportion (a_i), the proportion of that forest to be conserved (x_i), the percentage of the square to be conserved (a_i· x_i· 100%), the

Table 3 One optimal solution[a]

Square (#)	Area (km^2)	a_i	x_i	Area conserved		Existing reserves (≥ 1,100 km^2)
				(%)	(km^2)	
2	11,966	0.4702	0.21267	10.00	1,197	La Encrucijada & El Triunfo
5	11,973	0.2228	0.44883	10.00	1,197	–
7	11,861	0.2359	0.23018	5.43	644	Montes Azules
8	11,834	0.6334	0.08573	5.43	643	Montes Azules
9	11,909	0.5232	0.17775	9.30	1,107	–
10	11,859	0.5793	0.17262	10.00	1,186	La Sepultura
11	11,888	0.5315	0.10216	5.43	646	–
12	11,862	0.7871	0.00889	0.70	83	–
17	11,765	0.1561	0.59577	9.30	1,094	–
19	11,722	0.2193	0.45600	10.00	1,172	–
20	11,770	0.3199	0.16974	5.43	639	–
22	11,767	0.3898	0.25654	10.00	1,177	–
24	11,751	0.7778	0.06981	5.43	638	–
27	11,728	0.7597	0.13163	10.00	1,173	–
29	11,766	0.2299	0.03045	0.70	82	–
30	11,643	0.1482	0.67476	10.00	1,165	Banco Chinchorro
31	11,697	0.4704	0.09715	4.57	535	–
32	11,617	0.9054	0.11045	10.00	1,162	Calakmul
33	11,671	0.8536	0.11715	10.00	1,167	–
35	11,627	0.0820	1	8.20	953	Pantanos de Centla
37	11,649	0.0447	1	4.47	520	Los Tuxtlas
38	11,649	0.0457	1	4.57	533	Los Tuxtlas
40	11,637	0.2967	0.33704	10.00	1,164	–
42	11,686	0.4508	0.12045	5.43	635	–
46	11,666	0.3527	0.28353	10.00	1,167	–
48	11,532	0.3079	0.32478	10.00	1,153	Sian Ka'an
54	11,534	0.2944	0.18444	5.43	626	–
60	11,518	0.5953	0.16798	10.00	1,152	–
61	11,584	0.5430	0.18416	10.00	1,158	Sierra de Manantlán
62	11,553	0.0930	1	9.30	1,074	–
63	11,455	0.0643	1	6.43	737	–
64	11,448	0.6040	0.16556	10.00	1,145	–
65	11,499	0.5592	0.17883	10.00	1,150	–
66	11,432	0.5725	0.17467	10.00	1,143	–
67	11,443	0.3526	0.01985	0.70	80	–
71	11,440	0.1898	0.28609	5.43	621	–
78	11,357	0.3177	0.31476	10.00	1,136	Yum-Balam
84	11,366	0.5045	0.19822	10.00	1,137	Sierra Gorda
88	11,310	0.6228	0.16056	10.00	1,131	–
106	11,152	0.4814	0.20773	10.00	1,115	El Cielo
112	11,165	0.8654	0.11555	10.00	1,117	–
115	11,144	0.2507	0.39888	10.00	1,114	Sierra de Laguna
120	11,115	0.0843	0.59312	5.00	555	–

Table 3 continued

Square (#)	Area (km^2)	a_i	x_i	Area conserved		Existing reserves (≥ 1,100 km^2)
				(%)	(km^2)	
127	11,082	0.1611	0.04345	0.70	78	–
136	11,028	0.2214	0.22584	5.00	552	–
137	11,044	0.0655	0.76336	5.00	552	–
184	10,808	0.0595	1	5.95	643	–
191	10,853	0.8685	0.10708	9.30	1,009	
193	10,771	0.1628	0.61425	10.00	1,077	Cajón del Diablo
197	10,814	0.0006	1	0.06	6	–
200	10,716	0.0354	1	3.54	379	Maderas del Carmen
206	10,769	0.8366	0.00837	0.70	75	–
225	10,591	0.0392	1	3.92	415	–
231	10,628	0.0300	0.17000	0.51	54	–
237	10,616	0.0441	1	4.41	468	–
238	10,611	0.0208	1	2.08	220	–
241	10,453	0.0508	1	5.08	531	–
242	10,542	0.0389	0.13110	0.51	54	–
SUMA:				393.44	45,136	

[a] There are alternative solutions with the same minimum area, but varying combinations of included squares; a_i = the forested proportion of the square, x_i = the proportion of that forest to be conserved, "Area conserved (%)" = percentage of the square to be conserved ($a_i \cdot x_i \cdot 100\%$)

absolute area in km^2 corresponding to that percentage, and the name of existing reserves with an area of at least 1,100 km^2 in the square (a total of 17). Table 3 shows at the bottom the sum of the percentages being 393.44%, i.e., 3.9344 squares. The sum of the absolute area at the bottom of Table 3 is 45,136 km^2. The average area of the 58 involved squares is 11,472 km^2 (45,136/3.9344), and the average area of protection for each species, corresponding to 10% of a square, is 1,147.2 km^2 (rather than the mentioned 1,100 km^2). This corresponds to an average protected area for each species of 33.9 km by 33.9 km (1,147.2$^{0.5}$) in the optimal solution.

The reserve network proposed in Fig. 5 is not "the last word" on forest conservation. The Simplex algorithm indicates that there are at least 34 alternative solutions, in addition to the optimal solution described here: forcing the algorithm to include one of the following 34 squares in the solution does not cause a change in the objective function value (3.9344 squares), and only causes the combination of squares to change (technically the "reduced cost" is 0): #16, 23, 25, 28, 34, 45, 49, 53, 68, 70, 76–77, 79–80, 90, 103, 126, 142–143, 158, 175–176, 190, 196, 203, 207, 211, 215, 226–228, 230, 233, 236. There may also be combinations of these squares that could be included without increasing the value of the objective function. Furthermore, the "cost" of forcing other non-selected squares into the solution may be low. The maximum possible increase of the objective function, after forcing square #174 into the solution, is 0.87 squares, i.e., the minimum area would increase from 3.93 to 4.81 squares minimum area to conserve all 846 tree species. This provides room to consider other combinations of squares for a network, without an excessive "cost" of land under conservation. For example, one may want to include #28 (Sierra de Guerrero) or #39 (Chinantla region) in the optimal solution.

Note in this context that one can always augment the number of constraints in an optimization problem until the problem at some point becomes infeasible (i.e., the constraints contradict each other). It is fortunate that the problem of this study leaves room for additional constraints at zero or low cost, which provides a robust basis for refining the problem. While some additional squares may enter the optimal solution and other may leave it with such refinement, many of the squares of the optimal solution presented in Fig. 5 will always take part in a feasible solution.

Comparison with complementary analysis

As mentioned in the introduction, one of the most popular algorithms among conservation biologists for selecting reserve networks has been complementary analysis, rather than linear optimization. Therefore, it is of interest to compare our results with the ones obtained by complementary analysis for conserving species richness. Following this procedure, the first square that enters is #38, conserving 344 species or 40.7% of the 846 species. The next, second-priority square is #32, which conserves another 170 species, which together with the first-priority square accumulates 60.8% of the species. The squares that follow are #62 (+136 species), #9 (+42), #80 (+28), #7 (+24), #24 (+13), #78 (+13), #112 (+9), #84 (+8), #2 (+8), #237 (+6), #22 (+6), #33 (+4), #48 (+4), #61 (+3), #115 (+3), #136 (+3), #17 (+2), #64 (+2), #19 (+2), #126 (+1), #94 (+1), #184 (+1), #20 (+1), #25 (+1), #45 (+1), #60 (+1), #120 (+1), #196 (+1), #225 (+1), #5 (+1), #40 (+1), #63 (+1), #66 (+1), #88 (+1), and #238 (+1). In total there are 37 squares in the final solution. Now, there are two options. One could set aside all forested area in the chosen squares, even if for many squares that would imply more than 10% of the square's area. In that case the total conserved area for Mexico would correspond to 13.51 squares. Alternatively, one could conserve a proportion of 10% throughout the 37 squares, in order to take in some way the 10% constraints of the linear optimization model also here into account. In that case the total conserved area would correspond to 3.7 squares, compared with a larger conserved area of 3.93 squares of the linear optimization model of this article. There are, however, two crucial differences between the result of 3.7 squares for complementary analysis and 3.93 squares for linear optimization:

(1) The complementary analysis does not provide as a result that the sum of the conserved area for each species necessarily corresponds to a minimum of 10% of still-forested area. Even in the first option where all forested area in the squares would be set aside for conservation, resulting in a 3.4-times larger conservation area for Mexico (13.51/3.93), there are 33 species in the solution of complementary analysis that would *not* be conserved in at least 10% of still-forested area, though in the linear programming solution this is well feasible. In complementary analysis it is not possible to include such a constraint. For example, *Amphitecna breedlovei*, *A. steyermarkii*, *Cyphomandra rojasiana*, *Micropholis melinoniana*, and *Nectandra purpurea* would be conserved in an area corresponding to only 0.6% of a square.

(2) In the above solution of the complementary analysis, the following protected areas are missing: #8 (part of Montes Azules), #10 (La Sepultura), #30 (Banco Chinchorro), #32 (Calakmul), #35 (Pantanos de Centla), #37 (part of Los Tuxtlas), #106 (El Cielo), #192 (Cajón del Diablo), and #200 (Maderas del Carmen). An adapted version of the algorithm of complementary analysis would allow to include these squares in its solution. But if for comparison we simply run the linear optimization without the 15 protected-area constraints, the resulting total conserved area sums up to only 3.55 squares, 4.1% less than

found in the complementary analysis (3.7 squares): while this difference is not big, it nevertheless shows that the solution of the complementary analysis is finally also suboptimal.

Calculating irreplaceability and vulnerability indices

The evolution of a network of protected forest areas should give priority to those sites first that are most irreplaceable and most vulnerable (Margules and Pressey 2000). *Irreplaceability* measures the importance of the sites to achieve the conservation target, and *vulnerability* measures the likelihood that important features will be lost because of development or degradation. Sites with high irreplaceability and vulnerability should be protected first, given budget limitations, because loss of an irreplaceable site will compromise the achievement of the overall conservation target (Pressey and Taffs 2001; Noss et al. 2002; Lawler et al. 2003).

Knowing that all 846 tree species are contained in the 58 squares of the optimal solution proposed in Fig. 5, these squares could be prioritized for conservation efforts. Let us define the *irreplaceability index of a square* as the average vulnerability index of the species found within it, with the vulnerability index of a species being the inverse of the number of squares in which the species is found in Mexico. For example square #242 contains *Pinus quadrifolia* (found in 6 squares in Mexico), *Pinus jeffreyi* (5 squares), and *Pinus coulteri* (3 squares). The irreplaceability index of this square would then be 0.233 [(1/6 + 1/5 + 1/3)/3].

Furthermore, let us define the *vulnerability index of a square* as the inverse of the proportion of forest within it. For example square #242 would have a vulnerability index of 25.7 (1/0.0389). We could calculate the *priority index for forest conservation* of a given square as the product of the squares' irreplaceability index with its vulnerability index. The priority index of square #242 would then be 5.99 (0.233·25.7). If we carry this exercise out for all the 58 squares shown in the optimal solution in Fig. 5, then the following 14 squares have a priority index for forest conservation that is greater than 1 (in decreasing order): #197, 238, 225, 242, 237, 241, 231, 38, 200, 137, 37, 62, 63, and 184. Square #225 has the highest irreplaceability index, containing species that are found on the average in only 3.5 squares. Square #197 has the highest vulnerability index, because only 0.06% of the square contains forest. These indices provide additional information for deciding about further field work and analysis.

Another goal for protected area development has been compactness. For example Fischer and Church (2003) present an optimization model that minimizes the perimeter of the reserve network. In our country-wide approach, however, it is neither expected nor desirable to have one large conservation area, but rather many reserves should be distributed throughout the country. They are supposed to be accessible by visitors, scientists, and guards. Furthermore, the risk of destruction by fire or pests should be minimized, and genetic diversity should be conserved throughout the country.

Discussion

In Mexico there are at least 2,000 and possibly over 4,000 native tree species that reach a height of 3 m or more, in a flora of over 22,000 species of seed plants. In

🖄 Springer

"megadiverse" countries such as Mexico, scientists are racing against time to accumulate information on tree species and their geographical distribution, before forest habitats—with their biodiversity—fall victim to land use change. Originally, 52.0% of Mexico's land area was covered with forests. In 1999–2000, the total forest area in Mexico represented 33.3% of its land area, when excluding scrub land. Consequently, 36% of Mexico's land, covered originally with forest, has been deforested [(52.0–33.3)· 100%/52.0].

Given this situation, conservation strategies have to be taken in a situation of incomplete scientific information. In this study, 846 tree species (perhaps 37% of Mexico's total number of native tree species) were used to propose a reserve network for tree biodiversity conservation. While one can discuss to what extent the 846 species are representative for Mexico's overall tree diversity, we do neither attempt nor claim that the tree species of this study are a surrogate group (Gaston 1996; Margules and Pressey 2000) for all of Mexico's plants or even for its overall biodiversity or environmental services. Rather, the proposed reserve network is a first attempt to focus forest conservation efforts in Mexico optimally on some critical areas.

The 846 species are not clumped in a few squares in the optimal solution: the mean number of species per square in the optimal solution is 148.3, ranging from 2 species in square #197 (Cedros island) to 344 species in square #38 (Los Tuxtlas forest in Veracruz). Also, species are on the average not restricted to a single square in the optimal solution, thus reducing the risk of extinction: the median number of squares in which a species is found in the optimal solution is 8, ranging from 1 square for 48 species, to 41 squares for 1 species (*Tecoma stans*).

Tree species diversity in our sample is outstanding in southern Mexico. This is partly due to the species selection from sources predominantly from southern Mexico, but it is also expected as a general pattern. High biological diversity in southern Mexico has been reported for vascular plants by Villaseñor et al. (2003), exlusively for the 371 genera of the plant family Asteraceae by Villaseñor et al. (1998), and for mammals by Ceballos et al. (1998).

Using a standard optimization method from engineering and economics, developed by Dantzig (1963), we conclude that efficient land use planning on a national scale would allow to conserve the 846 tree species in a relatively small portion of the country, while taking into account currently existing vegetation and protected areas. An a priori assumption of our employed method is that point localities are representative for a larger forested area within a square, and that the detected species ideally are distributed in a homogeneous way throughout the forested portion of the square in question. For some tree species in some areas this may be true, for many others it will not. Consequently, we could discuss if the employed resolution (here squares of 1 by 1 degree) is the adequate one, and probably conclude that much more field work would be necessary to be able to work with a finer resolution. In practice, however, we will never be able to get to a resolution of single trees on a whole-country scale, and therefore we will never be completely satisfied with the employed resolution. On the other hand, for setting aside whole squares for conservation we would find little or no political support in the real world. Our method provides a way out of this dilemma. The idea to conserve a portion of a square leads us to discuss which portion should be assigned. It is important to emphasize that the optimization method of this paper does not determine where in a given square the conserved part has to be located!

Springer

Rather, analysis remains to be done subsequently, taking care of the previous uncertainty of the field data. It is here that one has to argue for intensive field work still to be carried out. A geographical square is an arbitrary unit, and may contain, for example, lowland rain forest on the one hand and mountainous pine forest on the other. Elevation above sea level with its associated climate variation is probably the most obvious factor that determines different species composition throughout the landscape within a given square. But there are also other important abiotic factors, such as soil and hydrological characteristics (Duivenvoorden 1995; Clark et al. 1999; Svenning et al. 2004).

Take for example square #38 (Los Tuxtlas area in Veracruz): in the lowland rain forest near sea level *Nectandra ambigens* is a typical tree element. Near the top of the volcano San Martín Tuxtla (1,400–1,700 m above sea level), *Nectandra ambigens* will not be found anymore, and a common tree element instead will be *Oreopanax xalapense*, which in turn is not found near sea level. Conserving both tree species within square #38 means assigning one conservation area to the elevated forest and another to the lowland rain forest, and that the two tree species *Nectandra ambigens* and *Oreopanax xalapense* will not be found together in any of the two. Violating in this way the a priori assumption of species being homogeneously distributed throughout the forested portion of the square, where they were detected, will result in a total area for each of the two species that is smaller than the supposed 10% of a square (approximately 1,100 km^2), or whatever value was chosen. In this way we violate a constraint of the optimization problem. While we can discuss case by case how severe this is, one needs to keep in mind that the value of approximately 1,100 km^2 is a reasonable but arbitrary value anyway, and some species will not even be found naturally in such an extended area.

How much conservation area a tree species needs in order to maintain minimum viable populations from a demographic and a genetic viewpoint, is a matter of ongoing discussion and analysis. Álvarez-Buylla et al. (1996) argue for thousands of individuals, even though the few measured populations of tropical rainforest trees revealed effective population sizes of less than 100 individuals. Considering our average conserved area per species of 1,147 km^2 or 114,700 hectares, and an average density within it as low as 0.1 adult tree individuals per hectare of a given species, there would still be a minimum pool of 11,470 adult individuals of each species in the optimal solution of Fig. 5, still enough to fulfill a minimum viable population size of thousands of individuals.

Furthermore, there is an ongoing discussion on how to adapt conservation strategies to global climate change (Bush 2002; Hannah et al. 2002). This consideration is particularly important when the strategy is to minimize the conservation area, because little native area would remain for the tree species to adapt their distributions. Villers-Ruíz and Trejo-Vázquez (1998) concluded that of 33 natural protected forest areas in Mexico, only 9 would remain in the same Holdridge life zone. For example the Sierra de Manantlán region's "subtropical dry forest" (#46, 60–62 in the maps) may change to "subtropical thorn woodland". This consideration on a changing vegetation-type, however, does not imply necessarily a big change of the species composition. Many of the same plant species found as tree species in dry forest, are also found as scrub species in thorn woodland (Rzedowski 1986).

Springer

The general consequence of a temperature increase will be a shift of the preferred distribution of the tree species to the North. Other factors will play in, in particular changes in precipitation patterns, and the current distribution and ecological niche of each tree species. In the Los Tuxtlas rainforest, for example, some species apparently do grow faster with increased temperature, as long as precipitation does not decrease too much (unpublished data from Martin Ricker). In Mexico there is also the interesting pattern that on the one hand there are northern species and genera that reach their southern limit (e.g., *Ulmus*), and on the other there are southern species and genera that reach their northern limit (e.g., *Pouteria*). The nearctic taxa may react different than the neotropical taxa to climate change.

In the optimal solution presented in Fig. 5, there are widely distributed species that remain widely distributed. The average of protected forested area in the optimal solution for all 846 species results to be 0.77 squares, but *Guazuma ulmifolia* is found in an area equivalent to 3.06 squares throughout the network of protected areas, followed by *Tecoma stans* in 2.99 squares. Correspondingly, *Guazuma ulmifolia* is found from 15° latitude to 29° latitude, spanning 14 squares from South to North, and *Tecoma stans* even over 15 squares. The maximum South–North range (18 squares) is reached by *Leucaena leucocephala*, while the average for all species is 5.4 squares. Of concern in climate change scenarios are obviously not these wide-ranging species, but the ones that restrict themselves only within one square from South to North. Calculating the South–North range for each of the 846 species in the optimal solution, reveals 71 species (8.4%) with such a restricted range of 1 square only. The list includes obviously the 18 species that apriori present a restricted distribution of one square only, but also others whose original distribution has been cut down in the optimal solution (e.g., *Attalea butyracea*). The 71 tree species are listed in Appendix 3. They should receive special vulnerability analysis and possibly management considerations, when implementing the optimal solution under the forecasted climate change.

Under the given constraints, the total minimum conservation area needed to conserve all 846 tree species, within the forest vegetation of the year 2000, is 45,136 km^2 (Table 3), corresponding to 2.3% of Mexico's total surface. How does that compare with the existing forest conservation area? From information provided by the Mexican Secretary of the Environment given for the 117 protected areas in 1999 on their web page (http://www.semarnat.gob.mx/sniarn), one can calculate that 36,496 km^2 are protected forest areas, corresponding to 1.9% of Mexico's land area. This area is composed of 41.9% tropical rain forest, 26.6% temperate forest (coniferous, oak–pine, oak), 15.7% dry forest, 12.0% mangrove, 3.6% moist montane forest, 0.1% palm forest, and 0.09% gallery forest. Compared to the existing 1.9% protected forest area, a conservation network corresponding to 2.3% of Mexico's land area for 846 tree species could be a realistic political goal.

Conserving all of Mexico's tree species will most likely require additional area. Furthermore, one has to keep in mind that tree diversity conservation is by no means the only objective when creating forest reserves. Providing recreation areas or protecting aesthetic landscapes are others. As a consequence, the percentage for forest conservation area demanded by society will be larger than 2.3% of Mexico's land area.

At the end, we like to point out some groups of neighboring squares as large geographic areas of interest for follow-up work on conserving tree diversity:

⌦ Springer

(1) The *Chiapas forests that extend into Tabasco, Oaxaca and Veracruz (Los Tuxtlas)*; this is the largest group, consisting of 14 squares: #2, 7–12, 19–20, 22, 24, 35, 37–38. Several of these areas coincide with the important areas suggested in the discussion of canopy tree species of the Atlantic slope rain forests by Wendt (1993). This includes the Los Tuxtlas area of Veracruz (squares #38 and #37), the Tuxtepec area of northern Oaxaca (#24 and not included in our optimal solution #39), the crescent area of extreme southeastern Veracruz (Uxpanapa) and adjacent Oaxaca (Chimalapas) to southern Tabasco and northern Chiapas (#22 and not included in our optimal solution #21), and the Lacandon area (#8 and #7).

(2) The *Yucatán Peninsula forests*, with 12 squares being the second-largest group: #17, 30–33, 48, 63–67, 78. See Ibarra-Manríquez et al. (1995) for a discussion on 437 tree species, of which 54 (12.3%) are endemic to the peninsula. Wendt (1993) also points to the southern Yucatán Peninsula (#17, 30–33) as important for conservation efforts. Especially significant in this area is the Calakmul Biosphere Reserve (#32).

(3) The forests of northern Baja California: #225, 237–238, 241–242.

(4) The Sierra de Manantlán region forest: #46, 60–62. And finally:

(5) The *forests of northeastern Mexico*, including the Biosphere Reserve "El Cielo" in Tamaulipas (#106), western Nuevo León (#120, 136), and the northern portion of the Sierra Madre Oriental in Southern Coahuila (#137).

Acknowledgements This project was initiated with funding from Petróleos Mexicanos (PEMEX) in 1998, and we are grateful for their support. Furthermore, we thank Laura Arriaga and the team from CONABIO for providing data on tree species distributions. Eladio Velasco Sinaca helped with the data processing. Finally, several anonymous reviewers made observations that led to improvements in the manuscript.

Appendix 1: List of 846 tree species and their distributions in 234 geographical squares

The appendix is too extensive to be published here, and therefore is available upon request from the authors (mricker@ibiologia.unam.mx, gibarra@oikos.unam.mx, ems@ibiologia.unam.mx, hmhm@ibiologia.unam.mx). In addition to the sources of the information, it lists the species alphabetically including the plant family, species author, some (important) synonyms, maximum tree height, and the code numbers (as in the maps of Figs. 1, 2, and 5) for the squares in which the species was detected.

Appendix 2: Formulation of the linear optimization problem

MINIMIZE

$0.0204x_1 + 0.0063x_{1a} + 0.4702x_2 + 0.1246x_3 + 0.0438x_4 + 0.2228x_5 +$
$0.0189x_6 + 0.2359x_7 + 0.6334x_8 + 0.5232x_9 + 0.5793x_{10} + 0.5315x_{11} +$
$0.7871x_{12} + 0.6836x_{13} + 0.7325x_{14} + 0.4283x_{15} + 0.1425x_{16} + 0.1561x_{17} +$
$0.1327x_{18} + 0.2193x_{19} + 0.3199x_{20} + 0.3189x_{21} + 0.3898x_{22} + 0.4057x_{23} +$
$0.7778x_{24} + 0.5394x_{25} + 0.7265x_{26} + 0.7597x_{27} + 0.6779x_{28} + 0.2299x_{29} +$
$0.1482x_{30} + 0.4704x_{31} + 0.9054x_{32} + 0.8536x_{33} + 0.2610x_{34} + 0.0820x_{35} +$
$0.0283x_{36} + 0.0447x_{37} + 0.0457x_{38} + 0.2799x_{39} + 0.2967x_{40} + 0.4812x_{41} +$
$0.4508x_{42} + 0.5155x_{43} + 0.6796x_{44} + 0.7513x_{45} + 0.3527x_{46} + 0.0032x_{47} +$
$0.3079x_{48} + 0.8225x_{49} + 0.8372x_{50} + 0.5119x_{51} + 0.0060x_{52} + 0.1605x_{53} +$
$0.2944x_{54} + 0.2005x_{55} + 0.2380x_{56} + 0.4949x_{57} + 0.4409x_{58} + 0.4991x_{59} +$
$0.5953x_{60} + 0.5430x_{61} + 0.0930x_{62} + 0.0643x_{63} + 0.6040x_{64} + 0.5592x_{65} +$
$0.5725x_{66} + 0.3526x_{67} + 0.0048x_{68} + 0.0488x_{69} + 0.3304x_{70} + 0.1898x_{71} +$
$0.1150x_{72} + 0.0497x_{73} + 0.0472x_{74} + 0.2627x_{75} + 0.6300x_{76} + 0.3702x_{77} +$
$0.0627x_{77a} + 0.3177x_{78} + 0.1603x_{79} + 0.1385x_{80} + 0.0302x_{81} + 0.0843x_{82} +$
$0.1270x_{83} + 0.5045x_{84} + 0.2432x_{85} + 0.1755x_{86} + 0.1398x_{87} + 0.6228x_{88} +$
$0.5415x_{89} + 0.1190x_{90} + 0.0199x_{91} + 0.0244x_{92} + 0.0944x_{93} + 0.4275x_{94} +$
$0.1041x_{95} + 0.0284x_{96} + 0.1053x_{97} + 0.4262x_{98} + 0.8297x_{99} + 0.4561x_{100} +$
$0.0014x_{101} + 0.0072x_{102} + 0.0000x_{103} + 0.0419x_{104} + 0.4204x_{105} + 0.4814x_{106} +$
$0.0216x_{107} + 0.0044x_{108} + 0.0100x_{109} + 0.2868x_{110} + 0.5681x_{111} + 0.8654x_{112} +$
$0.4084x_{113} + 0.0002x_{114} + 0.2507x_{115} + 0.1011x_{116} + 0.0071x_{117} + 0.0776x_{118} +$
$0.3070x_{119} + 0.0843x_{120} + 0.0456x_{121} + 0.0031x_{122} + 0.0488x_{123} + 0.1313x_{124} +$
$0.7674x_{125} + 0.8944x_{126} + 0.1611x_{127} + 0.0074x_{128} + 0.0011x_{129} + 0.0295x_{130} +$
$0.0213x_{131} + 0.0080x_{132} + 0.0064x_{133} + 0.0842x_{134} + 0.1299x_{135} + 0.2214x_{136} +$
$0.0655x_{137} + 0.0095x_{138} + 0.0060x_{139} + 0.0701x_{140} + 0.5418x_{141} + 0.8682x_{142} +$
$0.7915x_{143} + 0.0836x_{144} + 0.0144x_{145} + 0.0010x_{146} + 0.0173x_{147} +$
$0.0117x_{150} + 0.1548x_{151} + 0.0322x_{152} + 0.0180x_{153} + 0.0025x_{154} +$
$0.0034x_{155} + 0.0106x_{156} + 0.3308x_{157} + 0.8302x_{158} + 0.8720x_{159} + 0.6685x_{160} +$
$0.0770x_{161} + 0.0140x_{163} + 0.0311x_{164} + 0.0024x_{165} + 0.0351x_{166} +$
$0.0550x_{167} + 0.0347x_{168} + 0.0627x_{169} + 0.0034x_{170} + 0.0003x_{171} + 0.0216x_{172} +$
$0.4018x_{173} + 0.8731x_{174} + 0.8402x_{175} + 0.5650x_{176} + 0.0487x_{177} + 0.0003x_{178} +$
$0.0011x_{179} + 0.0041x_{180} + 0.1039x_{183} + 0.0595x_{184} +$
$0.0059x_{185} + 0.0056x_{186} + 0.0202x_{187} + 0.3364x_{188} + 0.4533x_{189} + 0.8775x_{190} +$
$0.8685x_{191} + 0.3139x_{192} + 0.1628x_{193} + 0.0006x_{194} + 0.0017x_{195} + 0.0000x_{196} +$
$0.0006x_{197} + 0.0007x_{198} + 0.0160x_{199} + 0.0354x_{200} + 0.0006x_{201} + 0.0037x_{202} +$
$0.0000x_{203} + 0.2109x_{204} + 0.4163x_{205} + 0.8366x_{206} + 0.4387x_{207} + 0.2724x_{208} +$

$0.3141x_{209} + 0.0110x_{210} + 0.0000x_{211} + 0.0003x_{212} +$

$0.0000x_{215} + 0.0026x_{216} + 0.0105x_{217} + 0.0946x_{218} + 0.4736x_{219} + 0.3060x_{220} +$

$0.3115x_{221} + 0.0822x_{222} + 0.0532x_{223} + 0.0392x_{225} + 0.0000x_{226} +$

$0.0000x_{227} + 0.0000x_{228} + 0.0072x_{229} + 0.0664x_{230} + 0.0300x_{231} + 0.1076x_{232} +$

$0.0997x_{233} + 0.0118x_{234} + 0.0013x_{235} + 0.0000x_{236} + 0.0441x_{237} + 0.0208x_{238} +$

$0.0508x_{241} + 0.0389x_{242}$

SUBJECT TO

{DISTRIBUTION CONSTRAINTS:}

{*Acacia acatlensis*}

$0.4702x_2 + 0.7871x_{12} + 0.7778x_{24} + 0.5394x_{25} + 0.7597x_{27} + 0.2967x_{40} + 0.4812x_{41} +$

$0.4508x_{42} + 0.5155x_{43} + 0.5953x_{60} + 0.0930x_{62} + 0.2627x_{75} + 0.1398x_{87} +$

$0.8654x_{112} \geq 0.1$

...826 constraint inequalities in accordance with the distributions in Appendix 1...

{*Zygia stevensonii*}

$0.2193x_{19} + 0.4704x_{31} + 0.2610x_{34} + 0.3079x_{48} + 0.8225x_{49} + 0.8372x_{50} + 0.0643x_{63} +$

$0.3177x_{78} \geq 0.1$

{CONSTRAINTS FOR SPECIES WITH A TOTAL EXTENSION SMALLER THAN 10% OF A SQUARE:}

{*Bernardia spongiosa*}	$0.0930\,x_{62} \geq 0.0930$
{*Ceiba grandiflora*}	$0.0930\,x_{62} \geq 0.0930$
{*Erythroxylum panamense*}	$0.0457\,x_{38} \geq 0.0457$
{*Eugenia sotoesparzae*}	$0.0457\,x_{38} \geq 0.0457$
{*Eupatorium galeottii*}	$0.0457\,x_{38} \geq 0.0457$
{*Jacquinia arborea*}	$0.0643\,x_{63} \geq 0.0643$
{*Jatropha bullockii*}	$0.0930\,x_{62} \geq 0.0930$
{*Lonchocarpus unifoliolatus*}	$0.0457\,x_{38} \geq 0.0457$
{*Parathesis conzattii*}	$0.0457\,x_{38} \geq 0.0457$
{*Parathesis psychotrioides*}	$0.0457\,x_{38} \geq 0.0457$
{*Pinus contorta*}	$0.0392\,x_{225} + 0.0508x_{241} \geq 0.0900$
{*Pinus lambertiana*}	$0.0392\,x_{225} + 0.0441x_{237} \geq 0.0833$
{*Pinus muricata*}	$0.0208\,x_{238} \geq 0.0208$
{*Pinus radiata*}	$0.0000\,x_{196} + 0.0006x_{197} \geq 0.0006$
{*Pouteria rhynchocarpa*}	$0.0457\,x_{38} \geq 0.0457$
{*Psychotria sarapiquensis*}	$0.0457\,x_{38} \geq 0.0457$
{*Sloanea medusula*}	$0.0457\,x_{38} \geq 0.0457$
{*Sloanea petenensis*}	$0.0457\,x_{38} \geq 0.0457$

{CONSTRAINTS FOR FORCING LARGE EXISTISTING CONSERVATION AREAS INTO THE OPTIMAL SOLUTION:}

{Banco Chinchorro}	$0.1482\ x_{30} \geq 0.1$
{Cajón del Diablo}	$0.1628\ x_{192} \geq 0.1$
{Calakmul}	$0.9054\ x_{32} \geq 0.1$
{El Cielo}	$0.4814\ x_{106} \geq 0.1$
{La Encrucijada + El Triunfo}	$0.4702\ x_2 \geq 0.1$
{La Sepultura}	$0.5793\ x_{10} \geq 0.1$
{Los Tuxtlas}	$0.0447\ x_{37} + 0.0457\ x_{38} \geq 0.0904$
{Maderas del Carmen}	$0.0354\ x_{200} \geq 0.0354$
{Montes Azules}	$0.2359\ x_7 + 0.6334\ x_8 \geq 0.1$
{Pantanos de Centla}	$0.0820\ x_{35} \geq 0.0820$
{Sian Ka'an}	$0.3079\ x_{48} \geq 0.1$
{Sierra de Laguna}	$0.2507\ x_{115} \geq 0.1$
{Sierra de Manantlán}	$0.5430\ x_{61} \geq 0.1$
{Sierra Gorda}	$0.5045\ x_{84} \geq 0.1$
{Yum-Balam}	$0.3177\ x_{78} \geq 0.1$

{CONSTRAINTS THAT THE COEFFICIENTS SHALL BE PROPORTIONS, TAKING ON VALUES ONLY BETWEEN 0 AND 1:}

$$x_1 \geq 0, x_{1a} \geq 0, x_2 \geq 0, \ldots, x_{242} \geq 0$$
$$x_1 \leq 1, x_{1a} \leq 1, x_2 \leq 1, \ldots, x_{242} \leq 1$$

Appendix 3: Tree species with restricted south-north range (1 square only)

Amphitecna regalis, A. steyermarkii, Anisocereus gaumeri, Antirhea lucida, Attalea butyracea, Bernardia spongiosa, B. wilburi, Bunchosia mcvaughii, Carpodiptera ameliae, Cassia hintonii, Ceiba grandiflora, Cordia bicolor, C. salvadorensis, Cyphomandra rojasiana, Ebenopsis ebano, Eremosis oolepis, Erythrina caribaea, Erythroxylum panamense, Esenbeckia flava, Eugenia sotoesparzae, Eupatorium galeottii, Hamelia longipes, Jacquinia arborea, Jatropha bullockii, Licania gonzalezii, L. mexicana, L. sparsipilis, Lonchocarpus cochleatus, L. hintonii, L. mutans, L. unifoliolatus, Malpighia novogaliciana, Mappia racemosa, Micropholis melinoniana, Mosiera contrerasii, Nectandra leucocome, N. purpurea, Nopalea inaperta, Opuntia excelsa, Parathesis conzattii, P. psychotrioides, Parmentiera parviflora, Pereskia lychnidiflora, Phyllanthus grandifolius, Pilosocereus gaumeri, Pinus jaliscana, P. maximartinezii, P. muricata, P. nelsoni, native germplasm of *P. radiata, P. rzedowskii, Piranhea mexicana, Pourouma bicolor, Pouteria rhynchocarpa, P. squamosa, Prosopis glandulosa, Psychotria sarapiquensis, Quararibea fieldii, Rhacoma puberula, Rinorea deflexiflora, R. uxpanapana, Sebastiania tikalana, Sideroxylon eucoriaceum, S. excavatum, S. lanuginosum, S. peninsulare, Sloanea medusula, S. petenensis, Stenocereus eichlamii, S. laevigatus*, and *Xylosma velutinum*.

References

Álvarez-Buylla ER, García-Barrios R, Lara-Moreno C, Martínez-Ramos M (1996) Demographic and genetic models in conservation biology: applications and perspectives for tropical rain forest tree species. Ann Rev Ecol Syst 27:387–421

Bedward M, Pressey RL, Keith DA (1992) A new approach for selecting fully representative reserve networks: addressing efficiency, reserve design and land suitability with an iterative analysis. Biol Conserv 62:115–125

Berg CC (1972, reprinted 1985) Olmedieae, Brosimeae (Moraceae), Flora Neotropica Monograph 7. The New York Botanical Garden, Bronx, New York, USA, 229 p

Bohs L (1994) Cyphomandra (Solanaceae), Flora Neotropica Monograph 63. The New York Botanical Garden, Bronx, New York, USA, 176 p

Bush MB (2002) Distributional change and conservation on the Andean flank: a palaeoecological perspective. Global Ecol Biogeogr 11:463–473

Cairns MA, Dirzo R, Zadroga F (1995) Forests of Mexico. J For 93(7):21–24

Ceballos G, Rodríguez P, Medellín RA (1998) Assessing conservation priorities in megadiverse Mexico: mammalian diversity, endemicity, and endangerment. Ecol Appl 8:8–17

Church R, ReVelle C (1974) The maximal covering location problem. Pap Reg Sci 32:101–118

Church RL, Stoms DM, Davis FW (1996) Reserve selection as a maximal covering location problem. Biol Conserv 76:105–112

Church RL, Gerrard R, Hollander A, Stoms DM (2000) Understanding the tradeoffs between site quality and species presence in reserve site selection. For Sci 46:157–167

Clark DB, Palmer MW, Clark DA (1999) Edaphic factors and the landscape-scale distributions of tropical rain forest trees. Ecology 80(8):2662–2675

Cowan RS (1967, reprinted 1987) Swartzia (Leguminosae, Caesalpinioideae, Swartzieae), Flora Neotropica Monograph 1. The New York Botanical Garden, Bronx, New York, USA, 228 p

Csuti B, Polasky S, Williams PH, Pressey RL, Camm JD, Kershaw M, Kiester AR, Downs B, Hamilton R, Huso M, Sahr K (1997) A comparison of reserve selection algorithms using data on terrestrial vertebrates in Oregon. Biol Conserv 80:83–97

Dantzig GB (1951) Maximization of a linear function of variables subject to linear inequalities. In: Koopmans TC (ed) Activity analysis of production and allocation. John Wiley & Sons, New York, USA, pp 339–347

Dantzig GB (1963) Linear programming and extensions. Princeton University Press, Princeton, NJ, USA, 625 p

Dirzo R, García MC (1992) Rates of deforestation in Los Tuxtlas, a neotropical area in Southeast Mexico. Conserv Biol 6:84–90

Dorfman R, Samuelson PA, Solow RM (1958, republished in 1987). Linear programming and economic analysis. Dover Publications, New York, USA, 525 p

Duivenvoorden JF (1995) Tree species composition and rain forest–environment relationships in the middle Caquetá area, Colombia, NW Amazonia. Vegetatio 120:91–113

Dykstra DP (1984) Mathematical programming for natural resource managament. McGraw-Hill, New York, USA, 318 p

Farjon A, Styles BT (1997) Pinus (Pinaceae), Flora Neotropica Monograph 75. The New York Botanical Garden, Bronx, New York, USA, 293 p

Fischer DT, Church RL (2003) Clustering and compactness in reserve site selection: an extension of the biodiversity management area selection model. For Sci 49: 555–565

Gaston KJ (1996) Species richness: measure and measurement. In: Gaston KJ (ed) Biodiversity: a biology of numbers and difference. Blackwell Science, Oxford, UK, pp 77–113

Gentry AH (1980) Bignoniaceae – Part I (Crescentieae and Tourrettieae), Flora Neotropica Monograph 25(I). The New York Botanical Garden, Bronx, New York, USA, 131 p

Gentry AH (1992) Bignoniaceae – Part II (Tribe Tecomeae), Flora Neotropica Monograph 25(II). The New York Botanical Garden, Bronx, New York, USA, 371 p

Gilbert KC, Holmes DD, Rosenthal RE (1985) A multiobjective discrete optimization model for land allocation. Manage Sci 31:1509–1522

Gómez-Hinostrosa C, Hernández HM (2000) Diversity, geographical distribution, and conservation of Cactaceae in the Mier y Noriega region, Mexico. Biodivers Conserv 9:403–418

Gómez-Rodríguez G, Palacio-Prieto JL (2001) The Mexican national forest inventory 2000–2001: updating land cover maps using Landsat 7 ETM+ images. In: Proceedings of the International Symposium on Spectral Sensing Research, 11–15 June 2001, Québec City, Canada. International Society for Photogrammetry and Remote Sensing (ISPRS), pp 393–401

[380]

Hannah L, Midgley GF, Millar D (2002) Climate change-integrated conservation strategies. Global Ecol Biogeogr 11:485–495

Hekking WHA (1988) Violaceae Part I – *Ronorea* and *Rinoreocarpus*, Flora Neotropica Monograph 46. The New York Botanical Garden, Bronx, New York, USA, 208 p

Hernández HM, Bárcenas RT (1996) Endangered cacti in the Chihuahuan desert: II. Biogeography and conservation. Conserv Biol 10:1200–1209

Hillier FS, Lieberman GJ (1990) Introduction to operations research. McGraw-Hill, New York, USA, 954 p

Ibarra-Manríquez G. 1996. Biogeografía de los árboles nativos de la Península de Yucatán: un enfoque para evaluar su grado de conservación. Tesis de Doctorado en Biología, Facultad de Ciencias, Universidad Nacional Autónoma de México, México D.F., Mexico, 189 p

Ibarra-Manríquez G, Sinaca S (1995) Lista florística comentada de la Estación de Biología Tropical "Los Tuxtlas", Veracruz, México. Revista Biol Trop 43:75–115

Ibarra-Manríquez G, Sinaca S (1996a) Estación de Biología Tropical "Los Tuxtlas", Veracruz, México: Lista florística comentada (Mimosaceae a Verbenaceae). Revista Biol Trop 44:41–60

Ibarra-Manríquez G, Sinaca S (1996b) Lista comentada de plantas de la Estación de Biología Tropical "Los Tuxtlas", Veracruz, México (Violaceae-Zingiberaceae). Revista Biol Trop 44:427–447

Ibarra-Manríquez G, Villaseñor JL, Durán-García R (1995) Riqueza de especies y endemismo del componente arbóreo de la Península de Yucatán, México. Bol Soc Bot México 57:49–77

Kaastra RC (1982) Pilocarpinae (Rutaceae), Flora Neotropica Monograph 33. The New York Botanical Garden, Bronx, New York, USA, 198 p

Landrum LR (1986) *Campomanesia, Pimenta, Blepharocalyx, Legrandia, Acca, Myrrhinium*, and *Luma* (Myrtaceae). Flora Neotropica Monograph 45. The New York Botanical Garden, Bronx, New York, USA, 179 p

Lawler JJ, White D, Master LL (2003) Integrating representation and vulnerability: two approaches for prioritizing areas for conservation. Ecol Appl 13:1762–1772

Lott EJ (1993) Annotated checklist of the vascular flora of the Chamela Bay region, Jalisco, Mexico. Occasional Pap Calif Acad Sci 148:1–60

Maas PJM, Westra LYT, collaborators (1992) *Rollinia*, Flora Neotropica Monograph 57. The New York Botanical Garden, Bronx, New York, USA, 190 p

Margules CR, Pressey RL (2000) Systematic conservation planning. Nature 405:243–253

Margules CR, Nicholls AO, Pressey RL (1988) Selecting networks of reserves to maximise biological diversity. Biol Conserv 43:63–76

Mas JF, Velázquez A, Palacio-Prieto JL, Bocco G, Peralta A, Prado J (2002) Assessing forest resources in Mexico: wall-to-wall land use/cover mapping. Photogramm Eng Remote Sens 68:966–968

Miranda F, Hernández- XE (1963) Los tipos de vegetación de México y su clasificación. Bol Soc Bot México 28:29–179

Nicholls AO, Margules CR (1993) An upgraded reserve selection algorithm. Biol Conserv 64:165–169

Noss RF, Carroll C, Vance-Borland K, Wuerthner G (2002) A multicriteria assessment of the irreplaceability and vulnerability of sites in the greater Yellowstone ecosystem. Conserv Biol 16:895–908

Palacio-Prieto JL, Bocco G, Velázquez A, Mas JF, Takaki-Takaki F, Victoria A, Luna-González L, Gómez-Rodríguez G, López-García J, Palma M, Trejo-Vázquez I, Peralta A, Prado-Molina J, Rodríguez-Aguilar A, Mayorga-Saucedo R, González F (2000) La condición actual de los recursos forestales en México: resultados del Inventario Forestal Nacional 2000. Bol Inst Geogr UNAM 43:183–203

Pennington TD (1990) Sapotaceae, Flora Neotropica Monograph 52. The New York Botanical Garden, Bronx, New York, USA, 771 p

Pennington TD, Sarukhán J (1998) Árboles tropicales de México. Universidad Nacional Autónoma de México, and Fondo de la Cultura Económica, México D.F., Mexico, 521 p

Pennington TD, Styles BT, Taylor DAH (1981) Meliaceae, Flora Neotropica Monograph 28. The New York Botanical Garden, Bronx, New York, USA, 472 p

Poppendieck HH (1981) Cochlospermaceae, Flora Neotropica Monograph 27. The New York Botanical Garden, Bronx, New York, USA, 34 p

Prance TP (1972) Chrysobalanaceae, Flora Neotropica Monograph 9. The New York Botanical Garden, Bronx, New York, USA, 410 p

Prance TP (1989) Chrysobalanaceae, Flora Neotropica Monograph 9-S. The New York Botanical Garden, Bronx, New York, USA, 268 p
Pressey RL, Taffs KH (2001) Scheduling conservation action in production landscapes: priority areas in western New South Wales defined by irreplaceability and vulnerability to vegetation loss. Biol Conserv 100:355–376
ReVelle CS, Williams JC, Boland JJ (2002) Counterpart models in facility location science and reserve selection science. Environ Model Assess 7:71–80
Rodrigues ASL, Gaston KJ (2002) Optimisation in reserve selection procedures – why not? Biol Conserv 107:123–129
Rohwer JG (1993) Lauraceae: *Nectandra*, Flora Neotropica Monograph 60. The New York Botanical Garden, Bronx, New York, USA, 333 p
Rzedowski J (1986) Vegetación de México. Limusa, México D.F., Mexico, 432 p
Rzedowski J (1991a) Diversidad y orígenes de la flora fanerogámica de México. Acta Bot Mex 14:3–21
Rzedowski J (1991b) El endemismo en la flora fanerogámica mexicana: una apreciación analítica preliminar. Acta Bot Mexicana 15:47–64
Rzedowski J (1993) Diversity and origins of the phanerogamic flora of Mexico. In: Ramamoorthy TP, Bye R, Lot A, Fa J (eds) Biological diversity of Mexico: origins and distributions. Oxford University Press, New York, USA, pp 129–144
Rzedowski J, Reyna-Trujillo T (1990) Vegetación potencial. Map IV.8.2 of the Atlas Nacional de México, Instituto de Geografía, Universidad Nacional Autónoma de México, México D.F., Mexico
Sætersdal M, Line JM, Birks HJB (1993) How to maximize biological diversity in nature reserve selection: vascular plants and breeding birds in deciduous woodlands, Western Norway. Biol Conserv 66:131–138
Sleumer HO (1980) Flacourtiaceae, Flora Neotropica Monograph 22. The New York Botanical Garden, Bronx, New York, USA, 499 p
Sleumer HO (1984) Olacaceae, Flora Neotropica Monograph 38. The New York Botanical Garden. Bronx, New York, USA, 159 p
Sneath PHA, Sokal RR (1973) Numerical taxonomy. W.H. Freeman and Company, San Francisco, USA, 573 p
Snyder SA, Tyrell LE, Haight RG (1999) An optimization approach to selecting research natural areas in national forests. For Sci 45:458–469
Sousa M, Ricker M, Hernández HM (2001) Tree species of the family Leguminosae in Mexico. Harvard Pap Bot 6:339–365
Sousa M, Ricker M, Hernández HM (2003) An index for the tree species of Leguminosae in Mexico. Harvard Pap Bot 7:381–398
Standley PC (1920–1926) Trees and shrubs of Mexico. Contrib US Natl Herbarium 23:1–1721
Svenning JC, Kinner DA, Stallard RF, Engelbrecht BMJ, Wright SJ (2004) Ecological determinism in plant community structure across a tropical forest landscape. Ecology 85(9):2526–2538
Todzia CA (1988) Chloranthaceae: *Hedyosmum*, Flora Neotropica Monograph 48. The New York Botanical Garden, Bronx, New York, USA, 139 p
Underhill LG (1994) Optimal and suboptimal reserve selection algorithms. Biol Conserv 70:85–87
Vane-Wright RI, Humphries CJ, Williams PH (1991) What to protect?—Systematics and the agony of choice. Biol Conser 55:235–254
Varela S, Reyes C, Becerra G (coordinators) (1999) Atlas forestal de México. Secretaría de Medio Ambiente, Recursos Naturales y Pesca, México D.F., and Universidad Autónoma de Chapingo, Texcoco, Mexico, 103 p
Villaseñor JL, Ibarra-Manríquez G (1998) La riqueza arbórea de México. Bol Inst Bot Univ Guadalajara 5:95–105
Villaseñor JL, Ibarra-Manríquez G, Ocaña D (1998) Strategies for the conservation of Asteraceae in Mexico. Conserv Biol 12:1066–1075
Villaseñor JL, Meave JA, Ortiz E, Ibarra-Manríquez G (2003). In: Morrone JJ, Llorente-Bousquets J (eds) Biogeografía y conservación de los bosques tropicales húmedos de México: una perspectiva latinoamericana de la biogeografía. Facultad de Ciencias, Universidad Nacional Autónoma de México, México D.F. Mexico, pp 209–216
Villers-Ruíz L, Trejo-Vázquez I (1998) Climate change on Mexican forests and natural protected areas. Global Environ Change 8(2):141–157

[382]

Wendt T (1993) Composition, floristic affinities, and origins of the canopy tree flora of the Mexican Atlantic slope rain forests. In: Ramamoorthy TP, Bye R, Lot A, Fa J (eds) Biological diversity of Mexico: origins and distributions. Oxford University Press, New York, USA, pp 595–680

Williams P, Gibbons D, Margules C, Rebelo A, Humphries C, Pressey R (1996) A comparison of richness hotspots, rarity hotspots, and complementary areas for conserving diversity of British birds. Conserv Biol 10:155–174

Biodivers Conserv (2007) 16:1959–1971
DOI 10.1007/s10531-006-9115-9

ORIGINAL PAPER

Seeing the wood for the trees: how conservation policies can place greater pressure on village forests in southwest China

David Melick · Xuefei Yang · Jianchu Xu

Received: 7 August 2006 / Accepted: 8 August 2006 / Published online: 27 October 2006
© Springer Science+Business Media B.V. 2006

Abstract In the last 6 years China has introduced a number of policies to try and conserve forests and protect watershed integrity; these include a ban on commercial logging, reforestation projects, restrictions on upland farming and burning, and controls on livestock grazing. The blanket nature of these impositions when combined with rapid socio-economic changes have increased pressures on many small rural communities. In this paper, we examine the case of Jisha Village in northwestern Yunnan, China—a typical rural Tibetan community sustained by traditional agriculture and livestock management. The cessation of commercial logging has seen the community turn to towards other income streams such as non-timber forest products (NTFP), increased livestock and attempts to foster tourism. However, timber quotas together with new road access have spurred the development of unofficial markets for village firewood and enhanced access to nearby forests. In addition, the decline of bamboo—a traditional fencing material—has resulted in an estimated 35-fold increase in demand for pine wood. Wood demands in this community are swiftly exceeding the sustainable harvest levels. Forest loss does not merely represent the depletion or degradation of future village timber resources, but also the loss of NTFP habitat. Moreover, due to proscriptions on rangeland burning, pasturelands are becoming degraded and grazing in forests is more intensive—reducing forest regeneration. These findings support calls to improve the flexibility and incorporate local needs into forest policy—the problems highlighted here seem indicative of the practical and philosophical challenges facing environmental planning and research in China.

Keywords Community forest · Firewood collection · Logging ban · Non-timber forest products · Northwest Yunnan · *Pinus densata* · Tibetan villagers

D. Melick (✉) · X. Yang · J. Xu
Department of Biogeography and Ecology, Kunming Institute of Botany (KIB), Heilongtan, Kunming, Yunnan 650204, China
e-mail: dmelick@mail.kib.ac.cn

Introduction

In August 1998, after catastrophic downstream flooding, China's government introduced an immediate ban on all commercial logging of state forests in 18 provinces and autonomous regions (*Natural Forest Protection Program*, NFPP). Although applied generally, the NFPP is especially targeted at the headwaters of the Yangtze and, to a lesser extent, Yellow rivers (Zhang et al. 2000). Despite the fact that the ban was designed for state forest regions of specific environmental sensitivity, in 2001 Xinjiang instituted a complete logging ban in all natural forests regardless of ownership—apparently to impress the central government. Sichuan and Yunnan soon followed suit and introduced similar logging bans, which were supported by the central government (Zuo 2001, cited in Miao and West 2004). Along with the logging ban, the government also introduced a policy to reafforest sloping farm lands (*Sloping Land Conversion Program*, SLCP), together with more stringent controls on shrub and pasture burning.

Six years on, the success of these policies is a matter of perspective. The large investment in reforestation and the commitment to stabilize mountain catchments has been widely applauded by international forestry groups such as CIFOR (2005). While there is no doubt that Chinese policies to increase forest cover are admirable, it is still too early to know if downstream water quality and flood control have improved, and the quality of the new forests is debatable. Many replanted forests are monocultures, and areas designated as forest may be scrub communities of limited biodiversity and natural resource value (Rozelle et al. 2003; Willson 2006). Moreover, while the NFPP and SCLP may deliver long-term national benefits, these broad-reaching environmental policy changes have potentially large impacts upon local communities. The most common concerns include the loss and/or confusion of community user's rights, reduction in agricultural and grazing land, and poor targeting of appropriate ecological areas (Miao and West 2004; Xu and Ribot 2004; Xu et al. 2004; Weyerhaeuser et al. 2005). Such changes, coupled with past over-exploitation and degradation of natural resources, together with rapid socioeconomic changes, could exacerbate existing poverty differentials and place the future of rural minority communities in a very vulnerable position.

Within affected local communities, the most obvious change from the logging ban is the drastic loss of income. Losses vary but are generally very significant in affected regions; in Sichuan official revenues reduced by up to 64–75% (Winkler 2003; Miao and West 2004), while in Diqing—the subject of this study—logging accounted for up to 80% of regional income (Hillman 2003). Although commercial logging was effectively a state monopoly carried out by provincial or county logging enterprises, many villagers were dependent upon this industry, either working directly for logging companies or investing heavily to provide transport and support services (Zheng 2004). In addition it has been widely discussed that though local government may have borne the brunt of massive revenue decreases, this resulted in the reduction or loss of services delivery to communities (e.g. compensation for income and/or resources loss, education services, development of alternative industries)—crucially at a time when many communities are in greatest need of local government assistance and direction. The central government was certainly aware of the challenges facing affected communities and outlined several goals, noting the need to reduce and meet demands for wood products, reduce rural firewood

[386]

dependency and rapidly develop alternative revenue sources such as tourism and animal husbandry (Zhang et al. 2000; Winkler 2003). Such aspirations to improve standards and reduce poverty are clearly behind the *Develop the West* policies, which aim to bring new industries and opportunities to these regions as rapidly as possible. Immediately obvious results of this in Diqing are the expansion of road and transport links, attempts to improve communications and considerable development of tourist infrastructure.

The desire to conserve while promoting development is at the crux of the problems facing China—it is neatly encapsulated by the rural village communities, which are being pulled simultaneously in different directions by competing needs and interests. In many cases, threats to rural livelihoods are exacerbated, not only by the logging ban but also by the reduction of traditional materials and resources because of past state-driven over-exploitation. Moreover, associated conservation policies, such as the drive to reforest and bans on burning are also conspiring against moves to increase animal husbandry, placing further pressure on the already stretched local agricultural and forest systems. Similarly, development and poverty alleviation projects, such as road access and exposure to market forces have a double barrel effect of creating a rapid need for disposable cash income, while also facilitating greater access to village forest areas for exploitation.

In a recent survey of the Tibetan community at Jisha village in northwest Yunnan, we found that local use of fuelwood forest had intensified very rapidly in the last few years—having potentially serious ramifications for the intermediate to long-term sustainability of these community pine forests (Melick and Yang, unpublished data). In this paper we trace the causes of these changes, further investigating and quantifying the changing demands and pressures on this community with respect to their use of their traditional forest resources. We track the recent rapid political and ecological changes affecting this village and examine the long-term consequences. Finally we look towards ways in which policy may be amended to help these communities to escape the destructive cycle in which they appear to be becoming trapped.

Methods

This study is based on a synthesis of data collected over field trips to the region since early 2004. This work builds upon an ongoing community livelihood program in the Xiaozhongdian Township run by the Kunming based *Center for Biodiversity and Indigenous Knowledge* (CBIK). Participatory mapping of this region to identify major land-use types and histories has been undertaken with villagers at Jisha over several years; for this study this information was augmented with high resolution QuickBird satellite imagery. In June 2005, detailed forest field surveys were made: different community pine forests were surveyed for stand characteristics, structural and floristic diversity (Melick and Yang, unpublished data). Site visits were made throughout the village and local households questioned about wood consumption needs and forest use. Observations were made and measurements collected of firewood supplies and fence lines to determine local wood consumption. Lengths of fence lines were determined from field observations and from the QuickBird satellite images. Basal areas and stand volumes were calculated from stand assessments made during the forests survey (Melick and Yang, unpublished data). *Pinus densata* volumes were calculated from the survey data using a metric variant of Honer's 1967 volume equation using coefficients for white pine (Honer et al. 1983). Volumes were calculated for all measured trees within a size class, which were then multiplied by the average stem numbers for that cohort. The stand characteristics of lightly utilized forest figures were used as most intensive logging expansion is occurring in this forest type.

Jisha case study

Study site

This study was based around Jisha Village (centered 27°21′56.56″–27°28′36.86″ N; 99°45′5.99″–99°51′43.86″ E), Shangri-La County, Diqing Prefecture, northwest Yunnan (Fig. 1). Jisha comprises two sub-villages, Upper and Lower Jisha, which contain 83 households with a population of over 400 people. All villagers are Tibetan and dependent upon agriculture and animal husbandry (Li 2003). Only one crop is grown per year; the primary staple is barley, which is intercropped with wheat, potato and brassica. At an altitude of 3,200 m, Jisha lies in a shallow valley of cleared pasture and agricultural land set amid forested hills. The surrounding forest varies, but near to the village at lower elevations, it is dominated by *Pinus densata*. The pine is close to its altitudinal limit here: on the slopes above the village this merges into mixed spruce and oak forest (*Picea likiangenis*, *Quercus* spp.), eventually giving way to fir-dominated forest (*Abies georgei*) at elevations above ~3,600 m. Many areas of these higher forests up to 3,700 m (the limit for commercial logging operations) have been severely degraded by past commercial logging activities and also by insect pest

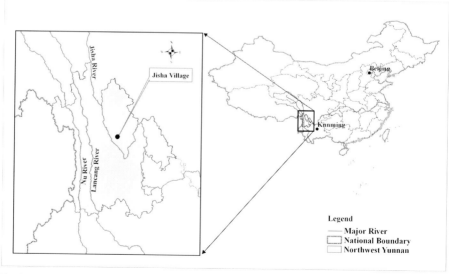

Fig. 1 Location of Jisha village, northwest Yunnan, China

🖄 Springer

attack. Spruce, and to a lesser degree, fir are prized timber while oak was the first choice for firewood—thus most merchantable timber trees have been removed.

Pinus densata dominates the fuelwood forests that now provide many of the daily requirements of Jisha village. This forest type is common on the lower mountain regions (i.e. below ~3,200 m) in northwest Yunnan and Sichuan. It is difficult to be precise, but villagers commonly use over 100 ha of nearby pine forest—their proximity to the village alone ensures these forests' importance. Although not particularly rich, these forests are a source of wood, animal bedding and fertilizer. Pine is used as a soft carving wood for lintels and feature work, and also, more extensively in recent times, for fencing materials and poor quality firewood. Although centrally dictated timber quotas purportedly exist for these communities, these are not well articulated and certainly there is little evidence of enforcement at a local level. In addition, pine forest can form important NTFP habitats, supporting matsutake, truffles and medicinal plants. Sacred forests at Jisha are dominated by small patches of remnant old growth spruce (seemingly in cold air basins or below katabatic flow gullies), but pine forests also have cultural significance, forming the cemetery forest, which has been long preserved as the place for traditional Tibetan sky burials.

Local socio-economics and politics

There have been very significant socio-economic changes to these communities in recent years—the most dominant of these has been the proscription of commercial logging. The absence of commercial logging has placed greater pressure on villagers to find alternate incomes, resulting in a rise in NTFP collection, attempts to increase livestock, off-farm work, and the collection of wood and timber from community forests, which is often (unofficially) used to supply township firewood and construction needs. In addition, these rising demands on village forests are accompanied by the construction of a new road; Jisha now has a large highway running through it (National Highway No. 214), effectively bisecting Upper and Lower Jisha, and driving straight through the cemetery forest south of the village (Fig. 2). This

Fig. 2 Looking north to upper Jisha along the new highway, which was cut through the cemetery forest. This road has granted easy access to large areas of pine forest to the south

road provides easy access to nearby towns; the trip to Zhongdian for instance has been reduced from an hour and half on rough unsealed roads to about 40 min on sealed highway. The road has improved access and also allowed an expansion of wood timber collection into nearby forests. It has also brought the development of a small village shop and a large externally funded emporium for tourist souvenirs. Development of tourism is the cause of angst within the village, splitting those who want to develop internal eco-tourism options supported by NGOs, against those who favor the government-backed large-scale development (including the proposed construction of a chairlift) (Li 2003). The full ramifications of this dispute will not be known for some time.

Politics in Jisha is through the elected village councils, which in turn sit below administrative village committees. Farmers also elect a natural village head. These are the bodies that directly administer to village needs. The politics are difficult to fathom, but at Jisha farmers are reportedly mistrustful of some village organizations following a lack of transparency during the 15 years of dealing with the government logging company. However, much daily village law is the direct result of long held village rules, taboos and mores (Li 2003).

Results and discussion

Recent forest use trends

The most obvious ecological trend at Jisha over the past few years is the increasingly high utilization of the community pine forests. Satellite imagery from November 2003 shows only about 7 ha of pine forest with clear signs of heavy use (i.e. obvious canopy gaps). Ground surveying shows this has expanded rapidly over the last 18 months. Now, the area of high-use looks to have increased by at least 3- or 4-fold (probably more). Much of this expansion is into the forest south of the village, beyond the cemetery forest, in areas that have been made more accessible by the new road system. Not only has the area of timber use expanded, but also intensity seems to have increased. Within these highly utilized forests, results show that basal areas and canopy covers are reduced by about 50% (Table 1, Fig. 3)—as the average canopy cover has dropped below 20%, technically, this should not be classified as 'forest' in China (Miao and West 2004). While rapid pine regeneration is likely, the structural survey suggested that these will be greatly simplified forests in terms of habitat diversity for many decades (Melick and Yang, unpublished data).

Table 1 The basal area and the average canopy cover (± SEM) for *Pinus densata* in the variously disturbed forest types near Jisha Village

Forest type	Basal area (m^2 ha^{-1})	Canopy cover (%)
Lightly utilized	34.63[a]	26.3[a] ± 2.5
Heavily utilized	15.99[b]	15.3[b] ± 3.4

Numbers denoted with different superscripted letters within the same column are significantly different (**$P < 0.01$). Adapted from Melick and Yang (unpublished data)

Fig. 3 The heavily utilized *Pinus densata* forest shows large gaps and the removal of over half basal areas and canopy covers

Estimates of timber needs and rate of tree removal

Estimating firewood use is problematic, nevertheless after examining annual storages and talking to locals we calculated an annual firewood volume of 5.96 m^3 per household. Given an average household of 4.82 people (i.e. population/households) this yields an annual per capita firewood demand of 1.24 m^3. It must be noted, however, that these volumes were for good quality oak. As this become scarcer more birch and pine are used and consumption rates of these are much higher. In addition, it is impossible to quantify the unofficial firewood trade—though, as firewood is sold in truckloads of 6 m^3, it seems likely that this is contributing significantly to wood removal. Although our annual per capita firewood results, are high compared with official figures for Tibetan communities in Litang County, Sichuan (0.5 m^3 per capita) (Winkler 2003), they are remarkably close to the 1.22 m^3 that was estimated for communities in northwest Yunnan's forested high mountain areas (Li 1993, cited in Winkler 2003). Similarly, higher values for Yunnan are also given by Xu and Wilkes (2004) who report a range of household consumption of 10–30 m^3 each year, resulting in an annual total of 600,000 m^3 of fuelwood collected in Diqing Prefecture.

Fencing needs were calculated from observations that split pine logs (in the size class 10–20 cm diameter) are the most common material. By contrast to pine, bamboo fencing uses only uprights and requires little or no cross bracing (Fig. 4). We recorded a total fence line of about 8,000 m at Jisha, so results show that on average each meter of pine or bamboo fence requires 0.067 m^3 and 0.004 m^3 of pine, respectively (Table 2). Given that pine fences need replacing every 4–5 years and bamboo every 10 years, this translates to a 35-fold increase in annual wood demand if all fences are converted to timber.

The removal of wood for building material is likely to be significant, but is notoriously difficult to quantify as local uses fluctuate. In their assessment of northwest Yunnan, Xu and Wilkes (2004) state that in Diqing Prefecture each year 1

🖄 Springer

Fig. 4 Traditional bamboo fencing is being replaced by less durable pine fences—this requires up to 35 times more wood

Table 2 The estimates for the comparative volume (m³) of pine wood required for pine and bamboo fencing at Jisha

Type of fence	Requirement/m	Total used	Annual requirement
Pine	0.067	533.1	118.5
Bamboo	0.004	31.8	3.4

in 30 households is allocated a quota for constructing a new house. The quota is 120 m³ per house in the highlands (though in fact, on average one house consumes a total of 150 m³, and exceptionally as much as 300 m³). Diqing Prefecture thus approves an annual total of 960,000 m³ of timber for local consumption. It is also difficult to trace where this timber is acquired—some is taken from nearby forest, but most is high quality large lumber, some of which is imported regionally.

In terms of tree removal from the Jisha village fuel forests, the most pertinent information is for fence construction—it uses pine wood and is clearly used regularly. Assuming similar tree densities and size classes as for the lightly utilized forest, it is apparent that almost a hectare of trees is removed per annum for fencing alone (Table 3). This is an underestimate, as total wood is not used and many trees with excessive side branch development are discarded. Moreover, this figure takes no account of the truckloads of low-grade firewood being sold to townships and road crews. Household firewood is also neglected as this is assumed to be oak; but blending

Table 3 Calculated wood volumes (m³) for usable *Pinus densata* in the lightly utilized forest at Jisha

Diameter cohort (dbh)	Average tree volume	Volume (ha⁻¹)
10–20 cm	0.088	86.2
20–30 cm	0.217	43.3
Total		129.5

with pine and birch is now rapidly impinging on the *Pinus densata* forests. Also the fence line measurement is conservative as we only recorded fences clearly evident from imagery—even though some obscured areas were likely to contain fences.

Causes of increased community forest exploitation

The causes of recent increased demands on fuelwood forest can be broadly divided into two areas: ecological and economic. Ecologically, past commercial logging has removed high quality timber (*Quercus, Picea*), simplifying the upper forests to such an extent that traditional management and forest use was effectively destroyed. A further factor is the demise of the bamboo understorey, due to the destruction of its shaded mountainside habitat. As a versatile and durable fencing material, bamboo is now being replaced by pine fencing, which, as explained earlier, requires an esti-mated 35 times more pine wood.

   Economically, several drivers are behind increased forest use. Until recently, barter systems were a significant part of trade in many of these rural towns; now cash economies dominate (Li 2003; Winkler 2003; Xu and Wilkes 2004). This is due to the general opening of external markets (NTFP for example are sold to outside inter-ests), but also to the literal exposure of the village to the new road, township fire-wood market, and new shops. In short, all these factors have increased the need to generate cash. The new road has also allowed greater access to forest. This is clear beyond the cemetery forest, where forest is accessed for timber cutting from the new highway and trails fan out from the road quarry. There now seems to be an unofficial firewood industry supplying surrounding townships and construction crews—the price for a truckload of pine and birch firewood is 500 RMB (USD 60).

Cascading problems

One of the most concerning aspects of the situation at Jisha is the self-perpetuating nature of the environmental problems. As outlined above, the loss of bamboo forest necessitates use of less durable pine fencing. Timber fencing is more expensive in terms of labor and wood: bamboo fences last about 10 years, moreover, the woven fences are sturdier have no cross bracing, use smaller supports and require less maintenance. In addition, the government-supported moves to increase livestock herds have seen even greater need for fencing; new fences were observed within the 18 months since the acquisition of the satellite image.

   The drive to increase livestock also leads to overgrazing resulting in increased weed problems and pasture degradation. These problems are compounded by government banning of traditional burning (a policy introduced to reduce forest fires). In addition, villagers say there is now more grazing in forests—particularly the heavily disturbed forest. This will inevitably lead to poorer regeneration and loss of forest quality. Moreover, it has negative impacts on wider replanting efforts as there is little incentive for grazers to protect plantation that will eventually reduce grazing habitat. Winkler (2003) reports a very similar scenario in Sichuan. This decreasing forest quality has ramifications for NTFP collection. The high value mushrooms are notoriously sensitive to habitat disturbance; villagers at Jisha no longer waste time collecting in heavily utilized forest, as the yields are so poor. Therefore, even should

🌱 Springer

immediate firewood demands decrease (i.e. with the completion of road building and perhaps with alternate fencing materials), the legacy of depleted forest habitat and reduction of NTFP will take decades to recover.

A similar vicious cycle has developed with new roads and enhanced vehicular access. The roads not only facilitate transport, making transfer of timber easier, but also allow access to more remote places. Naturally, the ample supply of transport and power tools (in part a legacy from past logging activities), further mitigates more widespread timber removal from community forests. Increasing use of vehicles increases running expenses. Thus as firewood collection becomes more far flung, villagers need to cut more to recoup costs. In addition, the road generates a market of its own, not only to townships who buy firewood (scarcer since the logging ban), but to road crews who use the pine wood to fuel bitumen melting. Villagers also say that road access makes traditional enforcement more difficult and "outsiders are known to remove wood". This was evident in the cemetery forest where some logging activity was apparent at the roadside edges of the site (despite community erected signs requesting respect for this forest).

Concerns with deepening wealth differentials in markets newly exposed to cash economies are well documented, but in Tibetan regions many successful villagers choose to invest in new and larger houses (Winkler 2003). At Jisha, new building activity is apparent, some spurred by the new attempts to attract and develop tourism. All of this activity is placing new demands on village wood resources (Xu and Wilkes 2004).

Institutional problems with conservation policy and research in China

It is important to note that while we have outlined specific problems for Upper and Lower Jisha, this case is likely to have widespread applicability to rural communities throughout mountainous regions of southwest China: most of these communities were dealt a financial blow by the logging ban and all have been targeted by the new policies to develop transport access and free market opportunities (Winkler 2003; Hillman 2003; Xu and Wilkes 2004). Moreover, concerns at Jisha epitomize how the expansion of village rights and the autonomy of forest users in China has been paradoxically accompanied by increased forest regulation at a central level. Generally, it seems, central policy overrides local concerns (e.g. the NFPP), however, the situation on the ground remains unclear in terms of the status of the tenure, usufruct rights and mutual obligations of community forestlands. In fact from a perusal of the relevant literature, the only thing clear about community forest rights is the state of confusion; a definition of community forest remains elusive (see for example, Miao and West 2004; Xu and Ribot 2004; Weyerhaeuser et al. 2006). Since a faculty of academics cannot pin down the details of community rights and obligations, it is unsurprising that villagers are generally non-plussed. At Jisha, villagers consider fuel forest to be their own resource for management as they see fit. It is clear that other forest areas are still utilized, but the relationship between village users and official government policies remains opaque to outsiders (and, we suspect, to many villagers).

At the local level, confusion towards forest use and timber rights stems not only from lack of codification of the rules, but also from an insecurity about tenure rights, because of recent fluctuations in forest policy. An uncertainty about the future is unlikely to foster responsible, far-sighted resource management. Moreover, any policing of levies, timber quotas and local regulations must be enforced by government, or, within villages, by designated forest guards. This system has been shown to be inefficient at best. In areas in Yunnan and Sichuan, reports comment on the compromised role of locally employed forest guards, who suffer from the absence of clear guidelines and a lack of incentive to enforce rules (Winkler 2003; Taveau and Wang 2005).

Clearly, individual cases of village resource management have unique aspects but poor communication and lack of information flows are often evident in northwest Yunnan. Forest bureaus and government departments, which have been severely disrupted by the logging ban, struggle with their new roles and sense of identity. New development projects with which they are involved often lack continuity and are frequently abandoned, having been placed unsuitably and/or suffering from a lack of support and maintenance materials. For example, in villages neighboring Jisha, the locals remain wryly amused by the poor quality concrete fence post that have failed after only a few years and which they claim, were never correctly located in the first place. Similarly, our field observations in Nuijiang show that *Animal Husbandry* or *Animal Fodder Bureau* personnel have little incentive to go to the field and, when they do, often lack set objectives.

As pointed out earlier, there are also apparent contradictions between—and within—conservation and resource policy. On the broad ideological scale recent reforms of village rights (i.e. Village Organic Law) have granted village committees greater land-use autonomy, while conservation policies such as NFPP, SLCP and fire proscriptions are nationally imposed and seemingly override these same community rights. Conflicting policy is exemplified in many villages where tree planting and fire restrictions clearly work against grazing interests, yet increasing livestock management is a stated aim of the new rural development program (Winkler 2003; Xu et al. 2004). Tracing the causes of such inconsistencies is difficult, but it is impossible for such sweeping conservation reforms to cater to local needs. Resource management is based on large-scale statistics and remote-sensed data—certainly to a farmer at Jisha this information can appear remote in every sense of the word. However, these technical limitations are further challenged by systemic problems or attitudes within Chinese institutions. Several authors have commented on the lack of information flow through various levels of government hierarchy, and while Chinese research has advanced quickly in terms of technical data collection, exchange and cooperation, it still generally lacks inclusive ecosystem analyses and stops short of involvement with social policy issues (Ruiz Pérez et al. 2004; Wu et al. 2004).

However, criticism of communication failures and lack of policy feedback can extend beyond Chinese institutions. Many projects involving international NGOs and scientists struggle to overcome similar obstacles. Generally, to an outsider there seems to be some replication and lack of communication in NGO projects. At Jisha, for instance, villagers had no knowledge of NGO programs being run in a neighboring village to try to establish thorn bush hedges to replace fences. On a broader level, the authors have seen overseas academics withhold relevant research from conservation planners, as they put their personal publishing ambitions ahead of greater needs. Similarly, global conservation organizations sometimes seem keener on appearance than substance, requesting immediate action plans with nebulous "demonstrable biodiversity outputs"—such plans no doubt read well in fundraising brochures, but, without meaningful population biology studies and a continuity of

Springer

baseline data, they are hollow. Highlighting these problems and trying to bring information out of the gray literature by presenting relevant examples and case studies is a good step towards developing some continuity and consistency for policy development.

Towards inclusive conservation policy at Jisha

With hindsight, it is easy to criticize Chinese conservation policy. In some ways, however, ours is not a criticism of conservation policies per se but rather an acknowledgement that minority communities are now paying the price for the past over-exploitation of natural resources. While township revenues were large, these community problems could be disguised—now they are becoming apparent. Government must act swiftly to grant exemptions to allow these communities to access their high value resources. This sort of inclusive policy will not be straight-forward; even for western governments with long experience of resource manage-ment the goal of balanced, socially inclusive conservation has long been advocated (e.g. Henning 1970), but remains largely unattained.

In line with the aim of promoting local community needs, it is up to researchers and NGOs to continue to press for policy changes. It is the role of researchers to take some responsibility to articulate specific policy options. Pragmatically, this can probably only be done on a case by case basis, since local situations vary, and, moreover, because local governments are most closely conscious of community attitudes. In the case of Jisha, our research suggests the following recommendations:

- greater village access to areas of higher-value forest for sustainable collection of firewood and timber, perhaps with acknowledgement of the supply market to townships;
- implementation of fencing alternatives;
- reintroduction of controlled pasture enrichment and rangeland burning, with cooperation of the *Forest Fire Bureau*;
- monitor the link between grazing and plantation success;
- definitive population biology and ecological studies to enable preparation of sci-entifically sound sustainable harvesting plans for NTFP.

Despite the challenges facing these rural communities some authors are optimistic and see the current situation in China as the opportunity to turn away from previous rapacious behavior and develop long-term ecologically sustainable industries (Winkler 2003; Xu et al. 2004). That the government is aware and attempting to respond to these challenges is clear—a key plank of the SLCP, for example, is an integration of environmental goals with those of agricultural restructuring and poverty reduction. It is the role of independent researchers to highlight the problems and inconsistencies that inevitably arise—such as the problems confronting Jisha. Equally important is our responsibility to outline practical solutions before the loss of community resources and cultures is irreversible.

Acknowledgements We are grateful for the assistance of Jisha villagers, particularly Mr. Wangzha and his family. We also appreciate the provision of advice and equipment by Drs. Horst Weyer-haeuser and Lance Whitnall. This work was in part supported by the Critical Ecosystem Partnership Fund (CEPF) Project on Applied Ethnoecology for Biodiversity Assessment, Monitoring and Management in NW Yunnan.

[396]

References

CIFOR (Center for International Forestry Research) (2005) Big bucks for the environment – China leads the way. May 25. Polex Alert. http://www.cifor.cgiar.org/docs/_ref/polex/english/2005/2005_05_25.htm. Cited 9 June 2006

Henning DH (1970) Comments on an interdisciplinary social science approach for conservation administration. BioScience 20:11–16

Hillman B (2003) Paradise under construction: minorities, myths and modernity in Northwest Yunnan. Asian Ethnic 4:175–188

Honer TG, Ker MF, Alemdag IS (1983) Metric timber tables for the commercial tree species of central and eastern Canada. Maritimes Forest Research Centre Information Report M-X-140

Li B (2003) Obstruction to local governance in natural resources management – a case study in Jisha Village northwest Yunnan. In: Xu JC, Mikesell S (eds) Landscapes of diversity: proceedings of the III symposium on MMSEA, 25–28 August 2002, in Lijiang, China. Yunnan Sciences and Technology Press, Kunming, pp 277–285

Miao GP, West RA (2004) Chinese collective forestlands: contributions and constraints. Int Forest Rev 6:282–298

Rozelle S, Huang J, Benziger V (2003) Forest exploitation and protection in reform China: assessing the impacts of policy and economic growth. In: Hyde W, Belcher B, Xu J (eds) China's forests, global lessons from market reforms. Resources for the Future, Washington, USA, pp 109–134

Ruiz Pérez M, Fu MY, Xie JZ et al (2004) The relationship between forest research and forest management in China: an analysis of four leading Chinese forestry journals. Int Forest Rev 6:341–345

Taveau S, Wang W (2005) Value of forest resources in a Miao community of Jindou Natural Village, Yunlong County, Yunnan Province. Research Report for Community Livelihood Program. Centre for Biodiversity and Indigenous Knowledge, Kunming, China. http://www.cbik.ac.cn/cbik-cn/cbik/our_work/download/Valueofforest.pdf. Cited 9 June 2006

Weyerhaeuser H, Kahrl K, Su Y (2006) Ensuring a future for collective forestry in China's southwest: adding human and social capital to policy reforms. Forest Policy Econ 8:375–385

Weyerhaeuser H, Wilkes A, Kahrl F (2005) Local impacts and responses to regional forest conservation and rehabilitation programs in China's northwest Yunnan province. Agric Syst 85:234–253

Willson A (2006) Forest conversion and land use changes in rural northwest Yunnan, China: implications for the 'big picture'. Mount Res Dev 26:227–236

Winkler D (2003) Forest use and implications of the 1998 logging ban in the Tibetan Prefectures of Sichuan: case study on forestry, reforestation and NTFP in Litang County, Ganzi Tap, China. Inform Bot Ital 35:116–125

Wu C-I, Shi SH, Zhang Y-P (2004) A case for conservation. Nature 428:213–214

Xu JC, Ribot J (2004) Decentralization and accountability in forest management: case studies from Yunnan, southwest China. Eur J Dev Res 16:153–173

Xu JC, Wilkes A (2004) Biodiversity impact analysis in northwest Yunnan, China. Biodivers Conserv 13:959–983

Xu ZG, Bennett MT, Tao R, Xu JT (2004) China's sloping land conversion programme four years on: current situation and pending issues. Int Forest Rev 6:317–326

Zhang PC, Shao GF, Zhao G et al (2000) China's forest policy for the 21st century. Science 288:2135–2136

Zheng L (2004) An analysis of household livelihoods in Tuomunan village, Xianggelila County, NW Yunnan. Center for Biodiversity and Indigenous Knowledge, Community Livelihoods Program, Working Paper 5. http://www.cbik.ac.cn/cbik-en/cbik/our_work/download/CBIK%20WP5%20ENGL.df. Cited 9 June 2006

🍬 Springer

Biodivers Conserv (2007) 16:1973–1981
DOI 10.1007/s10531-006-9125-7

ORIGINAL PAPER

Medicinal plant conservation and management: distribution of wild and cultivated species in eight countries

Mariel Aguilar-Støen · Stein R. Moe

Received: 6 December 2005 / Accepted: 11 September 2006 / Published online: 27 October 2006
© Springer Science+Business Media B.V. 2006

Abstract   In order to understand the particular challenges that medicinal plant conservation and management raise at the global level, it is necessary to address issues pertaining their distribution and the environments where they grow. When reviewing medicinal plant studies from eight countries in four regions we found that a high proportion of the reported medicinal plants had wide distributions across countries and continents. Most plants are found wild (40.5%) or naturalized (33.9%), while only 3.3% are cultivated. Since many species are distributed in wild conditions, cultivated and naturalized in several continents, conservation and management interventions would be best served through collaboration between host countries.

Keywords   Biodiversity · Agro-landscapes · Species diversity · Medicinal plants · Conservation

Introduction

Medicinal plants represent by the number of species used, the widespread nature of their use and their contribution to human health, perhaps one of the most significant ways in which humans directly reap the benefits provided by biodiversity (Farnsworth and Soejarto 1991; Hamilton 2004).

Most of the people depending on traditional medicine live in developing countries, and it has been estimated that between 60 and 80% of people worldwide rely mainly on traditional herbal medicine to meet their primary healthcare needs,

M. Aguilar-Støen
Center for Development and the Environment (SUM), University of Oslo, P.O. Box 1116, Blindern, Oslo NO-0316, Norway

S. R. Moe · M. Aguilar-Støen (✉)
Department of Ecology and Natural Resource Management, Norwegian University of Life Sciences, P.O. Box 5003, As NO 1432, Norway
e-mail: mariel.stoen@sum.uio.no

(Farnsworth and Soejarto 1991; Balick et al. 1996). Since ancient times medicinal plants have been harvested from the wild and cultivated in home gardens across the world (Halberstein 2005). Furthermore, many medicinal plant species have spread globally both via intentional and carefully planned transfers and as the unintentional outcome of people's movements (Fowler and Hodgkin 2004; Voeks 2004; Beinart and Middleton 2004).

A number of studies suggest that medicinal plant species are more often found in secondary growth or perturbed habitats, highly associated with human-modified landscapes (Voeks 1996; Caniago and Siebert 1998; Balasingh et al. 2000; Stepp and Moerman 2001; Voeks 2004). Perturbed habitats are sources of biologically active secondary compounds, as suggested by plant defense theory (Coley et al. 1985; Abe and Higashi 1991; Coley and Barone 1996).

The factors described above suggest that medicinal plant species should be a priority issue in conservation and sustainable development agendas. But contrary to plant species used for food and agriculture, medicinal plant species are a more diffuse category. The most important crops are cultivated and efforts exist towards the conservation of their genetic diversity (Fowler and Hodgkin 2004). Medicinal plant species are not always cultivated, they occur in a range of habitats, some are collected from the wild and many serve more than one function (Hamilton 2004). In order to understand the particular challenges that medicinal plant conservation and management raise at the global level, we need to reflect on; (a) how medicinal plants are distributed in the world; (b) to what extent medicinal plants are cultivated and to what extent do they grow in wild conditions and finally; (c) to what extent do the different countries in the world utilize the same plant species?

Methods

Eight ethnobotanical papers, concerned with medicinal plants in eight different countries were selected. The selection was based on an intention to cover two countries in each biogeographic region, excluding the Australian region (according to the classification suggested by Cox (2001). The countries selected were Mexico (Casas et al. 2000) and Guatemala (Girón et al. 1991) in the Central American region. Spain (González-Tejero et al. 1995) and Bulgaria (Ivancheva and Stancheva 2000) in the Holartic region. Uganda (Tabuti et al. 2003) and Ethiopia (Giday et al. 2003) in the African region. Nepal (Shresta and Dhillion 2003) and The Philippines (Lirio et al. 1998) in the Indo-Pacific region.

A database was created containing information on plant species, family names, parts of the plant used, geographical distribution (i.e. geographic areas where the plant occurs disregarding the factors involved in its dispersal or colonization) and cultural status (whether the plant is under cultivation or not). Eight categories were defined for geographical distribution and four for cultural status, (see description of the categories below).

Geographical distribution was determined using information available at Wiersema and Leon (1999); The International Plant Name Index (IPNI) (http://www.ipni.org/index.html) and the Plants Database from USDA (http://plants.usda.gov/index.html). As these sources report distribution in somewhat different forms, we homogenized geographical distribution of plants by creating eight categories. Hollis and Brummitt (1991) suggest nine geographical areas in 51 geographical regions for

🖉 Springer

recording plant distribution. Based on their classification, we categorized the plants into eight groups (see Table 1).

For cultural status we used information by Wiersema and Leon (1999) and the Plants Database from USDA. Plants that were not reported by those sources were assumed to be wild. Any information on cultural status reported by the studies analyzed was also used. The following categories were used: wild, naturalized, wild and cultivated and only cultivated.

Results

Species diversity and species overlap

From the papers selected a total of 611 plants species belonging to 132 families were reported, however the total of reports is 695 as 60 species were shared by two or more countries and therefore reported more than once (see Table 2). The families with higher percentage of species reported were Fabaceae (10.1%), Asteraceae (9.9%), Poaceae (5.5%), Solanaceae (4.6%) and Euphorbiaceae (3.5%).

A total of 60 plants were reported for more than one country (Table 2), while 17 plant species are reported for more than a pair of countries. Spain is the only country not to have any plant species in common with any of the other countries.

Table 1 Categories used for plant distribution in this study. The categories are based on areas and regions suggested by Hollis and Brummit (1991)

Category	Abbreviation	Criteria
Wide distribution within one area—Native	WD1AN	If the plant in question is reported in more than one region within one area and the country reporting falls within the area
Wide distribution within one area—Non Native	WD1ANN	If the plant in question is reported in more than one region within one area, and the country reporting does not fall within the area
Wide distribution within two areas—Native	WD2AN	If the plant in question is reported in more than one region in more than one area, and the country reporting falls within one of the areas
Wide distribution within two areas—Non Native	WD2ANN	If the plant in question is reported in more than one region in more than one area, and the country reporting does not fall within one of the areas
Limited distribution—Native	LDN	If the plant in question is reported in one region within one area and the country reporting is located in the same region
Limited distribution—Non Native	LDNN	If the plant in question is reported in one region within one area and the country reporting is not located in the same region
Endemic	E	For plants reported as endemic for one country
Cosmopolitan	C	For plants with distributions reported as such

Table 2 Plants reported in two or more countries ($n = 60$)

Family	Species	M	G	S	B	ET	U	PH	N
Araceae	*Acorus calamus* L.				x				x
Alliaceae	*Allium cepa* L.	x	x				x	x	
Alliaceae	*Allium sativum* L.	x	x			x		x	
Liliaceae	*Aloe vera* (L.) Burm. f.	x	x					x	x
Amaranthaceae	*Amaranthus hybridus* L.	x					x		
Papaveraceae	*Argemone mexicana* L.	x							x
Asteraceae	*Artemisia absinthium* L.		x			x			
Asteraceae	*Artemisia vulgaris* L.					x		x	
Liliaceae	*Asparagus racemosus* Willd.						x		x
Cannabaceae	*Cannabis sativa* L		x				x		x
Solanaceae	*Capsicum frutescens* L.						x	x	
Caricaceae	*Carica papaya* L.	x					x		
Apocynaceae	*Carissa edulis* (Forssk). Vahl					x	x		
Chenopodiaceae	*Chenopodium ambrosioides* L.	x	x				x		
Rutaceae	*Citrus aurantifolia* (Christm.) Swingle	x	x				x		
Rutaceae	*Citrus sinensis* (L.) Osb.	*x*					*x*		
Apiaceae	*Coriandrum sativum* L.		x		x				
Poaceae	*Cymbopogon citratus* (DC.) Stapf		*x*				*x*		*x*
Poaceae	*Cynodon dactylon* (L.) Pers.	x	x				x		x
Solanaceae	*Datura stramonium* L.	*x*					*x*		
Poaceae	*Eleusine coracana* Gaertn					*x*	*x*		
Myrtaceae	*Eucalyptus globulus* Labill.	x	*x*						
Myrtaceae	*Eucalyptus* sp.						*x*	x	
Euphorbiaceae	*Euphorbia heterophylla* L.	*x*					*x*		
Euphorbiaceae	*Euphorbia hirta* L.		x				*x*		
Fabaceae	*Glycyrrhiza glabra* L.				x				x
Malvaceae	*Gossypium hirsutum* L.	x	x						
Euphorbiaceae	*Jatropha curcas* L.		*x*				*x*		
Verbenaceae	*Lantana camara* L.	x	x				*x*		
Leguminosae	*Leucaena leucocephala* (L.) Benth.	x						x	
Linaceae	*Linum usitatissimum* L.				x	x			
Solanaceae	*Lycopersicon esculentum* Karst.	x	x						
Malvaceae	*Malva parviflora* L.	x	x						
Anacardiaceae	*Mangifera indica* L.	x	x				x		
Euphorbiaceae	*Manihot esculenta* Crantz.		x				x	x	
Asteraceae	*Matricaria chamomilla* L.		x		x				
Asteraceae	*Melampodium divaricatum* DC.	x	x						
Lamiaceae	*Mentha arvensis* L.							x	x
Leguminosae	*Mimosa pudica* L.		x				x	x	
Solanaceae	*Nicotiana tabacum* L.	x	x				x	x	
Lauraceae	*Persea americana* Mill.	x	x						
Piperaceae	*Piper auritum* HBK.	x	x						
Myrtaceae	*Psidium guajava* L.	x	x				x	x	
Euphorbiaceae	*Ricinus communis* L.	x					x	x	
Lamiaceae	*Rosmarinus officinalis* L.		x		x				
Rutaceae	*Ruta chalepensis* L.	x	x			x			
Poaceae	*Saccharum officinarum* L.	*x*					*x*		
Fabaceae	*Senna occidentalis* (L.) Link						x	x	
Malvaceae	*Sida rhombifolia* L.	x	x						
Solanaceae	*Solanum incanum* L.						x	x	
Solanaceae	*Solanum nigrescens* Mart. & Gal.	x	x						
Poaceae	*Sorghum bicolor* (L.) Moeench	x				x			
Asteraceae	*Tagetes erecta* L.	x						x	
Asteraceae	*Tagetes tenuifolia* Cav.	x	x						
Fabaceae	*Tamarindus indica* L.	*x*					*x*		

🕮 Springer

Table 2 continued

Family	Species	M	G	S	B	ET	U	PH	N
Bignoniaceae	*Tecoma stans* (L.) Juss. et Kunth	x	x						
Apocynaceae	*Thevetia peruviana* (Pers.) Schumann	*x*					*x*		
Asteraceae	*Vernonia amygdaliana* Del.					x	x		
Solanaceae	*Withania somnifera* L.					x	*x*		
Poaceae	*Zea mays* L.	x	x				x		

M = Mexico; G = Guatemala; S = Spain; B = Bulgaria; ET = Ethiopia; U = Uganda; PH = The Philippines; N = Nepal

Ethiopia has the highest number of plants in common with other countries (43.5%), followed by The Philippines (40%), Guatemala (34.7%), Uganda (18.6%), Mexico (17.3%), Nepal (16.1%) and Bulgaria (11.9%) (Fig. 1).

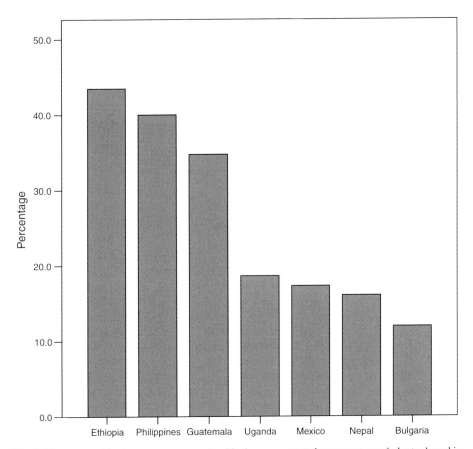

Fig. 1 Plants shared by two or more countries (the bars represent the percentage of plants shared in relation to the number of plants reported by country. Spain is excluded from the graph since it did not share any species with any other country)

Springer

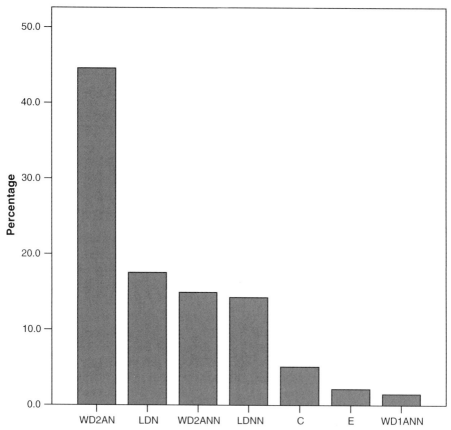

Fig. 2 Geographical distribution of plants reported for Guatemala, Mexico, Nepal, Spain, Bulgaria, The Philippines, Uganda and Ethiopia (as percentage n = 695). The descriptions of the different categories are given in Table 1

Geographical distribution and cultural status of plant species

As much as 43.9% of the recorded plants have wide distributions within two areas where they are native while only 16% of the native plants have limited distributions. Relatively few plant species (5.6%) have cosmopolitan distributions and only 1.9% are endemics (see Fig. 2). Only Mexico and Spain have recorded endemic plants. In Spain as much as 29.4 % of the recorded plants are endemics.

A large proportion of the medicinal plants (40.5%) are growing under wild conditions, only 3.3% of the recorded plants are exclusively cultivated while 33.9% are naturalized, 22.3% are found both cultivated and wild (Fig. 3). When focusing on the species that are shared between countries a large percentage (38.3%) are cultivated and wild, 31.7% naturalized, 16.7% only cultivated and 13.3% wild.

Discussion

A total of 132 families were reported in the cases we studied. The families with the highest number of species were Fabaceae, followed by Asteraceae, Poaceae,

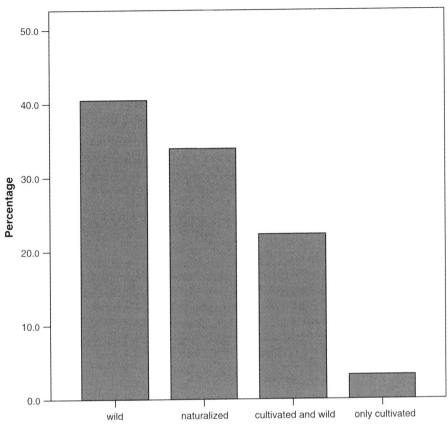

Fig. 3 Cultural status of plants reported for Guatemala, Mexico, Nepal, Spain, Bulgaria, The Philippines, Uganda and Ethiopia (as percentage $n = 695$)

Solanaceae and Euphorbiaceae (all >3%). In other previous studies these five families are reported to be among the ten most important medicinal plants families (Alcorn 1984; Kapur et al. 1992; Moerman et al. 1999). This indicates that special attention should be paid to conservation and management of plants belonging to these families.

Our results point to a high proportion of medicinal plants with wide distributions. This is very much in concordance with the situation of other plant resources used by people, particularly plant resources used for food and agriculture (Palacios 1999; Fowler and Hodgkin 2004).

The high proportion of medicinal plant species with wide distributions can be interpreted as the result of past and present plant movements resulting in improved medicinal floras around the world (Voeks 2004).

For the 60 plants reported in more than one country it is interesting to note that most of them are widely distributed, and most of them are either cultivated and wild (38.3%) or naturalized (31.7%). In contrast, when appreciating the total number of plant species considered (695), a high proportion is classified as growing in wild conditions or naturalized (40.5% and 33.3%, respectively) and a very low proportion is only cultivated (3.3%). This is particularly relevant for medicinal plant conservation efforts. Cooperation between different host countries is necessary to ensure

🕿 Springer

conservation and sustainable management of medicinal plant species. Our results show that many species are distributed wild, cultivated and naturalized in several places and accordingly their conservation and management would be better served by increased cooperation between the host countries, regardless of whether they occur naturally or have been naturalized or if the plants are domesticated.

For management we suggest increased focus on habitats where medicinal plant species grow. Other studies have stressed that many medicinal plant species inhabit secondary growth forests or disturbed forests. Voeks (2004) refers to 18 other studies (conducted in 10 countries) that have reported a higher incidence of medicinal plant species in disturbed or secondary growth forests. Stepp and Moerman (2001) concluded that weeds appear in much higher frequencies in the Tzeltal Maya and Native North American pharmacopoeias than would be predicted by the frequency of occurrence of weed species in general. Also other studies have found similar results (Posey 1984 in Brazil; Voeks 1996 in Brazil; Caniago and Siebert 1998 in Indonesia; Frei et al. 2000 in Mexico). Plant species thriving in disturbed and secondary growth habitats are species adapted to high competition and to increased herbivory (Coley et al. 1985) which make them produce higher proportion of secondary metabolites as defense mechanisms and thus also to be attractive as medicines.

In conclusion, many medicinal plants are widely distributed and used in many different countries, even across regions. Relatively few medicinal plants are cultivated. Thus, the conservation of medicinal plant species requires that efforts are focused on key habitats, including secondary forest, disturbed areas and agro-landscapes.

Acknowledgment M.Aguilar Støen has received support from the Norwegian Research Council project number 158984/510.

References

Abe T, Higashi M (1991) Cellulose centered perspective on terrestrial community structure. Oikos 60:127–133

Alcorn J (1984) Huastec Mayan ethnobotany. University of Texas Press, Austin Texas

Balasingh J, Thiruchelthil B, Jerlin B, Samuel A, Muthukumar A (2000) Medicinal flora of a tropical scrub jungle. J Econ Taxon Bot 24:737–745

Balick MJ, Elisabetsky E, Laird SA (1996) Medicinal Resources of the Tropical Forest: Biodiversity and its importance to Human Health. Columbia University Press, New York

Beinart W, Middleton K (2004) Plant transfer in historical perspective: A review article. Environ Hist 10:3–29

Caniago I, Siebert SF (1998) Medicinal plant ecology, knowledge and conservation in Kalimantan, Indonesia. Econ Bot 52:229–250

Casas A, Valiente-Banuet A, Viveros JL, Caballero J, Cortés L, Dávila P, Lira R, Rodríguez I (2000) Plant resources of the Tehuacan-Cuicatlán Valley, Mexico. Econ Bot 55(1):129–166

Coley PD, Barone JA (1996) Herbivory and plant defenses in tropical forests. Annu Rev Ecol Syst 27:305–335

Coley PD, Bryant JP, Chapin FS (1985) Resource availability and plant anti-herbivore defense. Science 230:895–899

Cox B (2001) The biogeographic regions reconsidered. J Biogeogr 28:511–523

Farnsworth NR, Soejarto DD (1991) Global importance of medicinal plants. In: Akerele O, Heywood V, Synge H (eds) The conservation of medicinal plants. Cambridge University Press, Cambridge, UK

Fowler C, Hodgkin T (2004) Plant genetic resources for food and agriculture: assessing global availability. Annu Rev Environ Resour 29:143–79

Frei B, Sticher O, Heinrich M (2000) Zapotec and Mixe use of tropical habitats for securing medicinal plants in Mexico. Econ Bot 54:73–81

Giday M, Asfaw Z, Elmqvist T, Woldu Z (2003) An ethnobotanical study of medicinal plants used by the Zay people in Ethiopia. J Ethnopharmacol 85:43–52

Girón LM, Freire V, Alonzo A, Cáceres A (1991) Ethnobotanical survey of the medicinal flora used by the Caribs of Guatemala. J Ethnopharmacol 34:173–187

González-Tejero MR, Molero-Mesa J, Casares-Porcel M, Martínez Lirola MJ (1995) New contributions to the ethnopharmacology of Spain. J Ethnopharmacol 45:157–165

Halberstein R (2005) Medicinal plants: Historical and cross cultural usage patterns. Annu Epidemiol 15:686–699

Hamilton AC (2004) Medicinal plants, conservation and livelihoods. Biodivers Conserv 13:1477–1517

Hollis S, Brummitt RK (1991) World geographical scheme for recording plant distributions. Hunt Institute for Botanical Documentation, Pittsburgh

Ivancheva S, Stantcheva B (2000) Ethnobotanical inventory of medicinal plants in Bulgaria. J Ethnopharmacol 69:165–172

Kapur SK, Shahi AK, Sarin YK, Moerman DE (1992) The medicinal flora of Majouri-Kirchi forests (Jammu and Kashmir State), India. J Ethnopharmacol 36:87–90

Lirio LG, Hermano ML, Fontanilla MQ (1998) Antibacterial activity of medicinal plants from The Philippines. Pharmaceutical Biol 36:357–359

Moerman DE, Pemberton RW, Kiefer D, Berlin B (1999) A comparative analysis of five medicinal floras. J Ethnobiol 19:49–67

Palacios XF (1999) Contribution to the estimation of countries' interdependence in the area of plant genetic resources. Food and Agriculture Organization of the United Nations. Background Study Paper No. 7

Posey DA (1984) A preliminary report on diversified management of tropical forest by the Kayapó Indians of the Brazilian Amazon. In: Prance GT, Kallunki JA (eds) Ethnobotany in the neotropics. New York Botanical Garden, New York, pp 112–126

Shrestha P, Dhillion SS (2003) Medicinal plant diversity and use in the highlands of Dolakha district, Nepal. J Ethnopharmacol 86:81–96

Stepp JR, Moerman DE (2001) The importance of weeds in ethnopharmacology. J Ethnopharmacol 75:19–23

Tabuti JRS, Lye KA, Dhillion SS (2003) Traditional herbal drugs of Bulamogi, Uganda: plants, use and administration. J Ethnopharmacol 88:19–44

USDA, NRCS (2004) The PLANTS Database, Version 3.5 (http://plants.usda.gov). National Plant Data Center, Baton Rouge, LA 70874–4490 USA

Voeks RA (1996) Tropical forest healers and habitat preference. Econ Bot 50:381–400

Voeks RA (2004) Disturbance pharmacopeias: Medicine and myth from the humid tropics. Ann Assoc Am Geogr 94(4):868–888

Wiersema JH, León B (1999) World economic plants. a standard reference. CC Press, Washington D.C

Biodivers Conserv (2007) 16:1983–1994
DOI 10.1007/s10531-006-9061-6

ORIGINAL PAPER

Biodiversity effects on biomass production and invasion resistance in annual versus perennial plant communities

Xiao Lei Jiang · Wei Guo Zhang · Gang Wang

Received: 6 April 2005 / Accepted: 24 April 2006 / Published online: 9 July 2006
© Springer Science+Business Media B.V. 2006

Abstract In a field experiment we constructed two different communities using both annual and perennial plant species, in which species diversity is experimentally manipulated. We want to test the relationships between species diversity and biomass production and invasibility and the possible mechanisms driving this relationships, especially, whether the identical mechanisms drive both diversity-production and diversity-invasibility relationships. Our results indicated that a positive diversity-production relationship and negative diversity-invasibility and production-invasibility relationships emerged in two different communities. However, the mechanisms underlying are different in two communities. In the annual communities, the observed positive diversity-production and negative diversity-invasibility relationships are linked by the sampling effect. In the perennial communities, however, the mechanism responsible for these observed relationships are the complementarity effect. Our results also found that, in addition to species diversity, species composition also play an important role in governing the observed relationship. The results of our study suggest that because species in different communities may differ in their life history, biological and physiological traits, mechanisms responsible for the observed relationship between diversity and biomass production and invasibiltiy are likely different.

Keywords Annual community · Biomass production · Complementarity effect · Invasibility · Perennial community · Sampling effect · Species diversity · Species composition

X. L. Jiang (✉) · W. G. Zhang
Key Laboratory of Grassland Agro-Ecosystem, Ministry of Agriculture, PRC, College of Pastoral Agricultural Science and Technology, Lanzhou University, Lanzhou 730020, China
e-mail: Jiangxl@lzu.edu.cn

G. Wang
State Key Laboratory of Arid Agroecology, Lanzhou University, Lanzhou, China

[409]

🙋 Springer

Introduction

The invasion of exotic species into assemblages of native plants is a pervasive and widespread phenomenon (Naeem et al. 2000; Dukes 2002; van Ruijven et al. 2003). Invasions can have impacts on ecosystem stability by modifying their diversity and their functioning (Mack and D'Antonio 1998). Therefore, the relationships between diversity and productivity and invasibility have attracted considerable recent attention (Wardle 2001). Researchers using plants, microbes and marine invertebrates studying the effect of diversity have generated substantial debate about both the patterns observed and the mechanisms underlying these patterns (Robinson et al. 1995; Palmer and Maurer 1997; Tilman 1997a; Crawley et al. 1999; Knops et al. 1999; Levine 2000; Naeem et al. 2000; Hector et al. 2001; Cardinale et al. 2002; Hodgson et al. 2002; Kennedy et al. 2002). Especially, whether identical mechanisms driven both diversity-productivity and diversity-invasibility relationships remain unclear (Hodgson et al. 2002).

There are many possible mechanisms for the observed diversity-productivity and diversity-invasibility patterns (Hooper 1998; Hodgson et al. 2002; Shea and Chesson 2002; Byers and Noonburg 2003; Jonsson and Malmquist 2003; Lambers et al. 2004), however, the most commonly used and accepted two mechanisms that could explain the positive effect of diversity on productivity and the negative effect of diversity on invasibility are the sampling effect (Aarssen 1997; Huston 1997; Tilman et al. 1997; Wardle 2001) and the complementarity effect (Knops et al. 1999; Tilman 1999; Naeem et al. 2000). The sampling effect is based on the premise that as more species are randomly drawn from a pool of species, there is an increased probability of including species which have a dominant role in driving the response variable being measured (Wardle 2001). Complementary effect is based on the premise that communities with more species could use a greater variety of resource capture characteristics, leading to greater use of limiting resources (complementarity), fewer resources being available for invading species, and therefore increasing productivity and reducing invasibility (Dukes 2002), Generally, these two mechanisms are not mutually exclusive and probably operated simultaneously. This made it difficult to test them as separated hypotheses (Lepš et al. 2001), and distinguishing between these mechanisms has attracted considerable recent attention (Wardle 2001). In this experiment we want to determine the relationship between diversity-biomass production and diversity-invasibility, and the contributions of sampling effect and complementarity effect in shaping these relationships using two different experimental plant communities.

Methods

The experiments were conducted during the growing season of 2002 at Grassland Experimental Station of College of Pastoral Agriculture Science and Technology, Lanzhou University, which located in Jingtai county (37°15′ N, 104°45′ E), Gansu province, northwest China. This area has an arid continental climate with the mean annual solar radiation of 6058 MJ m^{-2}, and mean annual sunshine duration of 2725.5 h. The mean annual temperature is 8.2°C and the mean annual accumulated temperatures above 10°C are 3038°C. The frost-free period is 160 days. The mean annual precipitation is 185 mm, and evaporation capacity is around 2300 mm. The soil is an Aridsol (serozem), with a pH value of 8.5.

The experiments were carried out on cultivated field. The last crop was harvested in 2001, the experimental sites were then ploughed. In spring of 2002, the experimental treatments were established using a randomized complete block design. The area of each plot is 3×4 m separated by 0.5 m buffer strips. Nineteen cultivated forage species for constructing resident communities were selected from local cultivation experiment field. The annual community consisted of nine species (Table 1), and nine diversity gradients: nine monocultures; four two-species mixtures; four three-species mixtures; four four-species mixtures; two five-species mixtures; two six-species mixtures, and one seven-, eight-, and nine-species mixtures. The perennial community consisted of ten species (Table 1), and ten diversity gradients: ten monocultures; eight two-species mixtures; seven three-species mixtures; five four-species mixtures; five five-species mixtures; two six-species mixtures, and two seven-species mixtures, one eight-, nine-, and ten-species mixtures. All treatments were replicated three times in a randomized complete block design, resulting in a total of 81 plots in annual communities and 129 plots in perennial communities. All mixtures were randomly allocated within one block and replicated in a second block with new randomization. To separate the species composition effect from the diversity effect, some diversity level treatments consisted of different assemblages of species (Tilman 1997b). Species composition of each diverse community was determined by a random draw from the species pool. In monocultures, seeds were sown at densities estimated to be higher than necessary to maximize aboveground biomass production and to ensure 100% cover (based on density and size per

Table 1 Species used in experiment and sown densities

Species	Life-history	Abbreviation	Sown density (no.seeds/m²)
1. *Avena sative*	Annual grass	AVSA	1125
2. *Lolium multiflorum*	Annual grass	LOMU	2000
3. *Panicum crusgalli*	Annual grass	PACR	450
4. *Sorghum sudanense*	Annual grass	SOSU	120
5. *S. vulgar × S. sudanense*	Annual grass	SVSS	135
6. *Melilotus albus*	Annual legume	MEAL	859
7. *Vicia sativa*	Annual legume	VISA	454
8. *Vicia villosa*	Annual legume	VIVI	407
9. *Amaranthus tricolor*	Annual forbs	AMTR	882
10. *Phleum pretense*	Perennial grass	PHPR	3650
11. *Dactylis glomerata*	Perennial grass	DAGL	1930
12. *Lolium perenne*	Perennial grass	LOPE	1667
13. *Elymus nutans*	Perennial grass	ELNU	1489
14. *Poa pratensis*	Perennial grass	POPR	3233
15. *Medicago sativa*	Perennial legume	MESA	1556
16. *Onobrychis viciaefolia*	Perennial legume	ONVI	802
17. *Trifolium repens*	Perennial legume	TRRE	2178
18. *Coronilla varia*	Perennial legume	COVA	250
19. *Silphium perfoliatum*	Perennial forbs	SIPE	12

individual of plants in local cultivation experiments and seed germination rate; Table 1). In polycultures, seeding densities of "resident" species were reduced according to the number of species in the community (e.g., for a 4-species community, a given species would be sown at one-quarter its density in monoculture).

Aboveground biomass was harvested at the peak of vegetation season (mid-August). Three 50 × 50 cm quadrats of randomly placed in each plot were clipped at ground level. The plant species number and aboveground biomass, cover, and density of each species were recorded in each quadrat. These sub-samples were then pooled, averaged, and separated by species, oven dried at 65°C, weighed and values expressed as dry mass of 1 m^2. The total biomass of realized species was used as a measure of the resident community productivity, and the total biomass of naturally colonizing weeds was used as a measure of invasibility. The Shannon–Wiener index was used for the index of species diversity:

$$H' = -\Sigma P_i \ln P_i$$

where P_i is the proportion of total aboveground biomass made up by species i (Dukes 2002).

The additive partitioning equation of Loreau and Hector (2001) was used for partitioning of selection (sampling) and complementarity effects. As shown in the additive partitioning equation, the net biodiversity effect can be partitioned as follows:

$$\Delta Y = Y_o - Y_e = [N \times \text{mean}(\Delta RY) \times \text{mean } M] + N \text{ cov}(\Delta RY, M)$$

where ΔRY_i is the deviation from expected relative yield of species i in the mixture, calculated as the difference between observed and expected relative yields. Observed relative yield of a species in the mixture is the ratio of its yield in the mixture and its yield in the monoculture. Expected relative yield is the proportion of the species planted. The first term $[N \times \text{mean } (\Delta RY) \times \text{mean } M]$ in the equation measures the complementarity effect, the second $[N \text{ cov } (\Delta RY, M)]$ the selection effect (Loreau and Hector 2001).

All data were analyzed as generalized linear models using SPSS 10.0 version. Biomass production and invasibility were regressed on species diversity, and the correlation between biomass production and invasibility determined. The correlation between productivity and sampling effects and complemetarity effects and invasibility was also determined. The effects of species diversity and species composition on biomass production and on invisibility were analyzed by using general linear models. Biomass production and weed invasion were log-transformed when necessary to improve compliance with equal variance and normality assumptions.

Results

On average, annual species performed better in productivity and invasion resistance than perennial species (Fig. 1). When planted in monoculture, species differed greatly in their biomass production and the ability to suppress weed invasion in both the annual communities (ANOVA, $F_{8, 18} = 207.66$, $P < 0.001$ and $F_{8, 18} = 94.899$, $P < 0.001$, respectively, Fig. 1a) and the perennial communities ($F_{9, 21} = 112.89$,

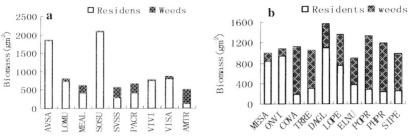

Fig. 1 Biomass production and invasibility of different monocultures (**a**) annual community; (**b**) perennial community

$P < 0.001$ and $F_{9,21} = 62.28$, $P < 0.001$, respectively; Fig. 1b). Of the annual monocultures, *Sorghum sudanense* plots were the most productive and the lest invasive communities, and *Amaranthus tricolor* and *Sorghum vulgare* × *S. sudanense* plots were the lest productive and the most invasive communities (Fig. 1a). Of the perennial monocultures, the two legumes, *Onobrychis* and *Medicago*, were the most productive and lest invasive communities, and *Poa* was the most invasive one, along with the grass *Phleum* and the forbs *Silphuium* (Fig. 1b).

On average, with the increase of species diversity, biomass production of resident species increased significantly, and community biomass production was positively correlated with species diversity in both the annual communities (Fig. 2a) and the perennial communities (Fig. 2b). Species diversity and community biomass production all had significant effect on suppression weed invasion: invasibility was negatively correlated with species diversity (Fig. 3) and biomass production (Fig. 4) in both the annual communities and the perennial communities.

Results of additive partitioning equation suggested that the two components of diversity effect, selection and complementarity, had strikingly different patterns in two communities (Figs. 5, 6). In the annual communities, regressing analysis indicated that sampling effects were positively correlated with species diversity (Fig. 6a), negatively correlated with community invasibility (Fig. 7a), while the correlations with complementarity effects were not significant (Figs. 5a, 8a). This indicated that, in the annual communities, the relationship between species diversity and productivity and invasibility was driven by sampling effect. In the perennial communities, however, the situation was different: complementarity effects were positively correlated with species diversity (Fig. 5b), negatively correlated with community invasibility (Fig. 8b), indicating that the mechanism operated was different from that of annual communities.

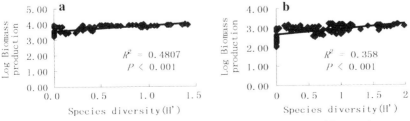

Fig. 2 Relationship between species diversity and biomass production (**a**) annual community; (**b**) perennial community

Springer

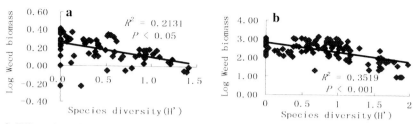

Fig. 3 Effect of species diversity on invasibility (**a**) annual community; (**b**) perennial community

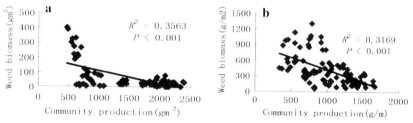

Fig. 4 Relationship between biomass production and invasibility (**a**) annual community; (**b**) perennial community

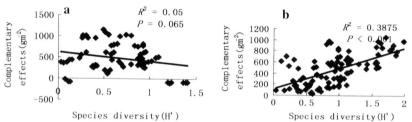

Fig. 5 Relationship between species diversity and complementary effects (**a**) annual community; (**b**) perennial community

The effect of species diversity may confound with species composition in our experiment. To separate such confounding effects, we analyzed the relative importance of species diversity and species composition using unique general linear models. Results indicate that species diversity and species composition had approximately equally important effects in our experiment (Table 2).

Discussion

Our results indicated that a positive diversity-productivity (Fig. 2) and negative diversity-invasibility (Fig. 3) and productivity–invasibility (Fig. 4) correlations emerged in two different communities, and that both the sampling effect and complementarity effect are likely to operate in driven a concomitant increase in productivity and decrease in invasibility with increasing species diversity (Figs. 5, 6). These results provide the experimental evidence for the resource-based hypothesis proposed to account for the positive diversity-productivity and negative diversity-

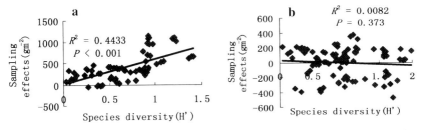

Fig. 6 Relationship between species diversity and sampling effects (**a**): annual community; (**b**): perennial community

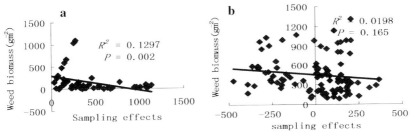

Fig. 7 Relationship between invisibility and sampling effects (**a**) annual community; (**b**) perennial community

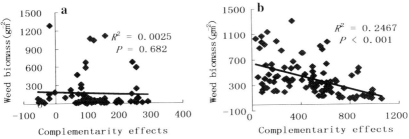

Fig. 8 Relationship between invisibility and complementarity effects (**a**) annual community; (**b**) perennial community

invasibility relationship observed in some experimental manipulated communities (Knops et al. 1999; Hector et al. 2001; Lepš et al. 2001; Tilman et al. 2001; Wardle 2001; Fridley 2002, 2003; Hodgson et al. 2002; Kennedy et al. 2002; van Ruijven et al. 2003; van Ruijven and Berendse 2005).

An increase in productivity (Aarssen 1997; Tilman et al. 1997; Cardinale et al. 2002) and decrease in invasibility (Levine 2000; Hector et al. 2001) with increased species diversity could be caused by either the sampling effect or the complementary effect. In the annual communities of our experiment, the significance of the sampling effect and its contribution to the observed responses to experimental manipulations of biodiversity indicated that sampling effect was responsible for the positive diversity-invasibility and negative diversity-invasibility relationship. This result consistents with other similar studies of experimental plant communities (Huston 1997; Tilman 1997b; Hector et al. 1999; Huston et al. 2000; Lepš et al. 2001; Fridley 2002), and confirms that a much larger effect of diversity on production occurred by

🖄 Springer

Table 2 Analyses of the effects of species diversity and of species composition on biomass production, weed invasion, sampling effects, and complementarity effects

Response variable	Annual communities								Perennial communities								
	Overall			Diversity		Composition			Overall			Diversity		Composition			
	R^2	F	P	F	P	F	P		R^2	F	P	F	P	F	P		
Biomass production	0.56	17.71	<0.001	21.13	<0.001	27.06	<0.001		0.61	11.61	<0.001	18.38	<0.001	19.74	<0.001		
Weed invasion	0.36	4.v81	0.027	5.17	0.008	0.81	0.371		0.48	6.84	<0.001	9.83	<0.001	14.56	<0.001		
Sampling effects	0.84	29.46	<0.001	52.26	<0.001	1.76	0.031		0.14	0.89	0.45	0.78	0.41	0.18	0.23		
Complementarity effects	0.127	0.81	0.35	1.76	0.191	1.03	0.24		0.65	12.01	<0.001	20.84	<0.001	14.19	<0.001		

means of sampling effects for particularly influential species. According to Aarssen (1997) and Huston (1997), the species that are most productive and least invisible in monoculture have the most positive effect on community productivity and invasion resistance. In our annual communities, however, the most productive (also most invasion resistant) species in monoculture, *Sorghum sudanense* (Fig. 1a), did not have the highest relative productivity and invasion resistance in mixtures. The second productive species in monoculture, *Avena sative*, performed better in mixture communities. All *Avena sativa* containing mixtures are all more productive and less invasible, suggesting that *Avena sativa* may dominate the community level response of productivity and invisibility. The dominant effect of *Avena sativa* may be due largely to its high resource use ability: has erect stems, grows rapidly during the seedling stage, monopolize both light and nutrients in the mixtures, thereby has higher efficiency of conversion of resources to biomass, and contributes more to community productivity (more than 85% of the mixture biomass is contributed by *Avena*) and remain less available resources for invasion weeds. Thus, a sampling effect can manifest itself through species that perform well in mixtures rather than through the most productive species in monocultures (Dukes 2001). Some short-term experiments with annual plant (Engelhardt and Ritchie 2001; Fridley 2002, 2003) also provide a strong evidence of sampling effect due to annual plants having limited resource use differentiation. Furthermore, species of annual community, although taxonomically diverse, are short-lived ruderals, and would not have the same range of morphological and physiological diversity of perennial community, consequently, have limited resource complementary resource use (Fridley 2002).

Contrast to the annual communities, however, a different mechanism operates in the perennial communities. An alternative explanation for the positive diversity-productivity and negative diversity-invasibility relationships is resource complementary use: complementarity of resource use between species results in higher productivity, and lower levels of available resources at species-rich communities, thus inhibiting invasion (Knops et al. 1999; Naeem et al. 2000). From the analysis, we found that a significant positive diversity-complementarity effects (Fig. 5b) and a negative complementarity–invasibility (Fig. 8b) correlation emerged in the perennial communities, indicating that both biomass production and resource complementary use increased with species diversity (Figs. 2b, 5b), consequently, resulted in a decreasing in invasibility with species diversity increasing (Fig. 3b). These provide strong evidence that the same mechanism, the complementary effect, is largely responsible for both the increased productivity and decreased invasibility of diverse mixtures in the perennial communities. These patterns are similar to results reported elsewhere which demonstrated that more diverse communities are both more productive (Tilman 1997b; Cardinale et al. 2002) and more resistant to invasion by other species (Levine 2000; Hector et al. 2001), and demonstrated that complementarity in resource use were the main driver of increased productivity at higher levels of species diversity in perennial communities (Tilman et al. 2001; Hooper and Dukes 2004; van Ruijven and Berendse 2005). A similar result was also obtained by Hodgson et al. (2002) with microcosms of bacterial communities, which demonstrated that identical mechanisms drive both diversity-productivity and diversity-invasibility relationships.

Theoretically, higher complementarity may be expected between species with great differences in functional traits (phonological, structural, and biogeochemical traits; Hooper 1998; Lepš et al. 2001). In particular, studies using species of greater functional diversity, conducted within larger plots, and of longer duration would be more likely to generate complementary species relationships, especially those based on phenology (Hooper 1998; Fridley 2002). The species used in our perennial communities, differed in their structural (such as canopy architecture, rooting depth), biogeochemical (such as N-fixing), and morphological (such as broad- and narrow-leaved species, erect or creeping shoot.) traits, so differ in seasonal (or phenology) and spatial patterns of resource (or light) use. Thus, we can extrapolate that mixtures with higher diversity would lead to complementary resource use and greater productivity, and consequently, remain fewer resources available for the invaders. Therefore, species-rich mixtures will be more productive and more resistant to invasion than the species-poor mixtures. However, the result of Hooper (1998) demonstrated that, even when species differ dramatically in phenology and morphological characteristics, complementary resource use was not necessarily positively correlated to the scales of diversity in mixture communities. Others (Hodgson et al. 2002) also reported that the occupation of distinct niches was not sufficient to produce a net complementarity effect. The possible explanation for this contrast results is that in addition to species diversity, resource availability can be influenced by many other factors (Davis et al. 2000; Dukes 2001; Byers and Noonburg 2003).

Consistent with some other studies of species diversity (Hooper and Vitousek 1997, 1998, Hooper and Dukes 2004; Lambers et al. 2004), our results suggest that species composition also play an important role in governing the relationship of diversity-productivity and diversity-invasibility. This can be demonstrated both by great variability in productivity and invasibility in monocultures (Fig. 1) and among mixtures with similar levels of diversity (Figs. 2, 3 and Table 2). This result suggests that some species could perform better in productivity and invasion resistance in some mixtures, but not in others. Clearly, species performances in mixtures are not simple functions of their performances in monocultures or less diverse mixtures (Hooper and Dukes 2004). Several studies (Grime 1997; Hooper and Vitousek 1997, 1998) have demonstrated that ecosystem function is mainly a consequence of the prevailing strategies of constituent species, in interaction with the abiotic environment. However, the fact that species composition are important in driving observed diversity-productivity or diversity-invasibility relationships does not mean that diversity effect can be neglected in such studies.

Our experiment clearly demonstrated that a similar positive diversity-productivity and negative diversity-invasibility correlations can emerged in two different communities. However, the mechanisms driving this concomitant increase in productivity and decrease in invasibility with increasing diversity of plant species are different. This suggests that because species in different communities (here, annual and perennial) may differ in their life history, biological and physiological traits, and may have different resource use style, and different interactions among them, mechanisms responsible for the observed relationships between diversity and both productivity and invasibility are likely different.

To summarize, from the results of our experiment, we can draw a conclusion that productivity and resistance to invasion can both increase with increasing species diversity, and these relationships can drown by identical mechanisms: in the annual communities, the positive diversity-productivity and negative diversity-invasibility relationship are linked by the sampling effect, while in the perennial communities, the mechanism driving a concomitant increase in productivity and decrease in invasibility with increasing diversity is the complementarity effect. This demonstrates that different mechanisms may drive diversity-productivity and diversity-vinvasibility relationships in different communities, where species diversity is experimentally manipulated. Because community processes is influenced by many factors, and interactions among them, the pattern and mechanisms of community invasibility are likely to be far complex than we found in this study. It is therefore important to disentangle and determine the importance of more generalized mechanisms operating in more complex communities and ecosystems.

Acknowledgements We would like to thank He Shi wei and Wu Deli for supporting field works. Two anonymous reviewers provided helpful criticisms of earlier drafts of this manuscript. This research was supported by the National Basic Research Program of China (Grant No. 2002CB 111505).

References

Aarssen LW (1997) High productivity in grassland ecosystems: effected by species diversity or productive species? Oikos 80:183–184
Byers JE, Noonburg EG (2003) Scale dependent effects of biotic resistance to biological invasion. Ecology 84:1428–1433
Cardinale BJ, Palmer MA, Collins SL (2002) Species diversity enhances ecosystem functioning through interspecific facilitation. Nature 415:426–429
Crawley MJ, Brown SL, Heard MS, Edwards GR (1999) Invasion-resistance in experimental grassland communities: species richness or species identity? Ecol Lett 2:140–148
Davis MA, Grime JP, Thompson K (2000) Fluctuating resources in plant communities: a general theory of invasibility. J Ecol 88:528–534
Dukes JS (2001) Productivity and complementarity in grassland microcosms of varying diversity. Oikos 94:468–480
Dukes JS (2002) Species composition and diversity affect grassland susceptibility and response to invasion. Ecol Monogr 12:602–617
Engelhardt KAM, Ritchie ME (2001) The effect of aquatic plant species richness on wetland ecosystem processes. Ecology 83:2911–2924
Fridley JD (2002) Resource availability dominates and alters the relationship between species diversity and ecosystem productivity in experimental plant communities. Oecologia 132:271–277
Fridley JD (2003) Diversity effects on production in different light and fertility environments: an experiment with communities of annual plant. J Ecol 91:396–406
Grime JP (1997) Biodiversity and ecosystem function: the debate deepens. Science 277:1260–1261
Hector A, Schmid B, Beierkuhnlein C et al (1999) Plant diversity and productivity experiments in European grasslands. Science 286:1123–1127
Hector A, Dobson K, Minns A, Bazeley-White E, Lawton JH (2001) Community diversity and invasion resistance: an experimental test in a grassland ecosystem and a review of comparable studies. Ecol Res 16:819–831
Hodgson D, Rainey PB, Buckling A (2002) Mechanisms linking diversity, productivity and invasibility in experimental bacterial communities. Proc Royal Soc London Ser B—Biol Sci 269:2277–2283
Hooper DU, Vitousek PM (1997) The effects of plant composition and diversity on ecosystem processes. Science 277:1302–1304
Hooper DU, Vitousek PM (1998) Effects of plant composition and diversity on nutrient cycling. Ecol Monogr 68:121–149
Hooper DU (1998) The role of complementarity and competition in ecosystem responses to variation in plant diversity. Ecology 79:704–719
Hooper DU, Dukes JS (2004) Overyielding among plant functional groups in a long-term experiment. Ecol Lett 7:95–105
Huston MA (1997) Hidden treatments in ecological experiments: re-evaluating the ecosystem function of biodiversity. Oecologia 110:449–460

Huston MA, Aarssen LW, Austin MP, Cade BS, Fridley JD, Garnier E, Grime JP, Hodgson J, Lauenroth WK, Thompson K, Vandermeer JH, Wardle DA (2000) No consistent effect of plant diversity on productivity. Science 289:1255a

Jonsson M, Malmquist B (2003) Mechanisms behind positive diversity effects on ecosystem functioning: testing the facilitation and interference hypotheses. Oecologia 134:554–559

Kennedy TA, Naeem S, Howe KM, Knops JMH, Tilman D, Reich P (2002) Biodiversity as a barrier to ecological invasion. Nature 417:636–638

Knops JMH, Tilman D, Haddad NM, Naeem S, Mitchell CE, Haarstad J, Ritchie ME, Howe KM, Reich PB, Siemann E, Groth J (1999) Effects of plant species richness on invasion dynamics, disease outbreaks, insect abundances and diversity. Ecol Lett 2:286–293

Lambers JHR, Harpole WS, Tilman D, Knops K, Reich PB (2004) Mechanisms responsible for the positive diversity-productivity relationship in Minnesota grasslands. Ecol Lett 7:661–668

Lepš J, Brown VK, Diaz Len TA, Gormsen D, Hedlund K, Kailová J, Korthals GW, Mortimer SR, Rodriguez-Barrueco C, Roy J, Santa Regina I, van Dijk C, van der Putten WH (2001) Separating the chance effect from other diversity effects in the functioning of plant communities. Oikos 92:123–134

Levine JM (2000) Species diversity and biological invasions: relating local process to community pattern. Science 288:852–854

Loreau M, Hector A (2001) Partitioning selection and complementarity in biodiversity experiments. Nature 412:72–76

Mack MC, D'Antonio CM (1998) Impacts of biological invasions on disturbance regimes. Trends Ecol Evol 13:195–198

Naeem S, Knops JMH, Tilman D, Howe KM, Kennedy T, Gale S (2000) Plant diversity increases resistance to invasion in the absence of co-varying extrinsic factors. Oikos 91:97–108

Palmer MW, Maurer T (1997) Does diversity beget diversity? A case study of crops and weeds. J Veg Sci 8:235–240

Robinson GR, Quinn JF, Stanton ML (1995) Invasibility of experimental habitat islands in a California winter annual grassland. Ecology 76:786–794

Shea K, Chesson P (2002) Community ecology theory as a framework for biological invasions. Trends Ecol Evol 17:170–176

Tilman D (1997a) Community invasibility, recruitment limitation, and grassland biodiversity. Ecology 78:81–92

Tilman D (1997b) Distinguishing between the effects of species diversity and species composition. Oikos 80:185

Tilman D, Lehman C, Thomson K (1997) Plant diversity and ecosystem productivity: theoretical considerations. Proc Natl Acad Sci USA 94:1857–1861

Tilman D (1999) The ecological consequences of changes in biodiversity: a search for general principles. Ecology 80:1455–1474

Tilman D, Reich PB, Knops J, Wedin D, Mielke T, Lehman C (2001) Diversity and productivity in a long-term grassland experiment. Science 294:843–845

van Ruijven J, De Deyn GB, Berendse F (2003) Diversity reduces invasibility in experimental plant communities: the role of plant species. Ecol Lett 6:910–918

van Ruijven J, Berendse F (2005) Biodiversity–productivity relationships: initial effects, long-term patterns, and underlying mechanisms. Ecology 102:695–700

Wardle DA (2001). Experimental demonstration that plant diversity reduces invasibility—evidence of a biological mechanism or a consequence of sampling effect? Oikos 95:161–170

⁄ Springer

TOPICS IN BIODIVERSITY AND CONSERVATION